The Petrels

their ecology and breeding systems

The Petrels

their ecology and breeding systems

JOHN WARHAM

Zoology Department
University of Canterbury
Christchurch, New Zealand

ACADEMIC PRESS
Harcourt Brace Jovanovich, Publishers

London · San Diego · New York · Boston · Sydney · Tokyo · Toronto

ACADEMIC PRESS LIMITED
24/28 Oval Road, London NW1 7DX

United States Edition published by
ACADEMIC PRESS, INC.
San Diego, CA 92101

British Library Cataloguing in Publication Data
Warham, John
The petrels.
1. Storm petrels
I. Title
598.42

ISBN 0–12–735420–4

Editorial and Production Services by Fisher Duncan
10 Barley Mow Passage, Chiswick, London W4 4PH

Printed and bound in Great Britain by
Mackays of Chatham PLC, Chatham, Kent

Contents

Preface

Research on seabirds in recent years has flourished, partly as a result of environmental concerns resulting in major programmes such as those of the Alaska Outer Continental Shelf Environmental Assessment conducted in the USA and similar investigations by Canadian and British organisations in the North Pacific, North Atlantic and the North Sea; by South Africa in the Benguela Current region of the South Atlantic as well as smaller projects by many countries. This work has been initiated mostly in response to proposed plans for exploitation of marine or sub-marine resources and has extended our understanding of the distribution of seabirds in relation to their foods, local fisheries, currents and other variables of the oceanic environment which affect the structures of marine communities.

In many of these communities tubenosed birds form important components at times dominating the above-water niche. Thus studies of petrels have figured prominently in these programmes, as they also have in recent censuses of breeding seabirds that have been carried out around the coasts and islands of the Northern Hemisphere. Parallel to this activity has been a surge in the study of petrels on their breeding grounds, particularly those in the Southern Ocean. New techniques for automatic monitoring of activity, for sex determination, for satellite surveillance of individual birds while at sea, now promise great refinements to our current knowledge of the petrels' place in nature.

The present work aims to fill the need for a concise account of these birds, taking in the mass of recent research and providing the means for gaining access to the literature for those requiring to delve further into particular topics. I hope that it will meet the needs both of the serious amateur and the professional ornithologist.

In this volume I have summarised the characteristics of the petrels genus by genus, followed by a more detailed account of their reproductive biology. Other aspects, such as behaviour and vocalisations, energetics and physiology, populations, foods and feeding, distribution, human use, and conservation and management form the subjects for a subsequent volume.

I am grateful to many colleagues who have kept me in touch with their current research and to them and their publishers for allowing me to use their illustrations: these are acknowledged in the text. Some have also commented on parts of the manuscript. Here I would particularly like to thank D.G. Ainley, L.B. Astheimer; J.A. Bartle, P.D. Boersma, W.R.P. Bourne, J.P. Croxall, K.W. Duncan, G.M. Dunnet, H.I. Fisher, C.R. Grau, P.C. Harper, C. Jouanin, P. Jouventin, I. McLean, S. Olson, R.L.C. Pilgrim, H. Rahn, R.E. Ricklefs, P.M. Sagar, B. Stonehouse, and G.C. Whittow.

With such a wide field to cover it is inevitable that some inexactitudes and errors of interpretation will have crept in: such errors of commission and omission are my own.

My colleagues in the Department of Zoology have also been supportive in many ways, as has my wife, Pat. After years of helping in the field she has been extremely patient at finding her spouse still tied to a desk during five years of retirement.

John Warham
Christchurch
July 1990

CHAPTER 1
Introduction: The Order Procellariiformes

Left to right: Laysan, Short-tailed and Black-footed Albatrosses.
Photo: author.

Contents

The Procellariiformes or petrels are a highly distinctive group of marine birds that have evolved from a single ancestor which they may have shared with the penguins (Sphenisciformes). Although the fossil record for both these orders includes many forms, and is rapidly improving with new finds, no obvious link between these groups has been unearthed. The early petrels of the Oligocene and Miocene seem to have been very similar to those of today, and for this, among other reasons, these birds have often been considered as an ancient group, perhaps derived from a late Cretaceous ancestor. Recent research using DNA–DNA hybridisation techniques, however, has suggested a much more recent origin, with the loons, frigate birds and penguins as their nearest relatives (Sibley *et al.*, 1988).

While the penguins lost their powers of flight to develop highly efficient means of swimming and diving to reach food, the petrels developed specialised systems for aerial locomotion that allowed their exploitation of pelagic food resources at or near the surface of the sea.

I External morphology

Petrels are readily identified on a single feature — their nostrils are sheathed in prominent horny tubes arising near the base of the bill (Fig. 1.1), hence the earlier name Tubinares for this order. The tubes apparently develop from growths of the cartilaginous walls of the nasal capsules. In the Diomedeidae or albatrosses these tubes are separated by the ridge of the culmen (Fig. 1.2), but in birds of the other three families they coalesce and may appear as a single tube divided by a median septum. The tubes tend to lie low along the culmen in shearwaters *Puffinus* spp., but in some of the storm petrels, e.g. *Fregetta*, they are prominent, even upturned.

Figure 1.1 *Head of Herald Petrel* Pterodroma arminjoniana heraldica*: the nasal tubes are fused together at the base of the culmen. Note the pale filoplumes protruding from the contour feathers of the crown. Photo: author.*

They are especially large in the fulmars and in the giant fulmars or giant petrels *Macronectes* spp. they occupy three-fifths of the length of the bill (Fig. 3.1).

Another special feature of the petrel bill is the division of the rhaphotheca, the horny coating to the beak, typical of birds, into 7–9 plates separated by deep grooves. These individual corneous elements (Fig. 1.2) are readily separated by maceration in a solution of sodium carbonate and they were named by a pioneer student of the group, Elliott Coues, in 1866.

Particularly in the larger species, the upper nail or maxillary unguis, is massive, decurved (reminding Coues of a cat's claw) and fits neatly into the lower mandibular unguis (unguicorn). This upper nail is stouter and thicker than any other of the bill plates, the extremely concave inner edges very sharp, and

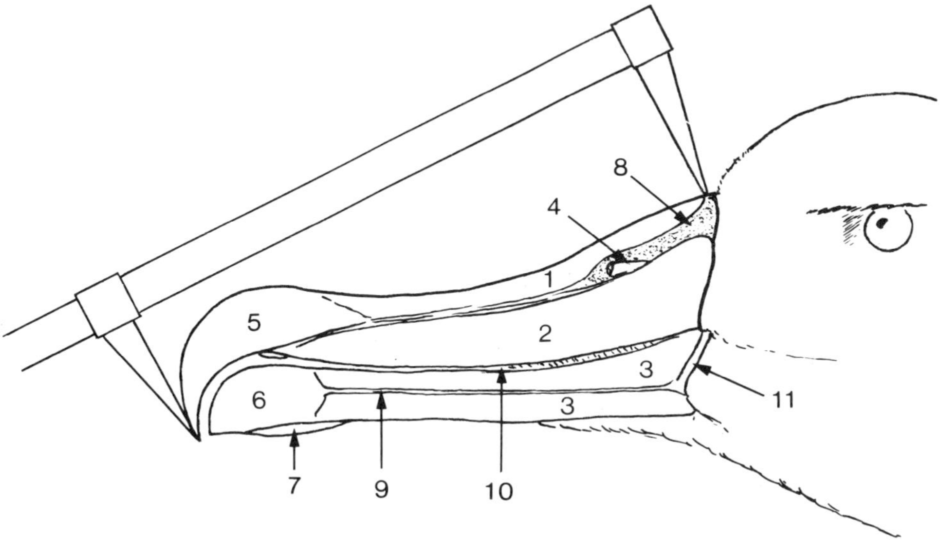

Figure 1.2 *The parts of a petrel's bill: 1. culminicorn; 2. latericorn; 3. ramicorn; 4. naricorn; 5. superior unguicorn (maxillary unguis); 6. inferior unguicorn (mandibular unguis); 7. inter-ramicorn (albatrosses only); 8. fleshy membrane; 9. sulcus; 10. tomia (cutting edges); 11. fleshy bar (albatrosses only). Dividers show method of measuring exposed culmen to tip of the hook of the nail. Bill to feathers would be slightly greater.*

evidently maintained in this condition by abrasion against the sharp inner faces of the lower nail. These nails, and the equally sharp cutting edges (tomia), form an excellent device for gripping slippery prey, for slicing through squid tentacles, and for tearing flesh by pulling backwards while the body is anchored by strong thrusts of the large webbed feet, a method often used by albatrosses and giant petrels. In fulmars the tomia are serrated to form combs for filter feeding. Bill colours range from fleshy horn (often suffused with pinkish) to black, with blue-grey shades in some prions *Pachyptila* spp.

The legs are rather long (in some storm petrels, very long), with the three front toes fully webbed but the hind one either absent or vestigial. Most have black or brown feet, sometimes these are flesh-pink or parti-coloured flesh and blackish. Prions, diving petrels *Pelecanoides* spp. and Little Shearwaters *Puffinus assimilis* have blue legs.

Like some other birds, e.g. gannets and ducks, petrel chicks grow two sets of down. Those of petrels (and penguins) differ in that the second set (usually called mesoptiles) grows up attached to the downs of the first or primary set (protoptiles) and the two may remain attached for some time. In such a 'double down' the junction shows as a constriction, apparently due to a slowing of growth, whereas the feathers proper (teleoptiles) push out attached to the downs of the second coat without any visible break.

The feather coat is thick, its insulatory powers usually augmented by subcutaneous fat deposits. The plumage often has a characteristic musty odour, sometimes stated, but not yet demonstrated, to derive from secretions of the preen glands. These are large and tufted. Assertions that each species has its own particular odour are also unsubstantiated, but some species smell more than others to the human nose, particularly the giant petrels, whose eggshells still smell strongly after 100 years in museum drawers.

The plumages are composed of black, brown, grey or white feathers, many species being whitish ventrally. A few have yellow feathers on the crown and many have pale filoplumes protruding here (Fig. 1.1). Species of high latitudes tend to be pale, those of low ones are often quite dark. Some have a checkered plumage and nocturnally active forms may have prominent underwing or tail patterns that may act as recognition marks when, as often happens, several congeners share a breeding place. A few species are polymorphic with dark, intermediate and pale birds which apparently interbreed at random.

Apart from pale edges to dark contour feathers, the juvenal plumage is generally very similar to that of the adult. The main exceptions are the giant petrels, most of which fledge in a wholly glossy blackish feather coat which soon browns with wear and lightens with age; and the Wandering and Short-tailed Albatrosses (*Diomedea exulans* and *D.albatrus*) whose bodies are dark brown at fledgling, but which lighten gradually during successive moults.

There are 11 primary quills in the petrel wing but the outermost is minute, so that there are always 10 functional primaries. The secondaries vary from 10 to about 40 (in the giant albatrosses *D.exulans* and *D.epomophora*), these differences being correlated with the proportionately much longer forearms in the bigger birds.

Recently fledged petrels have wholly fresh feathers, whereas the plumage of adults tends to be a mixture of old and new, that of dark-plumaged birds seldom lacking a few old brownish feathers. Moult of contour feathers often takes place continuously and during breeding, but that of the flight quills and secondaries is usually delayed until after that event, and for highly migratory kinds like Short-tailed and Great Shearwaters (*Puffinus tenuirostris* and *P.gravis*) is delayed until the contra-nuptial quarters are reached: the feathers are then shed rapidly. The juvenal coat is replaced during the young birds' first year at sea. Bill, iris and leg colours may also change with age, but apart from seasonal flushing in some species with pale bills or legs, colour change usually stops once maturity is reached.

A marked sexual dimorphism in plumage is rare among petrels. It is seen only in the two largest species, the Wandering and Royal Albatrosses, but sexual dimorphism in body dimensions is common. Usually the male is the larger, but with some storm petrels the reverse applies. Female petrels can be identified in the hand by the swollen lips to the cloaca for some time after laying. Otherwise sexing is difficult — see Chapter 11, Section IB.

II Anatomy

Identification of procellariiform skulls can be made even when the thick rhamphothecae are missing, by the characteristic shape of the ungues. The skulls also show a large, laterally expanded vomer in the palate which is fused to the palatines. As with most other seabirds, there are crescentic grooves above the orbits to house the extra-renal (salt) glands which are particularly large in petrels (Fig. 1.3), a feature to be expected of these highly marine birds.

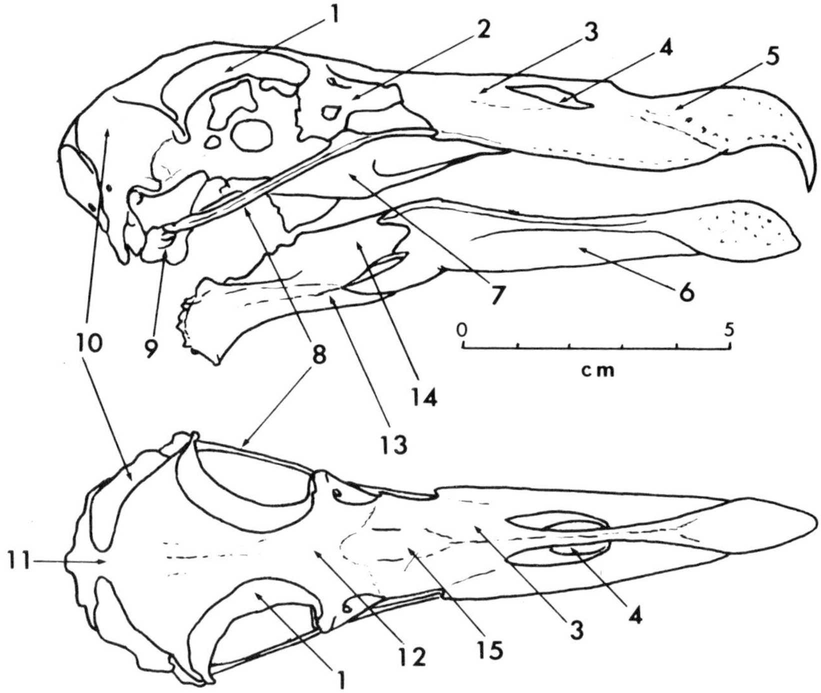

Figure 1.3 *Skull of giant petrel* Macronectes *in lateral view (above) and dorsal view. 1. Depression for supra-orbital salt glands; 2. pre-frontal (lacrimal); 3. nasal; 4. nares; 5. premaxilla; 6. dentary; 7. palatine; 8. quadrato-jugal; 9. quadrate; 10. temporal fossa; 11. sagittal crest (inter-fossal surface); 12. frontal; 13. angular; 14. surangular; 15. naso-frontal hinge.*

A notable aspect of the petrel brain is the large size of the olfactory bulbs. Recent studies have indicated that olfaction may be important in their lives, helping them in finding food, colonies and nest sites and perhaps in identifying partners, and, of course this sense works by day and by night. The musty smell of the feathers, nest materials and eggs may be involved. So may the tubular nostrils. These perhaps help to increase olfactory sensitivity by directing the inhaled air on to the olfactory epithelia, producing a 'nozzle' effect (Bang, 1966), but why the tubes should be huge in *Macronectes* and small in *Diomedea* is unexplained. In *Diomedea* the tube is divided internally so that the excurrent fluids from the salt glands, are kept separate from the incoming airstream (Bang, 1966).

Small petrels have relatively large brains for their body size, in large species the brain is relatively smaller but absolutely bigger. Presumably the difference helps account for the rather simple behaviours and restricted vocal repertoires of the small species compared with the complex and varied repertoires of large ones like the albatrosses.

A special feature of the digestive tract is the way in which the oesophagus passes without constriction into a very large proventriculus that fills most of the abdominal cavity (Fig. 1.4). The huge size of this bag-like organ allows very large meals to be stored or carried to the chick. The proventriculus twists back upon itself towards its apex before entering the small gizzard which is the stomach proper. This is twisted to the right of and above the proventriculus, with the pylorus opening from the posterior of the gizzard in the angle between the two parts of the bent proventriculus, resulting in a distinctive upwards curve to the duodenum which ascends to the right before forming the usual backwards-directed loop around the pancreas (Fig. 1.4). In some species the gut is peculiarly twisted, perhaps to increase the surface area for digestion.

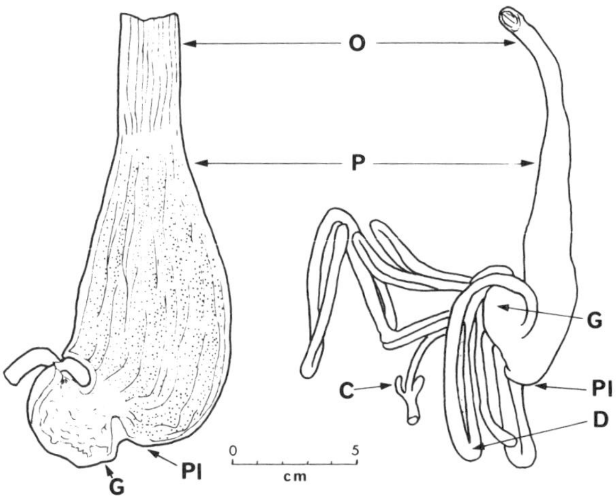

Figure 1.4 *Alimentary tract of Northern Fulmar* Fulmarus glacialis. *Left: oesophagus, proventriculus and gizzard opened longitudinally. Note the folded and highly glandular nature of the proventriculus which is huge compared to the gizzard. Right: alimentary tract showing the backwards-facing pylorus and the ascending duodenum as it leaves the gizzard. O oesophagus; P proventriculus; Pl lower level of proventriculus; G gizzard; C caecum; D duodenum. From Matthews (1949).*

The walls of the proventriculus are highly glandular, often ridged and very extensible, but the roles of the various glands have yet to be elucidated. They may be involved in the digestion of the stomach oil to be found in the proventriculi of most petrels. This oil originates from food and contains much wax ester, triglyceride and other substances. Petrels can metabolise wax esters (as an energy resource and for depot fat and water production), an ability rare among birds but not unique to petrels: at least one auk, *Aethia pusilla,* also includes wax esters in its food (Roby *et al.,* 1986). In surface nesting petrels stomach oil discharged through the mouth is used in defence, mostly by chicks. Contamination of their plumage by the oil may prove fatal to other birds but the petrels themselves appear to be immune.

III Body size

Procellariiformes range in body size from the Least Storm Petrel *Halocyptena microsoma* (wing span 32 cm; body weight 19.5 g) to the Royal Albatross (wing span 300 cm; body weight 8700 g). This is a range greater than in any other bird order.

Murphy (1936, p.475) pointed out that varied sizes and flight styles of Procellariiformes provided a nice example of adaptive radiation. He considered that, in a relatively uniform environment like the surface of the sea, once selection started to sort out different size types, their varying relationships to the available food and to their enemies would rapidly accelerate trends towards further diversification.

The species are not uniformly distributed over the size range. About 70% are of medium to small size (Fig. 1.5). In this figure the measure of body size used is the standard wing length, i.e. the length of the hand with attached primaries. This provides a convenient index of body size because wing length is highly correlated with the cube root of body weight (Warham, 1968). Body weights of individual petrels vary greatly due to deposition and metabolism of depot fat and oil, to variations in stomach contents and the like, and there are good figures for the mean weights of fewer species than there are for wing lengths, these being readily measured in museum skins.

The frequency polygon is of a bimodal distribution with no species of wing lengths 375–425 mm. However, the many special features shared by tubenosed

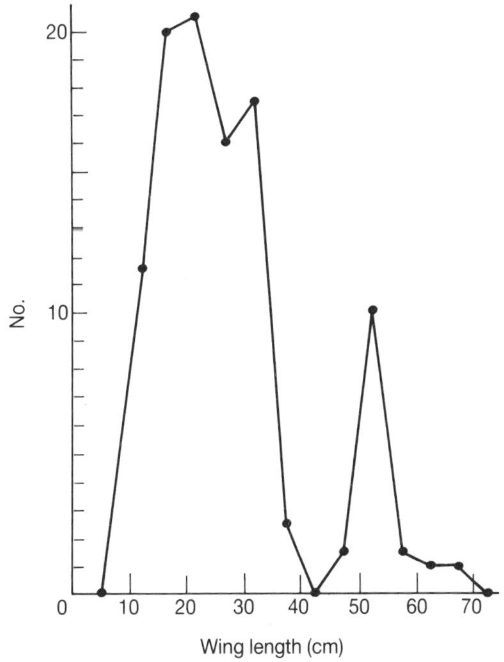

Figure 1.5 *Size variation for 103 species of Procellariiformes.*

birds indicate their common ancestry, so that during the evolution of species of the divers sizes that we see today, there must once have been birds of the missing sizes. A small piece of supporting evidence is the Middle Miocene albatross *Plotornis delfortrii* Milne-Edwards, 1878. The fossil is a tarsometatarsus 70.5 mm long which would have been characteristic of a petrel weighing about 2000 g, a size falling below that of the smallest present-day albatross and above that of the largest *Procellaria* petrel, i.e. in the distribution gap of Fig. 1.5.

Figure 1.5 has the typical negatively skewed form shown by the size distributions in many animal groups, distributions that Newell (1949) categorised as the result of 'phylogenetic size increase'. Stanley (1973) hypothesised that in the histories of many taxa, evolutionary novelties arose from comparatively unspecialised small to medium-sized ancestors which gave rise to smaller and larger forms better fitted for competition with other animals and for exploiting available and changing niches. In homeotherms like birds, problems of heat retention limit the extent to which a decrease in size is possible, whereas upper limits are usually set by the maximum practicable wingspans capable of supporting efficient flight.

Stanley suggested that many of the evolved forms, small or large, were more specialised, liable to extinction when environmental conditions changed, and unable to extract themselves from evolutionary 'dead ends'. In the Procellariiformes the extremes of specialisation are seen in the two giant albatrosses, the Royal and the Wanderer. These are incapable of sustained flight by beating their wings, their very existence depending on the continuation of the westerly wind flow of the Southern Ocean. Should wind patterns change and persistent calms prevail, this would lead to the demise of these birds unless they could shift elsewhere. On Stanley's reasoning the new niche created by the changed weather patterns might eventually be colonised by a new radiation evolved from a less specialised and smaller progenitor. Stanley's prescription for a history of phylogenetic growth would seem particularly fitting for the Procellariiformes, and there is evidence from the fossil record for the extinction of several albatrosses, mostly from the northern hemisphere, including the North Atlantic where none now breeds (Fig. 2.10).

IV Allometry

When examined as a whole many variables of petrel morphology, physiology and behaviour prove to be linked to body size, the latter usually measured by weight or mass. Easily measured variables are the dimensions of the body parts, egg weights and growth rates, the relationships being expressed by empirical equations. Some variables are linearly related to body size, but more usually the part is a non-linear monotonic function of the whole and the relationship best expressed by a logarithmic or power equation of the form of $y = ax^b$, where y is the dependent variable; x the body mass, a a constant (the intercept on the y axis or the value of y when $x = 1$), and b the value of the rate of change of size of organ, part or other variable with body size, i.e. the 'equilibrium constant' or 'partition coefficient' between the part and the body.

Some simple examples of these allometric relationships among petrels are given in Fig. 1.6. for female body weight versus egg weight (e), skeleton weight (s), culmen (c), tarsus (t), humerus (h) and manus lengths (m). The equations relating these variables to the body size (x) are:

$$y_e = 0.745x^{0.73} \qquad y_t = 5.7x^{0.31}$$
$$y_s = 0.035x^{1.1} \qquad y_h = 4.3x^{0.53}$$
$$y_c = 2.1x^{0.47} \qquad y_m = 8.6x^{0.40}$$

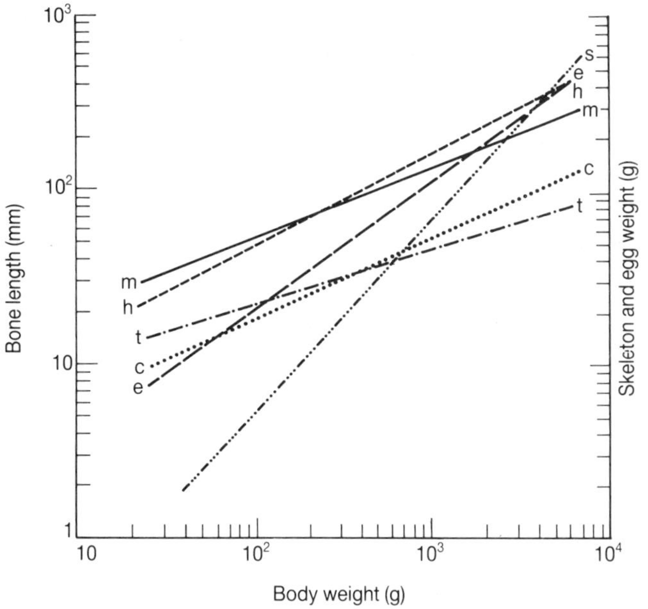

Figure 1.6 *Some allometrical relationships in petrels: (e) egg weight; (s) skeleton weight; (c) culmen length; (t) tarsus length; (h) humerus length; (m) manus length. From Warham (1968).*

Skeletal weight increases little relative to increasing body size, with $b = 1.1$, the exponent being only slightly above unity at which the part bears a constant ratio to the whole, i.e. the relationship is 'isometric'. Egg weight exhibits negative allometry ($b < 1$) and so does tarsal length ($b = 0.33$ for isometry for length and a three-dimensional variable like weight). The remainder show positive allometry ($b > 0.33$), i.e. while the rate of change of y with x remains constant, the part gets larger in relation to the body in larger petrels. The reverse obtains where the allometry is negative.

The variation about these regression lines differs considerably. For some, e.g. that relating egg weight to female body weight the correlation is very high ($r = 0.991$ in this instance); in others, like bill length, the variation is greater. For example, shearwaters have long, slender bills, gadfly petrels short stubby ones, so that for birds of the same body size in these two genera measuring bill size by length naturally results in a wider spread to the plotted points in graphs like Fig.

1.6 whereas if bill size were measured by some three-dimensional index the variation might well be reduced.

As all the variables of Fig. 1.6 are related to body weight, they are also related to one another. In a series of petrels graded according to body size the dimensions of the parts change according to a regular pattern: they are a function of body size itself and the form of the adult animal is set accordingly. A suite of curves such as this, to which can be added others for physiological and ecological variables, furnish a kind of blueprint of petrel organisation. They also allow predictions, within varying confidence limits, of the characteristics of little-known or extinct species and, where the points for a particular petrel diverge from the general trend lines, attention is drawn to the anomaly. If not due to faulty data, such divergences may often be explained on ecological or other grounds. For example, the fast development of fulmars discussed in Chapter 15, Section IVC shows up on curves for development rates and illustrates that the system has some flexibility.

The existence of 'rules' concerning the relationships between organ size or other variables and body size among Procellariiformes is not in itself a demonstration of an intimate phylogenetic relationship between the animals concerned, for there is a general relationship, usually of negative allometry, between organ weights and body weights of birds and mammals (sometimes called Welcker's Rule). Nonetheless, the general homogeneity of these allometric relationships of petrels is what might be expected of a group of animals with a common ancestor. Many other allometric relationships are discussed in the present work. Other factors do not vary with body size. Body temperatures, for example, being very similar for storm petrels and albatrosses.

V General ecology

With the exception of the diving petrels (Family Pelecanoididae) which, at least when nesting, seem to be inshore feeders, petrels mostly feed offshore or pelagically. They can travel for thousands of kilometres from their colonies to food-rich areas or can search tropical seas for patchy and shifting food concentrations along upwellings or local fronts. Their capacity to exploit such resources is helped by a series of energy-saving devices. Firstly, their body temperatures are lower than those of most other kinds of bird, averaging about 38°C (compared with about 41°C for birds in general) and, in some species at least, not rising more than a degree or so during flight. Thus less energy is required to maintain normal body temperature and less heat lost to cooler air and water. Secondly, they have a great ability to lay down subdermal fat and to store stomach oil, both of which can be drawn upon when food is limiting. Subdermal fat stores also aid insulation and help reduce heat loss. As a consequence, petrels can fast for extended periods and this helps them to nest, if necessary, far from their main food supply and also to penetrate seas where the food resource is unreliable and patchy and long spells between meals inevitable. Thirdly, most species conserve energy in flight either by dynamic or slope soaring like shearwaters and albatrosses or by other devices such as using the 'ground effect' and upthrust air above heaving seas as storm petrels do.

The petrel feeding system has its limitations. Most petrels must find their food in the top two or three metres of the sea, although some shearwaters have been seen feeding 20 m down. Even so, these probably compete rather little with the penguins whose feeding grounds they often share and which by their greater diving capability, can exploit a much deeper water column. To some extent this restriction to petrel feeding is eased by their ability to feed at night, taking advantage of the diurnal vertical migration of plankton and its associated predators.

Many petrels are solitary feeders, especially in the tropics; others forage in flocks, and even solitary ones band together around localised food sources. The classic example is that of the congregation that scavenges at a whale or seal carcass in the Southern Ocean; the albatrosses and giant petrels tearing the flesh and blubber while smaller birds such as Cape Petrels *Daption capense* grab medium-sized morsels and small fry like Wilsons's Storm Petrels *Oceanites oceanicus* snatch tiny scraps and fat globules without alighting.

Large petrels feed on large prey, albatrosses mostly on squid, but also, like giant petrels, killing and eating other seabirds when the chance arises. Many medium-sized species are also squid eaters but others take crustacea such as euphausids and copepods and many shearwaters rely mainly on fish. A good deal of petrel feeding is done in association with whales, tuna and other predators: by forcing their smaller prey towards the surface they make them available to the tubenoses. Storm petrels mostly pick up surface plankton (e.g. fish eggs, copepods and amphipods) by snipping up individual items while dancing close to the sea's surface. The feeding styles ued by petrels are summarised in Fig. 1.7, based on the categories of Ashmole (1971).

Flight styles vary with body size and wing shape. In general, the largest members of the order are highly developed for gliding and have wings of very high aspect-ratio whereas the small members have more conventional broader wings conferring reduced gliding powers but greater manoeuvrability. Fig. 1.8 shows typical wing outlines for representative members. The increasingly narrower wings with increasing body size is a consequence of the greater lengths of the humeral and radio-ulnal segments of the wings relative to those of the hand and its attached primary feathers.

On land many petrels walk or run only with difficulty. This is especially true of the shearwaters, birds with long pelves and whose femora articulate well to the rear of the body — modifications to increase swimming ability. These birds walk with a sinous action and often rest on their tarsi: if hurried they travel quite quickly, however, with beating wings, half running, half flying. They can stand erect only momentarily whereas albatrosses and, to a lesser extent, giant petrels, may do so for long periods. Shearwaters and albatrosses find difficulty in taking off in calms as their wings form highly effective aerofoils but only at high air speeds. Albatrosses therefore either need good runways or high points for take off under these conditions, while shearwaters then require elevated launching places from which they can dive and acquire the necessary lift for flight. Most medium-sized species among the shearwaters, gadfly petrels *Pterodroma* spp. and prions climb

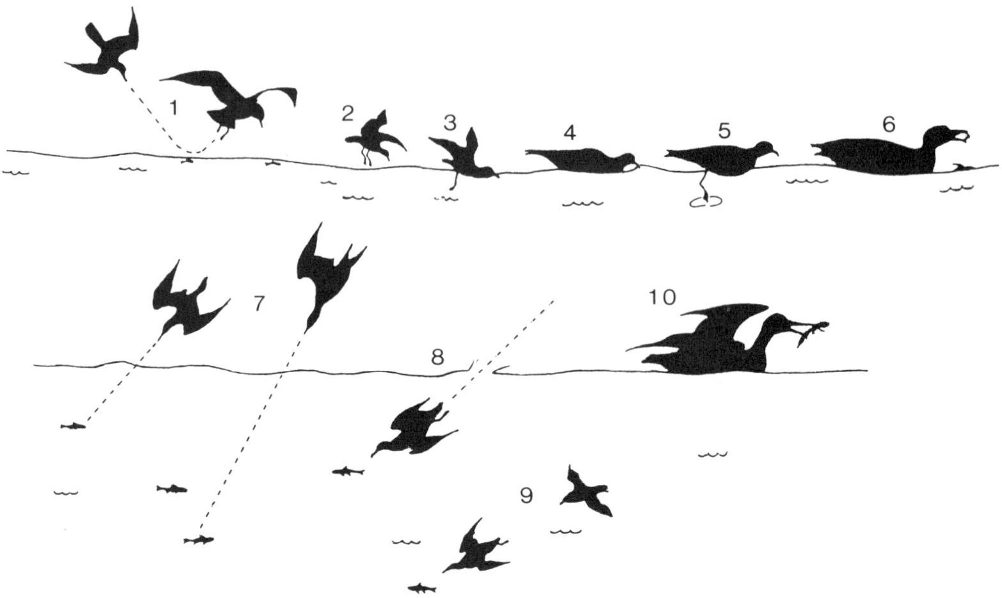

Figure 1.7 *Feeding styles of petrels: 1. dipping; 2. pattering; 3. hydroplaning; 4. surface filtering; 5. foot stirring; 6. scavenging; 7. pursuit plunging; 8. pursuit diving; 9. pursuit diving: wings; 10. surface seizing. After Ashmole (1971), courtesy of Academic Press, London.*

well and often use tree trunks and branches as launching platforms. Smaller, broad-winged petrels find little difficulty in taking off and alighting: their wing loadings are much less than those of the larger ones.

VI Breeding ecology

Petrels are social or colonial breeders, their nests often placed as close as possible without the owners infringing the individual distances of their neighbours. Breeding aggregations may be vast, with millions of pairs on a single island. The larger kinds — albatrosses, giant petrels and other fulmars — nest on the surface, the smaller ones mostly underground or in rock crevices, but, in some tropical locations where predators are scarce and/or the ground unsuitable for burrowing, even medium-sized birds may lay on the surface.

The age at first breeding ranges from 2 to 11 years according to species, the smaller kinds maturing sooner than the large ones. Reproductive rates are low but life expectancy high. All tend to return to breed at their birthplaces — philopatry — although some show a considerable talent for colonising new places. Most attempt to breed annually (usually at the same site and with the same partner) but, if successful, the largest albatrosses and a few of the medium-sized ones only nest biennially.

Successful breeding involves the coordinated efforts of both parents, and the time elapsing from arrival to the departure of the chicks is very long compared to

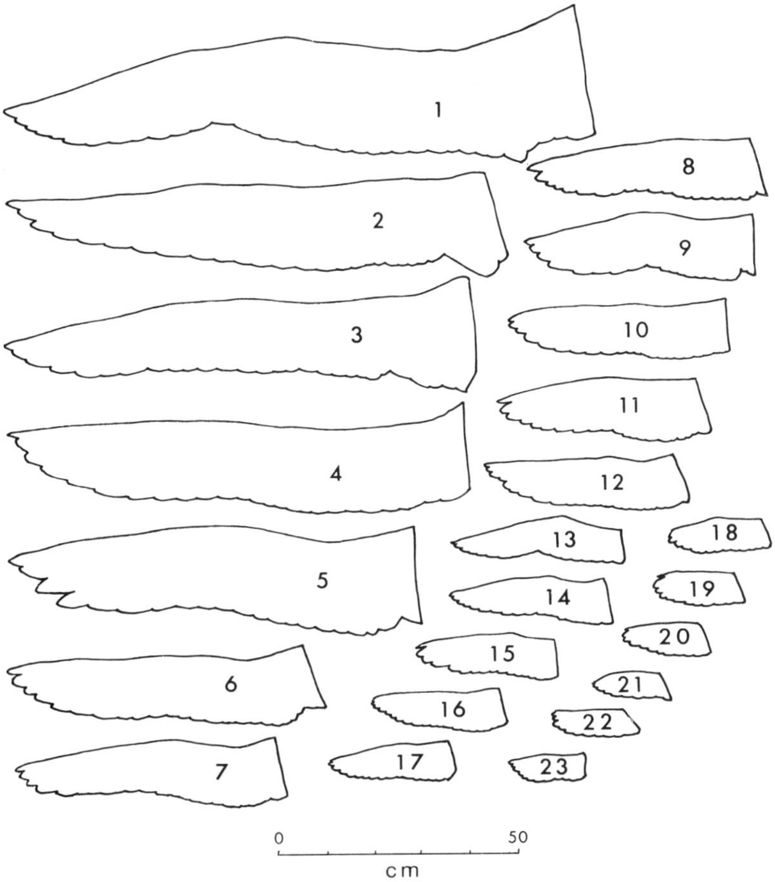

Figure 1.8 *Wing outlines of petrels: 1. Royal Albatross; 2. Light-mantled Sooty Albatross; 3. Grey-headed Albatross; 4. Black-footed Albatross; 5. Northern Giant Petrel; 6. Westland Petrel; 7. Grey Petrel; 8. Northern Fulmar; 9. Streaked Shearwater; 10. Providence Petrel; 11. Buller's Shearwater; 12. Cape Petrel; 13. Mottled Petrel; 14. Cook's Petrel; 15. Blue Petrel; 16. Little Shearwater; 17. Antarctic Prion; 18. Leach's Storm Petrel; 19. Black-bellied Storm Petrel; 20. White-faced Storm Petrel; 21. Grey-backed Storm Petrel; 22. British Storm Petrel; 23. Common Diving Petrel.*

other similar-sized birds. A typical breeding pattern is shown in Fig. 1.9.

The time spent ashore before the egg appears may be considerable and there is a pre-laying exodus of the female (and sometimes of the male too) to feed at sea while making her single, white, ovoid egg. This is very large for the size of the bird, weighing up to 28% of her body weight in the smallest forms.

Nesting places are on islands, headlines or mountains protected from mammalian predators. The few that breed on mainland continents do so in hot deserts (*Oceanodroma hornbyi* in Chile; *Bulweria fallax* in Arabia) or in cold ones like *Pagodroma nivea* in Antarctica, where few land animals of any sort can survive and predation is low. Breeding birds are at risk from predation when ashore as most have limited mobility on land and are easily caught. Moreover both incubation and

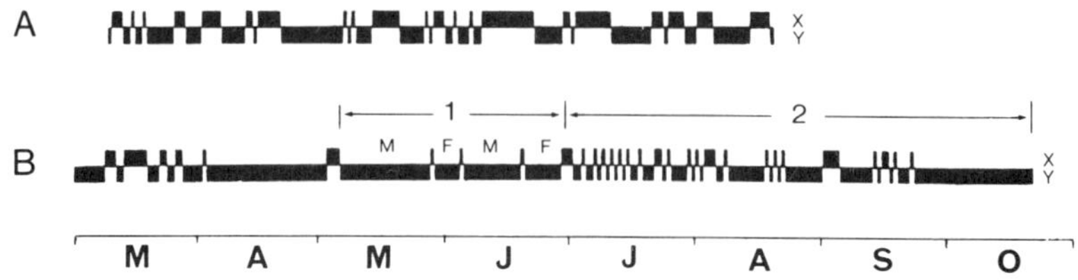

Figure 1.9 *Cycles of (A) non-breeding and (B) breeding Dark-rumped Petrels* Pterodroma phaepygia *from May to October. Upper marks (X) show days when burrows were entered, (Y) when not visited. 1. incubation; 2. chick stage. From Simons (1985).*

chick rearing are prolonged with both sexes sharing incubation in stints that may last a week or more. Eggs and young can withstand extensive chilling, cold and starved chicks sometimes becoming torpid and yet surviving to fledge. Both parents bring food for the young. Chicks of burrowing species are left alone by day within a day or two of hatching, having attained homeothermy, but the surface nesters guard their chicks for longer, natural predators like skuas or crows being ready to feed on untended young. These also need greater protection from the elements than those underground. Feeding is by regurgitation from the parent's proventriculus. Most chicks grow to exceed adult weight but shed weight before fledging. Nestlings of migratory species may be deserted by their parents before they fledge: with sedentary species there is no desertion period.

During the period of immaturity the younger age classes stay at sea for several years. With increasing age they start to return to the breeding places. The oldest, near-mature, birds are the first of the non-breeders to come back, often doing so at the same time as the breeders (Fig. 1.9); the younger non-breeders appear later, while yearlings typically are not seen ashore.

VII Global distribution

Petrels occur in all the oceans, but they are more numerous as species and individuals in the southern hemisphere where they may, like the penguins, have first evolved, although the fossil record is richer for the northern hemisphere. They penetrate to the polar regions and the Snow and Antarctic Petrels breed further south than any other bird except their parasite, McCormick's Skua *Stercorarius maccormicki*. Parents at one Snow Petrel colony at 80°S must fly about 350 km even to reach the coastline and further to reach open water and food.

Distribution at sea is governed by the distribution of food and, in the breeding season, by the location of nesting grounds in relation to food resource. This in turn depends on the whereabouts of currents, upwellings and fronts and other oceanic conditions. In contrast, choice of breeding locations seems governed more by the availability of safe islands and promontories which need not necessarily be very near to the feeding areas.

When not nesting migratory petrels are so highly marine that they tend to shun the land and are often only found there when storm-driven onto a lee shore. Then 'wrecks' may occur of stranded, dead and dying birds, some of which may be blown far inland. The main exceptions to this antipathy to land are the giant petrels, the northern species *Macronectes halli* being especially ready to feed on beaches or islands on whale and seal carcasses or to prey on penguins, albatross chicks and even introduced rabbits.

Because of their ability to withstand long fasts and to find food at sea far from their normal feeding range, petrels can turn up almost anywhere. Thus there are records of Laysan Albatrosses from the North Pacific in the Indian Ocean, of a Mottled Petrel (that breeds only in New Zealand) walking along a ploughed field in New York State, and of a Swinhoe's Petrel from the north-west corner of the Pacific Ocean singing from a burrow on Grand Salvage Island in the North Atlantic.

Sedentary Procellariiformes may visit their nesting places outside the breeding season. Others make vast annual migrations that are among the most extensive performed by birds. These are trans-equatorial movements and the participants travel fast, probably feeding little on the way. Most such post-nuptial movements take the birds from the south to the north, but some migrate in the reverse direction. Still others shift longitudinally, particularly in the Southern Ocean where many seem to travel down-wind in the prevailing westerlies. Whether they return against the wind or continue westwards back to their colonies has not been clearly established.

VIII Populations

Colony sizes may be huge and attempts to estimate world populations of the commoner species involve large assumptions. Effects of their vast numbers may be felt far from their breeding grounds. The Sooty Shearwater *Puffiness griseus*, that nests only in the southern hemisphere, was estimated by Wiens & Scott (1975) to be the major seabird consumer during its fall passage off the Oregon coast, taking some 24 000 tonnes of anchovies during a 2-month's stay.

Many populations have changed in size during historic times. Most have been decreased due to acts of man, in particular to his introduction of placental mammals to the breeding places. Only one, the Guadalupe Storm Petrel *Oceanodroma macrodactyla*, seems to have been wiped out, but many are threatened. The use of petrels as food has a long history, but except for a gadfly petrel, *Pterodroma hasitata caribbea*, excessive exploitation seems not to have led directly to extermination. Some have even increased in numbers and are extending their ranges, the most notable examples being the Northern Fulmar *Fulmarus glacialis*, the Black-winged Petrel *Pterodroma nigripennis*, and the Laysan Albatross *Diomedea immutabilis*.

IX Systematics and nomenclature

The classification and systematics of the Procellariiformes have long been the subject of controversy and a general agreement on species' limits in the near future seems unlikely. The status of many subspecies is even more controversial, many having been named when 'splitting' was in vogue. On the other hand the urge to 'lump' monotypic genera into a single genus with neither new evidence nor a detailed re-evaluation of existing evidence is equally unwise, particularly when much biology and behaviour awaits comparative examination.

The working list used here (Appendix 1) is based mainly on the conservative checklist of Jouanin & Mougin (1979) together with some recent proposals such as those of Bourne (1983a) on *Pterodroma mollis* and its allies and of Imber (1985a) on *Pterodroma*. Alexander *et al.* (1965) set out the consensus of workers on the group at that time on the families and genera and their names. Subsequently much new information has been gained and my checklist is a compromise: some proposals have been followed, others, that appear to need stronger data for acceptance, have not. The result is a list of 103 species divided between four well-defined families — the Diomedeidae (albatrosses); the Procellariidae (fulmars, gadfly petrels, prions and shearwaters); the Hydrobatidae (storm petrels), and the Pelecanoididae (diving petrels).

In this work the term 'petrel' is a synonym of Procellariiform, tubenose or Tubinare, as used by Forbes (1882), Godman (1907–10) and other early workers, rather than in the restricted and ill-defined sense of some recent authors and checklist compilers who refer to 'albatrosses and petrels' rather than to 'albatrosses and other petrels'.

The word 'petrel' itself, while popularly considered to derive from the way that storm petrels appear to walk on the water, like St Peter, was written as 'pitteral' as early as 1676 (Murphy, 1918) and may refer to the pattering action of the feet used by these birds when feeding, rather than to any link with the apostle. Dawson (1923) also pointed out that the voice of Leach's Storm Petrel *Oceanodroma leucorhoa* syllabilised as 'petteretteral' is instantly suggestive of the name and he thought this the derivation of the word. The name Procellariiformes itself was given by Furbringer (1888) in his classic work on the systematics and anatomy of birds.

CHAPTER 2

Family Diomedeidae — The Albatrosses

Light-mantled Sooty Albatross.
Photo: author.

Contents

I Introduction

Albatrosses, the largest of petrels, have been known to the Western world since the times of the early voyagers to the southern oceans. The first monograph on a group of petrels is the *Memoire sur Les Albatros* of Forster (1785). The name derives from the Arabic *al-cadous* after the Greek *kados*, a bucket, apparently first given to the European Pelican for its capacious pouch, and then applied to other water-birds. Linnaeus's name *Diomedea exulans* for the Wandering Albatross was perhaps an unfortunate choice, as the 'Bird of Diomede', based on the classical legend of Diomedes' transformation into a bird, and referred to by Pliny, St Augustine and others, was the smaller petrel we now call the Mediterranean Shearwater, *Calonectris d. diomedea*.

A good deal has been written about the folklore and superstitions relating to albatrosses, this huge bird naturally making a big impression on whalers and sealers envious of its mastery of gales. The Ancient Mariner's albatross is thought to have been based on one of the sooty albatrosses, or perhaps on a dark-plumaged giant petrel.

Despite the superstition about the bad luck following the killing of albatrosses, sailors often ate them. In Alaska the Kodiak islanders speared the Short-tailed Albatross *Diomeda albatrus* from boats during their summer migration and, understandably, the birds were wary of canoes. At the Chatham Islands east of New Zealand, Royal Albatross chicks were highly prized foods and specially designed canoes took the Moriori to the breeding islands (details in Skinner & Baucke (1928)). Albatrosses were also caught at sea by the Maori using specially shaped and baited bone hooks (Fig. 2.1). At Tristan de Cunha Wandering Albatrosses were snared by lines baited with penguin carcasses (Elliott, 1957).

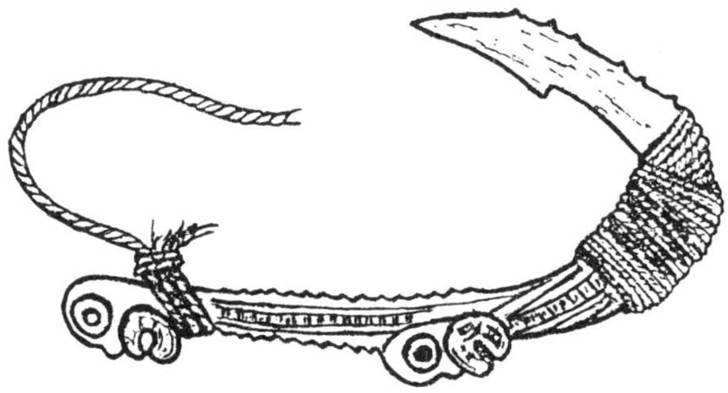

Figure 2.1 *Maori bone hook for catching albatrosses at sea. From Buck (1949).*

Albatrosses are among the few kinds of petrel that have survived captivity — *D. nigripes* for as long as 7 years (Plath, 1969).

Although most modern species live in the southern hemisphere, their fossil record is stronger in the north (Fig. 2.10). The very large *Gigantornis* from Nigeria is not now thought to be procellariiform but pelecaniform, and *Manu antiquus* from New Zealand is undeterminable (Olson, 1985d).

The earliest albatross-like bird so far described is *Diomedeoides minimus* based on femora from the middle Oligocene of Germany (Fischer, 1985). This author erects a new family Diomedeoididae to contain it. Otherwise the oldest fossil albatross is a small unnamed upper Oligocene bird (*c*. 25 x 10^6 years B.P.) from South Carolina (Olson, 1985d). Olson considers it related to the middle Miocene *Plotornis delfortrii*, from France (Milne-Edwards, 1874). Milne-Edwards noted albatross-like characters in his *Plotornis* material and Cheneval (1984), after re-examining the foot structure of the fossil, considers not only that *P. delfortrii* is a typical albatross but that the Miocene *Puffinus arvernensis* should also be placed in the Diomedeidae as *Plotornis arvernensis*. This was about the size of a fulmar and thus the smallest albatross so far recorded. There are also several middle Miocene species from California including one large (*D. milleri*) and one rather small (*D. californica*) form, and material probably of the latter, has also been unearthed in Oregon (Olson, 1985d; Howard & Barnes, 1987). Albatross material is also common in Lower Pliocene deposits at Lee Creek, North Carolina and there is a tibiotarsus of about the same age from Florida. *Diomedea anglica* was slightly smaller than a Wandering Albatross and came from Lower Pleistocene deposits in Suffolk, England. The best known southern fossil is probably *D. thyridata* (Wilkinson, 1969), a late Miocene bird from Victoria, Australia: fragmentary material from the Miocene of Argentina and the South African Pliocene has also been described (Olson, 1984, 1985a).

As Olson (1985d) points out, the picture is of a thriving and varied albatross fauna in the northern hemisphere in the later Tertiary and its disappearance must have been a relatively recent Quaternary event. The comparative richness of these northern remains does not necessarily imply a northern origin for albatrosses; rather that, like some other petrels, they were cosmopolitan. The relative abundance of their remains may be due to factors like the presence of suitable deposits and better conditions for fossilisation in the north, as well as more intensive searching there. That the North Atlantic could support an albatross population today is suggested by the 30-year residence of a Black-browed Mollymawk *D. melanophrys* with the Faeroese gannets (Andersen, 1895), and it is probably only the lack of a partner that prevents breeding by another bird of this species that appears in successive summers at Hermaness, Shetland (Bourne, 1977).

II General characteristics

A *General morphology*

Their great size is their most obvious feature, the smallest extant species, the Yellow-nosed *D. chlororhynchos*, with a weight of about 2.1 kg and wings spanning about 1.9 m compares with the Southern Giant Petrel *Macronectes giganteus*, weight 4.5 kg, wingspan 1.95 m. The 'great' albatrosses, the Royal *D. epomophora* and the Wandering *D. exulans*, with weights about 8.5 kg and wings spanning some 3 m, are the biggest seabirds. However, the size of the Wanderer has often been exaggerated and Dixon (1933), who measured 42 over 29 years at sea, found the mean span 3.10 m, the biggest 3.45 m and the smallest 2.92 m. Birds of the small subspecies nesting at Antipodes Island span about 2.71 m (Warham, 1977a).

Size apart, albatrosses are distinguished by the position of their nasal tubes. These are rather small and widely separated from each other on either side of the culminicorn, not fused together in the midline as in all other petrels.

The nostrils are complex. The naricorn, a thin, delicate scroll basically cone-shaped, protrudes forwards from the junction of the latericorn and culminicorn whose edge is emarginated allowing the protrusion. The structure is divided into upper and lower sections by an upward and inward projection that develops rapidly during the chick's first 2 months (Voisin, 1969). It is through the lower opening that the secretions of the supra-orbital salt glands flow, the upper orifice providing the main inlet for air for breathing and smelling. In many albatrosses the bases of the horny bill plates bear one or more transverse ridges: these may possibly be marks left by moults of the rhamphothecal covering, or growth lines of some sort.

A further unique feature of bill structure is the thin sliver of bone interposed between the rami of the lower mandible at their symphysis and sheathed by the corneous 'inter-ramicorn' (Fig. 1.2).

Coues (1866, p. 176) drew attention to the lobe in the concave cutting edge of the maxillary unguis in *D. exulans* which he thought might be an obsolete 'tooth'. This is clearly illustrated by Mathews (1912–13) for that species and for *D. albatrus* and in the earliest representations of albatrosses by Grew (1681) and Albin (1740). This indentation is different from the two small nicks in the cutting edges in some museum specimens which were made when albatrosses were caught by sailors with metal triangles having sharpened inner margins and baited with fat (Jones, 1936a, and Fig. 2.2).

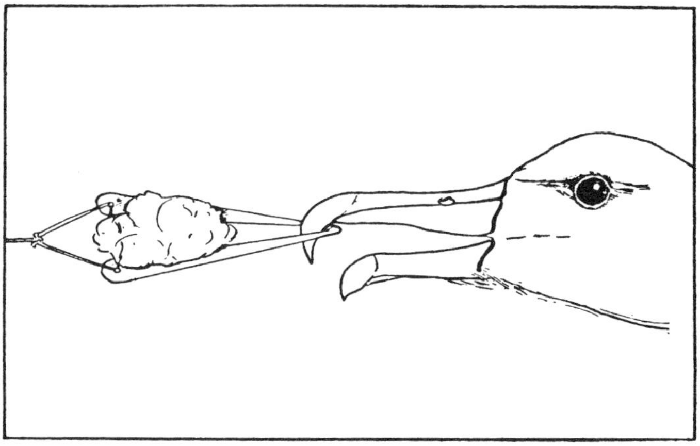

Figure 2.2 *Baited metal triangle used by whalers and sealers for taking albatrosses from sailing ships. A cork float was usually incorporated to keep the device near the surface. From Jones (1936a).*

The feet are strong and the webs large, the latter aiding take-off in calm weather (Forster, 1785). The hind toe is tiny, usually described as absent, represented externally only by a vestigial pimple but, in the larger species, just by a bony remnant revealed on dissection.

The albatross's ability to fly for long periods with fully extended wings has raised the question of how the strains on the shoulder and elbow joints are carried. Hector (1895) described an elbow-lock operating when the wing is expanded but Yudin (1957) disputed the existence of such a device. Instead he figured an arrangement where the head of the radius slides into a saddle on the head of the ulna when the wing is fully outstretched and held there by contracted entepicondyloradialis sublimus and profundus muscles. Mathews (1936) drew attention to ossifications of the patagial tendons in *D. cauta,* and supposed that the larger of these 'sesamoids' helped support the patagium when the wing was fully spread. However, neither the arrangement described by Yudin nor the sesamoid bones of Forbes (1882) and Mathews (1936), are restricted to albatrosses. Yudin found similar arrangements in *Puffinus, Oceanodroma* and *Fulmarus,* while Mathews found sesamoids in most genera apart from the fulmars.

Nonetheless, in the albatrosses, with their very long humeri and radii, the patagium is huge and an important part of the aerofoil so that its support when fully deployed would be specially useful. Pennycuick (1982) noted that in *Diomedea* its anterior part creates a drooping leading edge to the wing with a high camber and hollow underside giving the inner segments steep angles of attack even when the wing as a whole is horizontal. Because of the long inner segments of the wing and shorter terminal one, resting albatrosses fold their wings at the elbow and the wrist so that the primaries are almost completely obscured beneath the secondary feathers.

Pennycuick also described a 'shoulder lock' in albatrosses and giant petrels. When the wing is fully protracted the humerus is locked in place at the shoulder and cannot be raised but, on retraction by a few degrees, the wing is released and can be lifted. The effect is readily demonstrated on live or recently dead birds, and is due to a fan-shaped sheet of tendon of the superficial pectoralis muscle. Arising from the keel of the sternum, this converges to join the rest of the pectoralis insertion at the deltoid crest of the humerus. When the tendon fan is cut the locking effect is lost.

Except for the two sooty albatrosses, genus *Phoebetria,* which have good aerial dexterity, albatrosses have short tails, very short in the 'great' albatrosses.

The albatross skeleton has been described by Shufeldt (1888) using *D. albatrus;* Eyton (1867) illustrated that of *D. exulans* and Pycraft (1899) its skull. Kuroda (1954) illustrated and discussed the skeletons of the Laysan and Black-footed Albatrosses *D. immutabilis* and *D. nigripes*. Condon (1939) pointed out that albatross crania seen from behind, are only slightly convex, whereas in other petrels the dorsal border of the crania above the occiput are strongly arched (Fig. 2.3). This is seen well in the Waved Albatross *D. irrorata* whose head has a particularly flat-topped appearance

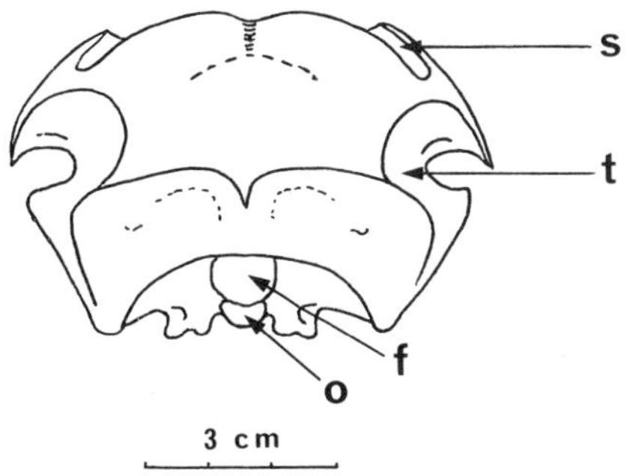

Figure 2.3 *Posterior view of cranium of the Short-tailed Albatross, mandibles removed: (s) depression for salt gland; (t) temporal fossa; (f) foramen magnum; (o) occipital condyle. From Shufeldt (1888).*

— see e.g. plate 8 in Loomis (1918). Albatrosses have deep lunate depressions above the orbits housing the salt glands but in *D. irrorata* they are unusually small and Watson & Divorky (1971) surmised a less salty diet for that semi-tropical species. In the palate the most distinctive feature is the keeled and complex vomer (Fig. 2.4) (Forbes, 1882) and there are no basipterygoid processes.

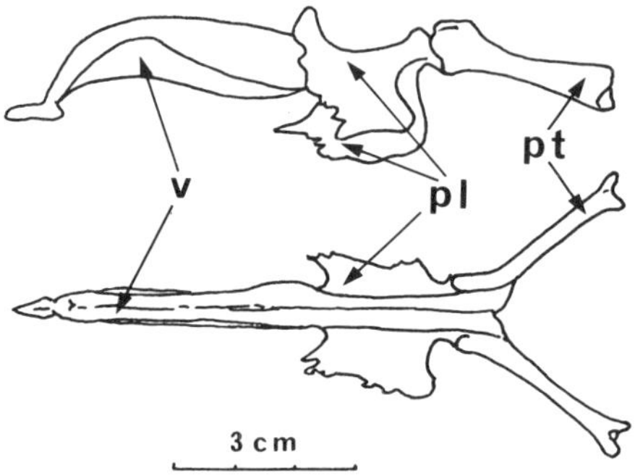

Figure 2.4 *Vomer and associated bones of Short-tailed Albatross. The albatross vomer (v) is peculiar in being abruptly downturned in midlength and keeled ventrally in lateral (upper) and dorsal view. It is ankylosed to the palatines (p1), here cut away from the skull (pt) pterygoids. From Shufeldt (1888).*

Forbes described and illustrated some of the musculature of *D. exulans* and *D. albatrus* while Klemm (1969) made a detailed study of the hind limb muscles of *D. immutabilis* and *nigripes*.

Other aspects of the internal anatomy that have been investigated include the complex air-sac systems of *D. exulans* and *Phoebetria* sp. (Ulrich, 1904) and of *D. immutabilis* (Hamlet & Fisher, 1967).

B *Plumage*

Most fully adult albatrosses are white-bodied with dark upper wings and tails. Only *D. nigripes, Phoebetria* and *D. irrorata* retain sooty or dark brown body plumage to maturity, but the young of *D. albatrus* and *D. exulans* fledge with mainly dark brown feathers, and some subspecies of *exulans* retain much of this juvenile pattern into adulthood.

Albatrosses separate clearly into two size groups — the 'great' albatrosses, the Wandering *D. exulans* and Royal *D. epomophora* (plus *D. amsterdamensis* if considered a full species); and the smaller north Pacific and Galapagos species together with the remaining southern kinds. The southern forms of the second group have white bodies, dark mantles and upperwings and grey tails and are commonly called 'mollymawks' in the countries where they breed. This was a term introduced to the south by the whalers, from the Dutch 'Mallemuck' meaning 'foolish gnat' and was originally applied to the Northern Fulmar which thronged around fishing boats like gnats and were so greedy for food as to be easily killed (Martens, 1675).

In the North Pacific the Laysan Albatross has a plumage pattern rather like that of the mollymawks and has often been thought closely related to them, but their behaviour and their ectoparasites suggest otherwise. The Short-tailed Albatross, the largest petrel of that region, also looks rather like an undersized male Wanderer but its bill is different in colour and structure, its crown wears golden feathers like those worn by gannets, and its behaviour too is much more like that of the other northern kinds than any of the southern ones. Golden feathers are also seen on the heads of adult Waved Albatrosses, birds that are otherwise notable by their pale mantles that are barred or waved with brown and tails that stay dark brown throughout life. In all these smaller albatrosses as well as the two sooty species the nasal tubes run rather further forward along the bill than they do in the 'great' albatrosses.

The mollymawks have a broken ring of white feathers behind and below the eye, but this is inconspicuous except with those that are grey on the face like Buller's *D. bulleri* and the Grey-headed species *D. chrysostoma*. The function of these eye rings is not known, nor of the much more conspicuous ones of *Phoebetria*: they do not appear to alter during displays, for example. The eyes themselves are rather recessed beneath the supraorbital ridges perhaps to help as eye-shades, and the depression above, behind and before the eye in mollymawks and *D. immutabilis* is lined with dark feathers which may also help to reduce light scatter into the eye.

Albatross chicks usually have pale grey down, but those of *immutabilis* and particularly *nigripes* grow distinctive dark grey, loose, ragged down with white tips that produce a silvery, peppered effect. The downy young of the Waved Albatross are also unusual being either a drab grey, with or without creamy tips, or light grey, subterminally creamy white but dark brown at their tips and the down also tends to be ragged or twirled together in tufts (Loomis, 1918). This colour polymorphism is evidently not sex-linked.

Sooty albatrosses remain dark throughout life but Wandering and Short-tailed species, that wear mainly blackish brown feathers on fledging, whiten extensively during successive moults. Black-footed Albatrosses also tend to get whiter ventrally in old age. The changes with age have not been described for the Short-tailed Albatross whereas thousands of *D. exulans* chicks have been ringed and hundreds of known-age birds recovered, particularly off New South Wales by the late J.D. Gibson and his team who scored the plumage states of captured birds according to a code (Fig. 2.5). Recaptures of previously scored birds made at intervals of up to 21 years show that for some populations of the subspecies *chionoptera* the progression to the whitest plumage takes about 20 years independent of sex (Robertson & Warham, unpubl.). The Gibson Code has been further elaborated by Jouventin *et al.* (1989).

The irides of all species are brown except for the subspecies *impavida* of the Black-browed Mollymawk *D. melanophrys*, where they lighten to a pale buff as adulthood approaches. The eyes of the Waved Albatross seem to be directed forwards more than in other species, perhaps enhancing binocularity.

Adult mollymawks have coloured bills, usually with a pink tinge to the upper nail, but they start life with blackish bills and fly when these are still dark and their body plumage is often darker on head, neck and underwings than that of the adults. Mollymawks differ from other albatrosses in having fillets of orange or yellow skin around the base of the mandibles and an orange or pinkish gape stripe

BACK		
1	All brown	
2	Mottled brown and white with some pencilling throughout.	
3	Mostly white and pencilled with some dark feathers on lower back.	
4	Pencilled all over (4H = heavy and distinct)	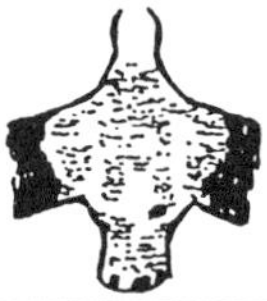
5	Faint pencilling, or traces thereof.	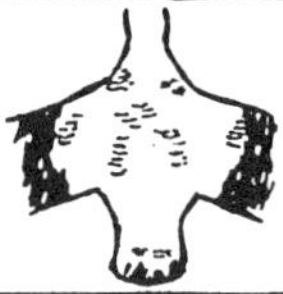
6	White	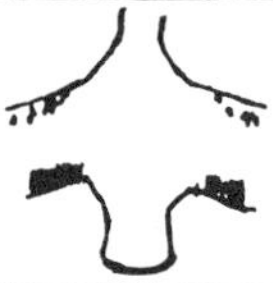

HEAD		
1	Sharply defined brown juvenile pattern	
2	Juvenile pattern still easily discernible	
3	Some dark feathers in crown & nape	
4	Separated dark crown	
5	Traces of dark feathers in crown	
6	Pure white crown	

INNER WING		
1	All dark	
2	A few white feathers	
3	Distinct white patch at elbow	
4	White patch merging with white of back	
5	Mostly white	

TAIL		
1	All dark	
2	White and dark	
3	Few dark tips	
4	Pure white	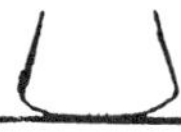

bared in display (Fig. 2.6). The Black-browed and Shy Mollymawks have a pencil-thin black line around the base of the beak, a feature seen also in one northern species — *D. albatrus*.

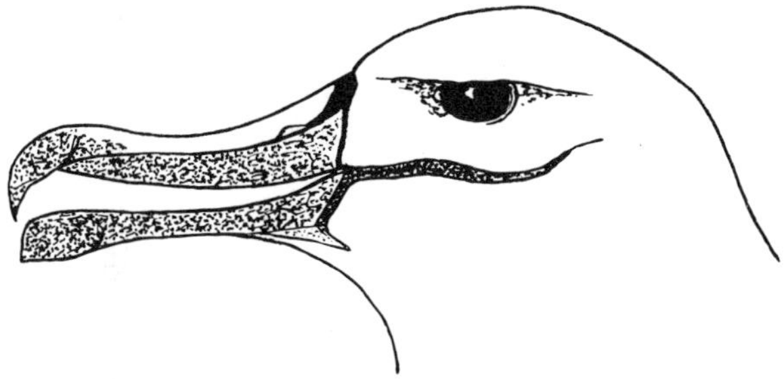

Figure 2.6 Diomedea cauta *displaying its orange cheek stripe. This is continued down the bases of the ramicorns as an orange bar. Note the thin black line at the base of the latericorn and the broad band of black skin separating the pale culminicorn from the forehead — a characteristic of subgenus* Thalassogeron.

The 'great' albatrosses have white or horn-coloured bills. These and the pale eyelids become pink when flushed with blood during exertion — when displaying, feeding chicks or when hauled from the sea to a ship's deck. The effect is perhaps greater during the pre-laying stage of the breeding cycle when the knee joints and rear of the tarsi may also acquire a pinkish flush (Richdale, 1952, p.81).

A peculiarity of the plumage of Wandering Albatrosses is the occurrence of pink stains on the sides of their necks. The stains appear to be superficial, may occur on one or both sides, and may go part-way down the nape or even across the back of the head forming a band from ear to ear (Warham & Bell, 1979; Tickell, 1980). This phenomenon has long been known and Murphy (1936, p.559) suggested that the chemical nature of the stain needed investigation. This has not yet been done, but the phenomenon occurs in all the populations examined, on birds of both sexes, captured at sea or on land, in breeders and non-breeders, and in winter and/or summer. Bennett (1860) said that the stain faded with time but feathers clipped from birds still show the colour 20 years on (pers. obs.).

Tickell (1980) found that about 75% of adults were so marked and carried the stain over long periods. It seemed confined to the outer vanes of the feathers with no signs of exudation from the skin and he pointed out that as the heaviest staining was on members of pairs that were separated, it seemed unlikely that it was euphausid pigment transferred during allopreening. Tickell suggested that the colour was from some externally acquired contamination, perhaps during feeding on *Euphausia*. His observations also tend to discount its being due to a scent gland or discharge from the ear. At South Georgia the males were the more heavily

---

Figure 2.5 *(opposite page) Gibson code for scoring Wandering Albatross plumage phases using a 4-figure combination. Circled numbers refer to arbitrary stages in plumage development, not to age, and in some populations much of the dark juvenal pattern is retained into adulthood. From Gibson (1967).*

stained but Macquarie Island males less so (Tompkins, 1983a). Rather similar stains on Little Gull *Larus minutus* have been attributed to pigments in their preen gland oils (see Gardner, 1987).

C *Moult*

Little is known of moult in albatrosses, no-one having regularly examined individual birds of known age over a long period. Some, like breeding *D. immutabilis*, start a slow moult of contour feathers around hatching time and continue it for several months (Fisher, 1967). Harris (1973) examined primary moult in *D. irrorata* known to be not less than 4–7 years old. As appears usual among petrels, wing and tail feathers were not shed during breeding. Old feathers among the inner primaries were the first to be replaced. Up to three generations of feathers were present among the primaries, secondaries and wing coverts and Harris considered that as intermediates were missing, the moult must occur over only a few months, his three types of feathers being assumed to be <1, 1 and 2 years old respectively. Of 1487 birds examined, 755 had replaced all their primaries since the previous season, the others had retained 1–3 old primaries, these usually being the three innermost. He thought that the sequence was descendant from the proximal primaries outwards, perhaps starting at two foci and possibly likewise in the secondaries, i.e. 'wave moult'. In this species Harris found that 30 birds that had successfully reared chicks had retained significantly more old primaries than 15 unsuccessful ones, implying that for successful birds, tied longer to the island, there was insufficient time for a complete replacement before the next season.

Brooke (1981) and Brooke & Furness (1982) examined the flight feather moult of some southern mollymawks. Their samples were small and were of birds of unknown age and breeding experience. Some of their examples of adult birds and most immatures seemed to show normal descendant moult, others showed wave moult in which a second descendant moult started before the first had finished. Brooke thought that most adults had interrupted their moult to breed as with *D. irrorata*, but non-breeding immatures seemed to show a similar pattern. The basic sequence with *D. melanophrys* was to shed the secondaries in an ascending series from two foci, one at the outermost (secondary 1) and one at numbers 19 or 20. At later ages replacement appeared to be rather random and the pattern on the two wings might differ, whereas that of the primaries was symmetrical. Tail feathers had three foci of replacement on either side.

The situation seems even more complicated, however, for Brooke & Furness (1982) found three adult and one immature Black-browed Albatrosses showing an ascending moult of their primaries and one a descending moult in its secondaries. More work on birds of known status, sex and age is clearly needed to determine typical patterns.

Albatrosses shedding their outer primaries may have significantly reduced wing spans, but at least in non-breeders, a reduction of 0.5 m in span evidently does not seriously reduce the safety margin for foraging in *D. melanophrys* and *D. cauta* (Brooke, 1981).

D *Sexual dimorphism*

Male albatrosses tend to be larger, weigh more and to have heavier bills than do females. In some, e.g. *D. nigripes* and *D. immutabilis*, the differences can be seen in the field when partners are standing together, but such differentiation is seldom possible with mollymawks. Frings & Frings (1961) found that by measuring head widths at their widest point just behind the eyes males *immutabilis* were separable by their greater dimensions. Some, perhaps all, adult mollymawks can be sexed by this means. Male *immutabilis* are also thicker in the neck (H.I. Fisher, pers. comm.).

Richdale (1952, p.81) noted that male Royal Albatrosses had bigger feet than their partners and Westerskov (1960) proposed sexing the species on the length of the middle toe nail; in males this was >24 mm, in females <24 mm.

Wandering, and to a lesser degree, Royal Albatrosses, are usually more easily sexed on plumage characters: males tend to have fewer, if any, dark feathers on the crown, body and tail than their partners, and this is true even of races that breed in dark plumage.

E *Hybridisation*

Hybrid petrels are rare and among albatrosses have only been described between the Laysan and Black-footed species. These often nest close together and in most recent seasons several such birds have been found on Midway Atoll. In 1980 there were five of them (pers. obs.). Fisher (1972) figured examples and described their behaviour. Those watched in 1980 often tried to 'dance' with Laysan Albatrosses but failed to complete the performance and there is no instance of successful breeding by a hybrid. They were never seen to show any interest in nearby Black-footed birds.

III The genera

Thirteen species are listed in Appendix 1, but species' limits are provisional. *D. (exulans) amsterdamensis* is regarded as a full species by its describers (Roux *et al.*, 1983) — see also Jouventin *et al.* (1989) and there are questions about the status of the three subspecies of the Shy Albatross *D. cauta*: are some or all of these full species? The status of the two forms of *D. melanophrys* is also doubtful: they may need to be raised to full species because pairs with the dark eyes of the nominate form have been found nesting in colonies of *D. m. impavida* (G. Taylor, pers. comm.).

Harper (1978) analysed plasma proteins from four species, finding the patterns of *D. exulans* and *D. epomophora* very similar but with *D. melanophrys* and *P. palpebrata* showing significant differences from the larger species in the mobility of some components (Fig. 2.7). For chromatograms from plasma of *D. immutabilis* and *D. nigripes* see Brown & Fisher (1966).

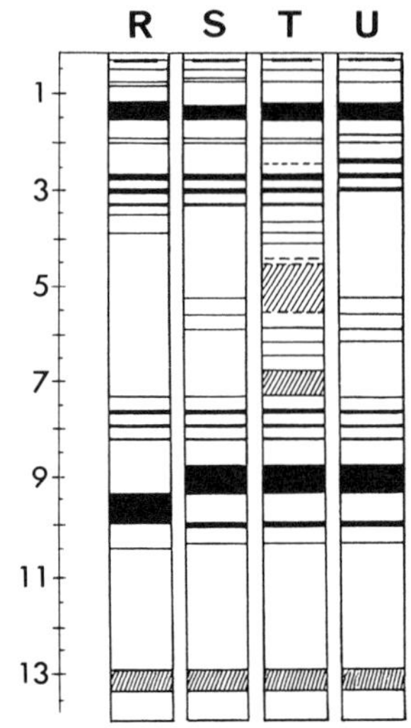

Figure 2.7 *Blood plasma proteins of (R)* P.palpebrata, *(S)* D.exulans, *(T)* D.epomophora, *and (U)* D.melanophrys. *A significant difference in the mobility of the albumin compared with the others is evident in* Phoebetria, *while the transferrin triplet and esterase bands of* D.melanophrys *also differ from the rest. For details of banding patterns see Appendix 2. From Harper (1978), courtesy of* New Zealand Journal of Zoology.

A *Genus Diomedea*

As early workers like Audubon (1838, p.329) realised, the different species can be identified by the shapes and arrangements of the bill plates (Fig. 2.8). Many of the differences show in nestlings, allowing chicks in mixed colonies to be identified.

Although most current classifications admit but two extant genera, the bill differences have provided the basis for three subgenera in addition to *Diomedea sensu stricto*:

1 *Phoebastria*

Phoebastria is characterised by the descent of the culminicorn behind the nasal tubes to overlap the latericorns so that the base of the culminicorn forms a massive boss before the eyes. This condition obtains in all three North Pacific species and in the Galapagos *D. irrorata* (Fig. 2.8,16).

2 *Thalassogeron*

In *Thalassogeron* the bill is rather compressed laterally and behind the nostril the culminicorn is separated from the latericorn by a strip of black skin. This is most apparent in *chlorohynchos* and *cauta*, less so in *bulleri* and *chrysostoma*. In all four, the bases of the latericorns are widened, their lower edges in advance of the upper, producing lateral plates in front of the eyes rather than a boss. All are mollymawks of the Southern Ocean (Fig. 2.8, 2–8).

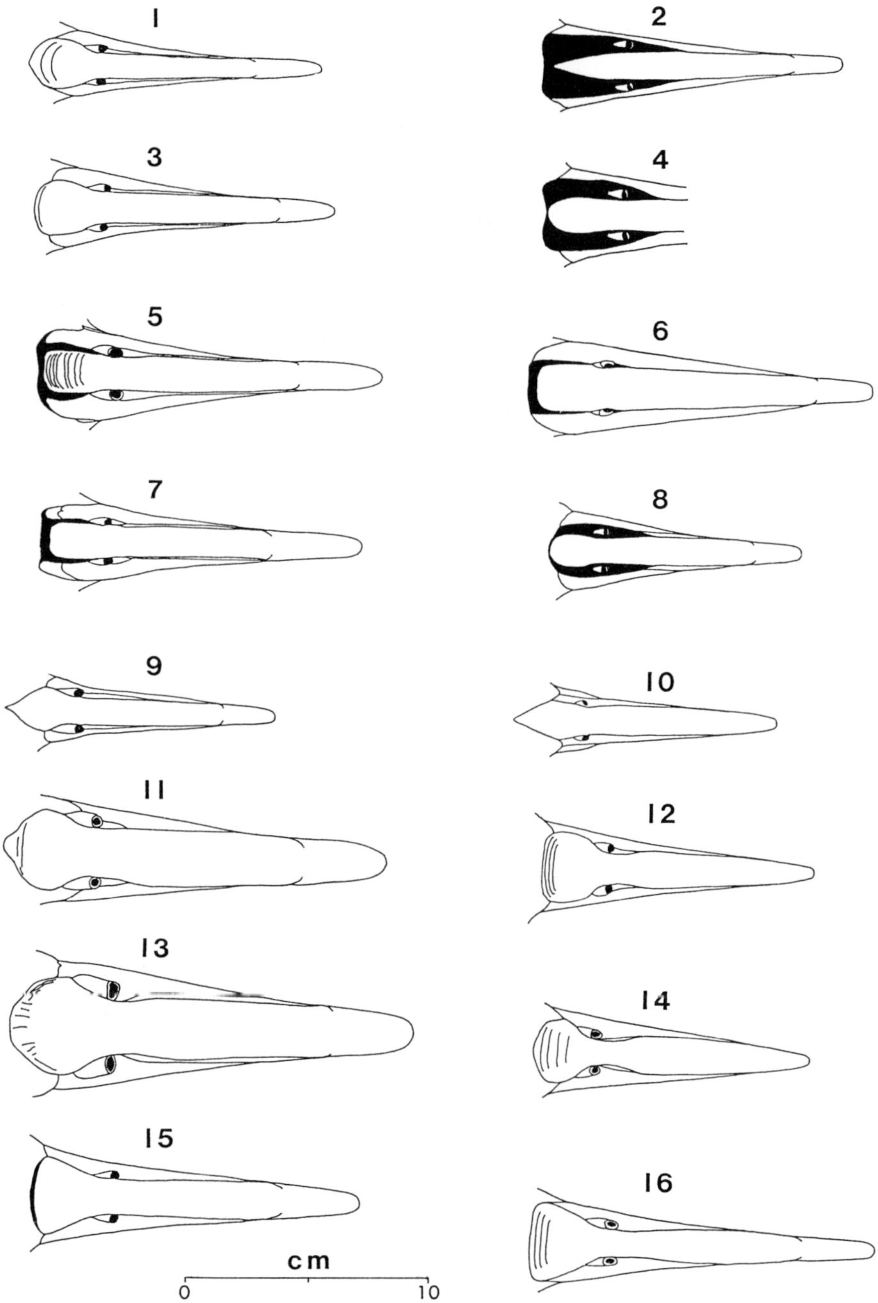

Figure 2.8 *Bills of albatrosses in dorsal view: 1.* D.melanophrys; *2.* D.chlororhynchos bassi; *3.* D.bulleri; *4.* D.c.chlorohynchos; *5.* D.c.cauta; *6.* D.c.salvini; *7.* D.c.eremita; *8;* D.chrysostoma; *9.* P.palpebrata; *10.* P.fusca; *11.* D.exulans; *12.* D.immutabilis; *13.* D.epomophora; *14.* D.nigripes; *15.* D.albatrus; *16.* D.irrorata. *del. R. Holdaway.*

3 *Thalassarche*

Thalassarche shows a condition of the bill plates intermediate to that of the mollymawks and the great albatrosses. The culminicorn is expanded and rounded at its base with no skin or membrane visible between the plates or before the frontal feathers. The Black-browed Albatross is the sole member of this taxon (Fig. 2.8,1).

Some current studies indicate the desirability of reviving some of these, e.g. *Phoebastria*, the four species involved seeming to stand out from the southern ones in behaviour and morphology. Palma & Pilgrim (1984) suggested that *D. irrorata* was most closely related to the southern mollymawks because it was the only northern albatross hosting the feather louse *Harrisoniella ferox*, otherwise found only on mollymawks. However, the structural and behavioural similarities relate that bird to the other northern species. Nevertheless, the Waved Albatross is unique in various facets of its biology and morphology, being the only equatorial species, the only one laying in May and June, the only one with bill longer than its tail and with very distinctive plumage. It also has four kinds of Mallophaga not found on other petrels suggesting that it may be the oldest of the extant northern albatrosses. It could have arisen from an ancestor that gave rise to the mollymawks and which, on invading the north, brought the feather lice with it.

B *Genus Phoebetria*

The two sooty albatrosses retain mainly brown or grey-brown plumage throughout life. Their bodies appear fusiform due to their pointed bills, and slender, wedge-shaped tail: the result is an aerodynamically smooth outline for these most aerial of albatrosses. The bill is laterally compressed, glossy black and has a deep sulcus along the side of the ramicorn filled with coloured flesh, bluish in *P. palpebrata*, yellowish in *P.fusca*. The nostrils are rather exposed and arise from the maxillary plates without an emargination of the culminicorn. The salt glands are very large and provide a diagnostic feature for the skull as the depressions housing the glands are separated by only a sharp ridge (Pycraft, 1899; Condon, 1939). According to Forbes (1882) the hallux has an external claw.

Phoebetria has often been regarded as the most primitive of albatrosses on account of its dark plumage — a juvenile character of some of the other species. Its behavioural repertoire is also smaller and simpler than that of other albatrosses. Mathews (1912–13, p.266) pointed out that the V-shaped sides of the bases of the ramicorns in *D.exulans* and some mollymawks, which allow the feathering of the face to extend forwards as a salient into the lower mandible, suggest their evolution from a groove-bearing ancestor. The fleshy bar at the base of the ramicorn in mollymawks may be a remnant of a sulcal membrane. In *Phoebetria* the feathering of the face also ends in a triangular salient along the ramicorn to give way to the sulcal fillets but on the culmen the frontal feathers make a sharp re-entrant.

The most prominent feature of the plumage is the bold ring of white feathers, interrupted in front, around the eye. This gives the bird a solemn, monk-like mien and presumably accounts for the generic name (from the Greek *phoibetria*, = soothsayer or prophetess).

IV Life styles

Although they are such aerial birds, and seen far from shore, adult albatrosses spend much of their time on land because the breeding cycles are very long. They are the most mobile of the petrels when on their feet.

A *Walking*

Albatrosses are surface nesters. They rest on their bellies and rise to their feet in two stages — first onto their tarsi, then onto their feet. They can stand around for some time without apparent tiring: their big flat feet no doubt help. The Laysan and Black-footed species are especially mobile, lifting onto the very tips of their toes at the height of their 'dancing' displays.

Albatrosses usually walk slowly as if planning each step, but the mollymawks and the northern species may be very sprightly and can run quickly, e.g. when displaying or attempting rape. The Waved Albatross even walks with its egg tucked into a brood 'pouch' in the style of an *Aptenodytes* penguin. Sooty albatrosses seldom walk, perhaps because they are ledge nesters with little room for perambulation, and Klemm (1969, p.141) noted the more proximal arrangement of the leg muscles in these birds which implied some restriction of movement and a shorter stride.

B *Swimming and diving*

Albatrosses swim well, riding quite high on the sea, and can move quickly with powerful strokes of their webbed feet. Like other petrels, when feeding or trying to maintain position on the water they will half-unfurl their wings and allow the wind to take their weight, sometimes completely, so that they walk on the water.

Few birds seem less designed for diving, but shallow dives for food, including complete immersion, have been recorded for *D.melanophrys*, *D.chlororhynchos*, *D.nigripes* and even *D.exulans* (references in Voisin, 1981).

C *Flight*

Due to their heavy wing loadings and high aspect-ratio wings, albatrosses fly fast but, as they usually follow a sinusoidal course, estimation of ground speeds is difficult. Pennycuick (1982) calculated ground speeds for three albatrosses at 10–11 m s^{-1} and a Wandering Albatross that kept company with a helicopter up to 31 m $^{s-1}$ broke off contact by diving ahead and below the plane (Robertson & Jenkins, 1981).

Take-off in winds is usually easy, whether from sea or land. Most nesting areas are within reach of cliff edges or ridges from which departure is facilitated during calms. Becoming airborne from flat seas involves a long run with wings beating the air and feet beating the water. Under windy conditions merely spreading the wings is enough for lift-off but the great albatrosses often nest among tussock and fern and walk to clearer places where they can unfurl their wings without obstruction. The Sooty Albatrosses are particularly adept: even in quite light airs they may rise directly in the breeze to hang in the air a few metres up, wings and tail being constantly trimmed to every nuance of the wind without being beaten. However, these birds do seem to need a cliff ledge for take-off.

Landings are usually more hazardous than departures because, with its heavy wing-loading, an albatross's approach speed is high and its short tail and high-aspect ratio wings allow little opportunity for course correction if wind conditions change suddenly, e.g. through gusting. In strong winds landings are usually easy but in light airs several circuits may be needed in which the bird almost stalls above its eventual alighting point before accelerating and trying again. This is the normal procedure with *Phoebetria* which touch down precisely. At Midway Atoll many Laysan Albatrosses misjudge wind conditions and land heavily, even turning somersaults before coming to rest, having failed to stall and reduce their ground speed to zero at the point of touch-down. Another hazard there is a forest of introduced *Casuarina equisitifolia*. Many pairs nest under these trees (probably some chose their sites before the trees appeared), and now have to land on roads or paths and then walk below the canopy to their nests. Such obstacles are alien to the albatrosses' natural world, and some die on striking tree trunks.

When landing, whether on water or on land, the webbed feet are spread to increase drag and the birds do not use vigorous 'back-pedalling' wing strokes when trying to control their stalls: instead a peculiar and rapid rotation of the wing about its axis effects a similar kind of control (Pennycuick, 1982).

The ability of albatrosses to fly with the minimum of exertion is proverbial. They can also fly in calms, beating their wings through shallow amplitude with most of the movement being of the distal wing segment that pivots on the wrist, but this sort of flight is not maintained for long.

Normal progression is achieved partly by 'dynamic soaring', a process that has come under repeated scrutiny and analysis. This relies on the gradient in horizontal wind speed that exists between that at the sea's surface and that higher up. Due to loss of energy by the breeze in moving the waves and in friction with the sea, the wind speed near the surface is lower than it is above it, until at a height varying with the wind speed, the gradient is zero.

The classical explanation of how the birds extract energy from a sheared wind can be separated into four components (Gray, 1968). Travelling near the surface at high speed and with much momentum due to its considerable weight, the bird banks hard and suddenly climbs into the wind — the windward climb (1). During this phase the bird gains potential energy due to the thrust of the wind on the cambered underwings and, more important, the suction effect of the high-speed flow over the aerofoil. The decelerating climb continues as long as there is enough wind gradient, but when drag and the backward component of the body weight slow the ascent, the bird makes a high altitude leeward turn (2) and accelerates down or slightly across the wind (3) and when nearly at the surface (often cutting into it with the wing tip) the albatross turns sharply across wind and, (4) slides along hugging the sea and still travelling fast. This is when most of the progression against the wind is made: the albatross expends its stored energy where the wind resistance is low.

Using this flight style the bird can tack across the wind and make substantial progress against it, or travel even faster downwind. The high aspect-ratio wing minimises energy loss through aerodynamic drag. Such soaring is most efficient in strong winds; when the winds are weak the birds often boost their progress with several flaps of their wings during the windward turn.

The sequence of movement has been described by Jameson (1958) and Gray (1968), in great detail by Cone (1964), and in a computer simulation by Wood (1973).

However, it is a commonplace observation that much albatross (and other petrel) movement is along a wave, the birds evidently taking advantage of the curtain of wind deflected up from its windward face. This is 'slope soaring' and has long been recognised as used by albatrosses, but it has only rather recently been appreciated that it may be more important than dynamic soaring. Wilson (1975) proposed that not only does slope soaring along waves provide the initial 'kick-off' for dynamic soaring, but that the birds can use the upcurrent to gain speed and, if this is sufficient, can then zoom up from the wave and travel in the desired direction. He tested this hypothesis in a computer simulation, and Pennycuick (1982) extended the idea during experimental studies at South Georgia where he could record albatross movements and wind variables with automatic instruments. He estimated that gradients useful for dynamic soaring only extend to about 3 m from the surface, whereas the great albatrosses often pull up to 15–20 m in strong winds. Pennycuick concluded that most of the energy for windward climbs comes from slope lift, gained at relatively slow speeds, and less from the wind gradient. The latter may perhaps be the major energy source in downward flight, but no direct data were obtained on that. In reality both slope and dynamic soaring seem to be used. In violent storms, however, so much spray is sheeted from the wave tops that soaring is impossible and wing beats are brought into play (Dixon, 1933).

The result of these manoeuvres is that the albatross generally moves along at an angle to the prevailing wind. The average direction tends to be down wind in non-breeding birds, hence the vast movements seen especially in the Southern Ocean. Nesting birds must obviously be able to make headway against contrary winds if these persistently blow from the same quarter.

D *Ship following*

Most kinds of albatross will follow ships, tacking across the wake ready to splash down beside discarded rubbish or offal and sailing around the ship from time to time under appropriate wind conditions. The birds' numbers fall off towards nightfall but individual birds may trail a ship for several days. Tschudi (1856) caught, marked, and released a Wandering Albatross which stayed with the ship for 6 days and Dixon (1933), identifying individuals by missing wing feathers or broken webs in their feet, also found some in attendance for several days.

Although the ship travelled 190 km on a dark night, a bird caught up with it the following morning. Dixon had no evidence that albatrosses remained with the ship after dark but could not say how these found it again in the morning. Dixon saw Wandering Albatrosses flying at sea on light nights, but they were not then attracted to ship's offal. They can also be seen flying over the breeding places in moonlight. Buller (1894) reported a marked sooty albatross (probably *palpebrata*) that followed a ship for 4800 km.

Most ship following is much briefer. Cock & Schneider (1982) analysed sightings of recognisable *D.exulans* between Cape Town and Marion Island. The mean time of following was 44 min (range 3–250 min), but the morning mean was only 9.4 ± 4.8 min compared with afternoon figure of 78.6 ± 6.4 min. These authors suggested that the birds were hungrier later in the day, having fed overnight.

Loye Miller (1940) reported that colour-marked *D.nigripes* did not follow his slow moving boat for more than 50 km, and wondered whether they dropped out to return to known feeding territories: the work of Gibson's team shows that individual birds cross great distances to return to known but restricted feeding areas.

E *Behaviour*

Albatrosses have larger brains than other petrels and their behaviours during breeding are probably the most complex of the tubenoses, the visual and vocal components presumably augmented by olfactory clues, although nothing is known of these.

Large size has favoured the evolution of this complex communication system. In the first place the birds are big enough to be active on land by day without interference from predators, so that visual signals can play a major role, whereas the nocturnally active smaller petrels may have to rely more on aural and olfactory clues. Secondly, their absolutely larger brains presumably give them a greater capacity to store, handle, remember and respond to complex information than can the smaller petrels. Thirdly, they live longer than the smaller kinds but must make repeated breeding attempts to replace themselves by two breeding adults (in stable populations, at any rate). Thus they live socially for a major part of their lives and this makes a complex communication system more valuable, and they have more time to learn it. Presumably this has been selected for by the good communicators having left more progeny than the poor ones.

In addition to mutual and reciprocal preening as seen among all petrels, albatrosses perform elaborate courtship and pair-bonding rituals that may involve widely spread or partly spread wings, head swinging and bobbing, bowing into nests, most accompanied by loud whistles, brayings and bill clappering. The elaborate posturings and vocalisations involved in territorial advertising and mate attraction may be presented in definite sequences or 'sets', the best-known example being the 'dance' of the Laysan Albatross studied by Meseth (1975), the vocalisations of this bird and the Black-footed species being described by Sparling (1977). Richdale (1949, 1950) described the repertoire of Buller's Mollymawk and of the Royal Albatross, and Tickell (1984) compared the behaviour of Black-browed and Grey-headed Mollymawks. Warham & Fitzsimons (1987) dealt with the vocalisations of Buller's and other albatrosses. The displays of the sooty albatrosses were examined by Berruti (1981b) and a very detailed analysis of the visual communication patterns of *P.fusca* by Jouventin *et al.* (1981).

Established and paired breeders display relatively little. Most activity on the nesting grounds is that of pre-breeders and of birds that have lost their mates. The pre-breeders may form 'clubs' and display among themselves on bare headlands or other areas adjacent to breeding colonies. The great albatrosses assemble in groups of three or more (the so-called 'gams') in open places in the general nesting terrain. Activity is greatest in the afternoons and evenings.

Pre-breeding sooty albatrosses spend a lot of time in dual flights over their nesting places (Fig. 2.9) and other species also use aerial displays at times. Their significance is not clear (Warham, 1976). Many albatross species have been seen performing 'ground' displays at sea and Laysan Albatrosses have even alighted on a ship's deck to perform mutual displays (Young, 1986).

Figure 2.9 *Dual display flights of* Phoebetria *albatrosses are facilitated by long tails which confer mobility exceptional among albatrosses.*

V Distribution

Figure 2.10 shows how the species are concentrated in the Southern Ocean. At the Prince Edward and Kerguelen Groups six species breed — a great albatross, three mollymawks and two sooty albatrosses. The recent discovery of *D.cauta salvini* meeting at the Crozets (Jouventin, 1990) means that seven albatrosses breed in that archipelago. Both giant albatrosses nest on Campbell and the Auckland Islands. At Heard Island a single Wanderer chick was found in a nest in 1980 (Johnstone, 1982). The only adult identified ashore was a male that had been ringed at

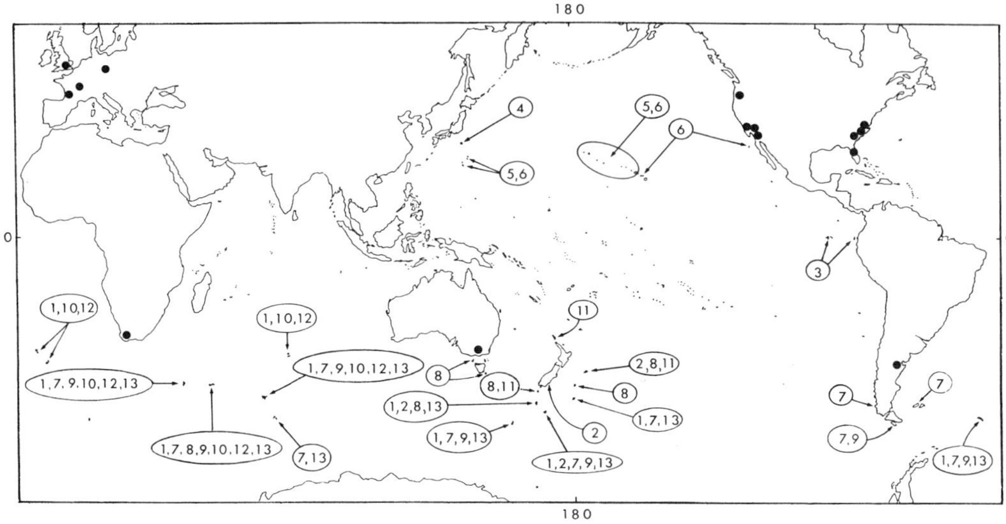

Figure 2.10 *Breeding distribution of albatrosses: 1. Wandering; 2. Royal; 3. Galapagos; 4. Short-tailed; 5. Black-footed; 6. Laysan; 7. Black-browed; 8. Shy; 9. Grey-headed; 10. Yellow-nosed; 11. Buller's; 12. Sooty; 13. Light-mantled Sooty. Solid circles show sites where fossils have been found.*

Macquarie Island 13 years before. Heard Island may once have been a regular breeding place, its climate being much like that of South Georgia, but although there were two old nest mounds in 1980, no further evidence of breeding has been obtained.

Rather surprising is the presence at Marion and Prince Edward Island of *D.chrysostoma* but not *D.melanophrys* and at Heard and McDonald Islands of *melanophrys* but not *chrysostoma*, because the last is typically more of a cold water bird than the former and its winter foraging range lies further to the south. Also absent is any albatross from even the most northerly islands of the Scotia Arc except South Georgia where four species breed.

Like other petrels, albatrosses tend to philopatry and some subspeciation has occurred, notably in *D.exulans* whose forms tend to follow eco-geographical rules for size, but not entirely for plumage colour. The most northerly populations at Tristan da Cunha, Gough, Amsterdam, Auckland and Antipodes Islands are smaller than the southerly populations, subspecies *chionoptera*. The whitest birds do breed furthest south, close to and on either side of the Antarctic Convergence, but the most northerly ones at Tristan da Cunha are intermediate in plumage when adult whereas those at Amsterdam Island, at almost the same latitude, are very dark. Further east, around New Zealand, all populations lie in the Subantarctic Zone, but the northernmost at Antipodes Island are dark birds, to the south those at the Auckland Islands are pale or of intermediate colour, further south again at Campbell Island they are dark, and furthest south in that sector those at Macquarie Island are light — typical *chionoptera*.

The southern albatrosses breed in or near the west wind zone, the recently discovered group of *D.bulleri* at the Three Kings Islands, north of New Zealand at 34°S being the most northerly in the hemisphere. North Pacific species use the prevailing westerlies there, particularly during the breeding season (the northern winter). However, calms are not rare around the Hawaiian islands but these do not appear to restrict the attendance of birds at their nests: evidently continuous strong winds are not essential for successful breeding.

The Waved or Galapagos Albatross, nesting the northern end of the cool Humboldt Current and in the south-east trade wind zone may experience less dependable wind conditions than any other of its family. Murphy (1936, p.531) commented on its laboured flight and frequent wing beats in light winds, with no dynamic or wave soaring and even appearing less buoyant than pelicans and frigate-birds flying nearby. However, in good winds the bird evidently flies well.

VI Feeding and foods

A *Feeding methods*

Albatrosses take most of their food by surface seizing. They alight to grasp their prey with hooked, razor-sharp beaks. They will scavenge for ships' discarded waste, particularly if fatty, but their tastes are catholic. Smaller sea birds may be killed and eaten, and as albatross gapes are enormous, such prey are often swallowed whole with bones and feathers regurgitated as castings through the mouth, but the birds seem to depend largely on squid. Murphy suggested that these were most likely caught after dark during the animal's daily migration to the surface and recent work based on the identification of squid 'beaks' from albatross gizzards has bolstered this idea. There are nevertheless few first-hand accounts of these birds taking squid. Gibson & Sefton (1955) reported Wandering Albatrosses swallowing pieces of dead cuttlefish *c.* 30 × 15 cm, tearing off chunks by raking backwards, bracing themselves with spread feet and half opened wings.

Smaller species can be quite agile. Barton (1979) saw *D.cauta* and *D.chlororhynchos* gliding slowly against the wind over mackerel and pilchard schools and taking the fish without alighting. The larger species, *cauta* landed to swallow its catch, the other swallowed it head first while on the wing. In the North Pacific *D.nigripes* takes flying fish in a similar manner (Thompson, 1951). Milledge (1977) noted that the smaller, more manoeuvrable *chlororhynchos* was very adept at taking offal, but was last in the albatross hierarchy when it came to competing for it.

Off New South Wales many Wandering Albatrosses are attracted to a regular but seasonal food supply provided by a cuttle fish *Amplisepia verreauxi* which dies in large numbers, partly from attacks by dolphins, and in Australian seas, as elsewhere, cetaceans and albatrosses and other petrels are often found together. It has even been suggested that some of the food found in albatross stomachs comes from whale vomit.

The analysis of the Wandering Albatross's diet made by Imber & Russ (1975) supports the suggestion by Hector (1895) of nocturnal feeding on cephalopods detected by their luminescent organs. Clarke & Prince (1981) disagreed, pointing out that while seven of the nine food species in their samples had photophores, these were screened to be invisible from above. This view has been strengthened by further work on the squid prey of *D.exulans* by Rodhouse *et al.* (1987).

B *Feeding range*

Breeding albatrosses can travel far in search of food and Wandering Albatrosses have been caught up to 1800 km from their nests while rearing a chick (Weimerskirch & Jouventin, 1987). Fisher (1967) found that Laysan Albatrosses foraged up to 1770 km away but, as they made 50–60 round trips when tending their chicks, they could seldom go beyond 2 days' travel from them (Fisher & Fisher, 1969, p.184).

Tickell (1968, p.62) spray-dyed 305 breeding *D.exulans* at South Georgia. Most of the 52 re-sightings at sea were within 230 km and to the north where whales are plentiful. He concluded that the birds could find adequate food within a few days' flight range from their nests. Croxall *et al.* (1984a), making an allowance of $\times$ 1.5 for zigzag progression, gave foraging ranges varying from 428 km for Black-browed Mollymawks, 615 km for Grey-headed Mollymawks, 709 km for Light-mantled Sooty Albatrosses and 1480 km for Wandering Albatrosses. These figures were obtained from activity recorders attached to the birds. There have been several other attempts to calculate foraging ranges, e.g. by Prince & Francis (1984) and Pennycuick *et al.* (1984). However, data from satellite-tracked *D.exulans* indicate much greater ranges than these. One male travelled 15 200 km from its nest during a 33-day foray (Jouventin & Weimerskirch, 1990). Weimerskirch & Jouventin (1987) provided evidence that male Wandering Albatrosses of the Indian Ocean tended to forage over Antarctic waters (50–60°S) while their partners chose tropical pelagic and offshore waters (35–45°S).

C *Foods*

There are numerous recent studies of albatross diets, most depending on the examination of stomach contents. These include those of *D.bulleri* (West & Imber, 1986), of *D.chrysostoma* and *melanophrys* by Prince (1980b) and Clarke & Prince (1981), of *D.exulans* chicks by Clarke *et al.* (1981) and Rodhouse *et al.* (1987) and of adults by Imber & Russ (1975) of *D.irrorata* by Harris (1973), of *D.immutabilis* and *D.nigripes* by Harrison *et al.* (1983), and of *P.palpebrata* by Thomas (1982). For all species squid and octopus, fish and crustacea comprise the main food types, with some differences in the sizes and species taken by the different albatrosses, particularly where they are sympatric, e.g. with *D.melanophrys*, and *D.chrysostoma* at South Georgia (Prince, 1980b).

VII Breeding

A *General*

The main characteristics of the petrel breeding system — slow development of egg and chick, care shared equally by both parents involving long stints on the egg, and fasts by the chicks interspersed with large meals — apply equally to albatrosses. Mate and nest-site retention and philopatry also tend to be the rule. The special features of the breeding system relate mainly to the birds' large size. The North Pacific species and those of the Southern Ocean lay at the same time — the northern winter and the southern summer — but the Galapagos species is again exceptional: it lays in the northern spring (April and May).

All nest on the ground, usually on flat areas or wide ledges, less often on narrow ones, the specialised niche of the sooty albatrosses. All are too big to burrow and need not do so because they have currently few significant enemies on land (though possible past enemies are unknown). The chicks are less immune from natural predators like sheathbills and giant petrels, and are guarded for several weeks. Nests of Wandering and Royal Albatrosses tend to be sited in depressions giving some shelter from the wind, but other sites are very exposed. The chicks of these species have to survive exposure throughout the southern winter and their down is extremely thick and matted, providing a waterproof insulation. Mollymawk chicks are also well insulated but those at the more southerly sites fly before the onset of winter. Their heavy coats inflict some stress on the chicks on unseasonally warm summer days.

Nests vary in shape (for details see Chapter 12, Section IIA), Mollymawks and sooty albatrosses build truncated cones of soil and vegetable matter, dished at the top for safer holding of egg and chick, the great albatrosses shallower cones of soil and vegetation.

The Black-footed Albatross's nest, a hollow at the top of a sandy beach, will have disappeared long before the next laying season, but some Laysan Albatross nests among coral sand become consolidated with growing vegetation and are more permanent. The extreme situation is that of the Galapagos species which builds no nest, lays its egg among lava breccia and shifts about, taking its egg with it.

Their large size, tameness, capacity for fasting, and surface-nesting make albatrosses attractive subjects for field experimentation and some kinds of physiological research. For example, in studying the metabolic cost of incubation, Grant & Whittow (1983) could enclose wild Laysan Albatrosses on their nests with airtight boxes with minimal stress to their subjects, whereas Croxall & Ricketts (1983), in similar work with Wandering Albatrosses, weighed each adult daily in a sack and, although no egg was lost, clearly some stress was sustained by the birds. To monitor changes in hormone levels some birds were cannulated for regular blood sampling (Hector & Goldsmith, 1985). Hector *et al.* (1986) performed laparotomies on breeding birds under general anaesthesia to measure changes in gonad size; only 7% of the birds failed to resume their normal routines and all returned to their nest areas in subsequent years.

B *Breeding frequency*

An unusual feature of the breeding system is the deviation from the usual petrel procedure of annual breeding. The two great albatrosses, the Grey-headed Mollymawk and the sooty albatrosses *P. fusca* and *P. palpebrata* breed biennially, the rest, as far as known, breed annually. For the great albatrosses biennial breeding is clearly tied in with their size, the incubation and nestling periods being size-linked. These birds take too long from egg laying to chick fledging to breed every year. The Wanderer, for example takes 78 days for incubation and the chicks average 278 days old when they fly (Tickell, 1968). If they succeed in raising a chick the birds miss the next year and return to breed the following one. Should nesting fail during incubation or the early chick stage they usually return the next year, but the longer a pair tends the chick the less likely their reappearance the ensuing season. Tickell tested the capacity of failed breeders to lay in year $\times + 1$ by killing chicks between March and October. He found that all who lost small chicks in March tried again the following year, as did 70% of those that failed in April, 33% in May, 50% in June, but none of those deprived from July onwards re-appeared the next year (Tickell, 1968, Table 36). One of Tickell's pairs laid every season for 6 years but never succeeded in rearing a chick.

Thus the typical situation with both great albatrosses is that there are two demi-populations such that in any particular year the breeders comprise mainly those that nested two seasons before and a lesser number that failed during the previous year. During Tickell's study 20% of the birds that laid but were unsuccessful tried again in the next year. By removing eggs he found that at least 83% would lay the following year.

Some ecological correlates of biennial breeding were discussed by Tickell (1970) who pointed out that at South Georgia the long chick stage of *D.exulans* allows laying in mid-summer and avoids the risk of having to build on frozen ground, a hazard sometimes faced by the annual-breeding mollymawks there.

Shortage of time as a reason for non-annual nesting seems less likely with the smaller biennial layers but at South Georgia, and evidently elsewhere, *D.melanophrys* nests annually while the similar-sized *D.chrysostoma* nesting nearby breeds biennially (Tickell & Pinder, 1967; Prince, 1985). Breeding *D.chrysostoma* arrive at the nesting grounds about 126 days after the previous years' fledglings have gone (165 days for *melanophrys*), the total development time (incubation + nestling period) being 213 days for *chrysostoma* but only 184 days for the other species. Tickell and Pinder thought that the 39-day difference (11% of the year) was enough to preclude annual breeding, but Fisher (1976, p.137) pointed out that incubating *immutabilis* regain a 20% weight loss in only 3 weeks at sea. Therefore biennial breeding in *chrysostoma* would seem to be due to factors other than length of parental care or body size. Food may be one of these factors. Prince (1980b) found that *chrysostoma* depended much more on squid than did melanophrys which mainly ate krill and these have a higher energy and calcium contents than squid (Clarke & Prince, 1980). Also, by means of inter-species cross-fostering experiments, Prince & Ricketts (1981) showed that there were species-specific constraints on growth, *chrysostoma* chicks growing faster than normal when raised by *melanophrys* and vice versa (Fig. 2.11). These authors believed that the slower growth of the *chrysostoma* chicks and the need for their parents to catch more of a poorer quality prey accounted for their biennial breeding, the differences in growth rates reflecting interspecific adaptations. Ricketts and Prince (1981) and Prince (1985) suggested that the nutrionally poorer diet of the *chrysostoma* had led to the evolution of a 6-week longer development time than that of *D.melanophrys*. In consequence, *D.chrysostoma* could not attain breeding condition during their post-nuptial winter at sea, when their diet was still relatively poor.

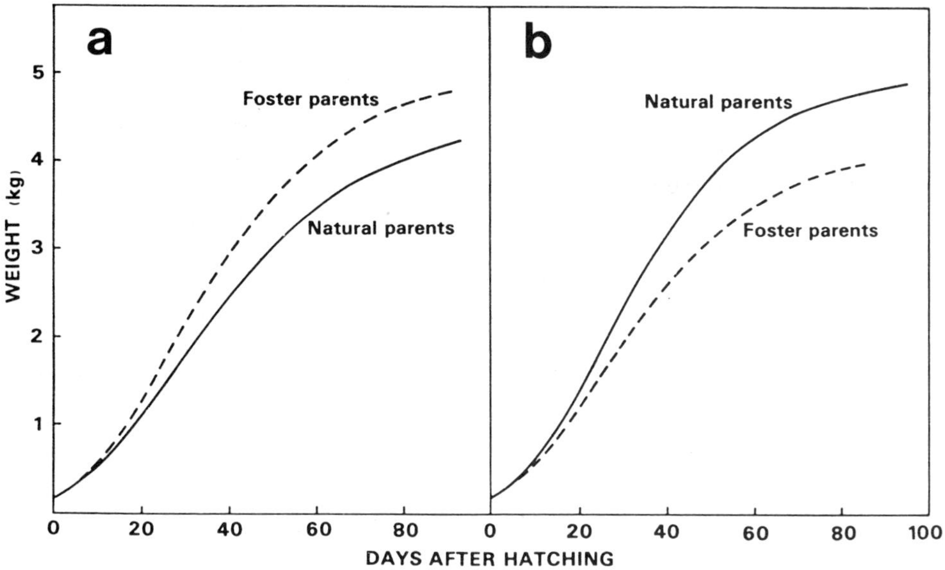

Figure 2.11 *Growth of Grey-headed Albatross chicks (a) and Black-browed Albatross chicks (b) reared by their natural parents (solid lines) or by parents of the other species (broken lines). From Prince & Ricketts (1981).*

Biennial breeding appears to be the norm with *P.palpebrata* (Kerry & Colback, 1972; Jouventin *et al.*, 1982) and *P.fusca* (Weimerskirch, 1982). These small albatrosses have very long reproductive cycles — total development time for *palpebrata* being about 230 days (Thomas *et al.*, 1983) and for *fusca* 250 days (Weimerskirch, 1982).

Thomas (1982) attributed the long nestling period of *palpebrata* to its breeding and feeding places being far apart so that it fed its chick less often than mollymawks like *D.melanophrys* nesting nearby. It seems likely that this extremely pelagic foraging, probably involving crossing the Antarctic Convergence, may be the main reason for alternate-year breeding of *P.palpebrata*, for, at least at South Georgia, the species takes much energy-rich krill (Thomas, 1982). More information is needed to establish all the factors involved, particularly for *P.fusca*, whose chicks evidently may fledge before being full-grown (Berruti, 1979). Jouventin and Weimerskirch (1984) emphasised that two-thirds of *P.fusca* ashore are non-breeders due to the delayed sexual maturity, the birds not returning until 8.4 years old or more and not breeding until at least 11.9 years old.

VIII Dispersal and migration

In the sense of a regular movement of a population between a breeding area and a wintering one, several albatrosses can be said to be true migrants. Otherwise the centres of the populations shift northwards during May to July, that is, after the breeding season for both northern and southern medium-sized species, the northern birds being winter and the southern ones summer breeders. The movement reverses in October to February. There is often a strong down-wind component, usually westwards, in the outward movement.

These movements are nonetheless rather complex, complicated by the biennial breeding and winter feeding of the great albatrosses, by differences in the movements of the different age groups, and by biasses in banding recoveries and of observations at sea. There is increasing but indirect evidence that the great albatrosses may circumnavigate the globe during their immature years, an idea mooted by John Gould some 150 years ago.

Documentation of population shifts by the two common North Pacific species is reasonably good thanks to the analyses of Shuntov (1974), Fisher & Fisher (1972), Sanger (1974a,b) and Robbins & Rice (1974). The latter analysed 324 recoveries of Laysan and 399 of Black-footed Albatrosses, many of known age, so could calculate some age-specific distributions.

Many analyses of movements based on recoveries and sightings of southern species have also been made. These include those of Amiet (1958), Barton (1979), Dixon (1933), Gibson (1967), Hagen (1982), Jehl (1973), Liversidge & Le Gras (1981), Shuntov (1974), Sudbury *et al.* (1985), Tickell (1967), Tickell & Gibson (1968) and Weimerskirch *et al.* (1985).

Perhaps the most remarkable movement is that of Wandering Albatrosses from all the major breeding places, that are drawn to the waters off New South Wales for the cuttle-fish harvest in late winter and early spring already mentioned. Many are regular visitors. For example, a Kerguelen bird first banded off New South Wales and retrapped there in 1966 and 1975 was caught again in July 1980, 24 years after its first capture. Analysis of 76 Crozet and Kerguelen birds retrapped off New South Wales shows that the birds are there either during their non-breeding year if a chick had been reared, or during the same season if they lost their egg (Weimerskirch *et al.*, 1985) and Fig. 2.12.

IX Populations and mortality

A *Population sizes*

Although it is often thought that the largest albatross populations are in the Southern Ocean, about 335 000 pairs of Laysan Albatrosses breed in the North-west Hawaiian Islands (Harrison *et al.*, 1984). This compares with the greatest concentration in the south — the 390 000 pairs of Black-browed Albatrosses around the Falkland Islands (Croxall *et al.*, 1984b).

Wandering Albatrosses are most plentiful at the Auckland Islands where about 13 500 pairs breed; the smallest populations are the two or three pairs of *exulans* that probably still nest at Nightingale Island, Tristan da Cunha and the relict group of *amsterdamensis* at Amsterdam Island where there are only about 40 breeding birds (Jouventin *et al.*, 1989).

Most southern mollymawks have large populations, others are more modest. Warham & Bennington (1983) mapped and counted the $4750 \pm 10\%$ pairs of Buller's Mollymawks with nests at the Snares Islands. The Waved Albatross of the

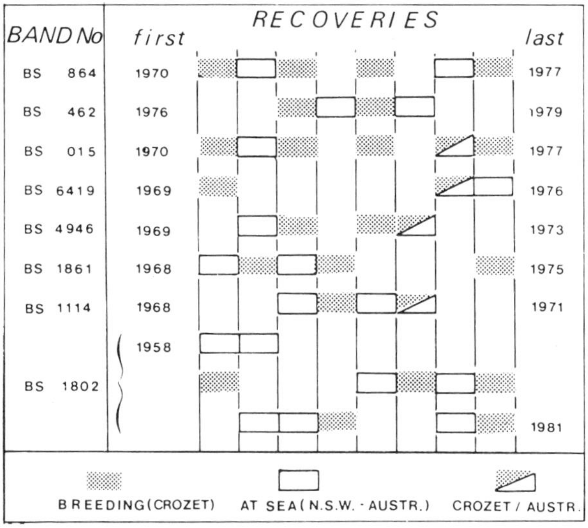

Figure 2.12 *Year to year movements of ringed Wandering Albatrosses between New South Wales and the Crozet islands. Each rectangle represents one or two recoveries during 1 year. From Weimerskirch* et al. *(1985).*

Galapagos Islands is one of the rarer species; Harris (1973) estimated that at least 12 000 pairs bred there.

The rarest species is the Short-tailed, *D.albatrus,* whose only known breeding place is on Torishima, some 575 km south of Tokyo where there were about 165 adults in the 1987 season (Hasegawa, pers. comm.).

With such large, long-lived birds population increase is necessarily slow at first, as exemplified by the Royal Albatross colony at Taiaroa Head, New Zealand which, although carefully guarded for more than 40 years, still had only 17 pairs that laid in the 1987/88 season and eight chicks reared.

On the other hand, the White-capped Mollymawks of Albatross Island, Bass Strait, depleted by sealers, have increased from about 400 nests in 1894 to have about 1500 chicks in 1973 and about 2000 pairs in 1983. The island has difficult access and the increase has apparently been in response to a relaxation of human pressure. The Wandering Albatrosses of Macquarie Island have never recovered from decimation by starving sealers who had to eat them. This population still seems to be declining for in 1959/60 16 pairs laid and very few failed to hatch their egg; in 1975 there were 27 pairs; in 1982 only 7 pairs and in 1985/86 no chick was reared (pers. obs.; Tompkins, 1985; Rounsevell & Brothers, 1984). The species has also steadily declined at South Georgia (Croxall, 1979) and at the Crozet Islands (Jouventin *et al.*, 1984). Increased entanglements in fishing nets may be a common factor here.

An albatross that is regaining a range lost to past depredations of Japanese plume and egg hunters is *D.immutabilis*. This has not only been displaying at many sites in the main Hawaiian islands and has established itself on Kaui, but has also been prospecting potential nesting places at the Revillagigedos Islands and on Alijos Rocks (Mexico) and has bred on Guadalupe Island in that country (Pitman, 1988).

B *Population structure*

Although hatching success can often be quite low, and many chicks die in their nests, adult survival is usually high. Richdale & Warham (1973) found some *D.bulleri* still breeding at the same sites and with the same partners as when they were first ringed as adults 23 years before. The oldest marked wild bird of any species is probably the female Royal Albatross banded as an adult in 1937 at Taiaroa Head which was still breeding there in 1988 when more than 58 years old. Yellow-nosed Mollymawks have been recovered at 32 and 37 years old (Hagen, 1982) and in 1972/73 at Midway Atoll there were at least 11 Laysan Albatrosses 40 or more years of age (Fisher, 1975a).

Being large, diurnally active, tame and easily marked, albatrosses have attracted the attentions of biological demographers and there are now good samples of birds of known ages. Annual survival rates for biennially breeding adult Royal and Wandering Albatrosses were estimated at 97.0 and 95.7% (Richdale, 1952; Tickell, 1968) and for annual breeders like *D.irrorata*, *D.immutabilis* and *D.melanophrys* and figures of 95%, 92% and 94.7% respectively, were obtained (Harris, 1973; Fisher, 1975b; Croxall, 1984). Data for five southern species were presented by Weimerskirch *et al.* (1987). Weimerskirch & Jouventin (1987) analysed the survival of Wandering Albatrosses from the Crozet Islands, where the population has also fallen since 1965. They found that males survived better than the females, possibly because the latter fed more in waters used by commercial fishermen than did the males.

Life tables for many species have been presented, e.g. for *D.immutabilis* (Fisher, 1975b), for *D.melanophrys* (Pascal, 1979), for *D.exulans* (Tickell, 1968; Barrat *et al.*, 1976) and for *P.fusca* (Weimerskirch, 1982); most of these indicate that about 0.5–1.0% of a cohort will attain the age of 80 years.

Albatrosses may come back to their natal islands at 2 years of age but few younger than 4 years old are seen there. Weimerskirch *et al.* (1987) gave 8.9 and 9.8 years as the mean ages at first breeding of *D.chlororhynchos* and *D.melanophrys*, 12.0 and 12.2 years for *P.palpebrata* and *P.fusca* and 11.3 years for *D.exulans*. From a sample of 428 inexperienced *D.immutabilis* paired together Van Ryzin & Fisher (1976) found that the average age of first breeding in the males was 8.3 years, in the females 8.9 years.

C *Mortality*

Deaths of chicks and eggs from natural causes vary considerably from year to year. Losses arise through rain and flooding, vulcanism, tidal waves, from unstable substrates (e.g. with *D.albatrus* — see below), by nests becoming buried by sand and by similar density-independent factors. The Waved Albatrosses studied by Harris (1973) suffered occasional complete failure when the whole or part of a colony deserted its eggs for unknown reasons, apparently not directly connected with El Nino effects. Their habit of shifting about with their egg leads to losses due to breakage or to the egg falling into crevices.

Other chicks die of starvation. After fledging more young birds die than older ones — Fisher (1974b) reported that his Laysan Albatrosses lost about 9% annually during years 1–4, but only 1.8% annually during years 4–8.

It is almost impossible to know just how these initial mortalities occur at sea but in calm weather many Laysan Albatross fledglings swim away from the nesting islands and Fisher (1975b) counted 37 large sharks in the Midway Atoll lagoon when the birds were leaving and found 13 fledglings in the stomach of one 4 m shark.

Apart from such man-made hazards as tree-trunks (and radio aerials) referred to earlier, *D.immutabilis* and *D.nigripes* pick up considerable quantities of the plastic rubbish of Japanese origin that floats across the North Pacific. Some may be fed to the chicks with their meals or regurgitated onto the ground (Fig. 2.13). What effect

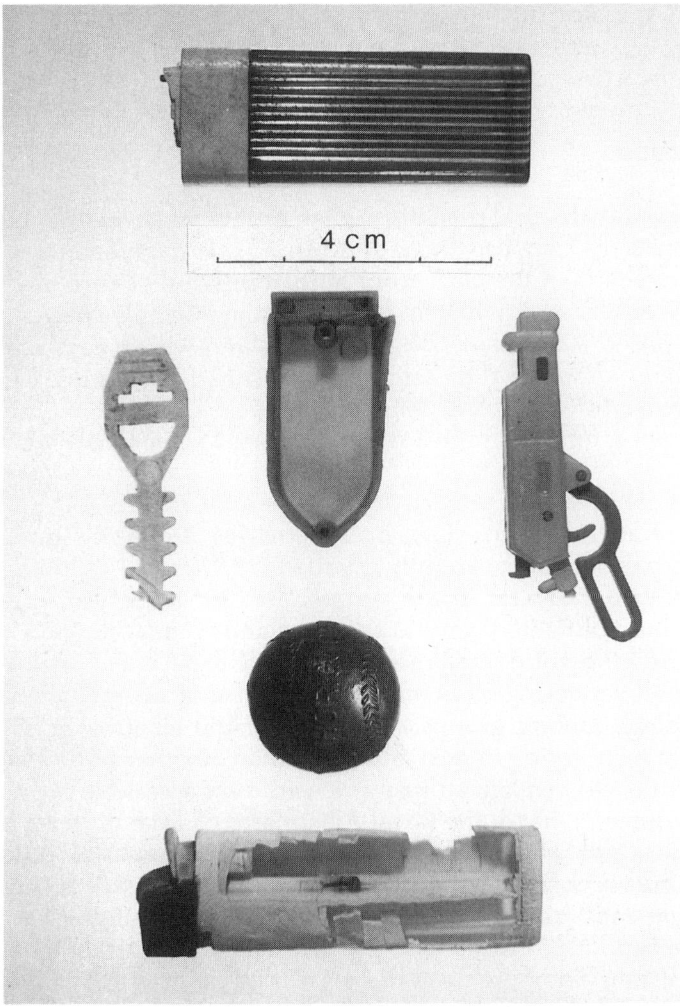

Figure 2.13 *Plastic rubbish of Japanese origin regurgitated by a Laysan Albatross on Midway Atoll. Photo: author.*

this has on the health of the birds is unclear, as quite large undigested squid beaks accumulate in gizzards, but it seems unlikely to be beneficial.

Other deaths indirectly due to man are those of the Laysan Albatrosses prospecting new sites for nests in Hawaii, many of which are killed by dogs. For example, 30 were lost in this way at a young colony on the Pacific Missile Range on Kauai on 22 February 1984.

A very unusual cause of death among Laysan Albatrosses comes from attack by Polynesian Rats *Rattus exulans* (Kepler, 1967). At Kure Atoll these climb onto the backs of incubating birds, gnaw through the feathers and eat the flesh below. After dark over 20 rats have been seen on a single albatross. The birds make only ineffectual efforts to fight off the rodents and most of those attacked die. Fortunately this appears to be a learned behaviour and has not been seen elsewhere, even on Midway Atoll where *R. rattus* is plentiful. Presumably this predation still continues on Kure as Harrison *et al.* in 1984 recommended the removal of the rats from that place.

X Conservation

Albatrosses have few natural predators when ashore and even the chicks seem able to defend themselves from cats by oil spitting, so that maintenance of the breeding habitat and protection of the birds from human interference are the main practical steps for their conservation while on land. Probably the only species still harvested is the Yellow-nosed Albatross of Nightingale Island whose eggs are taken by the Tristan islanders, although they temporarily banned harvesting when the birds' numbers decreased (Richardson, 1984).

Long-line netting for fish in the Tasman Sea and Pacific Ocean may well prove a major hazard.

Two current conservation projects may be mentioned here. At Kilauea Point on the island of Kauai, Hawaii, Laysan Albatrosses are being lured to take up territories and partners on a carefully guarded US National Wildlife Refuge. Birds had been attempting to nest in the general area for some years and what were presumably young birds flew around the headlands prospecting for nest sites, but none had entered the quite small reservation. The experiment began in 1982 to try and attract these wandering birds into the refuge using a combination of playbacks of tape recordings among groups of models of the albatrosses. By the 1987/88 season 11 pairs were trying to nest and more than 30 other birds visiting the area. This experiment could result in an easily viewed, very accessible population likely to rival for public interest that of the Royal Albatrosses of Taiaroa Head, New Zealand.

The second example concerns the Short-tailed Albatrosses of Torishima Island. The breeding area is on a bare slope of volcanic ash and losses occurred when this shifted, burying eggs and chicks. The planting of clumps of grass across the main nesting area began in 1981 to stabilise the ash. The birds quickly took advantage of the change, nesting along the rows of tussocks. Nesting success increased so that by 1987 over 50 chicks fledged whereas in the late '50s only 0–10 chicks were reared annually (Hasegawa, 1982, pers. comm.). The stage now seems set for a steady if not rapid expansion.

CHAPTER 3
The Fulmars

Cape Pigeon.
Photo: author.

Contents

I Introduction

This group is composed of seven species of robust, short-bodied, medium to large petrels in five genera: the Northern and Southern Giant Petrels *Macronectes halli* and *M.giganteus*, the Northern and Southern Fulmars *Fulmarus glacialis* and *F.glacialoides*, and the Antarctic and Snow Petrels and the Cape Pigeon in monotypic genera *Thalassoica, Pagodroma* and *Daption*.

The fossil record is sparse. *Fulmarus miocaenus* Howard, 1984, described from a humerus of middle Miocene age from California, was smaller than living *Fulmarus*, but *F.hammeri* of the Californian late Miocene, was rather larger than these (Olson, 1985d).

Cape Pigeons (or Cape Petrels) and giant petrels were well known to the early mariners in the Southern Ocean, the latter's vulture-like habits often described, and the Polynesian settlers on the Chatham Islands ate the northern species *M.halli* until very recently.

But the bird best known in the northern hemisphere is the Northern Fulmar which has been an important asset to some communities, its eggs and meat being eaten, its oil burnt in lamps, its feathers used in bedding, its bones as hooks, and its carcasses as bait.

Northern Fulmars are the only tubenosed birds known to transmit a disease fatal to man. A virulent strain of a viral ornithosis appeared among the birds at the Faeroes and in Iceland in 1933. The disease was contracted after the fulmar harvest, mainly by those splitting and salting young birds, and resembled a form of pneumonia. Up to 1937, 32 cases were fatal. The initial identifier Rasmussen (in Fisher, 1952) suggested that it had been transmitted by fulmars scavenging Argentinian *Amazona* parrots. Large numbers were shipped to Europe when psittacosis was rife in South America. Infected parrots that died were tossed overboard and Rasmussen believed that some carcasses were within the range of the Faeroese fulmars which thereby contracted the disease. Although psittacosis regulations stopped fulmar fowling in Faeroe and Iceland, and hence human fatalities, infected birds themselves still died, though without any apparent long-term effects on their populations. By 1966 fulmars were again being harvested at Faeroe, with greater precautions being used in their preparation as food.

'Fulmar' is a Norse name taken by Vikings to the Hebrides and subsequently carried over into the Gaelic, being in use there by the eleventh or twelfth centuries at the latest (Lockwood, 1954). It is a compound of *ful* + *mar*, or 'foul gull', doubtless in reference to the bird's oil spitting habit.

II General characteristics

The species have often been grouped in a sub-family Fulmarinae, but, unlike the Diomedeidae, that can be separated solely on the presence of an inter-ramicorn plate at the gonys (Fig. 1.2), the fulmarine petrels share most of their characters with other genera.

A *Anatomy*

The most prominent external features are the large nasal tubes, lower at their bases. These naricorns are fused together so as to appear like a single tube with a rather rounded, forwards opening, an impression heightened by the thin septum between them not extending to their aperture. These tubes reach their greatest development in *Macronectes* 'a sea bird of protean appearance' (Murphy, 1936), where they extend for three-fifths of the length of the bill, leaving no gap between them and the maxillary unguis (Fig. 3.1A).

The bill is strong with the maxillary unguis very convex and hooked and the mandibular one stout and obtuse and either unhooked or only very slightly so; the

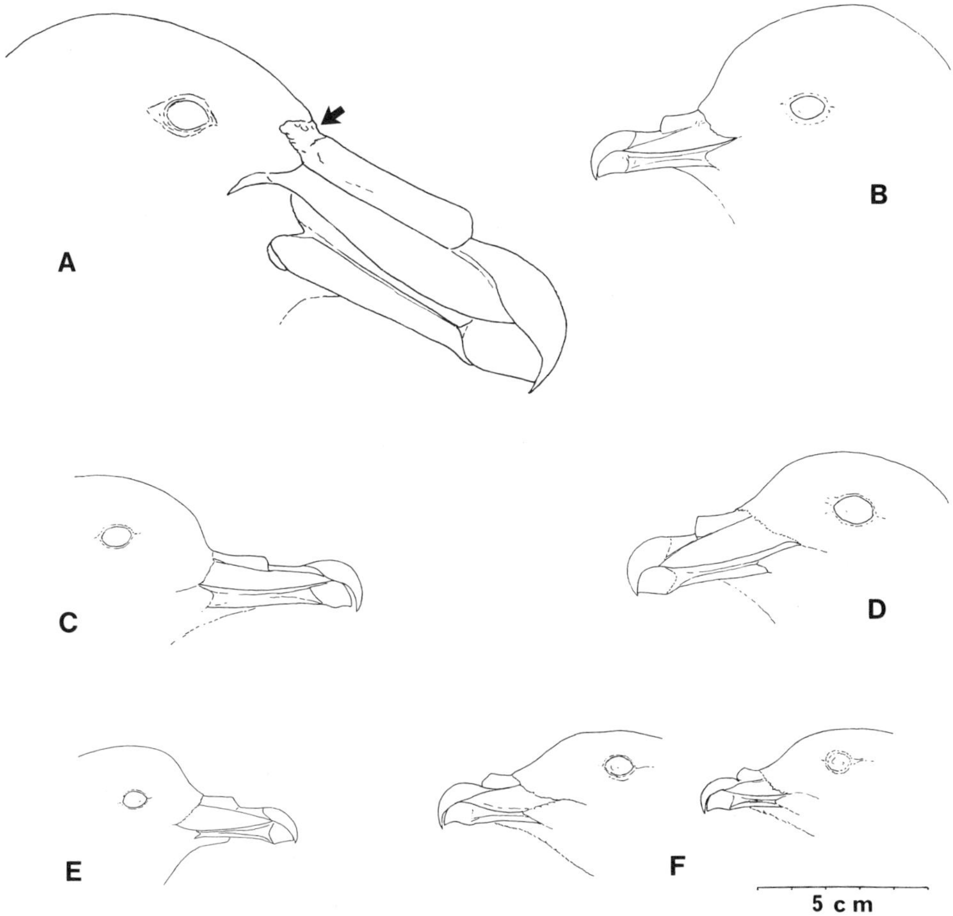

Figure 3.1 *Bills of fulmars. (A) Southern Giant Petrel* Macronectes giganteus*: and old (?) bird with encrustation (arrowed). (B) Antarctic petrel* Thalassoica antarctica. *(C) Southern Fulmar* Fulmarus glacialoides. *(D) Northern Fulmar* F.glacialis. *(E) Cape Pigeon* Daption capense. *(F) Snow Petrels* Pagodroma nivea *— large and small examples — see Mathews (1912–13, p.177),*

whole is specialised for shearing flesh and filter-feeding (Kay, 1953). Within the mouth the proximal edges of the rhamphothecal coating on the upper and lower mandibles bear a series of ridges or lamellae. Aided by the fleshy and blunted tongue, these help retain small food items (Voous, 1949). Such striae are particularly well developed in *Daption*, less so in *F.glacialis* (Fig. 3.2.) and poorly so in *F.glacioloides*. They are present in giant petrels though these are not usually thought of as filter-feeders. An important skeletal feature is the fusion of the pre-frontal (lachrimal) bones of the skull to the frontals (Pycraft, 1899; Bourne, 1975), a character shared with *Pterodroma*. Bourne (1985) suggested that this fusion strengthened the skull to facilitate surface-seizing of prey. Another fulmarine skull feature is the way in which the dorsal anterior arm of the quadrate bone extends in

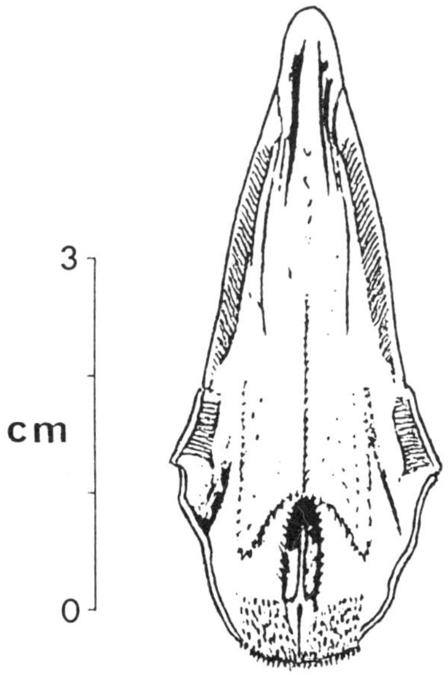

Figure 3.2 *Palate of Northern Fulmar showing oblique ridges on the rhamphotheca of the inside of the tomia. From Forbes (1882).*

front of the post-orbital process (Kuroda, 1954), a feature shared with a storm petrel *Oceanodroma*, however. In the palate there are well developed basipterygoid processes. According to Verheyen (1958, p.15), except for *Pterodroma*, the fulmars are the only petrels to have a patella.

Another feature of fulmar anatomy is the absence of the bony ossifications of the patagial tendons (the patagial ossicles or sesamoid bones) found in the muscles of the forearms of albatrosses, shearwaters and gadfly petrels. They are lacking also in prions, and in storm and diving petrels (Rheinhardt, 1873; Forbes, 1882; Mathews, 1936).

The tracheae are peculiar in that the lower portions of those of *Fulmarus* and *Thalossoica* are divided by a median septum, better developed in *Fulmarus* where it extends for *c.* 63 mm. Such septa are lacking in *Pagodroma* and *Daption*, but in *Macronectes* the arrangement is more complex and peculiar, the septa forming two separate tubes for nearly 89 mm (Forbes, 1882, p.37).

B *Plumage*

The plumage of fulmars consists of grey, white or brown feathers and both *M.giganteus* and *F.glacialis* are polymorphic in colour patterns when adult, with no sex-linked variation. Some *M.giganteus* are white from birth, the rest typically have dark grey-brown bodies usually with lighter feathers on cheeks and neck, some being almost white in those parts. There are no wholly white Northern Fulmars, but pale, dark and intermediate colour phases, the proportions of which vary between populations.

White-phase birds apart, fledglings of both giant petrels wear glossy blackish-brown feathers that soon become duller and browner, gradually lightening on face and neck until the adult condition is reached.

Macronectes has 16 rectrices (more than other tubenoses), *Fulmarus* and *Daption* 14, and *Thalassoica* and *Pagodroma* 12.

The musty smell so typical of the procellariiform integument is very pronounced in fulmars and especially in *Macronectes*. Colonies of *M.halli* are often first detected by their smell.

1 *Polymorphism*

The polymorphic plumage of the Northern Fulmar ranges from birds that are all white except for a dark eye-mark (Fisher's LL or double light, to those that are uniformly dark grey — Fisher's DD or double dark (Fisher, 1952, p.267). The Atlantic birds have been separated by Salomonsen (1965) into *F.g.glacialis*, a high-Arctic form with thick, blunt bill and mainly dark-plumaged birds, and *F.g.auduboni* of the low Arctic and boreal regions and mainly light-coloured birds. The Pacific form *F.g.rodgersi* has a rather slender bill and a high proportion of dark-phase birds (Fig. 3.3). Van Franeker & Wattell (1982) hypothesised that the two morphs originated in different oceans — the dark in the Pacific. During warm inter-glacials introgressions of dark-phase genes into the gene-pool of the Atlantic birds occurred, resulting in the present confusing 'mix'. These authors pointed out that sub-specific segregation on colour morph and bill shape is not precise, there being much intergradation. A multi-variate analysis based only on breeding birds analysed sex by sex might help to clarify the situation.

2 *Moult*

The major study is that of Hunter (1984a) on the two sympatric giant petrels at Bird Island, South Georgia, supplementing that of Conroy (1972). In both species the wing moult of breeding birds starts before or soon after the eggs hatch, the primaries being shed from the innermost outward, i.e. in continuous descendant sequence. Initially two or three feathers are shed at once. Males start before their partners and both sexes of *giganteus* moult earlier in their annual cycles than do *M.halli*. Furthermore, while male *giganteus* begin shedding primaries around egg laying, then usually suspend further moult while raising their chick, their partners do not shed flight feathers until some 60 days after the males, and have no suspense period.

Hunter suggests that the ability to undertake two highly energetic activities like moult and breeding simultaneously is possible because of the summer flush of food (mainly dead Macaroni Penguins *Eudyptes chrysolophus* and fur seals *Arctocephalus gazella*), which provide over 50% of the diet of giant petrels at South Georgia. The birds are able to feed chicks large meals more often than any other procellariiform there (Croxall & Prince, 1980).

Hunter explained the differences in starting times between the sexes as a consequence of the female's smaller size (and greater proportional weight loss during incubation), together with the extra energy and protein she needs for egg

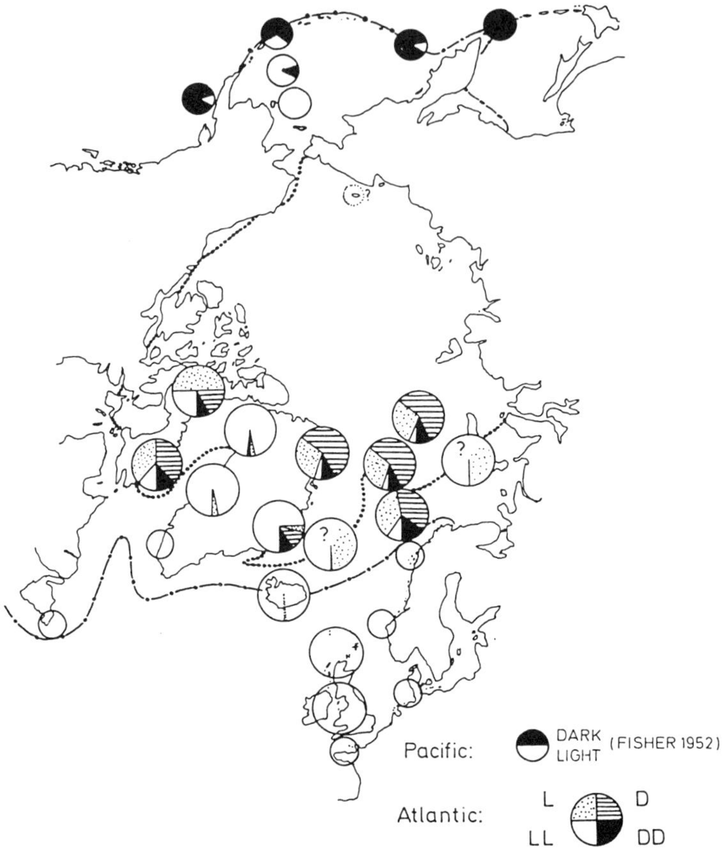

Figure 3.3 *Distribution of the colour phases of the Northern Fulmar. Dotted line encloses high-Arctic marine zonation, dashed line low-Arctic marine zonation. From van Franeker & Wattel (1982).*

production. As *giganteus* lays about 6 weeks later than *halli*, the former may have to start its moult sooner so as to complete it by the winter. Precise times taken by breeding giant petrels to complete their primary moult have not been recorded, but Hunter (1984a) suggests that it may take from 130 to 330 days. Such extended moults help account for the many birds seen in the southern summer with missing flight feathers.

Other southern fulmars studied do not shed primaries until their chicks are close to flying, e.g. in *Daption* and *Pagodroma*. Up to four primaries may re-grow simultaneously and a considerable loss in wing area may result — see, e.g. Fig. 8 in Biermann & Voous (1950) of *Daption* taken at sea. However, these, and both giant petrels, begin their body moults early, during incubation (Maher, 1962; Beck, 1969, 1970) and shed contour feathers accumulate around the nests of all four species.

Northern Fulmars also appear to begin body moult before that of the flight feathers, this latter being delayed until the chicks have fledged (Cramp & Simmons, 1977). And, although data on birds of known age are lacking, it seems likely that most pre- and failed-breeding fulmarine petrels moult earlier than birds that have reared chicks.

C *Sexual dimorphism*

Male fulmars tend to be heavier and longer in wing and beak than their partners (Table 3.1).

Table 3.1 Sexual size dimorphism in fulmars (male/female × 100)

Species	Bill length	Wing length	Weight	Authority
M.giganteus	115	106	126	Johnstone (1974)
M.halli	115	106	134	Johnstone (1974)
F.glacialoides	108	103	113	Mougin (1967)
F.glacialis	107	105	118	Mougin (1967)
T.antarctica	107	103	106	Biermann & Voous (1950)
D.capense	103	100	109	Pinder (1966)
P.nivea	109	104	123	Croxall (1982a) Jouventin & Viot (1985)

It will be seen that sexual size dimorphism in this group follows no orderly pattern: the fulmars with the greatest dimorphism are the largest and the smallest — the giant petrels and the Snow Petrel. The differences between the sexes of giant petrels allow paired birds to be readily sexed on bill size, and this dimorphism is also seen in large chicks (Conroy, 1972; Hunter 1984b) (Fig. 15.10). Furthermore, the 34% difference in weights between the sexes of *M.halli* (at Macquarie Island) is virtually identical to that from South Georgia of 33% (Hunter, 1984b) and from the Crozets of 32% (Voisin, 1968), so that in this species the difference is evidently greater than in *M.giganteus*. For *Pagodroma*, however, the considerable degree of difference between the sexes varies from place to place (Croxall, 1982a; Jouventin & Voit, 1985). Least dimorphic is *Daption* (Pinder, 1966; Sagar, 1986).

Croxall (1982a) plotted body weights against bill lengths to separate 85% of a sample of 43 of each sex of the Snow Petrel correctly, but some figures suggest that lengths of middle toe with claw could be more effective than weight — see Conroy (1972, Fig. 1) for *M.giganteus*. Dunnet & Anderson (1961) separated the sexes of *F.glacialis* using a statistical discriminant function based on bill dimensions and body weights.

III The genera

A Macronectes

The giant fulmars are as big as a mollymawk, weighing about 4.8 kg and with wings spanning some 1.9 m. They are characterised by their bulky bills whose nasal tubes are better developed than in any other bird (Fig. 3.1).

Macronectes has customarily been regarded as a very specialised form and it is the only tubenose feeding largely as a scavenger and predator on other birds and mammals and the only one to do this on land as well as at sea. Most petrels, from the albatrosses down to smaller species, will kill and eat smaller birds when these are available, but only *Macronectes* depends on predation for a major part of its food. Croxall & Prince (1980) suggested that the rather flattened skull of *Macronectes* might be an adaptation to facilitate thrusting its head into dead animals.

Another unusual feature of the bill in many birds is the development of an encrustation at the base of the culmen (Fig. 3.1A). The origin of this has not been determined but it is lacking in young birds and may be remnants of rhamphothecal material from previous bill moults. (In albatrosses, for example, series of transverse ridges are often seen at the base of the bill, perhaps formed in a similar way, but so far it has not been possible to use the ridges as growth lines for ageing.) Giant petrels also have exposed fleshy fillets at the bases of the latericorns and ramicorns (Fig. 3.1A).

The beak is well fashioned for tearing flesh. Like albatrosses, in dismembering carcasses, giant petrels bite the flesh and pull vigorously backwards so that the sharp inner edges of the superior unguis and the tomia help slice through muscle and ligament.

Macronectes consists of two sibling species of similar size and habits, whose distinctiveness was not established until recently (Bourne & Warham, 1966). Their many morphological similarities, such as the identical mobilities of some blood serum proteins (Shaughnessy, 1970) suggest that they became distinct in recent evolutionary times. Their distinctiveness was shown by the breeding at Macquarie Island of two separate populations laying about 6 weeks apart, the plumage of the early birds being monomorphic, the late nesters having some 7.8% of white birds.

Subsequently sympatric breeding was found to be the norm also at The Crozets, Marion and Prince Edward Islands and at South Georgia. It may also occur at Kerguelen and perhaps at Heard Island. The main differences between the species are set out in Table 3.2.

At some stations downy chicks are characteristically marked (Fig. 3.4) but, apart from those that are white from birth, the plumage of fledglings of both species is a glossy blackish brown. Their irides are also brown. Over successive moults both species lighten around the eye, cheeks and forehead, but this proceeds much faster in *giganteus*, some of which develop almost completely white heads and necks. This plumage lightening does not proceed more rapidly in males: adults cannot be sexed on plumage colour. The brown irides of all the young birds also lighten to grey or a whitish shade, but here the change is more rapid in *halli* and most breeding birds of this species are grey-eyed (often with residual brown spots), whereas many *giganteus* remain dark eyed into adulthood.

Bourne and Warham were unsure of the specific status of the giant petrels of Gough Island and the Falklands, pointing out that they seemed to be intermediate in some respects but with no white-phase birds. Recent work e.g. by Voisin & Bester (1981) and Devillers & Terschuren (1980) has confirmed their general findings but the taxonomic status is still unclear although the Falklands' birds are

Table 3.2 Differences between the two species of *Macronectes* (Modified from Bourne & Warham, 1966)

	Macronectes giganteus	*Macronectes halli*
Breeding range	Antarctic zone and convergence	Subantarctic zone and convergence
Summertime range	Mostly south of convergence	Mostly north of convergence
Body plumage	Dimorphic. Pale phase always white. Dark phase adult grey with pale grey or white head and throat. Pale feathers may extend down neck.	No pale phase. Adult has dark grey or brown body and head, face pale with freckled cheeks.
Chick's primary down	White or grey with top of head slightly darker.	Ash-grey above, white below: dark grey skull cap contrasts with pale cheeks.
Bill ungues	Grey-green to bright green.	Brown to pink, often with dark marks.
Iris colour of breeders	Mostly brown, some grey.	Mostly grey, occasionally brown.
Nesting behaviour	Colonial in open.	Solitary or social in sheltered places.
Laying date	Late: where both breed, about 6 weeks after *halli*	Early
Following ships	Seldom	Often

usually considered a population of *giganteus* and Voisin (1982) even proposed their being raised to subspecific status as *M.g.solanderi* Mathews, 1912. Further problems arise with birds found nesting in Argentina which appear to be typical *giganteus* with grey-green tips to their beaks but no white phase (Humphrey & Livezey, 1983; and photographs).

At the other side of the world, the *halli* nesting at the Chatham Islands, New Zealand, are rather small and many breed in dark, even black plumage, similar to that of juveniles elsewhere, and they often remain dark-eyed, though having the reddish ungues of typical *halli*.

There are differences in the foods and feeding of the two species (see below) and while the breeding phenology of *giganteus* varies with the severity of local conditions, that of *halli* is less variable. This is usually the earliest of Subantarctic petrels to lay, and at places where both breed there is a gap of about 3 weeks between the end of its laying and the start of that of *giganteus*.

Not surprisingly, as the two often nest in close proximity, hybridisation has been recorded (Johnstone, 1978; Burger, 1978; Hunter, 1983b). According to the last it is as high as 1.5% of the breeding population at South Georgia. All cases examined have involved male *giganteus* paired with female *halli*. Hunter pointed out that this probably arose because the only potential partners for unpaired female *halli* after the main body had mated, would be male *giganteus*. It has not been shown that hybrids are fertile.

Figure 3.4 *At some places like The Crozets and Macquarie Island, giant petrel chicks in primary down differ in colour patterns.* M.halli *(upper), tends to have a grey skull cap contrasting with the pale cheeks and underparts, and is pale grey on the back, whereas* M.giganteus *at the same stage is much paler, almost white, lacking contrasting patterns. From Voisin (1968).*

B Fulmarus

The two sibling species are very similar, rather gull-like birds, but with a stiff-winged sailing flight (making them look like miniature albatrosses) interspersed at times with rapid wing beats. They are of similar size, weighing some 750 g, wings spanning about 115 cm, and form a superspecies. A peculiarity of the southern species *F.glacialoides* is its long intestine, extending for nearly 2 m, twice that of *Daption* (Cunningham, 1871, p.224). Murphy (1936, p.601) wondered whether the difference was correlated with food specialisation.

The literature on Northern Fulmars is enormous. Fisher (1952) referred to his unpublished bibliography with 2204 citations. Fisher's monograph itself is mainly concerned with the history of the bird, its past and present distribution and annual cycle. Long-term research at Eynhallow, Orkney is the main source of our more recent and very detailed knowledge of the species' breeding — see Dunnet *et al.* (1979) and Section VIIA below — while Mougin (1967) made a comparative study of both species on their breeding grounds.

Unlike those of the sibling giant petrels, the habitats of *Fulmarus* spp. lie at opposite ends of the earth and contact between them has been lost. Once both were birds primarily of polar seas and pack-ice and Voous (1949) concluded that the southern bird is the more primitive and that a population of this species colonised the Pacific Arctic in the early Pleistocene, getting there when glacial conditions shifted the southern birds northwards, some crossing the Equator along the cool waters off the west coast of South America. *F.glacialoides* is still seen along the Humboldt Current today as far as 10°S. Voous postulated a spread from the North Pacific via the Arctic Ocean to the North Atlantic. Thus, on this view, the North Pacific fulmars represent the oldest stock of *F.glacialis*.

C Thalassoica

The Antarctic Petrel is very distinctive by reason of its bold black and white plumage, not chequered as with the Cape Petrel, the black fading to a buffish brown by the end of the summer and at the start of the moult. The black bill is compressed laterally, rather gull-like, and the grooves between the plates are very pronounced creating a deeply sculptured effect. The bird has a rather steep forehead and rests on the sea in a characteristically high attitude, well shown in E.A. Wilson's drawings, e.g. in Lowe & Kinnear (1930, plate VII).

This is the least known of the fulmars and doubtless more breeding sites await discovery. It nests mostly on rocks and stacks around Antarctica and, like the Snow Petrel, also lays on mountain ledges far inland.

Falla (1937) was the first to assemble much information on the birds. At Ardrey Islet near Wilkes Station Orton (1968) found them arriving about a week before other local breeders, taking up nesting sites on sheltered, sunny, north-facing cliffs, only to be ousted by the bigger, more numerous Southern Fulmars when these appeared. Some gave ground only after vigorous fighting, but all eventually laid in a small area of ledge with scattered rocks. Cowan (1979) noted that Antarctic Petrels were more agile on their legs than nearby fulmars and could take off and land on flat places better than them.

At the same place Luders (1977) studied the display repertoire, noting the similarity of many of the actions to those of other fulmars and recording a head-tossing movement apparently restricted to *Thalassoica*. Some breeding data were also given by Pryor (1968).

Mehlum *et al.* (1988) have described the breeding habitats of this petrel in Dronning Maud Land and censused the largest colony there about 200 km from the sea and about 1600 m above sea level. Here Bech *et al.* (1988) studied aspects of the development of the chicks under extreme cold conditions.

D Daption

The Cape Pigeon, Cape or Pintado Petrel is the best-known of Southern Ocean seabirds, its checkered black and white dorsal plumage making identification simple. The patterning and intensity of the dark markings vary with the age of the feathers and state of the moult (Murphy, 1936, p.606). These change the appearance and led an early writer into describing eight species for the genus (Stephens, in Shaw, 1826).

The bill is specialised, differing from that of other fulmars in being depressed, with the lateral outlines as viewed from above convex, particularly the mandibular rami. These enclose a space filled with vascularised bare skin forming a small distensible pouch. The degree of convexity of the bill varies from bird to bird (Biermann & Voous, 1950, p.66) but whether these differences are biologically significant seems unknown.

This petrel has a very extensive breeding and feeding range (Fig. 3.10), nesting, rather sparingly, on the mainland and offshore islets of Antarctica, thence north to the Crozets at 46°S. In the New Zealand sector the subspecies *australe* breeds as far as the Chatham Islands at 44°S. The birds are found among the pack-ice but their populations centre around the low Antarctic and the Subantarctic. Even more than the Southern Fulmar, Cape Pigeons range north along the Humboldt Current into the tropics. Presumably their tolerance of mild conditions has helped vagrants to reach northern seas occasionally, unaided by man.

Cape Petrels are great followers of ships and may sail so close that they can be caught with a hand net. Hutton (1865, p.292) recorded a marked bird that followed a sailing vessel for 2400 km.

E Pagodroma

Snow Petrels are the smallest of the fulmars and were first described by Georg Forster during Cook's second voyage at about 52°S (Forster, 1777). Except for its black bill, grey-blue legs and dark eyes, it appears as a wholly white bird, rather long in the tail for a fulmar. The bill is laterally compressed and the hallux unusually large and stout for a petrel. In the hand the eye socket is seen to contain a small patch of black feathers and there is a grey wash along the anterior margins of the two outermost primaries and often faint grey vermiculations on the contour feathers of the back. According to Brown (1966) these features are not restricted to young birds as thought by Murphy (1936, p.634).

Adaptations to counter the cold are the thick body plumage which extends well down over the base of the beak, nearly to the tibio-tarsal joint and fills the inter-ramal space; together with the downy after-shafts, these features lessen heat loss through the skin. The feathering of the forehead helps to give the bird its characteristic physiognomy (Coues, 1864–66) and Fig. 3.1F.

Snow Petrels are highly specialised for life in the high-Antarctic, finding food in the leads between the pack-ice and around icebergs. Their flight is buoyant, rather unpetrel-like and the longish tail confers an ability to change direction rapidly. They are the ecological counterparts of the Ivory Gulls *Pagophila eburnea* of the Arctic. Griffiths (1983) considered that the low wing loading (Warham, 1977a) was an adaptation for life among the pack-ice where dynamic soaring is impossible. That the birds accumulate adequate energy reserves, however, despite that disability, is shown by their nesting far inland where access to the sea and food must involve long journeys and much flapping flight, albeit in almost continuous daylight.

The downy chicks are unusual in having two colour phases: they are either a uniform light grey, paler ventrally, or, less often, light grey dorsally except for a creamy white forehead and belly (Brown, 1966). These variants are not known to carry through to the adult plumage.

A peculiarity of this petrel is that at most breeding stations large, intermediate-sized and small birds nest without assortive mating, mixed pairs breeding successfully. No one size type is restricted to any particular kind of nest site.

Jouventin & Viot (1985) gave body weights of 26 males from the Balleny Islands, where large birds predominate, as averaging 369 g, and of 13 small males from Davis Station as 278 g, and bill dimensions varied likewise (Fig. 3.5). This kind of variability has been repeatedly discussed since its first notice by Bonaparte in 1855. He listed the extremes as *P.nivea* var *major* and *minor* (Bonaparte, 1857) and the former has been used in some modern checklists.

A succession of recent workers (Isenmann, 1970; Cowan, 1981; Croxall, 1982a; Jouventin & Viot, 1985) have tried to unravel the details. The considerable sexual dimorphism (Table 3.1) above, hindered early analyses, but Jouventin and Viot made sex by sex comparisons and also ran starch-gel electrophoretic analyses of blood proteins from birds of Adelie Land where 85% are large. The morphological variability in this population was accompanied by great genetic heterogeneity and a high proportion of polymorphic loci, although there was no direct correlation between the genetic and morphological variabilities, whether the latter was measured by weight, bill length or wing length.

The proportions of large birds breeding in various populations are shown in Fig. 3.5. Only small birds nest at South Georgia or near Davis Station; only large ones nest at the Balleny Islands, but elsewhere both types are found in varying proportions.

More data from breeding pairs are needed, but Jouventin and Viot hypothesise that the current situation arose during the Quarternary glaciations when, in summer, the Antarctic ice was 100–150 km out from the present coastline and mainland sites mostly inaccessible. They envisage populations of small birds developing in the Scotia Arc region where the climate was relatively mild, while in the refugium of the unglaciated Balleny Islands a large form differentiated. With the later climatic warming, birds of both types could expand their ranges to produce a vast hybrid belt and a high proportion of polymorphic loci would be expected in such a mixture. These authors suggest that one test of their idea would be to determine the genetic variability of the Balleny and South Georgian birds.

F *Relationships*

The fulmars have usually been thought of as an ancient group, although Fleming (1982) pointed out that the dependence of *Thalassoica* and *Pagodroma* on the pack ice argues for their having evolved after the formation of this feature in the late Tertiary. He concurs with others in considering that a migration of a southern fulmar stock to the north gave rise to the Northern Fulmar as a much more recent phenomenon, the crossing of the tropics perhaps being made during a Pleistocene glaciation.

Apart from the comparison of the Northern and Southern Fulmars by Voous (1949), little work has been done to clarify relationships among the fulmars, although there have been suggestions for 'lumping' together some of the monotypic genera. Harper (1978) tentatively concluded from his analyses of blood plasma proteins, that *Macronectes, Daption* and *Fulmarus* were closely related. All

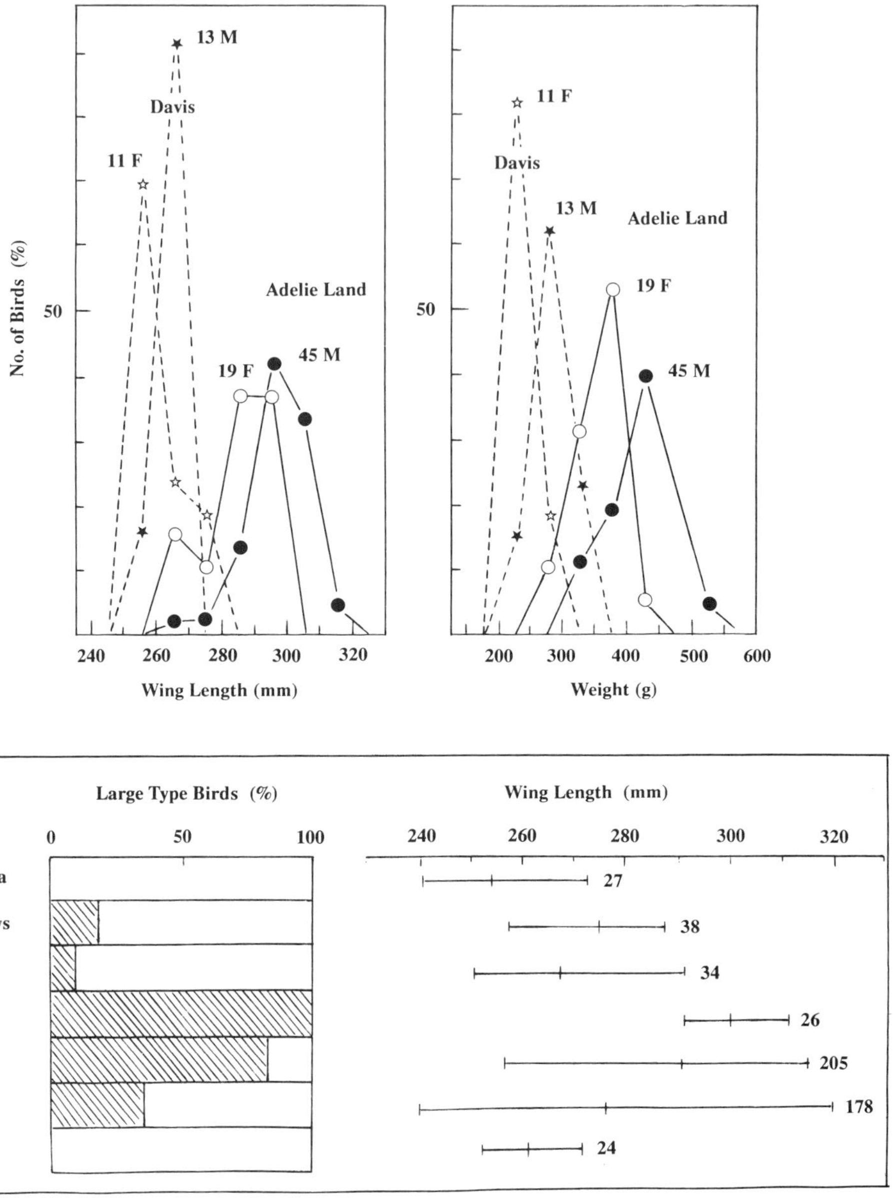

Figure 3.5 *Top: Wing lengths and weights of male and female Snow Petrels from Davis and Adelie Land compared. Bottom: Means and ranges of wing lengths of Snow Petrels from different colonies (right) and (left) the percentage of birds with wing lengths > 280 mm. From Jouventin & Viot (1985).*

had identical mobilities of transferrin 'triplets', but in some other features, like esterase activity, *Macronectes* was clearly distinct (Fig. 3.6). Barrowclough *et al.* (1981, Table 1), from another electrophoretic study, also found that *Pagodroma,*

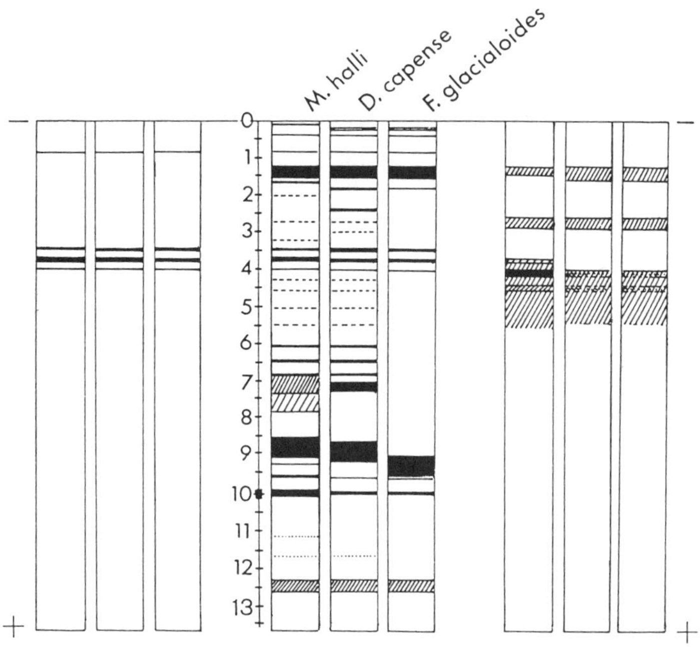

Figure 3.6 *Blood plasma proteins of three fulmars — note identical mobilities of transferrin 'triplets' (left), but varying esterase patterns (right). For key to components see Appendix 2. From Harper (1978), courtesy of* New Zealand Journal of Zoology.

Daption, Thalassoica and *Macronectes* were more closely related to one another than to other petrels they examined.

Imber (1985a) illustrated skulls of *Thalassoica* and *Pagodroma* and drew attention to features such as the lateral projection of their pre-frontal (lachrimal) bones and their large feet to support his ideas of a relationship with his genera *Lugensa* (= *Pterodroma brevirostris*) on the one hand and *Pseudobulweria* (= *Pterodroma rostrata* + *Pterodroma aterrima*) on the other.

IV Life styles

A *Locomotion*

All fulmars fly well. *Macronectes, Fulmarus* and *Daption* are stiff-winged and use much dynamic soaring, but in the pack-ice where there is little wind-shear, *Thalassoica* and *Pagodroma* must beat their wings frequently, the latter with a characteristic moth-like action. The flight of the Northern Fulmar has been studied in some detail, e.g. by Pennycuick (1960), its tameness and fondness for soaring along cliff faces using upcurrents created by the wind deflected from them facilitating observation. When alighting in calm weather or in up-drafts this fulmar, like other petrels in similar circumstances, uses a very rapid wing fluttering action apparently to aid stalling at touch-down.

Giant petrels fly very fast but, being short-tailed, lack the aerial dexterity of similar-sized sooty albatrosses, and they need reasonably flat ground for landing, so that their colonies tend also to be in fairly level places, though often near a slope to aid take-off. In calm weather they 'paddle' for some distance over land or sea, their webs fully spread, before they achieve enough lift to become airborne. The cliff-nesting species have no such problems, and in eastern Scotland *F.glacialis* has even been seen landing on and displaying from Rooks' nests in trees. However, fulmars do not climb trees as so many other petrels can.

The smaller fulmars avoid standing, but can do so briefly and Snow Petrels were described by Wilson as walking on ice 'with a very ungainly waddling'. *Macronectes* readily stands, walks and runs, but is less agile than an albatross and soon relaxes onto its tarsi.

Fulmars swim well and much prey is taken by surface seizure and by scavenging; swimming to maintain position is then important. Most fulmars are seldom seen diving, although all species probably do so occasionally. *Daption* dives regularly and went to about one metre down to take siphonophores off Heard Island (Downes *et al.*, 1959). Even *M.halli*, seemingly too buoyant for such activity, dived to 2 m to recover sinking food. It used half-closed wings to drive it along under water (Hemmings & Bailey, 1985). Harper (1987) saw similar behaviour whereas nearby *giganteus* remained on the surface and he wondered whether this was a constant difference in their foraging behaviours.

B *Behaviour*

Giant petrels often come ashore in loose groups to rest and sleep on beaches and grassy slopes far from their breeding sites. They also assemble at sea off penguin colonies waiting to capture newly fledged chicks. Antarctic Petrels rest and roost amid the pack-ice and large flocks may blacken floes and tabular icebergs.

Whether on land or at sea most fulmars take little notice of man and Northern Fulmars were easily taken by fishermen on the Grand Banks for bait. Antarctic Petrels have even been picked from the water by hand. But giant petrels are definitely timid birds, even at sea where, although familiar and regular followers of ships, they usually stay well back in the wake rather than come close in like Cape Pigeons. Most remain reluctantly on their egg when approached (flight means its loss to skuas or sheathbills) but some usually flee pell mell and their fear is contagious. Wholesale disturbance may result. Some early forays by bird ringers into colonies during incubation must have caused extensive desertions and when research bases are set up near existing giant petrel colonies the birds soon shift because of their intolerance to disturbance.

All fulmars defend their 'individual distances' by spitting stomach oil. Adult giant petrels seldom do this, although they disgorge their stomach contents (Chapter 15, Section IVB) but fulmar chicks are adept spitters and good marksmen, not splattering the oil wastefully like young albatrosses. The Northern Fulmar is particularly dexterous, shooting gobs of oil with sudden lunges of its opened beak at rates of two to three per second and giving a sharp hissing note at each lunge.

Behavioural repertoires have been little studied, even in the familiar Northern

Fulmar (summary in Cramp & Simmons (1977), mainly from M.A. Macdonald). In general, those of the four small species appear rather restricted; those of *Macronectes* more complex. A detailed analysis of the social behaviour of *M.giganteus* has been given by Bretagnolle (1988b). Actions common to all fulmars include mutual and reciprocal preening between partners or courting birds, and displays with opened bills and heads waved from side to side (Fig. 3.7). Such activities occur among paired birds, members of social groups and even at sea outside the breeding season, and seem to have both sexual and agonistic connotations depending on the circumstances and status of the participants. The accompanying vocalisations are complex acoustically and include series of cacklings and brayings rather unlike the calls of other tubenoses — see Guillotin & Jouventin (1980) for analyses of the voice of *Pagodroma*.

Figure 3.7 *Northern Giant Petrel in threat display — head waved, tail cocked, nape ruffled, wings spread. From a photograph.*

Copulation between mated birds usually involves little previous display. The mounted male strops his bill against that of his partner as occurs in albatrosses and other petrels. Hatch (1987) analysed many instances of copulations of Northern Fulmars with other than their own partners. He found that these were not the usual 'rapes' commonly seen among albatrosses (Chapter 12, Section VI), because most resulted from the solicitations of females.

Apart from oil spitting, threat involves extensive use of the spread wings, the spread and cocked tail, and often the raising of the feathers of the nape (Fig. 3.7). Hierarchies develop among *Macronectes* feeding communally at a dead whale or seal and much head waving, croaking and chasing occur between birds with established positions at the carcass and others, often in juvenal plumage, waiting to displace them (Warham, 1962; Voisin, 1978). In such situations the lighter, smaller females are disadvantaged.

Fulmars also call and display while on the wing, although aerial displays are easily overlooked. Typically the bird hesitates in its flight, braking with spread tail and vibrating wings while rolling its head and neck from side to side, bill wide open and giving a low-intensity braying cry (Fig. 3.8). Among Northern Fulmars such calls may involve threat (see, e.g. Pennycuick & Webbe, 1959) and perhaps this is true of the aerial displays of giant petrels (Warham, 1962, 1975, 1976). Similar actions by *F.glacialoides* have not been described but may have been overlooked.

Figure 3.8 *Aerial display of* M.halli: *the raised head and neck are swung from side to side, beak open to give a braying cry, wings held stiffly out and somewhat drooped, tail spread and feet lowered. From a photograph.*

V Distribution

A *General*

Fulmars are birds of high latitudes adapted to breeding during short summers of long days. All have circumpolar distributions (Figs 3.9, 3.10). The Northern Fulmar is the sole representative in the Northern Hemisphere and now nests, in small numbers, as far south as Britanny.

The southern fulmars are most numerous in the Antarctic Zone, with the most northerly breeders the giant petrels of Gough Island (40°S).

The Snow and Antarctic Petrels have the distinction of being the most southerly breeding of all birds apart from their attendant skuas *C.maccormicki*. Some lay on mountain ledges far inland from the Antarctic coastline. Snow Petrels nest at heights of up to 2100 m and 300 km from the sea (Lovenskiold, 1960; Somme, 1977) and have even been seen flying round mountains 415 km inland. Antarctic petrels nest on ledges in the Theron Mountains at 79°S (Brook & Beck, 1972), on the Pagodromen Mountains at 73°S (Somme, 1977), and in the Shackleton Mountains at 80°S (Croxall *et al.*, 1984b) where they are 350 km from the Weddell Sea. Whether the few pairs of McCormick's Skuas that breed near these sites are wholly dependent on the petrels or also make the 600+ km flights to and from the sea to feed, seems unknown.

Ainley *et al.* (1984) believed that Snow Petrels choose breeding places that allow summer access to areas of residual pack-ice where there is abundant food, and that they are reluctant to cross open sea to reach such areas, which would help explain the small population at South Georgia where there is limited access to the pack.

Bourne & Warham (1966) suggested that *M.halli* is adapted to feed to the north of

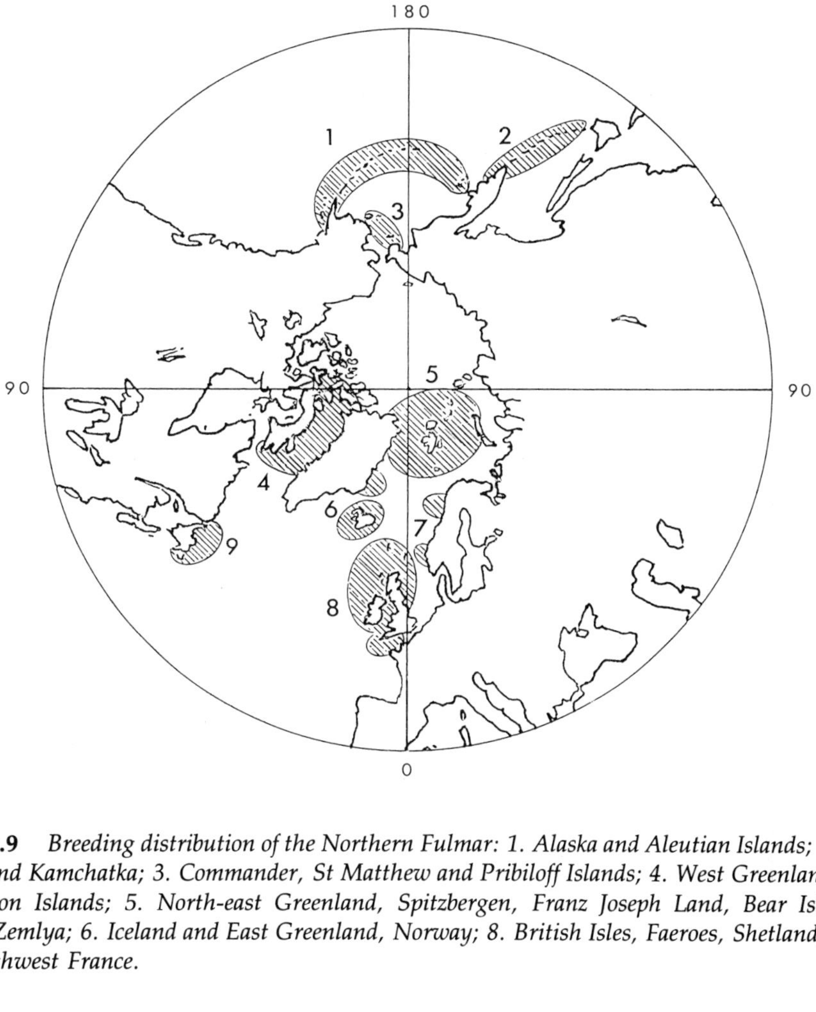

Figure 3.9 *Breeding distribution of the Northern Fulmar: 1. Alaska and Aleutian Islands; 2. Kurile Islands and Kamchatka; 3. Commander, St Matthew and Pribiloff Islands; 4. West Greenland, Baffin and Devon Islands; 5. North-east Greenland, Spitzbergen, Franz Joseph Land, Bear Island and Novaya Zemlya; 6. Iceland and East Greenland, Norway; 8. British Isles, Faeroes, Shetland, Orkney and Northwest France.*

the Antarctic Convergence, *giganteus* to the south of it, and this is supported by analyses of sightings at sea by Johnstone (1974) and others.

B *Range extensions*

The low-Arctic, eastern Atlantic, pale-plumaged form of the Northern Fulmar has undergone a major range expansion and population explosion in recent times, carefully documented by Fisher (1952). A similar increase has not occurred in the Western Atlantic or the North Pacific. The expansion of range and numbers seems to have begun about 200 years ago in Iceland, the colonisations spreading south the the Faeroes, Britain, Norway, northern France (in 1971) and Piccardy (in 1979) (Mercier, 1987). In Scotland the expansion continues despite a shortage of cliff sites, so eggs are laid on ledges in quarries, on window sills and castle walls, and

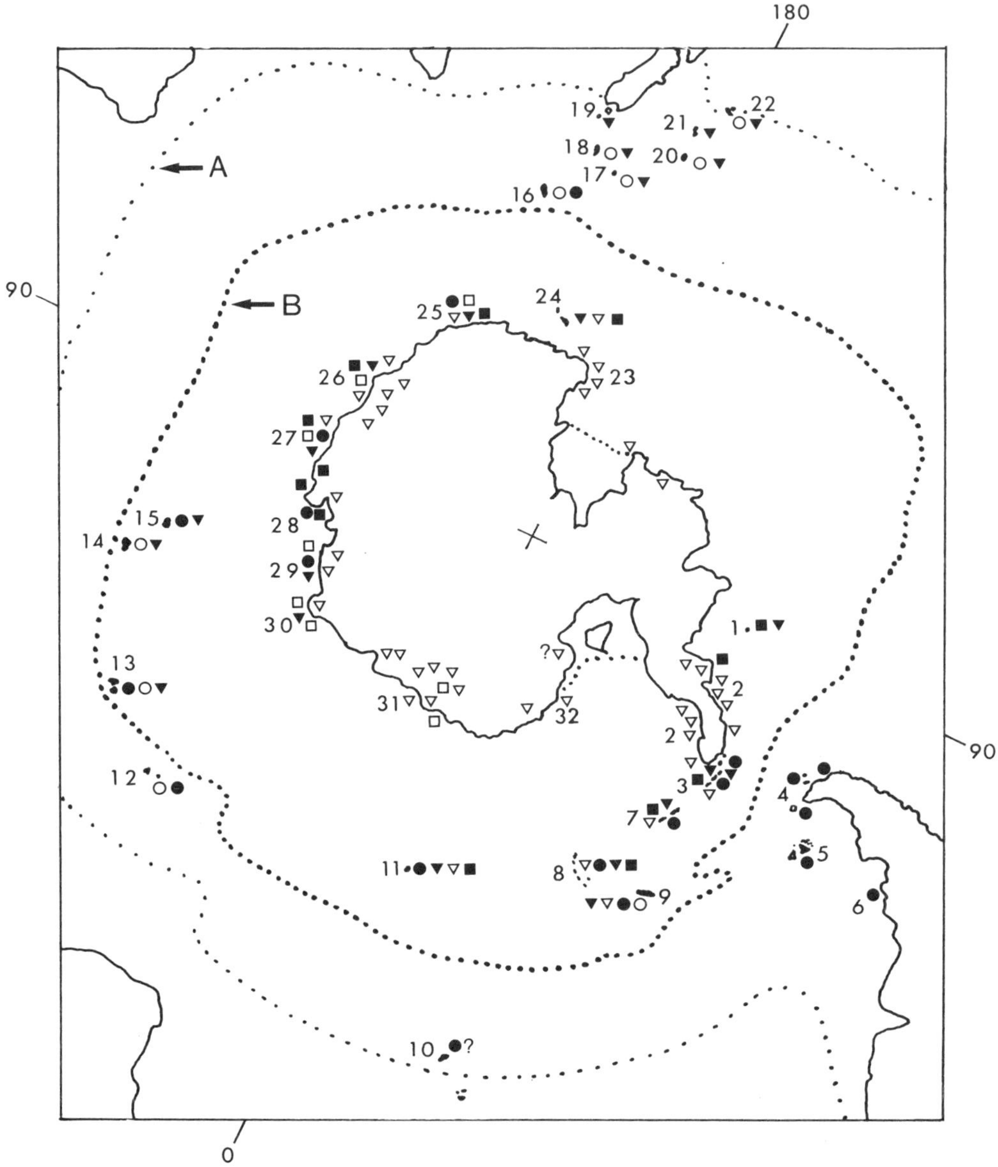

Figure 3.10 *Breeding distribution of southern fulmars:*
○ M.halli; ● M.giganteus; □ T.antarctica; ■ F.glacialoids; ▽ P.nivea; ▼ D.capense.
1. Peter I Island; 2. Antarctic Peninsula; 3. S. Shetland Islands; 4. Cape Horn, Isla Noir and Staten Island; 5. The Falklands; 6. Chubut; 7. S. Orkney Islands; 8. S. Sandwich Islands; 9. S. Georgia; 10. Gough Island; 11. Bouvet Island; 12. Marion and Prince Edward Islands; 13. Crozet Islands; 14. Kerguelen; 15. Heard Island; 16. Macquarie Island; 17. Campbell Island; 18. Auckland Islands; 19. The Snares; 20. Antipodes Islands; 21. Bounty Islands; 22. Chatham Islands; 23. Victoria Land; 24. Balleny Islands; 25. Terre Adelie; 26. Budd Coast; 27. Queen Mary Land; 28. Prydz Bay; 29. MacRobertson Land; 30. Enderby Land; 31. Dronning Maud Land; 32. Coats Land. A. Subtropical convergence; B. Antarctic convergence.

among low rocks on hillsides. Some have occupied Rooks' nests and have even laid on sand dunes near terns among the marram grass of the Ythan Estuary, Aberdeen (Anderson, 1982) and in the Monarch Isles, Outer Hebrides.

VI Feeding and foods

A *Feeding methods*

Fulmarine petrels have a wide variety of feeding styles. All will scavenge dead animals, behaviour particularly noticeable in Antarctic seas during the whaling era.

The lightest and most aerial of them, the Snow Petrel, however, picks most of its food from the surface by dipping, but often immerses its breast — 'contact dipping' — in so doing. Less often it dives but it seldom alights to seize prey. This and the Antarctic Petrel feed mainly in the leads among the pack-ice and at the bases of icebergs, often in flocks, the Antarctic Petrels usually being the more numerous. These also get much of their food by dipping but often dive 2–3 m in pursuit of prey. Both Northern and Southern Fulmars feed to some extent among the pack, as does the Cape Pigeon, but all three range far from the ice, which *Pagodroma* and *Thalassoica* seldom do. Southern Giant Petrels get some of their food among the pack-ice and in the open sea, but in the summer also scavenge on land and kill other seabirds both there and at sea.

Cape Pigeons swim confidently, riding waves buoyantly while picking individual food items from the surface like a pigeon pecking grain. At other times planktonic animals are taken at the tip of the beak in rapid succession while the expanded and contracted buccal pouch pumps out the imbibed water, leaving the food on the bill serrations (Matthews, 1929, p.576). Presumably this style is adopted when plankton 'blooms' are encountered and the fleshy tongue must also be involved. Matthews noted that when feeding like this the birds paddled vigorously with their feet to draw a current of water towards them from the front. Cape Pigeons also plunge-dive. Very rarely they may land to feed, having been seen to walk onto a beach attracted by dead notothenid fish.

Pagodroma, Daption and *Thalassoica* have black beaks, other fulmars pale ones: possibly dark beaks make better aiming guides when pecking small prey at the surface.

The feeding styles of the two *Macronectes* species are very similar, but some differences have been noted. For example, although both may follow ships to scavenge discarded garbage or food stirred up by propellers, *M.halli* is the more consistent ship-follower (Johnstone, 1974). The hierarchical arrangements accompanying scavenging on carcasses have been referred to and both species of giant petrels also dominate other potential feeders — skuas, gulls, wekas and sheathbills giving the larger birds a wide berth. Giant petrels have been known to side-swipe, kill and eat unwary birds as big as a large gull. During the southern summer some giant petrels live largely on penguins, taking chicks and adults of the smaller kinds for themselves and their young. They eat mollymawk chicks and smaller petrels as opportunity affords. Their victims may be alive or dead, on land or at sea, and carrion from seals and whales is searched for and devoured voraciously.

The striae on both upper and lower mandibles of both *Macronectes* species have been referred to (Fig. 3.2), but filter-feeding is not a method usually associated with these birds. However, Harper (1987) reports giant petrels (species unstated) filtering water for small planktonic prey by gulping mouthfuls and then extruding the excess water at the gape: this sounds very like filter feeding, Cape Pigeon style.

The distances from their nests at which food is found no doubt vary with the species and situation. Furness & Todd (1984) gathered direct information from 10 sightings at sea of colour-dyed *F.glacialis* from Foula, Shetland Islands. These were seen a median distance of 35 km from their chicks. Indirect evidence based on the median time of 6 h birds were away at sea (n = 67), gave an estimated potential range of 120 km.

B *Foods*

Crustacea and squid are important in the diet of all the southern fulmarine petrels. On the basis of the number of items in their stomachs, *Euphausia superba* and *E.crystallorophias* usually predominate, but on a weight basis squid may be more important; c.f. Ainley *et al.* (1984) and Ridoux & Offredo (1989). *Thalassoica* also captures many near-surface fish. *Pagodroma* in continental shelf waters of the Ross Sea depended largely on fish, whereas in deeper seas, squid were more important (Ainley *et al.*, 1984). Despite its size, *Macronectes* caught at sea contain large numbers of euphausids (taken by filter-feeding?), although squid may predominate on a weight basis. The picture is different if analysis is based on chick regurgitations when, measured by frequency-of-occurrence, remains of penguins and small petrels come first, seal flesh, krill, squid and fish being less important (Hunter, 1983a). Evidently, when rearing chicks, both forms of giant petrel are predators, using temporarily abundant local resources: their winter diets must be quite different.

Hunter (1983a) found that when krill was plentiful in summer, female South Georgian giant petrels of both species fed more on sea-caught prey (krill, squid and fish) than on dead seals, whereas the males fed mostly on seal flesh and gave this to their chicks. Indeed, he thought that at South Georgia the intersexual contrasts in diet (particularly marked with *M.halli*), were more important than interspecific dietary differences. Johnstone (1977) and Jouventin *et al.* (1982) reported that in summer *halli* is more addicted to carcasses and other food taken on land than is *giganteus*. Small prey, like prions and shearwaters, appear to be swallowed whole and even complete penguins have been found in stomachs.

Furness & Todd (1984) found that *F.glacialis* at Foula depended largely on sand eels during chick rearing. The birds of St Kilda, however, fed mainly on zooplankton, perhaps taken after dark.

VII Breeding

A *General*

Breeding studies of fulmars include those on *Macronectes* by Warham (1962), Conroy (1972), Jouventin *et al.* (1982), Hunter, 1983a, b; 1984a, b, c), and Voisin (1988); on *F.glacialis* those by Carrick & Dunnet (1954), Macdonald (1977a), Ollason & Dunnet (1978, 1983, 1986, 1988), Dunnet & Ollason (1978), Dunnet *et al.* (1979); of both *Fulmarus* those by Mougin (1967); of *Daption* those by Pinder (1966), Mougin (1968), Beck (1969), Isenmann (1970), Sagar (1979) and Mougin & Prevost (1980); of *Thalassoica* those by Orton (1968), Cowan (1979), Mougin & Prevost (1980) and Montague (1984); and of *Pagodroma* the studies by Brown (1966), Guillotin & Jouventin (1980), Cowan (1981) and Croxall (1982a).

B *Breeding patterns*

Fulmars do not burrow. They breed socially or colonially in the summer nesting either on flat ground (*Macronectes*) or using cliff ledges and sheltered crevices and recesses in which to lay their egg on bare rock or in shallow depressions lined with stones. Northern Giant Petrels often lay among tussock grass and ferns and make substantial nests from local materials. So does the southern species when such materials are to hand, but it mostly selects open, unvegetated sites when the nests are simple, saucer-shaped depressions. The shells of fulmars are thicker than those of other similar-sized petrels: presumably this reduces loss due to breakage (Chapter 13, Section VIB).

Tucked away in crevices and among talus debris the smallest of them, the Snow Petrel, is well placed to defend itself from skua attack and sites used by all cliff-nesters are often selected to give improved micro-climates notably giving some protection from the prevailing wind. Snow Petrels nesting inland come in flying high and then circle down the mountain sides until the nest level is reached (Somme, 1977).

Breeding systems follow usual petrel patterns, with nest-site and mate tenacity and a pre-laying exodus. However, changes of mate occur while the previous partner is still active more frequently than with most other petrels and giant petrels seem to shift their colony sites quite often, though over rather short distances. Nevertheless, some *giganteus* colonies have been in the same places for at least 45 years.

Fulmar chicks are guarded for much longer than those of burrowing species of similar size and the chicks have the advantage of defensive spitting when guarding ends. At South Georgia brooding and guarding by *M.halli* takes 11 days longer than by *giganteus* and male chicks of both species take 5–6 days longer to fledge than females (Hunter, 1984b). They are bigger than them and depart later.

A major difference from other tubenoses is that in adapting to short summers of long days and an abundance of food, with all fulmars the time from egg deposition to chick departure is significantly shorter than it is for petrels of similar sizes elsewhere (Chapter 15, Section VIC).

VIII Dispersal and migration

Notwithstanding contra-nuptial visiting (Chapter 12, Section IA), the centres of gravity of the Southern Ocean fulmar populations shift northwards during the winter. Some species must do that to escape the ice sheets closing around Antarctica and to remain in the pack-ice, and analyses of sightings of fulmarine petrels at sea, such as those of Shuntov (1974) confirm these general shifts even of species not dependent on the pack.

As many giant petrels and Cape Pigeons must remain within range of their breeding places in order to make their winter visits, the bulk of the northwards shift presumably consists of younger and non-breeding birds.

Recoveries of marked birds show that fledglings of both giant petrels tend to fly down-wind to the west, often reaching medium latitudes in the process. They do not breed until 6 or more years old, remaining at sea for several years before returning to their natal islands (usually), and recoveries of marked birds support the view that they circumnavigate the globe during that period. Similarly *F.glacialis* of low latitudes appear to be non-migratory as adults, the wide range of birds seen at sea in winter probably being due to the dispersal of immatures (Cramp & Simmons, 1977).

Southern Fulmars shift further north in some years than in others, e.g. in 1973 and in September and October 1975 more than 615 were found on the beaches of the North Island of New Zealand, while in the autumn of 1978 'wrecks' of this species occurred in southern Africa and southern Australasia. Some even reached Queensland. Antarctic Petrels also very occasionally reach as far as 36°S and in the winter of 1978 were plentiful around Stewart Island, with many washed ashore on western beaches of New Zealand. Yet the first record of the species in that country dates only from 1973.

Vagrants of southern species, particularly the Cape Pigeon, occasionally reach the Northern Hemisphere (see Bourne, 1967a). A giant petrel photographed off Ushant, France in November 1967 appears to have been *giganteus* but a sight record of an alleged giant petrel at Midway Atoll over three winters seems suspect as the possibility of confusion with immature *D.albatrus* (known to reach there) was not mentioned.

IX Populations and mortality

A *Population sizes*

The giant petrels have the smallest populations, some colonies being tiny, e.g. that of *M.giganteus* in Adelie Land with only 9 and 12 chicks in 1982 and 1984 (Jouventin *et al.*, 1984; Thomas, 1986). Williams (1984) estimated that about 310 pairs of *halli* and 1750 of *giganteus* breed at the Prince Edward Islands. Nearly half the world population of *halli*, *c.* 6500 pairs, is said to nest at South Georgia. At Macquarie Island in 1961 I ringed 2846 giant petrel chicks (all but about 150 of the total) and Johnstone (1977) estimated that 4000 ± 10% pairs of *giganteus* and 1000 ± 10% of *halli* bred there.

For a medium-sized, surface-nester, populations of Northern Fulmars are high. In Alaska there were computed to be about 2 million birds (Lensink, 1984), Greenland 200 000–400 000 pairs (Evans, 1984), with Britain and Ireland having 305 000 occupied nests (Cramp & Simons, 1977), the whole Atlantic population being of the order of 3 million birds (Brown, 1986). The regular censuses of this species done in Britain since 1912 are the most continued and thorough ever made of a petrel. For reviews see Fisher (1966) and Cramp *et al.* (1974).

The crevice nesters are virtually impossible to count with so many birds hidden or inaccessible and Mougin & Prevost's (1980) computation of a world population of 1 million Cape Pigeons can be little more than a guess. They also estimated a figure of 1.2 million Antarctic Petrels — very low compared with estimates for only two areas of East Antarctica of 3–5 million nesting around Prydz Bay (Montague, 1984) and of about 5.1 million birds along Marie Bird Land (Ainley *et al.*, 1984).

Counts of Snow Petrels have been attempted and Croxall *et al.* (1984b) surmised that 100 000 pairs bred around the Antarctic Peninsula and the Scotia Arc as well as 'several million' pairs of Southern Fulmars and perhaps a million Cape Pigeons. Ainley *et al.* (1984) estimated that some 2 million Snow Petrels were associated with colonies along the borders of the Ross Sea alone: the world population must be correspondingly large.

B *Population changes*

Most changes seem to be declines, particularly among giant petrels. For example, the tiny group of *M.halli* that formerly nested at Nelly Island off Stewart Island has disappeared and the colony of *M.giganteus* in Adelie Land near the French base, with 69 pairs in 1956, has now gone, apparently having moved to a less disturbed site. On the other hand the *halli* population at South Georgia has increased parallel to an increase in the fur seals breeding there. These prey upon penguins but fail to devour them completely and provide giant petrels with an increased food supply (Croxall & Prince, 1980).

However, the most remarkable change has been the increase in range and numbers already referred to of the Northern Fulmar. According to calculations by Cramp *et al.* (1974) the low-Arctic, pale-plumaged form continues to expand in Britain and Ireland at about 7% per annum, compound. The case of the increase is unknown and disputed. Fisher (1952, 1966) provided strong evidence for a correlation with the availability of offal from whalers and fishing trawlers. Ritchie (1930) supposed that it followed a decline in the human population on St Kilda, where he thought the birds had increased until lack of nest sites forced them to colonise elsewhere. However, there is no historical evidence that before the human occupation of such northern islands fulmars were abundant around Britain, so that some other change to the bird ot to its environment seems implicated. Wynne-Edwards (1962) proposed that a special genotype arose in Iceland particularly well adapted to the mild conditions of lower latitudes and that the expansion had little to do with human activities, while Salomonsen (1965) and Brown (1970) emphasised the probable importance of changes in oceanic conditions. The former believed that much of the eastern North Atlantic and that the original 'boreal' population, perhaps helped by a 'colonising' mutant, responded to the enlarged niche and greater availability of macroplankton. Bourne (1966) stressed that the availability of food in winter may be a crucial factor and offal could then be an important new resource. Brown (1970) suggested that to disentangle the feeding relationships of this omnivorous bird, although difficult, might offer the best way to assess the importance of food and oceanic factors in the population changes.

C *Population structure*

Mougin (1975) calculated survival and mortality rates for *F.glacialoides* at Pointe Geologie, Antarctica over five seasons based on recoveries of 95 ringed breeding birds; 93.1% survived annually, a mortality rate of 6.9%, giving a mean adult life expectancy of 14 years. This compares with an expectancy for *M.giganteus* at the same place of 25 years. Other figures calculated for southern species are for *Pagodroma* 4–7% annual mortality for adults (Hudson, 1966; Guillotin & Jouventin, 1980), whereas Brown (1966) calculated a rate of 18% for the birds at Mawson, Antarctica. South Orkney Island Cape Pigeons lost 5–6% of adults annually, giving a life expectancy of 16–20 years (Hudson, 1966) and for adult *halli* and *giganteus* at South Georgia annual mortality rates of 9.5 and 9.6% respectively were calculated by Hunter (1984b).

Over 17 seasons banded Northern Fulmars lost only 3% of breeding adults each year, their mean life expectancy being 34–35 years (Dunnet & Ollason, 1978) and a 'model' simulating annual changes to this population was used for predicting its size one year ahead (Ollason & Dunnet, 1983). In 1989 three birds were breeding which were banded as breeders in 1951, thus being at least 43 years old (Dunnet, pers. comm.).

D *Mortality*

While younger birds may die in storms, as evidenced by their bodies 'wrecked' on land, adults mostly die unrecorded, at sea. The possibility of mortality due to old age has arisen as the results of long-term marking studies mature (Dunnet & Ollason, 1978). The harsh environment of the polar regions involves losses of eggs and chicks due to nests being snowed in (parents often try to burrow through nonetheless), eggs becoming iced to ledges and other accidents. Few southern colonies are without attendant skuas ready to take eggs or unguarded and exposed chicks.

Mainly because of their inhospitable habitats and inaccessible nest sites, however, the fulmars are spared the attentions of alien predators. Even at places like Macquarie Island where cats are abundant, there is no evidence that these kill the chicks of the ground-nesting giant petrels. Northern Fulmars do suffer some loss through foxes, Peregrines and sea eagles, but these natural enemies have only local effects.

X Conservation

With fulmarine petrel populations so large, their nesting places so scattered and often remote and inaccessible, the future of this group of tubenoses seems more assured than most, barring major disasters such as oil spills among the pack-ice. The giant petrels are the most threatened. Their populations are small and their sensitiveness to disturbance will continue to involve reductions in nesting success as new Antarctic and Subantarctic bases are established with the consequent human interference, often unintentional and uncontrolled. There is thus a need to pin-point the key populations and colonies so that adequate protective measures can be instigated — measures that have already been taken at some of the long-established stations.

This need for undisturbed conditions is not restricted to giant petrels. Ollason & Dunnet (1980, 1986) showed that Northern Fulmars are also sensitive during nesting, disturbance reducing hatching success particularly among birds breeding for the first time.

CHAPTER 4

The Gadfly Petrels — Genera Lugensa *and* Pterodoma

Soft-plumaged Petrel.
Photo: author.

Contents

I Introduction

The gadfly petrels form the largest group of tubenosed birds, with 25 species in *Pterodroma* (*Pteros*, wing; *dromos*, running='winged runner'). *Lugensa*, a subgenus of Mathews (1942) has recently been revived for the Kerguelen Petrel, a bird more usually called *Pterodroma brevirostris*. *Lugensa* comes from *lugens*, mourning —

referring to the dark plumage — and is derived from a specific name *lugens* formerly used for that petrel until shown to be based on a drawing of another species, and so indeterminate (Bourne & Elliott, 1965).

Gadfly petrels are highly pelagic birds widespread in tropical and subtropical seas but with several breeding in the subantarctic zone. They breed socially or colonially, laying their eggs in burrows or on the ground and these are more rounded and thinner-shelled than eggs of other petrels of similar size (Chapter 13, Section VIB). The birds are also very vocal, with a wide repertoire. All studied use high-speed dual flights apparently as part of courtship and pair-formation. Many species have parti-coloured feet.

The fossil record provides some insight into past distributions but none of the material is pre-Pleistocene. Two new *Pterodroma* have come to light. Olson (1975a) described a medium-sized form, *P.rupinarum*, from St Helena (apparently related to the Tahiti Petrel *P.rostrata* of the Central Pacific) (Fig. 6.2) and Harrison & Walker (1978), and Harrison (1986), a small form, *P.kurodai*, from Aldabra Atoll in the Indian Ocean. Shufeldt (1916, 1922) described the skeleton of the Cahow *P.cahow* from a collection of subfossil petrel bones from a deep cave in Bermuda, a bird then thought extinct. He named this species *Aestrelata vociferans*, based on reports of the bird's clamorous voice; it had just been named *Aestrelata cahow* by Nichols & Mowbray (1916),

Other gadfly petrel material found in various deposits at the Chatham Islands and Amsterdam Island is, where identifiable, all of extant species. But such material may be useful when immature bone, identified by unformed articular ends and granular surfaces, is present: these may indicate breeding and, at times, harvesting of fledglings by local human populations.

II General characteristics

A *Size*

The taxa are composed of small to medium sized birds, ranging from Pycroft's Petrel *P.l.pycrofti* weighing 163 g, wingspan 74 cm to the Great-winged Petrel *P.macroptera* weighing about 525 g, wingspan 104 cm. The size distribution of 29 of the species is shown in Fig. 4.1. The smaller group of the histogram includes all the '*Cookilaria*' species (see below) and also *mollis, feae, madeira, macgillivrayi, inexpectata* and *brevirostris*.

B *Bill*

Gadfly petrels differ from other procellariids in their deep bills, always black in colour. The bill is laterally compressed, particularly in *Lugensa*. The maxillary nails are strongly hooked, convex above and concave below, the tomia of the upper mandible following a sinuous line when viewed from the side (Fig. 4.2). The tomia of the lower mandible are almost straight, the gonys a little concave. Well-marked sulci separate the lateral plates of both mandibles. The nasal tubes are short, rather

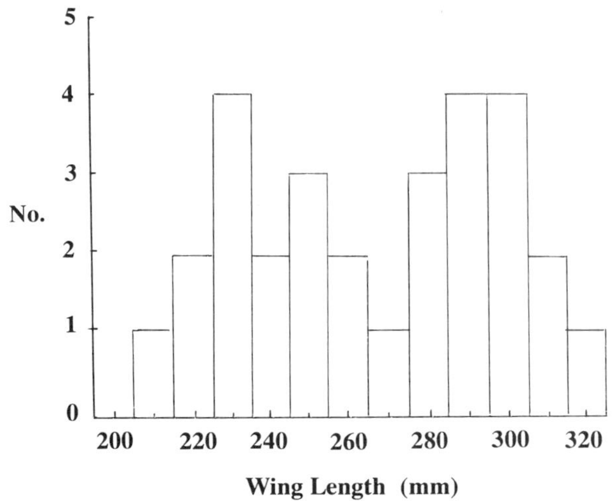

Figure 4.1 *Size distribution (by standard wing-lengths) of 29 species of gadfly petrel (sexes combined).*

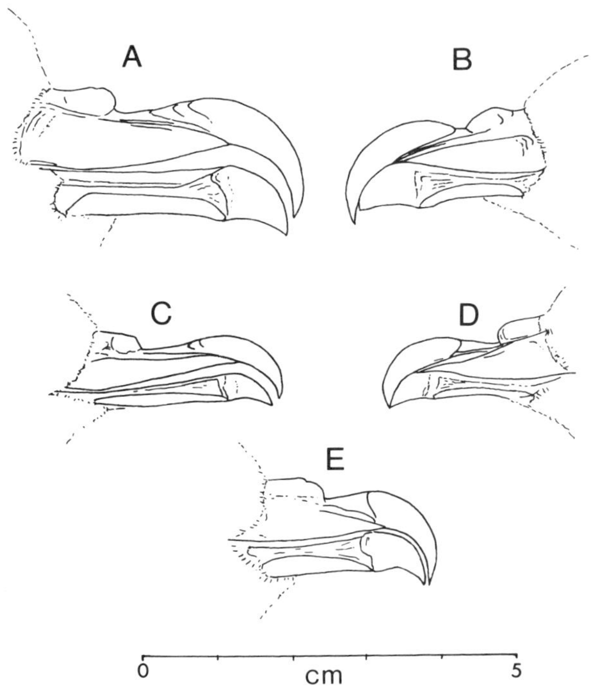

Figure 4.2 *Bills of gadfly petrels:*
A. incerta; *B.* brevirostris; *C.* cookii; *D.* hypoleuca; *E.* inexpectata.

elevated, with a thin central septum and apertures that usually open upwards. There is a little culmen exposed between the nasal openings and the upper nail.

Jouanin (1970) drew attention to differences in bill structure among *Pterodroma*. He noted that the bill of the Reunion Petrel *P.aterrima* was shaped differently from that of *P.macroptera* and the Trindade Petrel *P.arminjoniana*, and so were the nostrils. In *macroptera* the openings lie in a vertical plane, in *aterrima* they open on an inclined plane which contains the anterior edge of the median septum (Fig. 4.3).

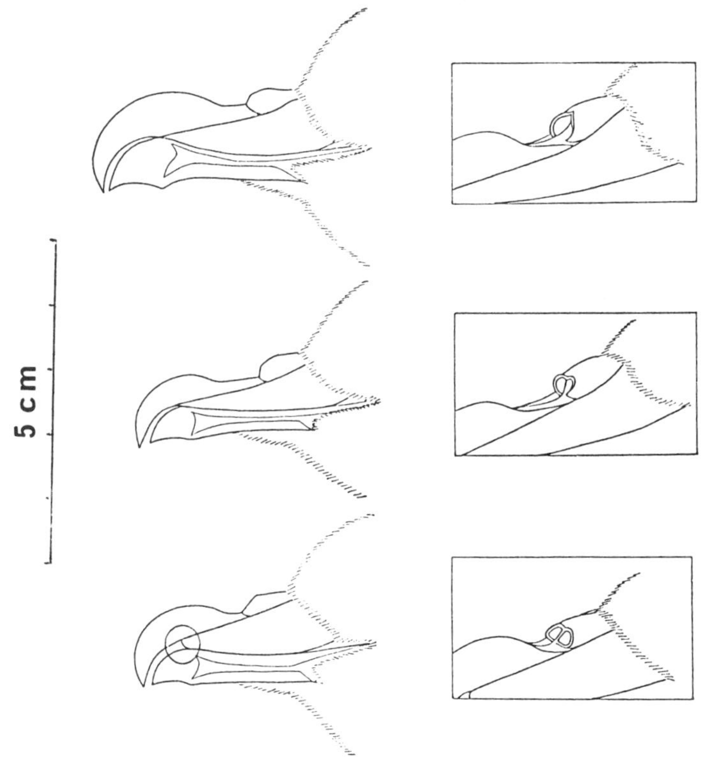

Figure 4.3 *Bills and details of nostrils of (top)* P.macroptera, *(centre)* P.arminjoniana, *and (bottom)* P.aterrima. *Note rhamphothecal notch of* aterrima *(circled). From Jouanin (1970).*

This figure also shows the notch in the rhamphotheca at the junction of the latericorn with the maxillary unguis of that bird. Jouanin noted that the bill of the Tahiti Petrel *P.rostrata* had a similar structure — shown well by Brasil (1917).

C *Foot colour*

In gadfly petrels the tarsi are rounded anteriorly and only slightly compressed, unlike those of diving species such as Sooty Shearwaters. Chicks of light- or dark-plumaged species are born with flesh-coloured feet that darken distally while they are in the nest (Fig. 4.4). Sometimes this does not occur and dark-plumaged adults may rarely have pale feet but species with all-dark plumage typically have black legs and feet as adults. *P.aterrima* is an exception, having the basal two-thirds of the webs, toes and legs flesh-coloured, the rest blackish.

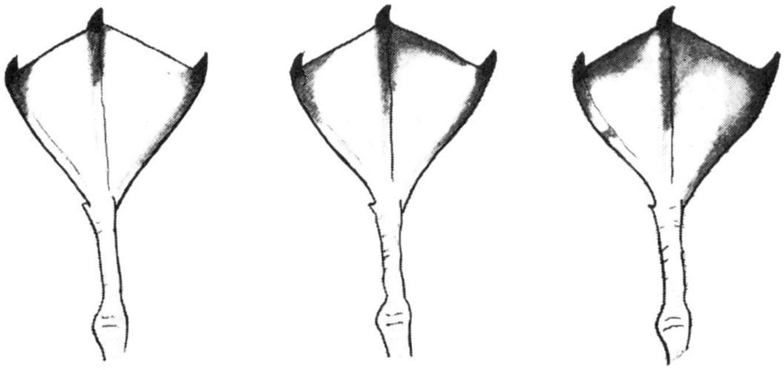

(a) (b) (c)

Figure 4.4 *Gadfly petrel chicks are born with pink feet. In* P.lessonii *darkening of the distal toes starts at about 40 days (a) and by 85–95 days (c) has almost reached adult condition.*

In polymorphic species light-phase birds have bi-coloured black and pink feet, dark-phase ones black feet, and intermediate-plumaged birds variable amounts of blackish on their tarsi, outer toes and inner webs (Oliver, 1955; Gill *et al.*, 1970). Dark edgings to the webs may confer greater strength, the better to resist abrasion, than if the webs were wholly flesh-coloured.

D *Plumage*

1 *General*

The plumage of five species is blackish brown, although there may be some pale feathers on the underwing. In most of the others the body feathers are brown or grey on the upperparts, with some white on the head, forehead and underparts, usually on the belly. A few species — *arminjoniana, neglecta* and probably *hasitata,* and one subspecies *P.l.brevipes*, are polymorphic, with light, dark and intermediate phases. The various morphs interbreed freely — Fig. 4.5. The Soft-plumaged Petrel *P.mollis* is less polymorphic than these, but does vary considerably in plumage. It has a rare dark phase but no extremely pale ones.

Apart from the Collared Petrel *P.l.brevipes*, the seven smallest gadfly petrels have upperparts that are grey or blackish, underparts white. This group is often designated as forming the subgenus *Cookilaria*, erected as a genus by Bonaparte in 1856 but imprecisely defined by him. Current practice is to sink *Cookilaria* into

Figure 4.5 *Dark and intermediate morphs of* P.neglecta, *Kermadec Islands. Photo: M.F. Soper.*

Pterodroma because, apart from their tendency to have legs of some shade of blue, there is rather little to separate them from smaller gadfly petrels like *P.mollis*.

Pterodroma tend to have steep foreheads with the pale-tipped feathers creating a scaled effect (Fig. 1.1). There are dark feathers in the eye sockets of nearly all the species and Harper (1978) suggested that these might help reduce sea glare. In pale-headed birds like the White-headed Petrel *P.lessonii* these dark patches extend on either side of the eye socket. Only in white-phase Kermadec Petrels *P.neglecta* are such patches lacking (see Serventy *et al.*, 1971, p.98).

Underwing patterns vary (Fig. 4.6) and all tend to have a wedge of darker feathers spreading diagonally from near the carpal joint back towards the body. This includes *P.macgillivrayi* not shown in Fig. 4.6. The patterns provide useful identifying features for birds seen at sea.

2 *Sexual dimorphism*

While plumage polymorphism is not sex-linked, sexual dimorphism in body size and related features has been shown in at least three species — *arminjoniana, macroptera* and *rostrata*. In *P.a.arminjoniana* (but not in *P.a.heraldica*) Murphy & Pennoyer (1952) found males significantly larger than females. This may also be true of Gould's Petrel *P.l.leucoptera* (Hindwood & Serventy, 1941). Falla (1937) and Imber (1971) established that male *P.macroptera gouldi* have significantly longer and deeper bills and weigh more than females and Johnstone & Niven (1989) used a discriminant function to sex this species. Reinstatement of the subspecies *P.rostrata trouessarti* Brasil, 1971, has been proposed by de Naurois & Erard (1979) because,

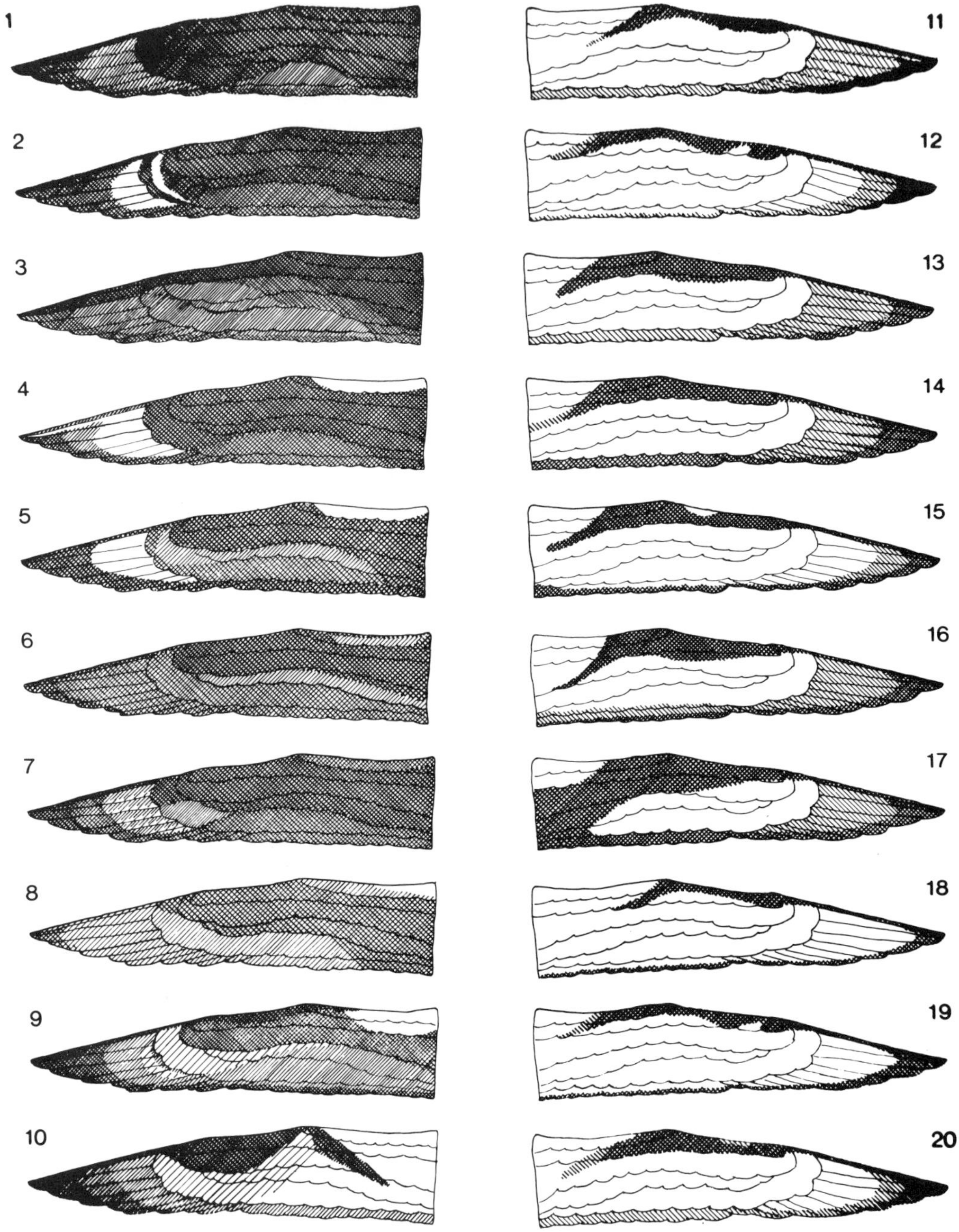

Figure 4.6 *Underwing patterns of gadfly petrels (not to scale):*
1. macroptera, lessonii, incerta *and* rostrata; *2.* solandri; *3.* magentae; *4.* neglecta; *5.* arminjoniana; *6.* alba; *7.* ultima; *8.* brevirostris; *9.* mollis, feae *and* madeira; *10.* hypoleuca; *11.* cahow *and* hasitata; *12.* externa; *13.* baraui; *14.* phaeopygia; *15.* inexpectata; *16.* nigripennis; *17.* axillaris; *18.* cookii *and* defilippiana; *19.* longirostris; *20.* leucoptera.

sex for sex, the latter has a bigger, heavier and deeper bill than the typical form (Fig. 4.7).

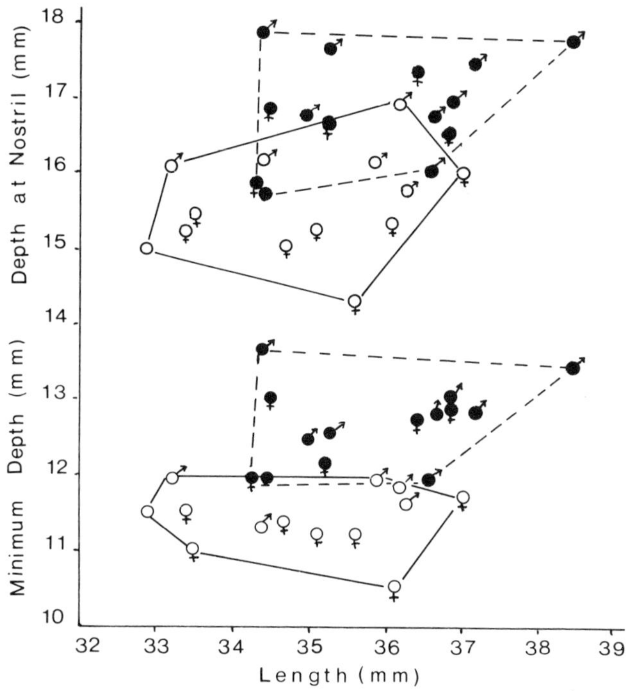

Figure 4.7 *Bill lengths and depths of* Pterodroma rostrata. *Closed symbols:* r.trouessarti; *open symbols:* r.rostrata. *From de Naurois & Erard (1979).*

Imber (1971) reported that males of 16 species of gadfly petrel have more filoplumes on their crowns, napes and hind-necks than the females have. The pale, downy tufts ending these simple structures may protrude from the contour feathers of the head (Fig. 1.1), in some birds almost forming a crest. Although thought by this author to have been previously unnoticed, prominent head filoplumes of *P.leucoptera*, *P.externa*, and *P.aterrima* were commented on by Stejneger (1893), Loomis (1918), Lonnberg (1921) and Godman (1907–10, p.180) respectively.

Imber hypothesised that the filoplumes aided sex recognition during mate selection, and he pointed out their absence in white-headed species like *P.lessonii*, where they would be invisible against the white contour feathers. Be that as it may, much courtship on the ground takes place on dark nights when such plumes would be invisible, while courting birds, even of *P.macroptera* (the species that first attracted Imber's attention to the matter), may lack protruding filoplumes. Perhaps the sexual differences in head filoplume development arose when gadfly petrels were solely diurnally active.

3 *Polymorphism*

Murphy (1936, p.693–697) pointed out that in polychromatic petrels, the proportions of the various phases is likely to change with time. He quoted Labat's description of the Black-capped Petrel *P.hasitata* on Guadeloupe Island, West Indies as being a black bird (Labat, 1722), whereas du Tetre (1667) had reported it as being white-breasted. Murphy & Pennoyer (1952) also scored the colour phases of skins of *P.arminjoniana* from the South Pacific and found that the proportions of the colour morphs varied greatly from island to island in the Pitcairn Group. Iredale (1914) classified the colour phases of *P.neglecta* at the Kermadec Islands, and Merton (1970) obtained rather different ratios from the same island nearly 60 years later, but he commented on the intergrading nature of the variation which created difficulties in categorising some birds.

E *Internal anatomy*

The gadfly petrel skeleton was described by Forbes (1882), Pycraft (1899), Shufeldt (1922) and Kuroda (1955), Shufeldt comparing it with *Puffinus* and Kuroda with *Puffinus* and *Calonectris*. The skeleton of the gadfly petrel is that of a short-necked bird compared with the shearwaters, one with a deep rib cage and deep sternal carnia to carry the big pectoral muscles. The pelvis is also rather short and deep. The skull is bulky and short, rounded downwards above the foramen magnum (Fig. 4.8), thus resembling that of the fulmars and not the more delicate and elongated skull of the shearwaters (Fig. 7.1). The depression marking the temporal fossa is shallow (deep in shearwaters), and the openings in the frontal and orbito-olfactory septa from the orbit to the nasal chambers are small — big in shearwaters. Like *Puffinus* the skulls of adult *Pterodroma* and *Lugensa* have ankylosed pre-frontal (lachrimal) bones, a feature separating gadfly petrels from *Bulweria* (Chapter 6, Section II), whose skull is otherwise very pterodromine. According to Harper (1978) young and fledgling *Pterodroma* have free pre-frontal bones and a perforated anti-orbital wall. In *Lugensa* fenestrated extensions of skull bone lateral to the supra-orbital depressions may help to protect the large eyes (Imber, 1985a).

Murphy & Pennoyer (1952) pointed out the great similarities between the skulls of gadfly petrels, finding those of *neglecta*, *arminjoniana* and *ultima* identical: yet all breed independently and sympatrically. Nor could those authors detect any constant differences in skull morphology between surface and burrow-nesters.

Harper (1973) compared the skulls of *P.mollis* and *L.brevirostris* and found that the latter had 20% longer orbital openings than *mollis*. He thought this difference a consequence of the larger eyes of *brevirostris* associated with nocturnal feeding.

As expected of birds that seldom swim under water, neither the humerus nor the tibiotarsus is flattened to the degree seen in shearwaters and, as shown in Fig. 4.9, in *Pterodroma* and *Lugensa* the cnemial process (crista cnemialis cranialis) of the tibiotarsus is short and spatulate, not elongate as in *Puffinus* (Shufeldt, 1922; Klemm, 1969, p.182). There are two small sesamoid wing ossicles in *Pterodroma*, one in *Puffinus* and none in fulmars.

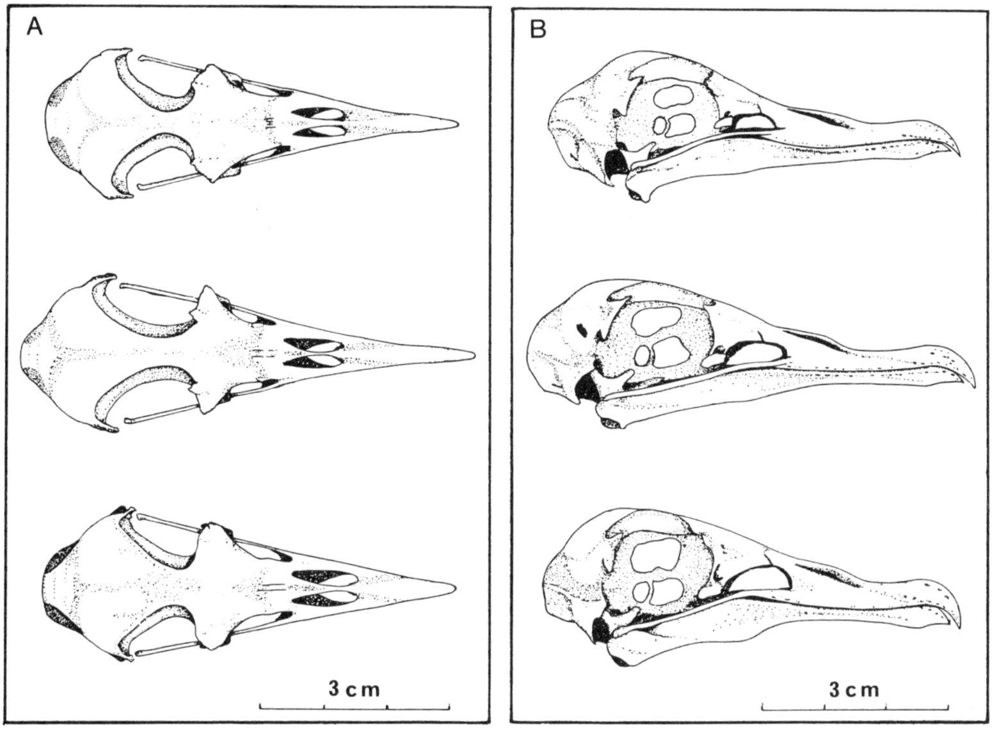

Figure 4.8 *Dorsal (A) and lateral views (B) of* 'Cookilaria' *skulls: top* P.l.pycrofti, *centre* P.cookii, *and bottom* P.axillaris. *From Fleming (1941b).*

In the viscera a peculiarity of *Pterodroma* is the bead-like form of the lower intestine in many species (Kuroda, 1955), which is due to a helicoidal twisting (de Naurois & Prevost, 1972; Thibault, 1977) and Fig. 4.10. Imber (1985a) has used the degree of twisting as a taxonomic character. He separated *Lugensa* from *Pterodroma* partly on the basis of its possessing simple, untwisted upper intestines similar to those of fulmars. Kuroda hypothesised that the twisting gave a greater surface area for digestion and he has elaborated his hypothesis (Kuroda, 1986).

Aspects of the musculature of White-headed and Kerguelen Petrels were examined by Forbes (1982) and of the Bonin Petrel by Klemm (1969).

III The species

A *General*

The species of gadfly petrels are ill-defined, current classifications being mainly based on breeding distributions, on external morphology like size, on plumage patterns and on skeletal features. Murphy & Pennoyer (1952) drew attention to one of the problems; that despite their different plumages and distributions the

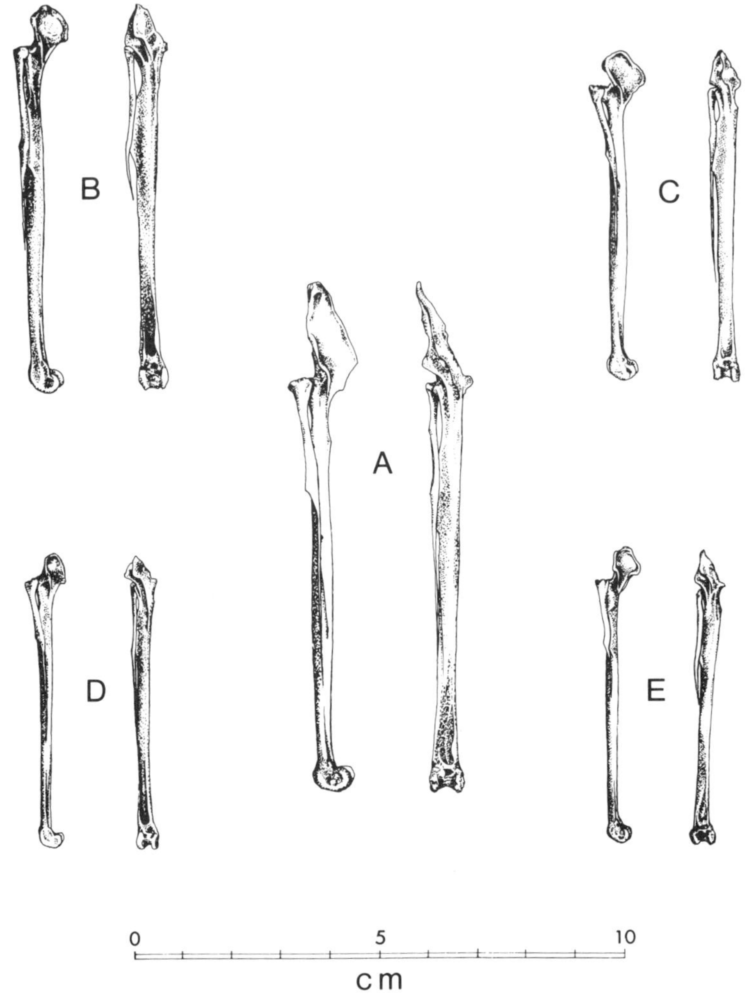

Figure 4.9 *Right tibiotarsi of gadfly petrels in lateral (L) and front (R) view to show crista cnemialis cranialis compared with A.* Puffinus griseus; *B.* P.lessonii; *C.* L.brevirostris; *D.* P.mollis; *E.* P.inexpectata.

skeletons are often virtually identical. They cited Schlegel's Petrel *P.incerta* and *P.lessonii*, considering them sibling species, the first adapted to subtropical and the second to subantarctic seas.

Various attempts have been made to group together species supposedly most closely related into superspecies. Jouanin and Mougin (1979), for example, suggested that the large forms *macroptera, lessonii, incerta*, and possibly also *solandri* and *magentae* form one such superspecies group. Others may be *neglecta, arminjoniana*, and *alba*; *ultima, mollis* and *inexpectata*; *cahow, hasitata, externa, baraui* and *phaeopygia*.

That these very similar species are distinct is shown by up to four kinds nesting on the same island without interbreeding even when their nesting seasons are

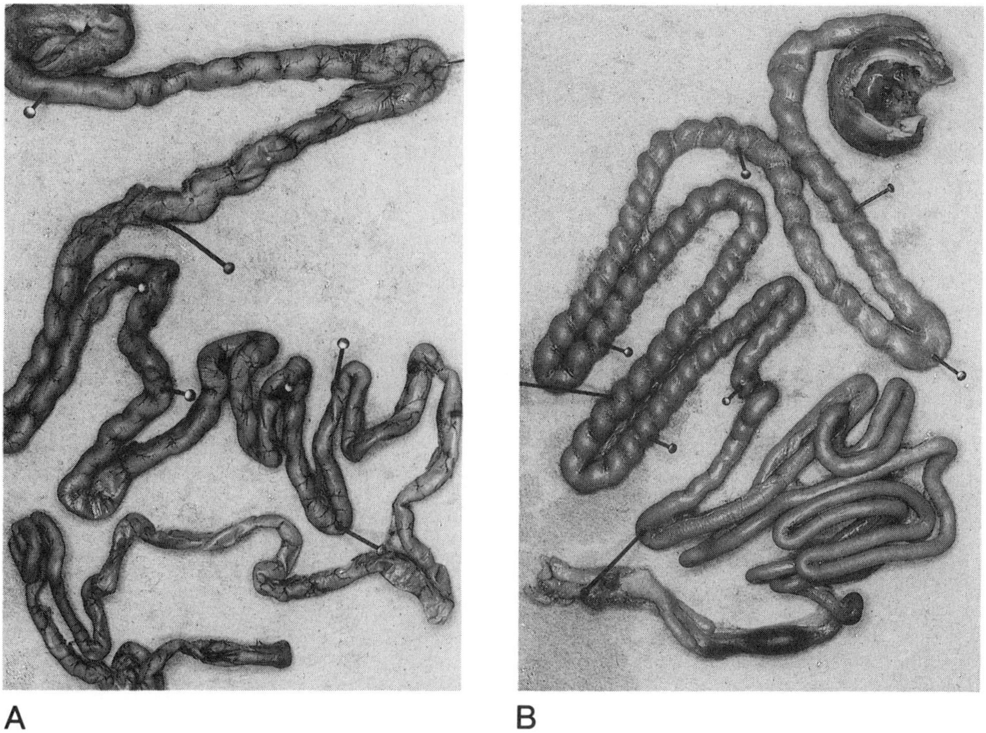

Figure 4.10 *Intestines of (A)* P.inexpectata *('transitional' condition) and (B)* P.m.gouldi *('continuous' helicoidal twisting). The gizzard lies at the top in each photograph. From Imber (1985a).*

congruent. For example, at Gough Island *macroptera*, *mollis*, *incerta* and *brevirostris* breed. Other instances are shown in Fig. 4.14.

The White-headed and Great-winged Petrels are similar-sized birds with almost indistinguishable skeletons and have sometimes been considered conspecific. Both breed at Kerguelen and on The Crozets. The first is a summer and the second a winter breeder, so that interbreeding would be unlikely even if they were summer and winter forms of the same species, but the cycle of *macroptera* is so long that their times ashore overlap considerably (Barre 1976, Table IX). Nevertheless, mixed pairs have never been reported and the two birds appear to be quite distinct probably, like the two giant petrels, *macroptera* (a melanistic form) finds food for its chick to the north in subtropical seas while *lessonii* feeds to the south in subantarctic ones.

Bourne & David (1983) suggested that a few of the many specimens of gadfly petrels collected from sympatrically breeding populations at Ducie and Oeno Islands, Pitcairn, may be hybrids, but there are no field observations of two different species raising the same chick, nor authenticated hybrids among the *arminjoniana*, *alba*, *neglecta* and *ultima* that breed there. Nonetheless, the opportunities for inter-breeding would seem particularly high where breeding seasons are extended. It has also been postulated that all-dark *mollis* (see below) might be hybrids of that species with *brevirostris*, but this seems to be unlikely as dark-phase *mollis* at Marion Island may have normal-plumaged partners (Schramm, 1983).

Although sympatry without significant interbreeding establishes the specific credentials of some tropical species, others of similar size whose ranges do not overlap have less clear claims to specific status. A good example is the *hasitata, externa, phaeopygia* complex. Indeed, Bourne (in Palmer, 1962) lumped *e.externa, e.cervicalis, cahow, p.phaeopygia, p.sandwichensis,* and *hasitata caribbea* as subspecies of *hasitata* on the basis of their distinctive underwings (Fig. 4.6), large size, and long tails. He considered the Pacific forms of this complex as replacing each other around the margins of that ocean.

From considerations of external morphology Jouanin (1970) concluded that the Mascarene Black Petrel *P.aterrima* is the Indian Ocean representative of the Tahiti Petrel *P.rostrata,* a widespread form of the tropical Pacific, and from skeletal evidence Olson (1975a) came to similar conclusions and drew attention to similarities between all *rostrata* subspecies and the extinct *P.rupinarum* of St Helena (Fig. 6.2).

B Pterodroma mollis *and allies*

The taxonomy and relationships of the Soft-plumaged Petrel complex are complicated by breeding in both hemispheres (Fig. 4.14). In the North Atlantic nest the distinct forms, *feae* and *madeira,* possibly the result of a double invasion from the south. *P.(m).feae* is the larger and breeds either on the arid Desertas Islands near Madeira in the autumn, or in warmer seas on the mountains of the Cape Verde Islands in mid-winter. *P.(m).madeira* is the smaller and paler, nesting in early summer around the humid mountain tops of Madeira. The Southern Ocean populations of *P.mollis* are morphologically less diverse (except for the occasional dark morphs) and all burrow in the summer on damp hillsides.

Bourne (1983a) proposed that the northern forms be given full specific status — *P.madeira* on Madeira proper and *P.feae* on the Desertas and Cape Verde Islands — restricting *mollis* for all the southern populations — an arrangement followed in Appendix 1. See also Bourne (1987a).

Birds of these various populations tend to vary a good deal in the depth of their pigmentation on back, breast and outer tail feathers. Very occasionally completely dark birds have been collected at Gough and Marion Islands, birds that have been confused with the Kerguelen Petrel, another dark-plumaged species. This bird is, however, quite distinct.

The Soft-plumaged Petrel and *P.inexpectata* are of similar size and have often been thought closely related. Nevertheless their plumages are rather different, particularly their underwing patterns (Fig. 4.6), as are their voices (Warham, 1979) while only *inexpectata* is highly migratory. Bourne (1983a) considered *mollis* a more generalised species than *inexpectata,* the latter having evolved in the New Zealand region and developed a highly migratory regime that takes it to the North Pacific, to escape the southern winter, whereas *mollis* has remained a sedentary summer breeder throughout its wide southern range.

C 'Cookilaria'

This group comprises about seven rather small species presently confined to the Pacific Ocean — *cookii, axillaris, nigripennis, leucoptera, hypoleuca, defilippiana* and *longirostris* of Appendix 1. These birds have been treated in many different ways by different revisers and the above separation follows Jouanin & Mougin (1979) and will surely need modifying when these birds have been studied in more detail. Jouanin & Mougin suggested that *nigripennis* and *axillaris* might form one superspecies (excluding *hypoleuca* because of its cuneate tail) and *cookii, defilippiana, longirostris* and *leucoptera* another.

Murphy (1929) divided them into two species — *P.cookii,* with five subspecies, mainly on the colouring of their upperparts. Fleming (1941b) believed that Murphy's *P.cookii* was a polyphyletic assemblage, noted similarities between the skulls of *nigripennis* and *axillaris* and their distinctiveness from both *cookii* and *pycrofti* (Fig. 4.8). Others who put forward different groupings were Falla (1942), Austin (1952) and Bourne (in Palmer 1962). Bourne reiterated that the *'Cookilaria'* petrels have distinctive blue feet and than *axillaris* and *nigripennis* do not belong to *P.cookii.*

Since then *nigripennis* and *axillaris* have been found to nest sympatrically at South-east Island in the Chathams Group without inter-breeding (Kinsky, 1970). These therefore warrant full specific status. One of Murphy's subspecies, *P.c.orientalis,* is also now considered as probably based on examples of typical *cookii* from New Zealand collected during their contra-nuptial migration to the eastern South Pacific (Falla, 1942).

Jouanin and Mougin (1979) followed Falla in sinking another subspecies, *P.c.masafuerae* (Lonnberg) into *P.longirostris* Stejneger and in placing *pycrofti* as a southern form of *longirostris.* They also gave full specific to the heavy-billed and long-tailed *P.defilippiana* of Mas a Tierra, San Felix and San Ambrosio Islands, and treated *brevipes* of the southern tropical Pacific as a form of *leucoptera* — see also Bourne (1983b).

All seven species have parti-coloured feet. In *nigripennis, axillaris* and *hypoleuca* the legs and basal thirds of the webs and toes are flesh-coloured, their distal parts, especially the outer toes, blackish. With *P.cookii, defilippiana* and *longirostris,* on the other hand, the feet are bluish proximally, blackish on the distal two-thirds of webs and toes, while in *leucoptera* (often stated to have blue and black feet), the legs of birds of the typical race are a dull grey-flesh, darker distally as in the other species. *P.l.brevipes* does, however, have bluish legs; and evidently leg colours are not absolutely fixed as some *nigripennis* from Norfolk Island and elsewhere have been found with blue legs (Serventy *et al.,* 1971, p.103). Perhaps, as in some albatrosses, leg colour changes with changes in breeding condition.

D *Subspecies*

Many subspecies have been proposed of which Beck's Petrel, *P.rostrata becki* (Murphy, 1928) is particularly interesting. It is a miniature edition of the typical form. Its breeding place has never been found (if it still exists), and the only specimens are the two collected by Rollo Beck at sea, east of New Ireland in 1928. Other small examples of gadfly petrels are known, e.g. of *lessonii*, but whether these are merely 'sports' of birds from unknown populations or subpopulations is unclear. In 1978 de Naurois reported morphological differences between *P.leucoptera* breeding in the mountains of New Caledonia and those of the typical form from Cabbage Tree Island, New South Wales. Imber & Jenkins (1981) confirmed these differences, supported the erection of a subspecies *caledonica*, and showed that it is this form that tends to be 'wrecked' on New Zealand North Island beaches.

E *Recent classifications*

Imber (1985a) proposed a radical re-arrangement based primarily on the intestinal twisting, the feather-lice, some skull characters, the status of the rhamphothecal notch, and on foot size and underwing patterns. He removed *brevirostris* from *Pterodroma* to monotypic *Lugensa* because he considered this bird to be closer to the fulmars. Also removed were *rostrata, macgillivrayi, aterrima* and extinct *rupinarum* into another Mathewsian taxon, *Pseudobulweria*, having the latericorns forming notches with the upper nail, simple intestines, fused pre-frontal bones that do not project beyond the quadratojugals, large feet and somewhat laterally compressed tarsi. His genus *Pterodroma* consists of 29 species, re-grouped into four subgenera (Fig. 4.11 and Table 4.1). His proposals involve raising several subspecies to full specific status.

Apart from following his use of *Lugensa* for the Kerguelen Petrel (in several respects an odd bird), I have preferred a more conservative classification and retain this bird among the gadfly petrels. Imber's ideas have yet to receive detailed attention and there are doubts about the suitability of gut morphology as a taxonomic character (Kuroda, 1986) and problems in interpreting evidence from ectoparasite–host distributions Bourne (1987a).

Some further light on species' limits should result when new characters based on behavioural and biochemical data are available. To the present time the behaviour of gadfly petrels has received little attention, their vocalisations are mostly unknown and unevaluated, and the only examination of biochemical characters is Harper's 1978 analysis of some of the plasma proteins of *macroptera, lessonii, inexpectata, l.pycrofti* and *cookii* (Fig. 4.12). He found many common features in their electrophoretic patterns, with those of *macroptera* and *lessonii* being very similar and those of *l.pycrofti* and *cookii* almost identical. Patterns from *P.inexpectata* showed features common to both those pairs of birds. The pre-albumins at 'relative mobility units' 102 and 113 (Fig. 4.12) only in *l.pycrofti* and *cookii* were interpreted by Harper as support for the distinctiveness of the subgenus *Cookilaria*. Unfortunately Harper's samples included neither *brevirostris* nor any of the species Imber shifts to *Pseudobulweria*.

IV Life styles

Gadfly petrels, while usually colonial when breeding, are often solitary at sea, feeding on surface or near-surface prey, and often far from the continental shelves.

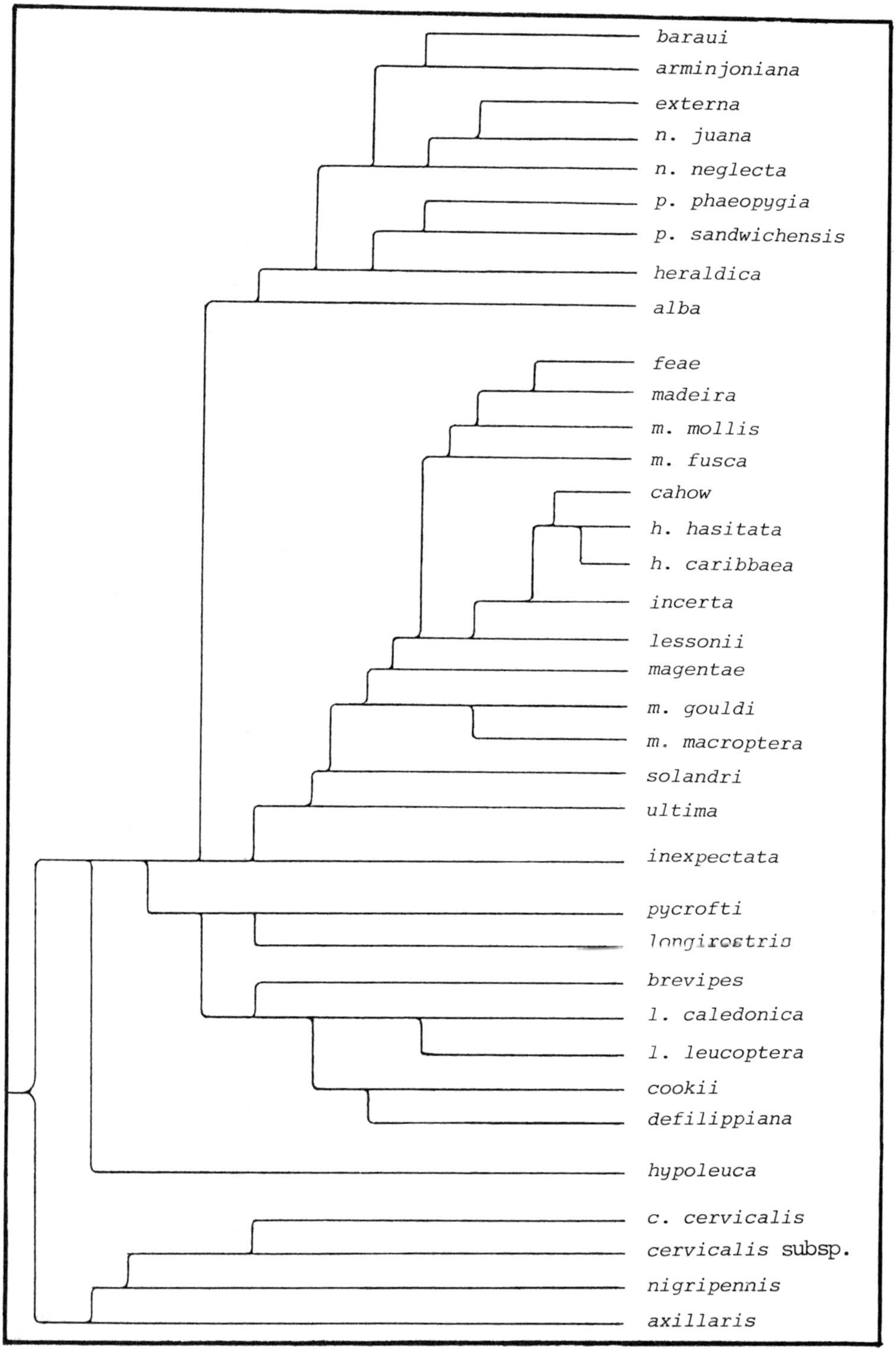

Figure 4.11 *A phylogeny of* Pterodroma. *From Imber (1985a).*

Table 4.1 A classification of the gadfly petrels

Family Procellariidae Subfamily Fulmarinae Genus *Pterodroma*		
Subgenus *Proaestrelata*	Subgenus *Hallstroma*	Subgenus *Pterodroma*
Pterodroma axillaris	*Pterodroma alba*	*Pterodroma ultima*
P.nigripennis	*P.heraldica*	*P.solandri*
P.cervicalis (2 subspecies)	*P.phaeopygia* (2 subspecies)	*P.macroptera* (2 subspecies)
P.inexpectata	*P.neglecta* (2 subspecies)	*P.magentae*
P.hypoleuca	*P.externa*	*P.lessonii*
Subgenus *Cookilaria*	*P.arminjoniana*	*P.mollis* (2 subspecies)
Pterodroma pycrofti	*P.baraui*	*P.madeira*
P.longirostris		*P.feae*
P.leucoptera (2 subspecies)		*P.incerta*
P.brevipes (2 subspecies?)		*P.hasitata* (2 subspecies)
P.cookii		*P.cahow*
P.defilippiana		

From Imber (1985a)

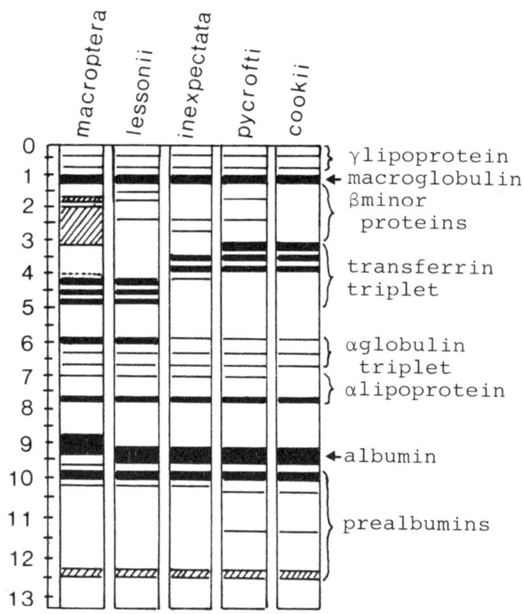

Figure 4.12 *Plasma proteins of 5* Pterodroma. *From Harper (1978), courtesy of* New Zealand Journal of Zoology.

A *Locomotion*

1 *At sea*

Their powerful and very fast flight is well displayed at sea, the name 'gadfly petrel' referring to their impetuous stooping manoeuvres, the old name *Oestrelata* for the genus coming from the Greek *oistrelatos* — goaded by a gadfly. The larger species, in particular, travel in great expansive spiralling curves, arcing high into the sky then swooping down like miniature albatrosses. Because of this towering progression, the birds are often first detected against the sky, being hard to pick out against the water. At the height of the movement they can command a wide field of view. Gadfly petrels tend to be over-dispersed at sea and it is possible that their towering flight style lets them monitor the activities of their neighbours. This may explain why several birds, and even several species, tend to concentrate round small, localised food sources, the downward movement of one bird having drawn in others to investigate.

These are extremely aerial petrels and settle on the sea rather less than do most other species. However, they have to alight when scavenging, e.g. when taking tentacles of the large squid so often found in their stomachs. They are also found resting on the water both near nesting places and on the contra-nuptial feeding grounds. Such behaviour is reported for the Phoenix Petrel *P.alba* (Gallagher, 1960) and for *P.mollis* (=*feae*), seen by Murphy near the Cape Verde Islands sitting in flocks facing the wind, with one bird pecking at the carcass of a large squid (Bannerman, 1968, p.179). When they were still plentiful on Little Barrier Island, New Zealand, Sibson (1949) saw a ring of *P.cookii* resting by day on the water around the island. He noted how they rode high in the water (cf. Wilson's drawing of *P.lessonii* in Lowe & Kinnear, 1930, plate v.).

Nevertheless, while awaiting the onset of darkness gadfly petrels seldom form tight-knit 'rafts' offshore like shearwaters do. For example, *P.hypoleuca* has not been reported massing on the sea near its large colonies in the North Pacific. Some such flocking has been reported of the Galapagos Petrel *P.phaeopygia*. Harris (1970) saw small flocks forming off Santa Cruz during the breeding season and Jouanin & Gill (1967) were told that Barau's Petrel *P.baraui* appears on the sea off Reunion about an hour before flying inland, but that it does not feed in those coastal waters.

2 *On land*

Grounded gadfly petrels tend to shuffle along on their tarsi, but when running raise their wings and most are more agile than shearwaters of similar size. Gadfly petrels can also climb trees, usually to assist take-off when nesting in forests, and this facility is not restricted to small species like *P.pycrofti* (Bartle, 1968), as is sometimes thought. That larger kinds can do this was noted by Governor Hunter back in 1793 writing of the 'Providence Petrel' *P.solandri* at Norfolk Island. On Lord Howe Island this bird still uses coastal bushes as launching pads, as does *P.rostrata* on tropical islands of the South Pacific. Even *P.hypoleuca* which, under entirely natural conditions would be unlikely to encounter anything higher than a *Scaevola* bush, can scramble up the trunks of trees introduced to Midway Atoll.

This petrel is typical of the smaller species, very agile on the wing, swirling around into pockets on the fringes of the *Casuarina* copses at Midway and touching down gently and precisely. It is skittish on land and if alarmed darts off from level ground and is gone with a quick leap into the air. Bigger species are less agile. To some extent this agility is reflected in the wing loadings. *P.mollis* has a wing loading of about 45 N m^{-2}; a shearwater like *P.assimilis* of similar weight 61 N m^{-2}, but the smaller of the *Cookilaria* only 33–38 N m^{-2}.

Like most tubenoses gadfly petrels tend to make repeated passes over the breeding area, hesitating on each circuit over a particular place, often the nest, where they eventually land. The smaller species hesitate least and touch down with greater control after fewer circles than the bigger ones (Bartle, 1968; pers. obs.). In open situations landings may be made quite close to the burrow and Bonin Petrels at Midway Atoll and Great-winged Petrels at Eclipse Island, Western Australia often alight only 3–7 m away; where nests are in forest, however, they may have to descend through gaps in the canopy or come down on the fringes of the woods and walk to their burrows.

B *Voice*

Gadfly petrels have loud voices, call much from the ground and air and have large repertoires. Common to many species is a repeated 'ti-ti' call — hence the Maori name 'titi' for some *Pterodroma*. Other calls are long, low-frequency moans and persistent churring sounds. These repertoires have not been adequately worked out, the most detailed to date being that of *P.hypoleuca* (Grant *et al.*, 1983) which provides a good example of a typical repertoire for a *Cookilaria* petrel.

Despite many common features, the voices of the different species seem distinctive, although studies of those of sympatric species where vocal identification at night could be important, have not been published. Murphy & Pennoyer (1952) pointed out that the varied underwing patterns could also provide identification clues for the birds.

More than most tubenoses, gadfly petrels will often respond to human imitations of their calls or to rhythmical noises like hand-clapping. This behaviour was reported of the Cahow *P.cahow* as early as the fifteenth century. Birds in flight can be lured to the ground and into the hands of the caller: for a history of recent observations see Warham (1988a). Further work has shown that *P.externa*, *P.nigripennis*, *P.l.pycrofti* and *P.macroptera* can all be called down and the technique has been used to locate remnant colonies of the last species (Tennyson & Taylor, 1990).

C *Behaviour*

Apart from their aerial displays, courtship and allied behaviour, although unstudied, show many facets common to petrels in general, with much mutual preening, copulation on the ground near the burrow, and territorial behaviour. The larger species are particularly tame and take little notice of man (Fig. 4.13). The smaller ones are more prone to fly when approached or when disturbed in the open by day or by night. Diurnally active kinds call and perform aerial chases by

Figure 4.13 *Most gadfly petrels are tame and defend themselves by grasping and tearing with their powerful bills. This is* P.macroptera *nesting on the surface at Eclipse Island, Western Australia. Photo: author.*

day, although their activity tends to peak towards dusk (Gardner *et al.*, 1985; Warham, 1988a).

When flying after dark some species are readily attracted to lights. *Pterodroma mollis* at Antipodes Island was brought to the ground by the beam of a small hand-torch whereas *P.lessonii* at the same place and time could be enticed down only with difficulty even with a powerful spotlight. Swales (1965) also found *brevirostris* highly phototactic. In the Hawaiian islands *P.phaeopygia*, particularly fledglings on their maiden flights, may be distracted by lights around sports facilities and land in nearby streets. Misty nights are particularly dangerous for them. Much the same applies to *P.baraui* at Reunion and at Midway Atoll some *hypoleuca* die after colliding with street lights. Fires also attract gadfly petrels. Hunter's men on Norfolk Island lit fires to decoy Providence Petrels (Hunter, 1793; Iredale, 1911), a device also used to catch Cahows (Beebe, 1935), while on Haiti peasants practise 'Sen Sel' — their name for luring *P.hasitata* on foggy nights into fires set on the mountains above the nesting places (Wingate, 1964).

V Distribution

A *General*

Breeding distributions are shown in Fig. 4.14 (see also Table 4.2), showing that

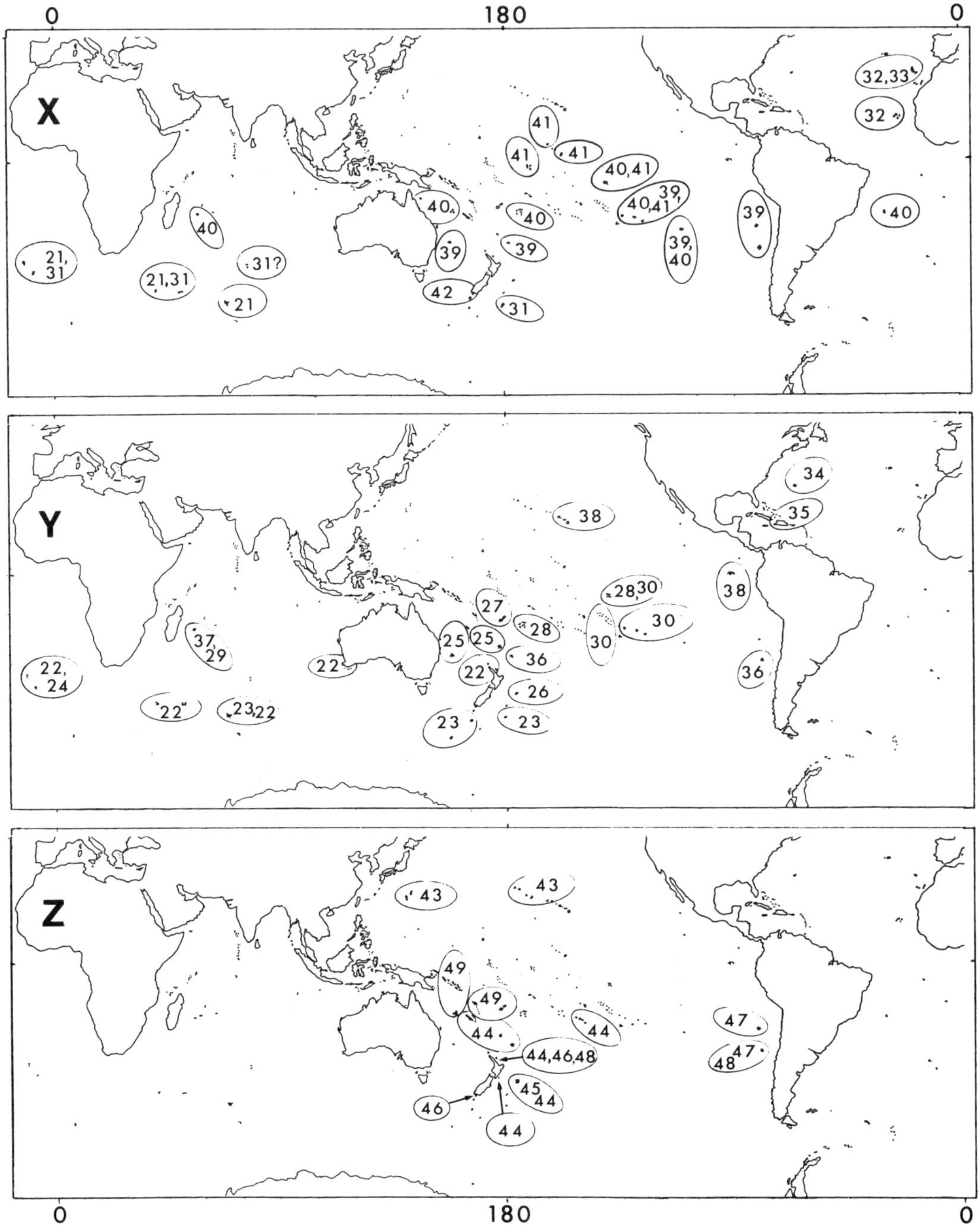

Figure 4.14 *The breeding distribution of gadfly petrels:* **X** *21* brevirostris; *31* mollis; *32* feae; *33* madeira; *39* neglecta; *40* arminjoniana; *41* alba; *42* inexpectata. **Y** *22* macroptera; *23* lessonii; *24* incerta; *25* solandri; *26* magentae; *27* macgillivrayi; *28* rostrata; *29* aterrima; *30* ultima; *34* cahow; *35* hasitata; *36* externa; *37* baraui; *38* phaeopygia. **Z** *43* hypoleuca; *44* nigripennis; *45* axillaris; *46* cookii; *47* defilippiana; *48* longirostris; *49* leucoptera.

Table 4.2 Some morphological and behavioural characteristics of gadfly petrels

Characteristic	*macroptera*	*lessonii*	*incerta*	*solandri*	*magentae*	*rostrata*	*macgillivrayi*	*neglecta*	*arminjoniana*	*alba*	*ultima*	*brevirostris*	*mollis*	*inexpectata*	*cahow*	*hasitata*	*externa*	*baraui*	*phaeopygia*	*hypoleuca*	*nigripennis*	*axillaris*	*cookii*	*difilippiana*	*longirostris*	*leucoptera*
Monomorphic plumage	X	X	X	X	X	X				X	X	X	(X)	X	X	(X)	X	X	X	X	X	X	X	X	X	
Polymorphic plumage								X	X																	X
Sexually dimorphic	X					X			(X)																	
Tropical						X	X	X	X	X	X		X			X			X		X					X
Subtropical	X		X	X	X	X		X	X				X		X		X	X	X	X	X	X	X	X	X	X
Subantarctic	(X)	X	X									X	X	X												
Highly migratory				X				?X						X			X			?X	X		X		?X	
Relatively sedentary	X								X	X	?X	?X			X	X			X	?X						
Summer breeder		X	X		X							X	X	X			X	X	X		X	X	X		X	X
Winter breeder	X			X		?X	?X				X		X		X	X				X				X		(X)
Seasonless breeder								X	X	X																
Surface nester	(X)							X	X	X	X													X		X
Burrow nester	X	X	X	X	X	X						X	X	X	X	X	X	X	X	X	X	X	X		X	X
Diurnal on land				(X)				X	X	X	X										(X)			X		
Nocturnal on land	X	X	X	X	X	X						X	X	X	X	X	X	X	X	X	X	X	X		X	X
'ti-ti' Call	X	X		X				X				X	X	X	X	X	X	X	X	X	X		X	X	X	X

(X) = less common characteristic or situation: ?X = needs confirmation.

while some gadfly petrels are adapted to cool temperate subantarctic waters, none breeds in the equivalent zone of the Northern Hemisphere, although *P.inexpectata*, exploits subarctic resources during its northern migration.

These petrels nest on islands, particularly on small atolls, but some inhabit large islands like those of Hawaii and New Caledonia, with some relict groups on mountains and cliffs in New Zealand. A few choose forested mountains, e.g. *P.hasitata* in the Caribbean and *P.rostrata* and *P.leucoptera* in the Pacific.

Murphy and Pennoyer (1952) emphasised that different species or groups of species characterise the tropical, subtropical and subantarctic zones of surface water, and that representatives of each may have circumpolar distributions. A good example is *P.arminjoniana*. This nests in the tropics of the Pacific, Indian and Atlantic Oceans (Fig. 4.14X, 40) and there is no evidence that these three groups now have any effective contact. *P.(r).aterrima* of the Indian Ocean and *P.r.rostrata* of the tropical Pacific are instances of populations now separated by continental barriers.

The patterns of Fig. 4.14 include some other disjunct distributions, evidently relicts of previously more continuous ones. For example, *P.cookii* now nests only at Little Barrier and Codfish Islands, places some 1350 km apart and in seas with quite different characteristics. This pattern must result from a loss of nesting areas elsewhere, perhaps due to tectonic sea-level changes and/or vulcanism. Others, like *P.baraui*, have very few breeding places, and are presumably remnants of formerly more widespread and successful populations. Only *P.mollis* and perhaps *P.hypoleuca* have populations of very close allies in both hemispheres.

From the wealth of species represented in the Chatham Islands' subfossil material collected by H.O. Forbes, Bourne (1967b) argued that these islands were once a major breeding station. He thought that in addition to the Taiko *P.magentae* (Bourne, 1964), the collection contained bones of *P.axillaris* and *P.inexpectata* and of a smaller form, perhaps Murphy's Petrel *P.ultima*. Only the first two are found there today, and the bones he then attributed to *inexpectata* seem more likely to have been those of *mollis* (Bourne, 1983a). However, as the local Polynesians, the Morioris, ate both resident birds and strays that reached the island's beaches (Sutton, 1979), identification of fledgling bone seems needed to prove past breeding.

Other bone deposits on Amsterdam Island in the Indian Ocean contain many remains of a large *Pterodroma*, possibly *neglecta*, and of a small one, possibly *mollis*: neither is known to have bred there in historical times (Jouanin & Paulian, 1960).

B *Range expansions*

While most gadfly petrels have declined in numbers in recent years, two have begun to expand their ranges and presumably their total populations. The most dramatic of these expansions has been that of the Black-winged Petrel. This has established (or re-established?) itself at Philip Island off Norfolk Island, and has also attempted to nest on the main island. It has reappeared at Lord Howe Island too in recent years. Subfossil bones show that it, or a very closely related form, was present in the past (Fullagar *et al.*, 1974), but early accounts by local naturalists do not mention it. The bird evidently began the re-colonisation in the 1960s and it is now quite conspicuous, although introduced Masked Owls *Tyto novaehollandiae* and cats may be restricting its increase. In 1971 probably not more than 10 pairs were nesting there.

Further west, birds have been found in burrows on Heron Island, Queensland, and at Coff's Harbour, New South Wales, but none of these exploratory episodes has yet resulted in an established Australian population.

Black-winged Petrels are often seen flying around headlands of northern New Zealand and, as at Norfolk Island, corpses killed by cats show that landings have taken place. Most such attempts seem doomed to fail, but those observed flying round the Poor Knights Islands, where there are no placental predators, could eventually succeed. Further south still, at Portland Island, a small population became established in 1979/80 (Eagle, 1980) and by 1981 had grown appreciably with about 25 active burrows and an apparently healthy non-breeding group (Warham, unpubl.). The population on South-east Island, Chathams Group, which may also be a recent colonisation, seems static, however.

The source of this expansion of Black-winged Petrels is unknown. The nearest colony to northern New Zealand is at the Three Kings Islands, whose population has never been censused, but which is thought to be quite healthy. A more likely source perhaps is Macauley Island in the Kermadec Group where the bird is abundant, and living space there may now be short.

The second gadfly petrel to have experienced a revival is the Providence Petrel *P.solandri*. This was formerly a bird of Norfolk Island, but was completely eaten out by the first settlers who were starving — hence the vernacular name — (Whitley, 1934), leaving only the birds at Lord Howe Island to perpetuate the species. In 1985 however, this petrel returned to nest on Philip Island, off the main Norfolk Island (Hermes *et al.*, 1986), and although the numbers are evidently quite small, the species should be fairly safe there. These birds seem to have effected their colonisation very quickly as they had not been reported prospecting before being found breeding.

Harper (1973) and Bourne (1983a) suggested that *P.mollis* is expanding its range in the South-west Pacific but the evidence for such an expansion is rather weak. *P.mollis* was unknown in New Zealand until discovered at Antipodes Island in 1969 (Warham & Bell, 1979) but at least eight specimens have since turned up in New Zealand after gales. Nevertheless, as no scientific party had stayed on Antipodes Island before that year, small populations of a nocturnally active petrel such as this could easily have been overlooked. Some support for the notion of a new colonisation comes from the small numbers seemingly present and by the subsequent finding of much burrowing activity and a good non-breeding segment in the community (Imber, 1983), which would suggest that this is a young, recently established population.

VI Feeding and foods

A *Feeding methods*

Much of the gadfly petrel's prey is captured when the birds swoop down to snatch food by 'dipping' (Fig. 1.5), often pausing on rapidly beaten wings to grab their prey. Or they scavenge dead or dying prey floating near the surface. Harper (1987) described *P.macroptera* hovering while picking crustacea from the water, then rising briefly into the air to swallow them, and he watched this species tackling squid by biting and lunging. One bird, struggling with a large squid, dipped its spread wings into the sea to act as brakes and hinder the victim's escape. Pitman (1982) noted both styles used by *P.phaeopygia*: feeding on carrion, usually dead squid, was less important than taking live squid and flying fish driven to the surface by tuna and other large marine predators which possibly included porpoises. Flying fish were caught in mid-air — 'aerial dipping' — a technique shared with *P.externa* and probably with most tropical gadfly petrels. Pitman's observations were made by day and he did not believe that it was necessary to infer that nocturnal feeding is important to *P.phaeopygia*.

Such localised feeding in tropical seas is often associated with large concourses of birds milling around above fish schools, with boobies like *Sula sula* and *S.leucogaster* and terns like *Sterna fuscata* diving into the sea while frigate-birds circle overhead to snatch flying fish in mid-air and to harass heavily laden boobies to disgorge.

Most *Pterodroma* show little interest in ships. The Soft-plumaged Petrel is an exception and Murphy (1936, p.712) caught it with a baited hook. In the South Atlantic (Harper, 1973) saw this species alighting in the ship's wake to investigate garbage and taking plankton churned up by the propellers, whereas *P.lessonii* and *P.inexpectata* ignored the ship. Other gadfly petrels can, however, be attracted to baits of hot fat.

Gill (1967) found Soft-plumaged Petrels feeding at night alongside his ship, but whether they would have done so beyond the lights is uncertain. According to Harper (1973) this bird is a diurnal feeder whereas *brevirostris* feeds at night, as does *macroptera* (Imber, 1973; Harper, 1987).

Ashmole & Ashmole (1967) estimated that some of the squid tentacles taken by *P.alba* had come from animals weighing up to 6 kg, and so were presumably dead when attacked. In contrast, the presence in their stomachs of many sea skaters *Halobates* sp. only 6 mm long showed that they could also pick tiny animals from the surface. They were not seen to do this, but presumably these insects were taken by dipping and by day in the manner of another *Halobates* predator — the Blue-grey Noddy *Procelsterna caerulea*.

Another technique was described for *Puffinus* and *Pterodroma* by Murphy (1936, p.487). This is to raise and hold the body just above the water using thrusts of the feet and beaten wings to hover while picking up surface items — 'pattering' (Fig. 1.5). The Ashmoles pointed out that the large area of webbing in the foot of *P.alba* could be an adaptation to this feeding style, rather than to enhance swimming ability.

B *Foods*

Gadfly petrels' gizzards almost invariably contain squid 'beaks'. Ashmole & Ashmole (1967) found that 52% of the stomachs of Phoenix Petrels held squid beaks and 66% their eye lenses. Squid made up three-quarters of the food in the samples by volume, a higher proportion than that in any of the other seabirds they investigated. The rest of the food of *P.alba* consisted of other invertebrates, fish averaging only about 14% of the total volume and the fish were smaller animals than the squid. For the Bonin Petrel *P.hypoleuca* the reverse may be true, as a wide range of fish are taken and these may, in toto, be more important than squid (Harrison *et al.*, 1983). As with *P.alba*, a sea skater, *Halobates sericeus* is also eaten. Squid featured prominently in detailed analyses of the stomach contents of chicks of *m.macroptera, mollis* and *brevirostris* reported by Schramm (1986).

From his analysis of the food of *P.m.gouldi* Imber (1973) concluded that the prey were mainly species living beyond the continental shelf and that 80% were of kinds associated with acoustic scattering layers. This implied that the birds fed by night and, as 90% of the prey were squid and fish that bear light-producing organs, he suggested that bioluminescence may help the petrels find their prey. Imber thought that these petrels were not concentrating on areas of upwelling, but taking scattered foods, and that their daytime feeding was mostly restricted to scavenging dead animals, including the coelenterate *Argonauta*.

An analysis of sightings of Black-capped Petrels off the south-eastern United States by Haney (1987) showed that these were concentrated seaward of the continental shelf break and associated with meanders among the Gulf Stream fronts. Further out the birds occurred mainly where local turbulences were created by currents deflected from under sea ridges and peaks.

The Ashmoles made the interesting suggestion that the rather soft plumage of gadfly petrels helps them to avoid the embraces of living squid and octopus, making it difficult for these to get a grip of birds swimming on the surface!

C *Feeding ranges*

The behaviour of *P.baraui* referred to above is typical of gadfly petrels in that they are seldom seen feeding or even flying close inshore by day (Falla, 1934; Warham, 1967; Ashmole & Ashmole, 1967; Warham, 1977a), and indirect evidence, such as the long incubation spells, points to the breeders feeding far from land. The Ashmoles suggested that the advanced state of digestion of the foods they found in *P.alba* implied distant feeding, and there is evidence that stomach oil retards digestion (Warham, 1977b).

In theory, *P.inexpectata* can range from its New Zealand breeding places to feed in Antarctic waters 2200–4000 km away during the 12–14 day spans off the egg (Warham *et al.*, 1977). Ainley & Manolis (1979) confirmed earlier reports by Falla (1937) and others of the abundance of this bird just outside the pack-ice in the sector 95°E to 140°W during the southern summer. They considered that, from the numbers involved, most of the off-duty birds must be present. Nakamura (1982) also saw large numbers off Antarctica between 100°W and 170°W at a time when incubation would have been in progress.

Pterodroma lessonii is another species that has been seen in Antarctic seas during the nesting season, but whether the sightings are of breeders off-duty or of non-breeders is not known. Nakamura (1982) showed that this species and *P.inexpectata* are zoogeographically segregated in the Australian sector of the Southern Ocean in summer. *P.lessonii* was common at lower latitudes in warm seas (*c.* 14°C),

P.inexpectata at higher latitudes and in seas <2°C. The Kerguelen Petrel *L.brevirostris* has also been sighted far to the south off Antarctica in the summer, but not necessarily beyond the range of those nesting at the Indian Ocean islands. At that season it has been found in Drake Passage and in the South Pacific at about 60°S as far as 145°W, and much more commonly near the South Sandwich Islands (Harper *et al.*, 1972).

From the species' composition of food fed to the chicks Imber (1976b) deduced that adult *P.macroptera gouldi* were foraging about 650 km from their nests, but that they may have been feeding closer during the early stages of chick rearing when parental visits were more frequent.

VII Breeding

The breeding patterns and phenology of gadfly petrels are rather similar in most respects to those of other procellariids, their most unusual features being the very long breeding cycles of some non-migratory forms and the prevalence of surface-nesting. There are also rather many winter-breeders (Table 4.2) and the eggs are more rounder and thinner-shelled than those of other petrels of similar size (Chapter 13, Section VIB).

The chief studies of gadfly petrel breeding biology are those on *lessonii* (Warham, 1967; Barre, 1976), on *macroptera* (Imber, 1976b), *brevirostris* (Mougin, 1969); *phaeopygia* (Harris, 1970; Simons, 1985; Cruz & Cruz, 1990), *inexpectata* (Warham *et al.*, 1977), *hypoleuca* (Grant *et al.*, 1983), *l.pycrofti* (Bartle, 1968; Dunnet, 1985) and on *macroptera*, *mollis* and *brevirostris* by Schramm (1983a).

A *The nest*

Surface nesting is particularly favoured by tropical species (Table 4.2). A few, like *P.macroptera* in Western Australia, may lay both in burrows and on the surface, but burrowing predominates: others: like *P.leucoptera* at Cabbage Tree Island, while generally laying on the surface in rock crevices and niches, may burrow elsewhere (subspecies *brevipes*). Some make no attempt at burrowing, notably *alba*, *ultima*, *arminjoniana* and *neglecta*. But few nest completely in the open in the style of an albatross: instead they lay in the shade of a root, in an embrasure among eroded coral or lava, or in the shelter of vegetation. Ashmole & Ashmole (1967) found that Christmas Island *P.alba* chose sheltered places for their egg, but they thought that in midsummer, when the vegetation was thin, exposed adults may have left them during the heat of the day. In that event the embryos would have been killed.

At many of these tropical sites the hard substrates prevent burrowing but whereas several kinds of surface-nesting *Pterodroma* may breed sympatrically, burrowing and surface-nesting gadfly petrels do not. Other burrowing species may be present nevertheless. At Cabbage Tree Island surface-nesting *leucoptera* have three kinds of burrowing shearwater as neighbours: possibly surface-nesting was the gadfly petrel's response to past competition for nesting space.

Surface-nesting *Pterodroma* have not colonised Antarctica, the medium-sized fulmars that lay on ledges and in crevices there can defend themselves with barrages of stomach oil, a protective device not used by adult gadfly petrels.

Nest densities achieved by gadfly petrels are seldom high, and lone pairs may breed successfully. The Bonin Petrels on Laysan Island probably lay at higher densities than other species, partly because the coral sand permits burrows in close proximity. Densities of nests of the Black-winged Petrel *P.nigripennis* on smaller islands in the Kermadec Group may also be high, all suitable ground being closely burrowed. At South Meyer in 1967 occupied burrows ranged from 0.3 m^{-2} to 2.2 m^{-2} (Merton, 1970). On the same island *P.neglecta* was nesting on the surface at only about 0.3 m^{-2}.

Where several species breed close together, as they do on islands in the Pitcairn Group, there may be some segregation by species. On Ducie Island Beck and Quayle found much intermingling of the four gadfly petrels nesting there, but with some evidence of local clustering in each species (Murphy & Pennoyer, 1952). Segregation may also arise due to selection by different species of different types of substrate. For example, at Gough Island *P.mollis* burrows into dry sheltered and steep hillsides, whereas *brevirostris* chooses waterlogged holes in upland situations (Bourne, 1983a). At Marion Island this latter bird also nests in marshy places, laying its egg on a low mound in the nesting chamber. The mound is surrounded by a drainage channel carrying excess water out through the burrow entrance (Fig. 12.3).

Gadfly petrels that lay on the surface are also active over their colonies by day and may visit their nests then. An exception is Gould's Petrel *P.leucoptera* which, though technically nesting on the surface, albeit cryptically, is strictly nocturnal over land, as is its burrowing subspecies *brevipes*.

Burrowers behave differently. Even where there are no predators these petrels usually arrive and depart only under cover of darkness. There are some exceptions among the tropical species. At Macauley Island, the Kermadecs, for example, *P.nigripennis* flies around in broad daylight and may enter its burrows then. Elsewhere it also flies over nesting places by day yet, at the newly established Portland Island colony, it appears overhead at dusk and none lands in daylight.

B *Annual cycles*

When several gadfly petrels share a single island their breeding seasons may differ as well as the habitats they use. At Marion Island *P.macroptera* lays in May, *L.brevirostris* in early October and *P.mollis* in mid-December (Schramm, 1983a). Hence, as Schramm points out, the birds are raising their chicks at different times so that competition for food may be reduced.

Best documented of the annual cycles is that of *P.macroptera gouldi* by Imber (1976b). This is a non-migratory, winter-nesting bird which, if successful, spends 9–10 months in breeding activities, a few single birds, usually males, being found ashore within a month of the chick's departure (Falla, 1934). According to Imber, this lengthy season is partly the result of a very long pre-laying exodus (Chapter 12, Section VII) of 53–80 days: the nestlings are fed for a similarly extended period of 108–128 days. The same is true of the typical subspecies in Western Australia where some adults return to burrows only 2 months after the chicks have fledged and with one nestling period of 134 days recorded (Warham, 1956, 1957).

Kerguelen Petrels *L.brevirostris* are also rather sedentary — absent from Crozet Island only during February and March — but summer-breeding, and in marked contrast to *macroptera*, these have quite short incubation and nestling periods of 49 and 60 days respectively (Mougin, 1969). This accelerated development seems not to be associated with any migratory drive but may be a response to a short summer of long days, as with fulmars. Yet *P.lessonii*, a bird sharing similar environmental conditions on land and sea with *brevirostris*, has no such accelerated development, the chicks leaving at about 102 days old. Winter breeding would probably be impossible at these comparatively high latitudes, the ground too hard for digging, the plankton descended to deeper water and the microclimates in the nesting chambers perhaps too harsh for chick survival. As it is, by laying in the summer, the young of both species can be safely unattended under ground when 2 days old, despite the high chill factors obtaining outside.

Smaller species take less time than larger ones to raise chicks. Bonin Petrels (weight 180 g) hatch their eggs after 48.7 days (Grant *et al.*, 1982c) and their chicks fly when about 82 days old (Pettit *et al.*, 1982a). Highly migratory kinds like *P.inexpectata* have more compressed breeding cycles and this bird takes about 50.5 days for incubation and about 75 days to rear its chick. This allows the breeding birds about 18 weeks for their migration to and from the North Pacific, involving flights of at least 24 000 km (Warham *et al.*, 1977).

Most of the small '*Cookilaria*' species breed on remote islands and only short-term observations have been made on their breeding systems. The main exception is *P.hypoleuca*, whose Midway Atoll population, although currently being decimated by rats (see below), has been readily accessible thanks to regular air services operated for the US Navy. At that island many birds burrow around the houses and installations and mains electricity is readily to hand. This has allowed a series of physiological studies to be pursued on the bird and its egg, e.g. by Grant & Whittow (1983, 1984).

C *Laying patterns*

Despite its extended breeding cycle, the spread of laying of *P.macroptera* extends over only 3 weeks and at any particular nesting place occurs at the same time annually, although varying with locality and latitude. More restricted laying periods are the rule among southern cold-water species like *brevirostris* and *lessonii*, as well as with some subtropical ones like *solandri* and *nigripennis*. But at Lord Howe Island *P.nigripennis* chicks in February were of very variable sizes, suggesting rather asynchronous layings. This could have been due to new colonists laying at different dates or many young, naive breeders may have been involved in a newly established colony, birds that in older populations might not have attempted to nest. These would be expected to lay late and have lighter chicks than more experienced pairs, due to parental inefficiencies.

Laying patterns are quite different for most of the tropical gadfly petrels. Vinson (1976) found *P.arminjoniana* on eggs at Round Island, Mauritius during 10 months of the year, and on each visit usually saw chicks at various stages of development.

Prolonged laying is also known of *P.neglecta* at Ducie Island, Pitcairn (Murphy & Pennoyer, 1952). At Christmas Island (Kiribati) Ashmole & Ashmole (1967) found *P.alba* incubating in every month and obtained evidence that there were two peaks of laying — May to July and December to January (Fig. 11.7) — as was also noted by Gallagher (1960). What is not clear is whether the same pairs nest annually at particular times, shift their timing from year to year, or lay at less than 12-monthly intervals like Galapagos Audubon's Shearwaters (Chapter 11, Section VC).

D *Chick desertion*

Also needing clarification is whether gadfly petrels usually feed their chicks until they leave, or, in the migratory kinds like *inexpectata*, desert them to start their northwards flights. Few observers have monitored nests after the chicks have left. Direct observation or automatic photo- or video-recording is needed.

The best data are for the Cahow, a bird usually regarded as non-migratory. Wingate (in litt.) has often found the chicks being fed during their last week ashore and some parents visiting nests after their chick has gone, but this is rare. Fifteen days from abandonment to the chick's departure was common and 30 days the maximum recorded, but the chick was weak and walked to the sea instead of flying there. Wingate suggests that chicks that grow fast are tended up to their fledging; those that grow slowly are more likely to be abandoned.

Other records of desertion periods from gadfly petrels are for *P.phaeopygia* — about 7 days at the Galapagos (Harris, 1970); 2–3 weeks in Hawaii (Larsen, 1967) — and for *P.neglecta* at the Kermadecs 10–14 days (Oliver, 1955). There seems to be no chick desertion by *P.macroptera* (Warham, 1956; Imber, 1976b). More precise figures from all species are needed.

VIII Dispersal and migration

Some gadfly petrels are trapped by on-shore winds and 'wrecked' on the beaches of South America, South Africa, and Australasia where these land masses extend into the west wind zone. For example, *P.mollis* is fairly common in winter among flotsam on the shores of south-western Australia. Presumably these are birds from the Indian Ocean colonies and, significantly, further east in Australia and in New Zealand such birds are rare (Serventy *et al.*, 1971). Much of the movement in the Southern Ocean during the non-breeding season may be down-wind dispersal of the younger age-classes and it is these that tend to predominate in the beach mortalities. Fleming (1950) found numbers of White-headed Petrels at sea to the north-east of northern New Zealand in winter; these could have been from the breeding places to the south of that country and suggest a northwards and down-wind shift after breeding. Sightings in mid-winter of flocks and 'rafts' of up to 150 of these birds in the north-west Tasman Sea (Jenkins, 1982) may have been of birds from Macquarie Island and, again, a north-east shift would have occurred.

Powlesland (1983, 1987) plotted the monthly recoveries of six species of gadfly petrel found dead on New Zealand beaches. The great majority of the 748

Kerguelen Petrels counted between 1939 and 1981 were found in September, so, with laying about 10 October, these were more likely to be non-breeders rather than breeders caught during their pre-laying exodus. Figure 4.15 shows that five other species counted between 1960 and 1984 had marked peaks in beach mortality. Virtually no *inexpectata* were collected in the winter, the majority then

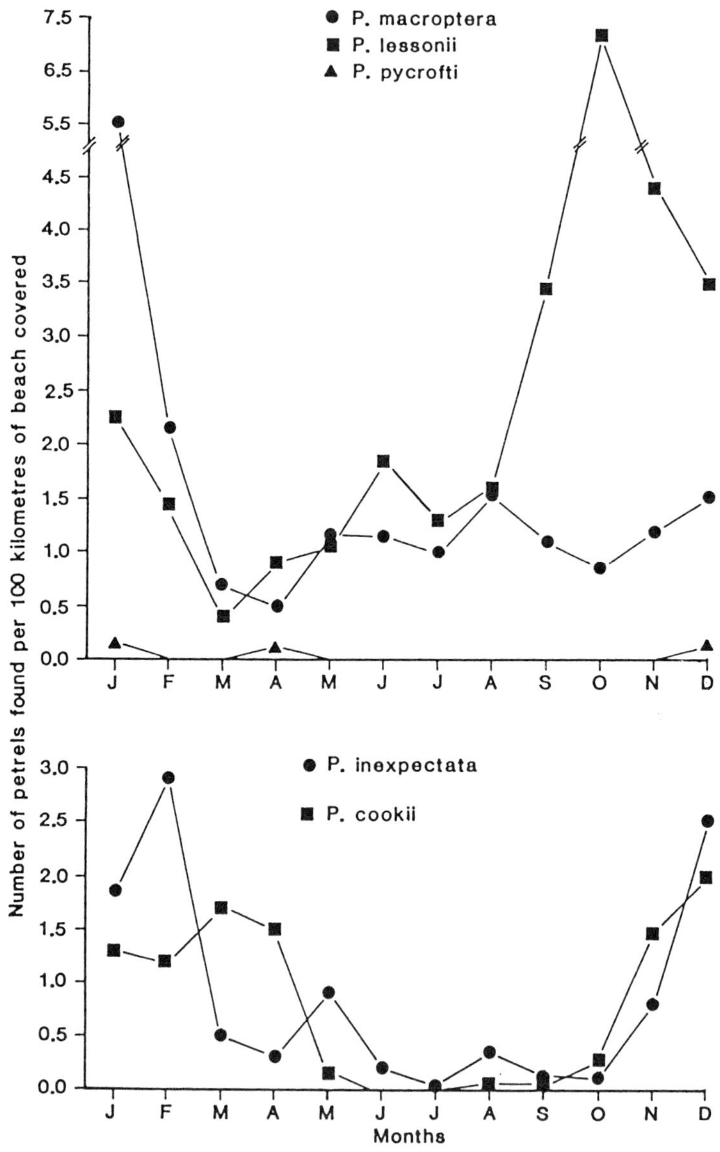

Figure 4.15 *Monthly rates of recovery in New Zealand (no. found dead 100 km^{-2} of beach searched) of five gadfly petrels. From Powlesland (1987).*

being in the North Pacific. Their spring and summer mortality may, as Powlesland suggests, be of returned and exhausted non-breeders having difficulty in feeding in New Zealand seas. Great-winged Petrels (winter breeders) mainly die in the summer and again, these are presumably non-breeders while the white-headed Petrels whose deaths peak in late spring and early summer are mostly in primary moult: these too must be non-breeders for laying is in December. The small numbers of *P.l.pycrofti* (23) and of *P.cookii* (434) found over 24 years reflect their small populations.

At least 11 of the 29 species of gadfly petrel are migratory whereas four seem to be essentially sedentary (Table 4.2). The most extensive movements are made by the trans-equatorial migrants — *inexpectata, cookii, solandri* and *nigripennis*, and possibly *neglecta* and *longirostris*.

Sightings in the North Pacific have been collated by Nakamura & Tanaka (1977), Ainley & Manolis (1979), Tanaka & Inaba (1981) and Tanaka *et al.* (1985). Mottled Petrels seem to penetrate the furthest north and select the coolest waters (*c.* 10°C), being particularly common along the subarctic front just south of the Aleutian Islands and the Alaska Peninsula. *Pterodroma externa* is common to the south-east of Japan in July to September, particularly in waters at 27–30°C, *P.nigripennis* in rather similar areas and waters at *c.* 25–28°C. *P.longirostris* also shifts to waters to the east of Japan, as far as 180°E, and feeds along the fronts of the Kuroshio Current and in seas at 24–27°C. The sole resident North Pacific gadfly petrel *P.hypoleuca*, which nests in the northern winter, also feeds east of Honshu in warm seas in the summer. These various species may occur singly or in mixed flocks and together with *B.bulwerii* and *P.pacificus* (Tanaka & Kaneko, 1983).

It is rather surprising to find that the Kerguelen Petrel seems to be sedentary, as this is a southern species that feeds mostly in the Antarctic Zone, and its fellow nester, the White-headed Petrel is not sedentary. The Kerguelen Petrel is a summer breeder but Mougin (1969) found it wholly absent from The Crozets only during February and March, reappearing in early April although the peak of laying was not until 10 October. It occasionally flies over land in daylight during the winter, at which season part of the population moves to the south (Bourne, pers. comm.). Another small species whose adults may be sedentary is *P.mollis*: birds have been found in burrows outside the breeding season (Wilkins, 1923, p.508). Possibly, while the younger age classes disperse from the nesting places, the older birds, being more proficient feeders, remain closer to home: it is unlikely that young, naive birds are able to return to the colonies during the non-breeding season. Where the other sedentary gadfly petrels go after nesting — the various tropical species and the North Atlantic representatives of the Soft-plumaged Petrel is unknown.

There is also a good deal of apparently random wandering, probably by immature birds. For example, the specimen of *P.arminjoniana* taken off North Carolina (Lee, 1979) and of *neglecta* seen flying over Hawk Mountain, Pennsylvania (Palmer, 1962), and, perhaps the most bizarre of all, the *P.inexpectata* found in New York State in April 1880 (Brewster, 1881). This bird had reached completely the 'wrong' ocean system by some unknown route, and, unlike some albatrosses, is not the sort of bird that a whaler might carry to the Northern Hemisphere.

IX Populations and mortality

A *Population sizes*

Populations of gadfly petrels tend to be rather small, at least in comparison with those of southern shearwaters, storm petrels and prions. Many have been reduced through predation by alien placental mammals, although the situation is not as critical as Harris (1970) made out, for he overlooked some thriving species.

Possibly *P.hypoleuca* of the Hawaiian North-West Chain is the most abundant for Harrison *et al.* (1984) estimated that up to 250 000 pairs lived on Lisianski Island alone. However, that of *P.externa* at its sole breeding place, Alexander Selkirk Island, Juan Fernandez Group, is very large. Brooke (1987) estimated that there were about one million pairs there as well as 131 000 pairs of *P.l.longipennis*. At Ducie Island, Pitcairn, a population of some 230 000 birds of *P.a.heraldica* was mooted by Beck & Quayle (Murphy & Pennoyer, 1952) and at 140 ha Whale Island, New Zealand, between 30 000 and 40 000 breeding pairs of *P.m.gouldi* (Imber, 1976b).

The numbers of Mottled Petrels *P.inexpectata* nesting around southern New Zealand must run to the hundreds of thousands, but any credible census seems impracticable at present because many nests on steep slopes are inaccessible and hard to differentiate from those of the usually more abundant Sooty Shearwater *Puffinus griseus*. One small population of this gadfly petrel nests on an islet in an inland freshwater lake — apparently the only such site known for a petrel.

Pterodroma includes some very rare species, particularly in the south-west Pacific. *P.macgillivrayi* is a small, all-dark form, one example of which was collected in Fiji in 1855. The species was not seen again until 1983 (Watling & Lewanavanua, 1985), when an adult was attracted to a spotlight and a fledgling seen in 1985. The re-discovered *P.(r).aterrima*, the Reunion Petrel, is also evidently very rare (Jouanin, 1970), and the breeding sites of *P.magentae*, another recent re-discovery, were only found in 1987 despite extensive searching over the previous 10 years (Crockett, 1984). By December 1989 41 adults and one fledgling had been banded (Crockett, pers. comm.). Other New Zealand rarities include the Chatham Island Petrel *P.axillaris* with a population in the very low hundreds and Pycroft's Petrel *P.l.pycrofti* which, although nesting on several islands, has but small numbers anywhere. At one of its major breeding places, Hen Island, an estimate of 100 pairs was made by Bartle (1986).

Elsewhere in the Pacific both the Galapagos and Hawaiian races of *P.phaeopygia* are small and endangered, the biggest colony in Hawaii, on the island of Maui, having an estimated 400–600 pairs by Harrison *et al.* (1984). In the south-west Gould's Petrel *P.leucoptera* has an apparently static population of perhaps 200–500 pairs (Fullagar, 1976).

In the Atlantic the Bermudan Cahow still survives thanks to imaginative protective measures (Wingate, 1977). In 1983, 35 nesting pairs fledged 20 chicks (Wingate, pers. comm.). Most other northern populations are similarly small, with the Jamaican Petrel *P.h.caribbea* usually considered extinct, but which could conceivably still linger on in montane forests of the Caribbean. The typical species is also rare but as yet uncounted. Further north the populations of *P.madeira* survive only by having retreated into inaccessible crevices on high inland cliffs.

Simons (1984) developed a population model for *P.p.sandwichensis* based on his 3-year study in Hawaii.

B *Mortality*

Natural mortality of gadfly petrels on land arise from causes similar to those of the petrels — to flooding or collapse of burrows, to predators like reptiles, snakes, skuas and perhaps frigate-birds. Rails like *Rallus philippensis* may suck unguarded eggs and Woodhens *Gallirallus sylvestris* take some chicks and eggs of Providence Petrels at Lord Howe Island and Wekas *G.australis* killed *P.cookii* and *P.inexpectata* at Codfish Island.

The surface-nesting gadfly petrels of the tropics and subtropics have few natural enemies on land: their losses from rats — *Rattus rattus* and *R.exulans* — may be serious, but all these seem to be man-made infestations. In contrast, only two of the subtropical and tropical *Cookilaria* species nest on the surface, despite the absence of predators. These are *P.l.leucoptera* and *P.defilippiana*, neither of which normally lays without some cover from aerial view. The Bonin Petrels of the Hawaiian Northwestern Islands, for example, are burrowers and yet have no gulls, hawks or skuas to harass them, so either they nest underground in response to some past predatory pressure, to competition, or, being small, need the ameliorated micro-climate of the nesting chamber to help chick thermoregulation on hot days. Another possibility is that frigate birds *Fregata ariel* and *F.minor* would take unguarded chicks. However, although frigate birds do take tern chicks, there seem to be no records of them eating gadfly petrels, even when both nest on the same islands. It might also be expected that tropical surface nesting petrels would guard their chicks for longer than the 1 or 2 days usual among the burrowing forms, but we lack data on this aspect. The chicks may be mobile enough to reach the protection of shade by their own efforts.

An unusual kind of mortality is suffered by Gould's Petrel at Cabbage Tree Island. It is caused by the 'Bird-lime Tree' *Pisonia umbellifera* whose long, ripe seeds fall to the ground in late summer and exude a viscous substance. As the birds move to and from their nests the seeds may become attached to their plumage. If heavily infested the feathers become matted, the birds become completely disabled and eventually die (Hindwood & Serventy, 1941).

As with other petrels, the real dangers for gadfly petrels when ashore are man-introduced pests, for example cats and dogs, rats, mongooses (in Hawaii), Wekas (on Macquarie Island). Direct predation by man may have paid a part in the past and *P.hasitata* is eaten in the Caribbean (Wingate, 1964) while *P.phaeopygia* was taken by the Polynesians in Hawaii (their 'UA U') and has only barely survived there. The Maori still use *P.macroptera* as a 'mutton-bird' and no doubt take other species like *P.inexpectata* and *l.pycrofti* when available. Even at the Kermadecs early European settlers almost exterminated nesting *P.neglecta*. Iredale (1914) assessed that there were 'about half a million individuals' of this bird on Raoul Island there, and in 1889 the settlers harvested about 12 000 chicks. There are probably none nesting on Raoul today, though they do so on other islands in the group, Merton (1970) calculating a total of about 5000 pairs on the two Meyer Islets.

X Conservation

Conservation of gadfly petrels usually involves protection from alien predators, by their control or removal. On some of the Galapagos Islands dogs and pigs are the main hazards for successful breeding of the Galapagos Petrel and pigs will need to be controlled if the newly discovered Chatham Islands Taiko is to be rehabilitated. Fortunately the eradication of cats from mountainous, heavily forested Little Barrier Island, New Zealand (Veitch, 1985) and the removal of Wekas from Codfish Island should lead to the recovery of the decimated populations of Cook's and Mottled Petrels at those sites.

The best documented work has been the conservation of the Cahow on Bermuda (another 'Bird of Providence' — see Butler, 1882). Not only were the numbers extremely low but the petrels were being evicted from their burrows by tropic birds. Wingate (1977) was able to induce the petrels to use artificial burrows and devised baffles that prevented the tropic birds from gaining access.

In addition to control of predators, successful breeding depends on the maintenance of suitable breeding habitat: as for other burrowing species, this may require protection from trampling by man and domestic animals. Sometimes simple solutions fail. When the Maori owners of Stephenson Island off the North Island of New Zealand found that their harvests of *P.macroptera* were declining, apparently due to damage to burrows by cattle, they had the cattle removed. Unfortunately this led to a great development of introduced kikuyu grass which has covered the island with an impenetrable mat, seems to have depressed the Great-winged Petrels still further, and driven the small Pycroft's Petrel from the area. At another site, the Poor Knights Islands, Pycroft's Petrel has almost disappeared, parallel with the increase of the aggressive and larger Buller's Shearwater.

CHAPTER 5

The Blue Petrel Halobaena *and the Prions* Pachyptila

Antarctic Prion.
Photo: author.

Contents

I Introduction

These are small petrels of the subantarctic and antarctic zones of surface water. The genera are considered together because they are often thought to be closely related.

Coues (1866) noted their similar plumage patterns of grey-blue above and white below, that their legs and the side plates to their bills are bluish in both, and that the serrations on the sides of the upper mandible of *Halobaena* are of similar construction to those of *Pachyptila*. In the former these fine striae run from the rear to about half way along the palatal edge of the mouth and are rather less developed than in most Fairy Prions *P.turtur*. In both genera the lamellae are in the lining of the mouth and evidently of ectodermal origin, not in the rhamphotheca as in fulmars. The tongue is simple and prion-like. In the myology of the hind limb and the proportions of the leg bones the two genera are similar (Klemm, 1969).

Likewise, both the prions and *Halobaena* have a dark band across the mantle linked to a shallow V-shaped bar of dark feathers along the wings that produce a broad 'M' across the upperparts in the flying birds.

Pachyptila and *Halobaena* share very similar *Naubates* and *Saemundssonia* feather lice (Timmerman, 1965; Pilgrim & Palma, 1982), while Jacob & Hoerschlemann (1982), on the evidence of the composition of their preen gland waxes, also concluded that Blue Petrels and the prions were closely related.

Bretagnolle (1990) compared aspects of the behaviour and vocalisations of *Halobaena* with those of prions, fulmars and gadfly petrels. He concluded that the links with *Pachyptila* are real and not due to secondary convergence.

Coues (1866) also saw resemblances between the Blue Petrel and gadfly petrels like *Pterodroma mollis* and *P.cookii*, notably in their laterally compressed bills and heavy maxillary ungues, in the proportions of toe to tarsal length, and in the pale edges to the forehead feathers of *Halobaena* which create a scalloped effect as in many *Pterodroma*. In wing shape, aspect ratio and wing loading the Blue Petrel resembles the gadfly petrels more than the prions (Warham, 1977a). Murphy (1936, p.724) also regarded the two genera as closely related, as did Harper (1978), who considered the similarities of plumage and structure to the prions due to convergence. He also found that the plasma protein patterns were much closer to those of *Pterodroma* than to *Pachyptila* (pers. comm.).

Checklists (e.g. Alexander *et al.*, 1965; Jouanin & Mougin, 1979) usually place (*Halobaena* between *Pterodroma* and *Pachyptila*. Imber (1985a) even hypothesised that all three genera had a common ancestor, *Halobaena* being the least specialised.

However, fulmarine links have also been proposed for both the Blue Petrel and the prions, and in some earlier lists both were placed in the subfamily Fulmarine, e.g. by Sharpe (1879). Klemm (1969), on myological evidence regarded *Halobaena* as closer to the fulmars than to the gadfly petrels. Finally Kuroda (1983), from detailed examination of the skeletons, has concluded that the Blue Petrel and the prions are indeed closely related, but placed both in his tribe Pachytilini of the subfamily Fulmarinae.

II *Halobaena caerulea* — the Blue Petrel

The sole member of this genus is a small seabird of the Subantarctic and Antarctic zones of surface water. It is unrepresented in the fossil material. Unlike any other petrel, its square-cut tail is prominently tipped with white, broadest in the centre. Otherwise, apart from its dark crown and dark suffusion from the nape that spreads downwards to form a broad but broken breast bend, it greatly resembles a prion in colouring. Its bill is rather slender, but with prominent nails (Fig. 5.16).

Halobaena is one of the less studied petrels. It is sexually monomorphic and slightly bigger than most prions, 34 South Georgian birds weighed 193 ± 18.6 g (Prince, 1980). Wing spans average *c.* 68 cm.

A *Life style*

At sea this petrel occurs in loose flocks of perhaps a hundred birds, sometimes in company with Southern Fulmars and prions, but tends to have a less erratic flight

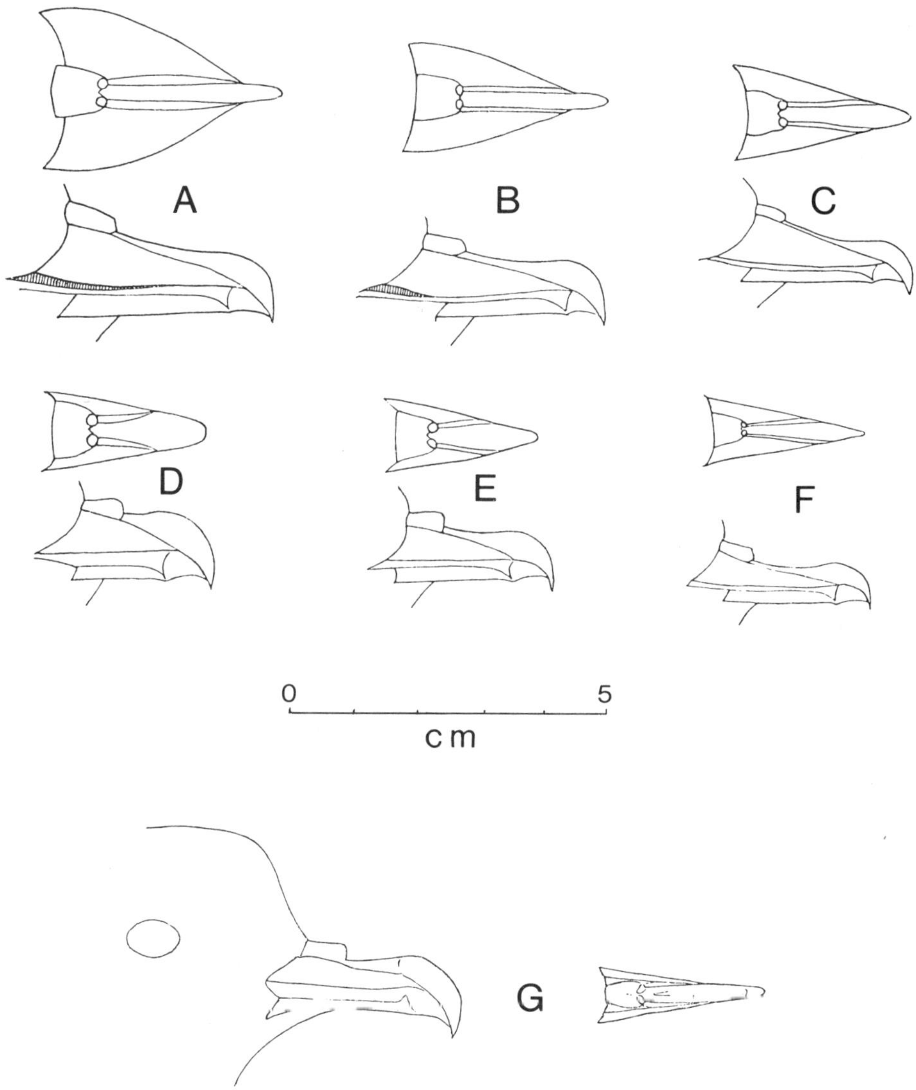

Figure 5.1 *Bills of prions (A–F) and the Blue Petrel (G), in dorsal and lateral view: A. Broad-billed* P.vittata; *B. Salvin's* P.salvini; *C. Antarctic* P.desolata; *D. Fulmar* P.crassirostris: *E. Fairy* P.turtur; *F. Thin-billed* P.belcheri; *G. Blue Petrel* H. caerulea. *Note exposed lamellae in (A) and (B). Mainly from Fleming (1941a).*

style than the latter, usually keeping closer to the water. Routh (1949) described the ease with which Blue Petrels skim along the leading (leeward) side of a wave during gales, and, rising slowly up its face using the eddy currents there, flip over a pass joining two waves, keeping as low as possible and gaining momentum as they slide down into the oncoming trough. On calm days, however, the flight was relatively laboured and the birds had to flap hard to keep up with Routh's ship.

These birds are quite agile on sea and on land, although in calm weather take-off at sea involves a hurried paddle across the surface like a fulmar.

B *Distribution*

The Blue Petrel is a circumpolar breeding species but is only known to nest at six locations, all on or close to the Antarctic Convergence (Fig. 5.2).

In summer Routh (1949) found prions (presumably Antarctic Prions (*P.desolata*))

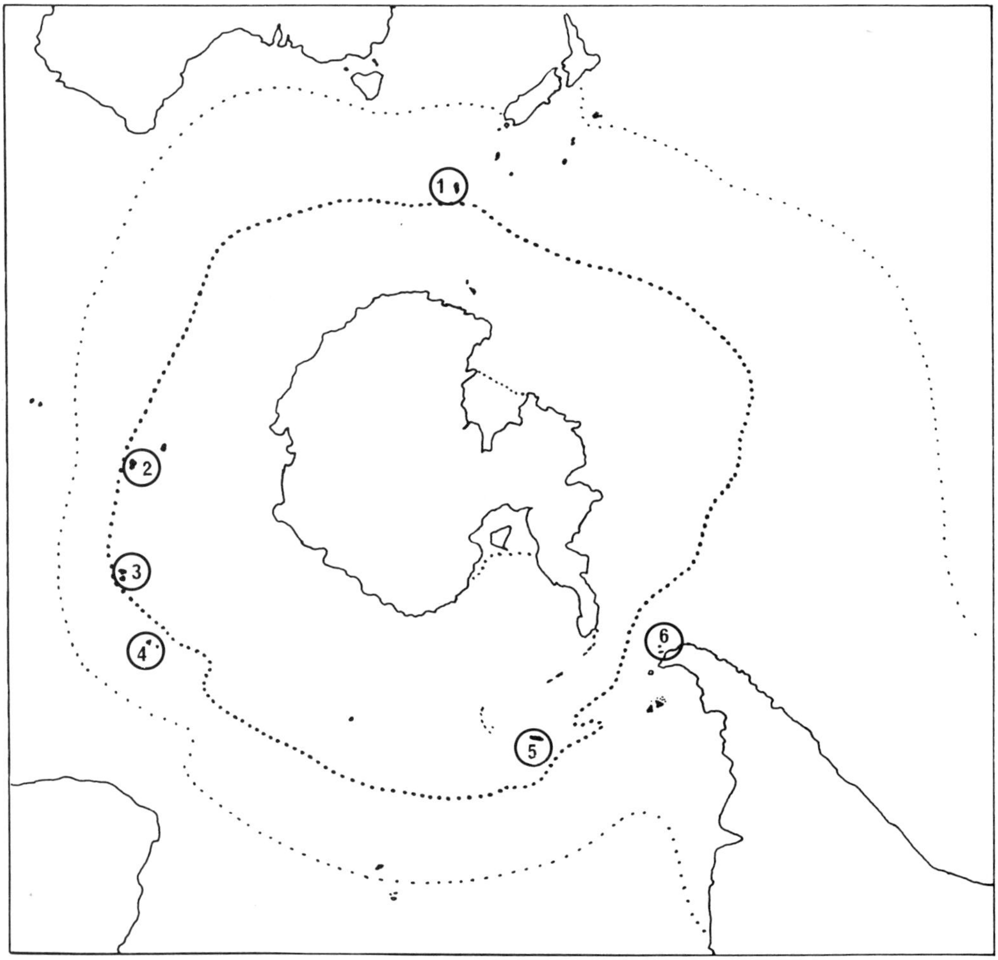

Figure 5.2 *Breeding distribution of Blue Petrels, Heavy dotted line = Antarctic Convergence; lightly dotted line = Subtropical Convergence. 1. Macquarie I.; 2. Kerguelen; 3. The Crozets; 4. Marion and Prince Edward Is.; 5. South Georgia; 6. Diego Ramirez, Cape Horn and nearby islands.*

and Blue Petrels in mixed flocks north of the pack ice and along the ice edge, but as his ship drew further south and east the Blue Petrels became more numerous while the prions disappeared. Still further south and the Blue Petrels too were absent.

The summer range of the species seems to lie in seas with surface temperatures of −1° to +6°C, the winter range in seas from −1° to −10°C, much the same as the preference of the Antarctic Prion (Mougin, 1975, Fig. 40). This Antarctic distribution was also emphasised by Biermann & Voous (1950, p.90) who reported many summertime flocks in waters at 0° to 1°C, with large numbers among the icebergs feeding on shoals of krill with Antarctic Petrels and Southern Fulmars. In February Blue Petrels in the Crozet sector of the Southern Ocean were also in seas at about +1°C, over water some 3000 m deep and 2000 km from the nesting places (Stahl *et al.*, 1985). Egg laying was but a month away. Zink (1981) saw a flock of 55 birds in open water of the Ross Sea on 22 January at 70°S — possibly the most southerly sighting of the species.

C *Feeding and foods*

There are few reports of the feeding style of these petrels. Biermann & Voous (1950) saw them swimming while taking prey by surface-seizing and also by diving from a swimming position. Unlike most prions, Blue Petrels followed their ship to take prey close by, often in association with whales.

At Marion Island *Halobaena* fed mainly on *Euphausia vallentini* with fish and squid being of lesser importance in the diet. Noctuid moths and assasin bugs *Nabis* sp. were also found in their stomachs: evidently these insects had been blown out to sea (apparently from Africa) and picked up from the surface (Steele & Klages, 1986). Brown *et al.* (1986) also recovered an average of 3.3 insects per gizzard in 45 Blue Petrels salvaged from a 'wreck' in Victoria in 1984. These were mainly common coleoptera of south-eastern Australia, and evidently taken from the surface and not far offshore. Insects (apart from *Halobates*) are unexpected prey for a pelagic seabird and suggest that the petrels were hungry and feeding unselectively on surface items which included indigestible pumice and plastic particles. Otherwise the food remains from these 45 birds were almost entirely of cephalopods. Stomach analyses of South Georgian *Halobaena* by Prince (1980a) showed that these were depending more on crustacea, but fish and squid were also important. Prince also found that when Antarctic Prions and Blue Petrels fed on the same prey the latter took larger animals than did the prions. Routh (1949) saw prions and Blue Petrels feeding together around the bases of icebergs, presumably taking advantage of the local flushes of plankton due to melt-water mixing effects — as the Northern Fulmar does around icebergs in the Arctic Sea.

Pumice in the stomachs of the South Georgian birds was thought to come from the South Shetland Islands and this and other indirect evidence led Prince to suggest that these Blue Petrels were feeding to the south west, perhaps 1350 km from their nests. The Marion Island birds must have been feeding closer in, because *E.vallentini* occurs north of the Polar Front which is about 330 km south of that island (Steel & Klages, 1986).

D *Breeding*

Paulian (1953) worked out the annual cycle at Kerguelen and was also the first to report the post-nuptial return to the burrows. Jouventin *et al.* (1985) added more data from the Crozets, Fugler *et al.* (1987) more for the Marion Island population.

The birds are burrowers. On Kerguelen they delve into soft ground below *Azorella* or *Acaena*; on Ile de l'Est, Crozet Islands, beneath a dense cover of low grass; and at South Georgia they dig below clumps of tussock grass. Paulian (1953) also measured the burrow temperature, finding that this varied only slightly despite considerable fluctuations in the air temperature outside.

The noise of Blue Petrels calling from the sky and from the ground may dominate the night-time chorus on the breeding grounds, for nests may be very concentrated. Burrow densities of about 0.72 m^{-2} were measured at South Georgia (Croxall, 1984).

These birds lay rather early, e.g. at the Crozet Islands their eggs appear at the end of October whereas *P.salvini, belcheri* and *turtur* lay at the end of November and early December, *desolata* not until the end of that month (Jouventin *et al.*, 1985).

E *Dispersal and migration*

Adult Blue Petrels appear not to migrate. The autumnal protogamic behaviour reported by Paulian was confirmed for Marion Island by Fulger *et al.* (1987). There the adults left when the chicks fledged, stayed at sea for some 6 weeks while replacing their flight feathers, to reappear at their nests after dark for about 2 months before leaving for the winter. At Macquarie Island Blue Petrels can be found in burrows throughout the year (Brothers, 1984). Others, probably mostly immatures, travel far, and large numbers may be driven onto lee shores in winter and early spring, often in association with cyclones. Derelicts on Western Australian beaches are usually thought to come from Kerguelen but no ringed bird has been found to establish provenance. Thousands were cast ashore dead or dying on those beaches around 6 August 1984 and by 11 August had turned up in Victoria, 2600 km to the east (Carter, 1984; Brown *et al.*, 1986). The latter suggested that Blue Petrels had been well distributed south of Australia and New Zealand in mid-winter, almost to the pack-ice, but had been forced north by south-westerly gales and, weakened through being unable to feed properly, had perished close to land. 'Wrecks' in New Zealand tend to occur slightly later, peaking in September (Fig. 5.3) and here, as in Australia, large numbers of Kerguelen Petrels are also involved (Powlesland, 1983).

Blue Petrels have been recorded throughout the south Pacific Ocean in winter and early spring, off the Chilean coast as far as 35°S and off Peru at 20°S (Szijj, 1967; Meeth & Meeth, 1977).

F *Populations and mortality*

Although Blue Petrels nest in few places, breeding aggregations can be large. Schlatter (1984) suggested that about 2 million birds might breed on the Diego

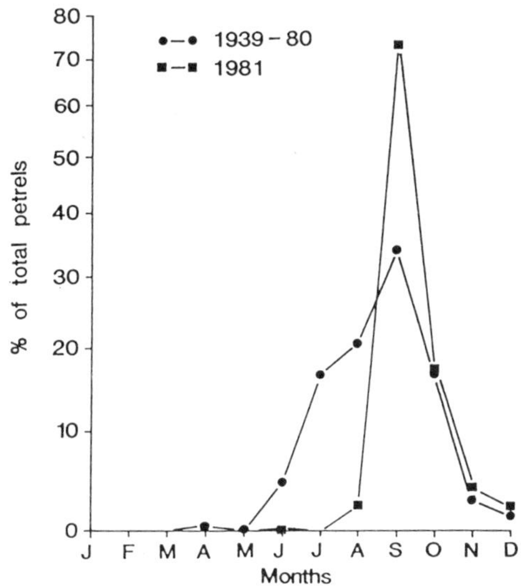

Figure 5.3 *Monthly distribution (% of totals) of beach-washed Blue Petrels in New Zealand, 1939–1980* (n = *360) and in 1981* (n = *343). From Powlesland (1983).*

Ramirez Islands near Cape Horn; 'tens of thousands' on three islands in The Crozets (Jouventin *et al.*, 1984), and 70 000 pairs on South Georgia (Prince & Croxall, 1983).

Like the prions, Blue Petrels suffer heavy predation from skuas which kill them on the ground at night. They seem safer at sea but mortalities there are unknown. Paulian (1953), however, took eight of them from the stomachs of six Leopard Seals.

Cats have reduced the numbers at Marion and Macquarie Islands and the remnant population of 500 to 1000 pairs at the latter station has withdrawn to safer stacks offshore. The mortality due to cats at Marion Island was estimated at 16 500 birds annually, amounting to 3.7% of the total energy needs of the estimated 2100 cats there (Williams, 1978) — see also van Rensberg (1985).

III The prions: genus *Pachyptila*

A *Introduction*

Prions are small petrels restricted to, and often abundant in, the Southern Ocean. The attention of science was first drawn to them by explorers like Cook and Weddell, but owing to their great similarity, precise identification of the 'blue petrels' of these early records is often impossible. Later, whalers and sealers told of their abundance, and from their association with cetacea called them 'whale birds'. They sometimes crowded round these mammals in such numbers that many were killed by blasts from the whale-guns (Murphy, 1936, p.624).

Until recently it has often been thought that most speciation within the genus took place during the late Tertiary and the Pleistocene (Fleming, 1941a), but the recent discovery of fossil material of several species from the early Pliocene of Cape Province, South Africa by Olson (1985a) shows that the radiation was well under way before the Pleistocene glaciations of Antarctica began to make major changes to the hydrology and climate of the Southern Ocean.

The fossils include plentiful post-cranial material of *Pachyptila salax*, a species considerably larger than any extant form (Fig. 5.4). The deposits contain incompletely ossified material which Olson believes were from nestlings of *P.salax*, as well as material probably of two other kinds of prion of similar sizes to those of today's forms. None nests in the warm waters of Cape Province today.

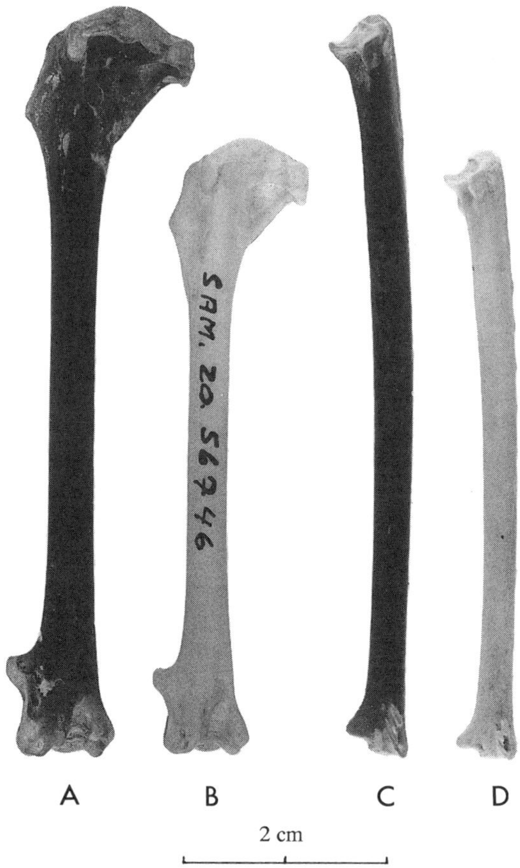

Figure 5.4 *Left humeri in caudal view (A, B) and left ulnae in dorsal view (C, D) of (A) and (C)* Pachyptila salax *Olson 1985 and (B) and (D)* P.vittata. *From Olson (1985a).*

B *General characteristics*

Prions are blue or blue-grey above, white below, and the darker mantle and wing coverts form a prominent M-shaped mark across the upperparts when the wings are spread. This pattern is not unique to prions and Blue Petrels, being seen in e.g *Puffinus bulleri* and *Pterodroma cookii*, but it forms a useful field character and was commented on by Cook (1777). Prions' tails are wedge-shaped with a blackish terminal band, the underparts white except for greyish undertail coverts. Their bills are blue, blue-grey or black, the legs blue. Wing spans and body weights range from about 56 cm and 130 g (Fairy Prion) to about 68 cm and 195 g (Broad-billed Prion). Males tend to weigh more than females (Tickell, 1962, p.46), otherwise prions are not sexually dimorphic apart from the Fulmar Prion *P.crassirostris*, the males of which have significantly longer bills and wider maxillary nails than the females (Harper, 1980). Twenty males averaged 23.4 $\pm$ 0.2 mm in bill length, 20 females 22.2 $\pm$ 0.2 mm. $P = 0.05$.

Facial patterns and widths of tail bands vary from species to species but the main differences are in bill structure (Figs 5.1 and 5.6). These range from the slender bill of the Thin-billed Prion *P.belcheri* to the stout one of *P.crassirostris* and the very inflated, specialised bill of *P.vittata*.

Comb-like lamellae fringe the sides of the palate and, as noted by Darwin (1872, p.239), resemble the filter plates of baleen whales. In the bigger species, *vittata*, *salvini* and in some *desolata*) these lamellae run the whole length of the maxilla and are visible when the bill is closed. Between the rows of lamellae and the cutting edges of the maxillae there are grooves into which the tomia of the mandibles fit (Fig. 5.5). In *vittata* the lamellae are lemon-yellow. With the other species of prion, particularly *belcheri*, the striations are less well developed, entirely absent distally, and with no grooves for the reception of the mandibular tomia.

Details of moulting patterns have not been worked out but in most, a body moult of the breeders starts towards the end of chick-rearing, the flight feathers being replaced later. Non-breeders possibly moult earlier (Harper, 1980).

There are few reports of prions calling at sea, but at the nesting places they are very vocal, using strings of low emphatic cooing notes — those of the Fairy Prion *P.turtur* being syllabilised as 'popper popper pop', and so on — but the calls of *P.vittata* are harsh, staccato barks. Most singing takes place after dark although both *crassirostris* and *vittata* call by day at some breeding places. Some of these calls have been analysed by Bretagnolle (1990).

C *The species*

The prions form one of the most difficult petrel groups for the taxonomist because of their many similarities. Specific separation is mainly on the size, shape, structure and colour of their bills. The colour of the lores, the extent of the superciliary stripe and the width of the tail band have also been used (Harper, 1980). Modern classifications still depend largely on the diagnoses of Coues (1866).

Unfortunately the bill characters show a good deal of intraspecific variability, the extent of which has not been clearly established, and there are overlaps in

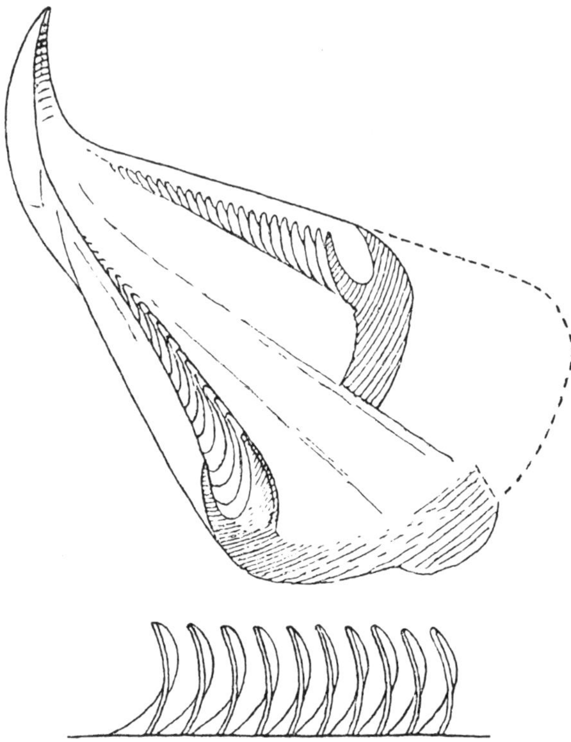

Figure 5.5 *Maxillary lamellae of* P.vittata *(much enlarged), Upper drawing is in palatal view; the lower shows 10 plates on the right side, as seen from the mid-line. From Murphy (1936).*

dimensions between birds of different forms; added to which bill shrinkage during the drying of skins is considerable, amounting to 1–13%, and unpredictable (Falla, 1940; Kinsky & Harper, 1968). Much of the museum material on which classifications have been based is of birds of unknown sex and age, so that immatures, whose bills are smaller than when full grown, have often been confused with adults of the smaller-billed species. The result of all these variables is that many subspecies have been named on inadequate evidence. Jouanin & Mougin (1979) recognised five species — *vittata, desolata, belcheri, turtur* and *crassirostris* — considering Salvin's Prion *salvini,* a lesser broad-billed form, as a subspecies of *vittata*. But the bill of *vittata* differs from that of other prions in being black, not blue (Falla, 1940), whereas that of Salvin's Prion is bluish like the rest. It has also been shown that the broad-billed prions of St Paul Island, usually classified as *P.v.macguillivrayi,* are closer to *salvini* than to *vittata,* not only on bill colour and bill size, but also in breeding 3 months later than *vittata* like *salvini* (Roux *et al.*, 1986). These authors therefore treat *salvini* as a full species, with subspecies *salvini* and *macgillivrayi*.

Fleming (1941a), in a landmark study, recognised these six species but regarded them as falling into three subgenera, each with a pair of species — *Pseudoprion (turtur, crassirostris); Heteroprion (belcheri, desolata)* and *Pachyptila, sensu stricto (salvini, vittata)*. He argued that this arrangement was supported not only by similarities in adult bill morphology, but also on that of the skull, palate, tongue and hyoid apparatus.

Murphy (1936, p.611, 615) saw the sequence of mouth structure as the result of the disharmonic growth of the bill in relation to the rest of the animal and the fringe of lamellae in *vittata* as the extreme development from the simple condition of minute striations as seen in *P.belcheri*. Fleming (1941a) developed this concept further. He showed that in ontogeny the bills of the larger-billed species pass through embryonic stages during which their structure resembles those of the smaller billed forms. He illustrated this idea of differential growth by comparing bill forms using D'Arcy Thompson's system of Cartesian coordinates (Thompson, 1917). Fleming suggested that the smallest species, *turtur* and *crassirostris*, are closest to the original stock. Using the outline of the bill of the latter species as a standard, he showed how the matrices needed to represent the shapes of the other species supported his subgeneric grouping (Fig. 5.6).

Cox (1980) noted the intergradation of many characters, and reduced the species to three (*vittata, turtur* and *belcheri*). However, he lacked adequate specimens of known provenance and made the novel suggestion that *salvini* was the result of extensive hybridisation between *desolata* and *vittata* and that *belcheri* and *desolata* interbred at Kerguelen. With very large numbers of prions, often of several species, nesting on the same island, occasional hybridisation might be expected, and possible hybrids have been reported, e.g. by Harper, (1980, p.275). Mixed pairs have not been found and Kerguelen *belcheri* and *desolata* nest at different times in different places (Weimerskirch *et al.*, 1989).

Harper (1980) examined 839 specimens from breeding places and 9247 from beach wrecks and supported Falla's separation into six species. He described differences between immatures and adults and showed that *vittata* and *salvini* are differentiated on the shape, size and colour of their bills and by their breeding times. He also thought that *salvini* is migratory and *vittata* not, whereas, at least at Marion Island, some *salvini* are also sedentary (see below).

Chromatograms of the blood plasma proteins of four of the six species showed many similarities and unusual pattern of double albumens (Fig. 5.7). The patterns for the Antarctic Prion *P.desolata* were distinctly different from the rest, whereas three populations of *P.turtur* proved to have almost identical protein complements and esterase patterns, as would be expected (Harper, 1978). Bretagnolle *et al.* (1990) report on the genetics, foods and vocalisations of four species of prion from the Indian Ocean.

D *Life styles*

Prions are social birds and at sea are usually seen in flocks, often of great size. In calms they float buoyantly on the surface, but in the air they have a distinctively erratic and fast flight. They rock and twist, often in unison, as they speed along, now close to the waves, now thrown high in the air, so that in sunny weather their white underparts flash out to attract attention and then, as they flick over to present their backs to the viewer, all vanish against a grey sea.

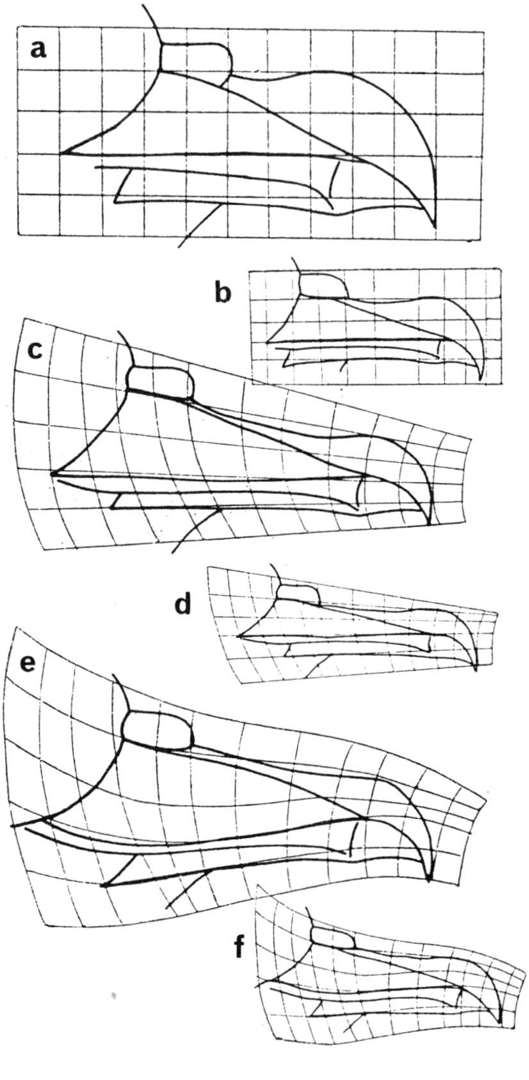

Figure 5.6 *Outline drawings of the bills of prions. From Fleming (1941a).* P.crassirostris *(a), considered the least specialised, is represented on a network of Cartesian coordinates and those coordinates imposed on outline drawings of the other species.*
(a) P.(Pseudoprion) crassirostris, *(b)* P.(P.) turtur, *(c)* P.(Heteroprion) desolata, *(d)* P.(H.)belcheri, *(e)* P.(Pachyptila) vittata, *(f)* P.(P.) salvini.

Murphy (1936, p.626) recounts that during the 1914–18 war, when the US navy was considering the best colour to paint naval vessels to lessen their detection by enemy submarines, he drew the authorities' attention to the prion's disappearing trick. His suggestions were ignored, but after their own optical experiments, the Navy came up with 'omega grey', a colour that proved to have identical reflective properties to those of the dorsal surface of *P.desolata*! Harper (1978) suggested that

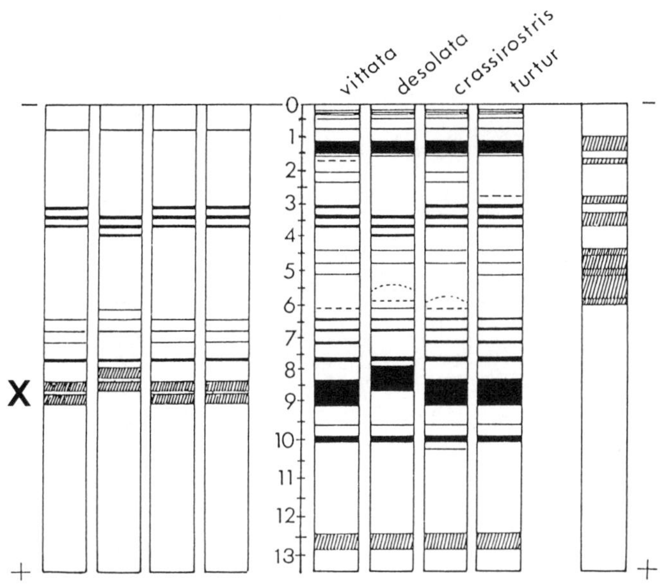

Figure 5.7 *Protein bands from blood plasma of four species of prion. Left-hand columns show the albumens resolved into two distinct bands at X — a pattern not seen in other petrels. The right-hand column gives the esterase pattern common to all four species. From Harper (1978), courtesy of* New Zealand Journal of Zoology. *For key to banding see Appendix 2.*

this cryptic colouration evolved as an adaptation to reduce detection by some diurnal predator, most probably the Tertiary equivalent of the Southern Skua *Stercorarius lonnbergi*.

Prions find little difficulty in taking off from level ground or water, even in calms. Birds resting in heavy seas lift at the last moment before a comber, fly through the spray to alight on the calm surface of the downward slope. Murphy (1936, p.622) likened them as rising like a school of flying fish. Harper (1972) noted that *P.belcheri* has a particularly buoyant and agile flight compared with which the slower, more deliberate wing action of *P.desolata* creates an impression of rather laboured progression. Prions sometimes dive for food, propelled by their wings, and have been reported going down to about 3 m. They normally ignore ships, but *vittata* is an occasional ship's follower.

Although they tend to shuffle around on their bellies when on the ground, Fairy Prions, can move nimbly on their feet with their bodies quite erect if necessary (pers. obs.). The large-billed *vittata* is less agile, and Milne-Edwards' drawing (Fig. 5.8) exaggerates their standing ability although providing a dramatic picture of the great development of the bill in this species.

In New Zealand Fairy and Broad-billed Prions often nest on forested islands, and, like many other tubenoses climb trees to facilitate take-off: one adult Fairy Prion took only 2 min to climb a 12 m pohutukawa tree. Fledglings were less capable (Harper, 1976).

Figure 5.8 *Skeleton of Broad-billed Prion, showing the enormous development of the beak in this species. From Milne-Edwards (1867–1868).*

E *Distribution*

Prions nest on suitable islands throughout the Southern Ocean (Fig. 5.9). They are most concentrated in the subantarctic zone, but *P.desolata* also breeds close to the Antarctic Continent, while *P.vittata* nests at low latitudes close to the subtropical convergence. The map shows the known situation, but with increasing attention to the southern islands, more colonies of these small and rather cryptic birds continue to be discovered. For example, since Harper's review in 1980, *P.turtur* has been found nesting at South Georgia, Kerguelen, St Paul, Macquarie and the Antipodes Islands. One small population of the Antarctic Prion at Cape Denison on the Antarctic mainland (Falla, 1937) appears to have disappeared, but other mainland colonies there may have escaped detection.

Some of the Fairy Prion colonies of the Atlantic and Indian Oceans are rather small, although not as small as initially thought, e.g. Berruti (1981a) reported those at Marion Island as numbering at least in the thousands of pairs rather than in the hundreds as earlier believed, and the Fairy Prions recently discovered on Kerguelen number 100–10 000 breeding pairs (Weimerskirch *et al.*, 1989). There are also more *P.turtur* on Beauchene Island in the Falklands than was earlier thought (Smith & Prince, 1985).

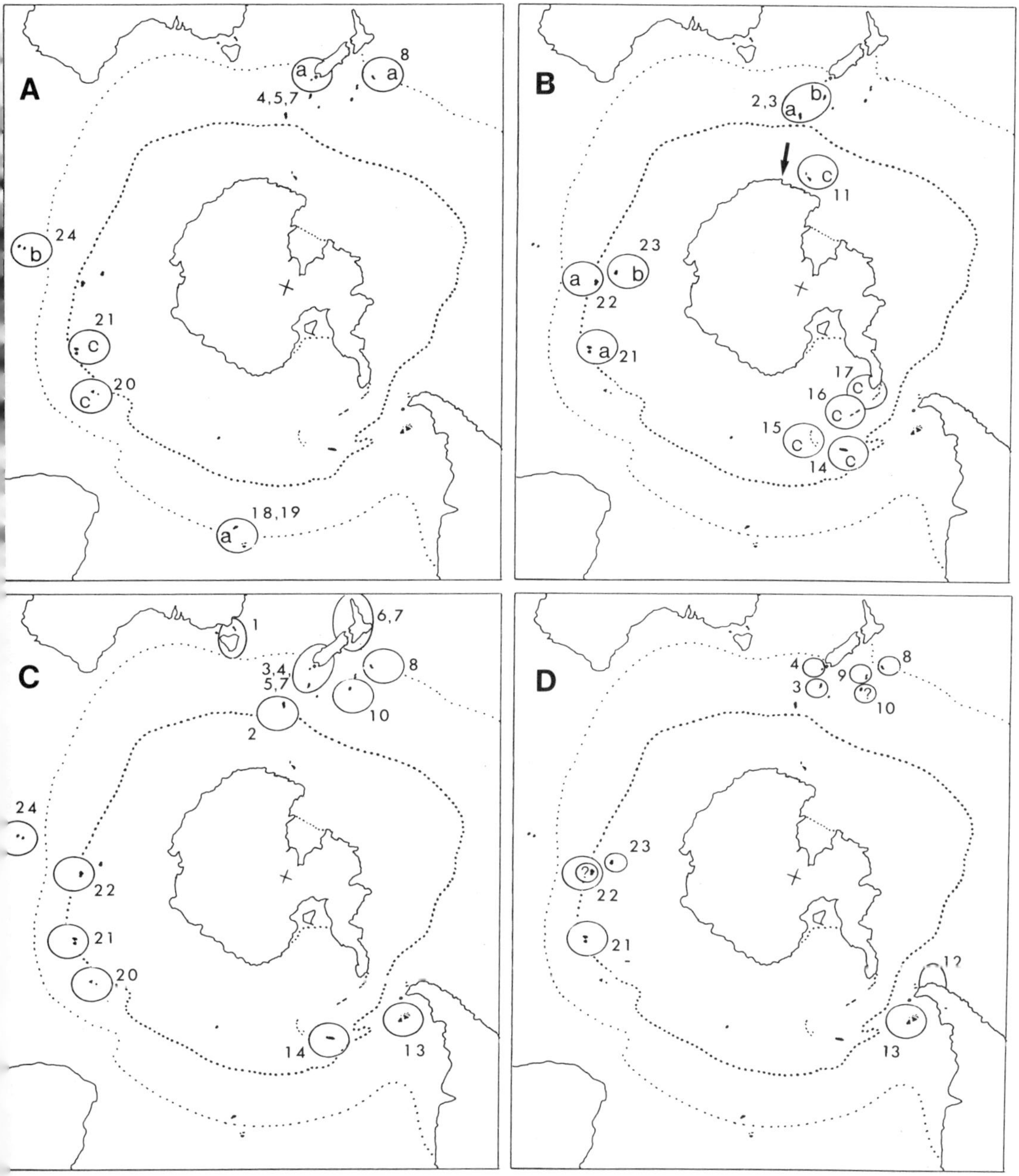

Figure 5.9 *Distribution of breeding colonies of prions:* **A.** *Broad-billed and Salvin's Prions: (a)* vittata*: (b)* s.macqillivrayi*; (c)* s.salvini. **B.** *Antarctic Prion: (a)* d.desgolata*; (b)* d.alter*; (c)* d.banksi. *Arrow points to Cape Denison, former breeding place.* **C.** *Fairy Prion* P.turtur. **D.** *Thin-billed Prion* P.belcheri *(large circles) and Fulmar Prion* P.crassirostris *(small circles). 1. Tasmania/Bass Strait, 2. Macquarie I., 3. Auckland Is.; 4. Snares Is., 5. Stewart I. and Foveaux Strait, 6. N.Z. North I., 7. N.Z. South I., 8. Chatham Is., 9. Bounty Is., 10. Antipodes Is., 11. Scott I., 12. Cape Horn and Is., 13. Falkland Is., 14. South Georgia, 15. S.Sandwich Is., 16. S.Orkney Is., 17. S.Shetland Is., 18. Tristan de Cunha, 19. Gough I., 20. Marion and Prince Edward Is., 21. Crozet Is., 22. Kerguelen, 23. Heard Is., 24. St Paul and Amsterdam Is.*

Figure 5.9 shows that at many island groups several species breed, the Chatham Islands and Kerguelen each having three and The Crozets four. Their coexistence does not necessarily mean that they share a particular habitat — the colonies may be well separated. But mixed colonies also occur, e.g. *P.vittata* and *P.turtur* at several New Zealand locations. In the Crozet Archipelago four species — *salvini* and *turtur* in great numbers and *belcheri* and *desolata* in very small numbers — together with abundant Blue Petrels, share the habitats of the Ile de l'Est (Despin *et al.*, 1972).

Fleming (1941a) drew attention to the geographical segregation of the species in his subgenera *Heteroprion* and *Pachyptila* as having no overlap in their breeding distributions. This view is no longer tenable because *turtur*, a bird of the Subtropical zone, now has a foothold in the Antarctic zone at South Georgia while at the same time nesting well north of the Subantarctic zone at St Paul Island and in northern New Zealand. Indeed, both *desolata* and *turtur* have circumpolar breeding distributions, whereas *crassirostris* is restricted to New Zealand and one Indian Ocean location, *belcheri* to the South Atlantic and Indian Ocean sectors and *salvini* to the Indian Ocean alone.

F *Feeding and foods*

Prions feed on concentrations of zooplankton and crustacea near the surface. The finer-billed species like *belcheri* and *turtur* pick up food items directly by surface-seizing or may fly low to take prey by dipping, keeping their wings outstretched so that contact with the water is only momentary. This technique has been particularly noted for *belcheri* by Strange (1980), while Harper (1972) described the birds as gliding on stiffened wings and deftly picking food from the surface. They did not alight but patted the water with their feet to gain extra lift and momentum.

Antarctic Prions use a special method termed 'hydroplaning' (Murphy, 1936, p.621) when straining small zooplankton (Fig. 1.7). The birds scurry along into the wind driven by their feet, their bodies resting on the surface but partly supported by the aerofoils of their extended wings and, with their heads submerged, scoop the food into their bills. Or they may swim while swinging their heads from side to side, sweeping the water with its contained food into their mouths. The skin between the rami of the mandibles in this and in other prions is elastic and extensible and is used as a storage pouch while the tongue is smooth and muscular and appears to provide the force to expel the water, leaving the animals trapped on the vanes of the sieve (Fig. 5.10). The pouch itself may also have a pumping function. According to Murphy this style of feeding was the basis for the whaler's term 'scooper' for these birds.

Opportunities for close observation of feeding are rare. From a pier Robinson (1961) watched Fairy Prions feeding on individual crustacea floating in clear water at various depths. The birds either reached down to grab the prey or dived to get it, but always brought the catch to the surface, swallowing it there intact or breaking it into smaller pieces.

The best review of prion diets is that by Imber (1981) who showed how the foods vary, as expected, in accordance with beak form, with no closely related species

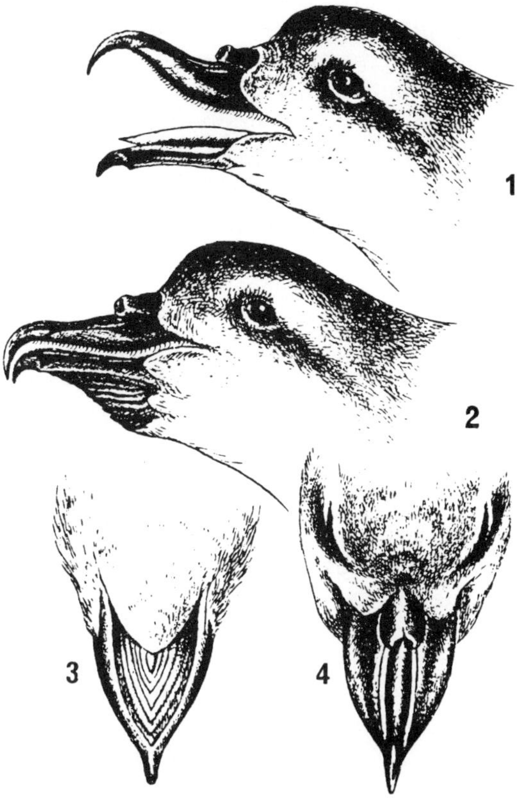

Figure 5.10 *Broad-billed Prion showing (1) 'strainers' and simple tongue, (2) buccal pouch expanded, (3) ventral surface and (4) dorsal surface of bill. From Wilson (1907).*

living sympatrically having the same diet. Like previous workers he viewed adaptations to exploit different prey as being a major force behind prion speciation.

The items taken are of small size. The Thin-billed Prion is recorded as feeding mainly on amphipods, euphausids and on squid no more than 2 cm long, and this bird evidently does not filter feed as do those with lamellated bills (Harper, 1972). The amphipod *Themisto gaudichaudii*, which swarms at the surface by night throughout the main range of *belcheri*, is particularly important (Harper, 1972; Strange, 1980) and the bird's tweezer-like beak would facilitate capture of this prey. Harper believed that much of its feeding is done at night, although some is done by day (Strange, 1980). Fairy Prions in the Australasian region depend mainly on the euphausid *Nyctiphanes australis*. Little is known about the diet of the Fulmar Prion but remains of stalked barnacles *Lepas* sp. have been found in stomachs (Imber, 1981), unusual prey for petrels. Imber considered that this bird's heavy bill would be suitable for tearing barnacles from flotsam and for crushing their shells.

Antarctic Prions will scavenge for food but much comes from filter-feeding without the need for seeing each item. A 16 g sample from a single stomach contained 41 000 copepods (Prince, 1980a). However, the main food during his sampling period was *Euphausia superba* — 97% by weight. Salvin's Prion diets are little studied, but its large filtering apparatus implies that it takes surface prey, probably copepods, several species of which occur near the nesting islands. In the New Zealand region the Broad-billed Prion depends largely on the copepod *Calanus tonsus*, abundant in summer when the chicks are being raised and fledged. Beach wrecks of this prion are most common during or soon after June and July when *C.tonsus* is least plentiful in surface waters (Imber, 1981).

G *Breeding*

Fairy, Broad-billed and Fulmar Prions tend to be sedentary and, at some breeding places at least, visit their nests outside the nesting season. Richdale (1965b) gave a dramatic description of a mass return of *P.vittata* to his study colony on Whero Island, southern New Zealand on 7 February, after a post-nuptial absence of about 5 weeks. The birds suddenly reappeared about 10 pm on a calm, dark evening and began a vocal chorus, with birds singing from the group, from within rock crevices, from the air and even from perches in trees. But by 11.30 pm the din had ceased and all had vanished underground.

The breeding cycles of prions have received a good deal of attention, partly because, despite their nocturnal and fossorial habits, they can withstand a fair amount of attention without deserting egg or chick. The chief studies have been those of Richdale (1965b) on *P.turtur* and *P.vittata*, of Harper (1976) on *P.turtur*, Strange (1980) on *P.belcheri*, Tickell (1962) on *P.desolata*, Derenne & Mougin (1976) and Berruti & Hunter (1986) on *P.salvini*.

Prions breed colonially, laying in chambers at the end of burrows driven into soft soil, or in rock crevices and rock piles. Fulmar Prions seem always to lay in crevices and these are usually a tight fit, preventing eviction by larger species such as shearwaters. However, crevice nesting does not distinguish *crassirostris* from *turtur* as Robertson & van Tets (1982) supposed, because *turtur* also nests in crevices, among rock debris, and in burrows, all on the same island.

When several species nest on one breeding place they are ecologically separated by the kind of terrain they choose. Thus at Ile Croy, Kerguelen, *P.desolata* is abundant and nests everywhere except in wet ground or among coastal rock debris, leaving the less numerous *turtur* to breed in humid inland sites and coastal rocks. Significantly however, *P.desolata* and Common Diving Petrels *P.urinatrix* use similar substrates here, and despite their rather different breeding timetables, the diving petrels suffer considerable loss of chicks due to direct competition from the larger prions (Mougin, 1985). At Ile de L'Est, Crozet Islands, *P.belcheri*, while nesting apart from other prions, does so in synchrony with that of *P.salvini* (Despin *et al.*, 1972).

Apart from a clear difference in laying dates of Fairy Prions in northern and southern New Zealand of 4 to 6 weeks (Harper, pers. comm.), the birds of the different populations of each species show little variation in the timing of breeding throughout their range. This probably also applies to the late-laying southern populations of *turtur* from New Zealand which seem to lay about the same time as those at the Falkland Islands.

The breeding system is typical of that of small petrels. The established breeders nest annually generally at the same sites and with the same partners as in the previous year. There is a re-occupation period and a pre-laying exodus by the females. Nest densities are often high and sites defended vigorously from other prions and in high-density areas some eggs get laid on the surface and are abandoned. Offshore the birds may assemble in 'rafts' to await nightfall. Incubation shifts are short — about 2.4 days for *turtur* (Harper, 1976), 2.0 ± 1.1 days for *desolata* (Tickell, 1962) but perhaps longer, 4–7 days, for *belcheri* (Strange, 1980).

In northern New Zealand Fairy Prion chicks are fed every night, often by both parents (Harper, 1976). He found that the average intake overnight of 33.7 g is about 30% more than one prion brings at a visit. Similarly, at South Georgia, Antarctic Prion chicks got meals on at least 83% of nights (Prince, 1980a), and high attendance rates appear to apply also with Salvin's Prion. Evidently many birds get their food locally when rearing chicks, and, in view of their rather short spells on the egg, may not go much further when off duty during the incubation stage.

Their small size and the presence of predators makes nocturnality and burrow or crevice nesting obligatory for prions. Only *P.crassirostris*, laying in the crevices on the rocky Snares Western Chain and the Bounty Islands, is active there by day. These birds perform mutual displays while perched in the open and in daylight. This is possible because of the almost complete absence of predators: the occasional skua or gull (*Larus dominicanus*) passing by may be mobbed by a tight group of prions in close pursuit (Robertson & van Tets, 1982), and even single birds will pursue skuas. At the Bounties these authors found that courtship and the accompanying vocalisations occurred by day, whereas at night calling stopped — an unusual reversal of normal patterns. At Heard Island the same species is almost entirely nocturnal, although birds may fly along the cliffs by day throughout the year. Downes *et al.* (1959) saw Fulmar Prions landing there at dusk in mid-winter (20 June) at a narrow entrance to a cave where large numbers roosted: on that date they counted 1500 birds leaving just before daybreak. It is not clear why this species should be nocturnal and active vocally after dark at Heard Island, where the nests are quite inaccessible to skuas, and diurnal elsewhere.

H *Dispersal and migration*

The movements of prions after breeding may be complex and recoveries of ringed birds away from the ringing places are rare. The sometimes spectacular mortalities of prions on lee shores during winter storms consist mainly of young and immature birds which can be identified by their thin crania and parietal 'windows' which are easily depressed by finger pressure in fresh specimens (Harper, 1980). These 'wrecks' occur frequently on the western shores of South America, southern Africa, and Australasia. The species found vary in abundance according to the continent involved and suggest a generally northwards movement with an easterly drift downwind during the winter. Thus among birds washed ashore in South America *P.belcheri* predominates; on southern African beaches this and *P.desolata* are the commonest prions, the latter being found in larger numbers in some years

(Avery, 1982); on Western Australian beaches *P.salvini* and *P.desolata* are the usual victims,the first probably being from both the Marion and Crozet Island groups (Severity & Whittell, 1962), the second probably from Kerguelen as Harper (1972) confirmed from an examination of 20 skins. Antarctic Prions form a major component of 'wrecks' in winter on southern Australian beaches — probably mainly of birds from Macquarie and Auckland Islands. Harper (1972) found few Antarctic Prions in the central South Pacific and believed that neither the New Zealand nor the South Atlantic birds are really circumpolar travellers in this sector, as had been suspected.

On New Zealand beaches *P.turtur* is the most abundant casualty, the birds being presumably from local colonies, but Harper (1980) identified some among them of the subantarctic race *subantarctica* (from the New Zealand southern islands) among the corpses of the west coast, indicating a northwards shift after breeding. Some *P.desolata* also reach New Zealand and may die only hours after their first flights (Harper, 1980). Although adults are involved in these beach casualties, it seems likely that most are too experienced to be caught and also that they range less far from their breeding places.

Mendelsohn (1981) related winter returns of Salvin's and Fairy Prions to Marion Island with the passing of cold fronts. The birds appeared sporadically in huge flocks offshore, some flying overland and some visiting burrows after dark. Such visits ended abruptly just before severe cold fronts crossed the island (Fig. 5.11). Mendelsohn suggested that the prions moved south to avoid bad weather and poor feeding conditions behind the front so bringing them close in. After the front had passed through they probably moved north again to preferred feeding grounds.

Northwards movements of various species of prions from the Falkland Islands and those of the Scotia Arc have been reported by Harper (1972) and others. From examples he collected, Harper deduced that the summertime flocks of *P.belcheri* in Antarctic seas are mainly of immature birds taking advantage of the swarming *Themisto gaudichaudii*. These cram the birds' stomachs.

Although *P.turtur* occasionally ranges north to Queensland and even to New Guinea and *P.belcheri* to about 17°S off western South America, none has been reported north of the Equator. One *desolata* was collected off the south coast of Java at 6°49′S (Becking, 1976). In the Indian Ocean there are records from as far north as the Comoro Islands and near the Equator off East Africa (Bourne, 1960). One at 9°40′S off Tanzania appears to have been *P.vittata*.

As Harper (1980) pointed out, young petrels are not constrained by the need for a yearly return to nest and in their 3–5 years' wanderings have time to explore marine habitats not used later in life and to cross marine barriers that may restrict the adults. He saw the migration of prions as fitting the 'exploratory model' of Baker (1978) with an adult home range crystallising out from a pre-breeding area of familiarity.

I *Populations and mortality*

Prion populations are often huge with 'millions' of pairs of *P.salvini* breeding on the Crozets (Jouventin *et al.*, 1984), about 22 million pairs of *desolata* on islands of

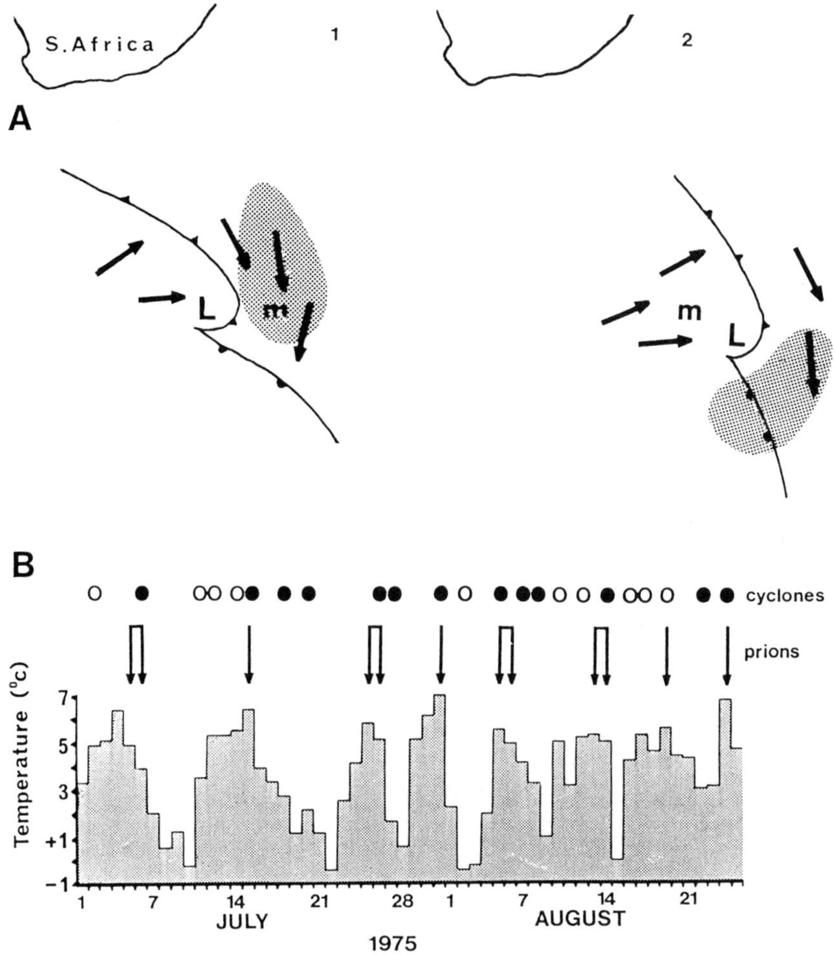

Figure 5.11 *A. Suggested movements of prions near Marion Island (m), as a low front passed through on two consecutive days. Prion flocks stippled, predominant winds arrowed, and conventional symbols show fronts. B. Mean daily temperatures and prion influxes (arrows) and cyclones (circles) passing Marion Island. Cyclones with pressures estimated as < 1000 mb solid circles, those estimated >1000 mb open circles. From Mendelsohn (1981).*

the Scotia Arc (Croxall *et al.*, 1984b), vast numbers of *vittata* at Gough Island (Swales, 1965) and untold numbers of *turtur* around New Zealand and Tasmania. Few detailed counts have been made but a colony of the latter species on Tasman Island was estimated to hold 25 000 pairs (Brothers, 1979a).

Other populations are small, from the tiny relict colony of *P.turtur* at Macquarie Island (only one egg found) whose numbers perhaps do not exceed 100 pairs (Brothers, 1984), to the 440 pairs of *desolata* estimated for Elephant Island near the South Shetlands and only 'hundreds' of *turtur* on South Georgia (Croxall *et al.*, 1984b).

On the breeding grounds prions suffer considerable mortality through flooding of burrows and other accidents and from predation by natural and alien predators. Fairy Prions have also been found with trematode 'anklets' round their legs (Richdale, 1944, p.179; Imber, 1984b), an infection that kills storm petrels (Chapter 9, Section IX). Deaths of prions from this cause have not been reported. Also, like many pelagic seabirds, prions ingest the small plastic pellets that now pollute the surface of the sea. The prions probably take them mostly when starved and such contamination is common among birds cast ashore dead or dying (Harper & Fowler, 1987). Whether this pollution harms the birds (or their chicks) is unclear.

The most widespread predator is the Southern Skua, *S.lonnbergi*. This seems unable to take prions flying by day, even over land (Brothers, 1984) and flocks of *desolata* and *vittata* off the breeding places seem immune to their attacks, but at night skuas catch adults on the ground, particularly in the pre-breeding period when the noisy younger birds are seeking partners. Skuas also get many fledglings. At some colonies they dig down with their bills to take birds in shallow nests, e.g. at those of *P.belcheri* in the Falkland Islands.

The intensity of predation by *lonnbergi* skuas at a big colony of Broad-billed Prions on the Chatham Islands was assessed by Young (1978). He found that skuas on territories among prions took two to six of these each night, those in areas with few prions or with alternative prey, took less. Most of the victims were adult and many were caught in the open and on dark as well as on moonlit nights. These skuas did not excavate prion burrows.

At some New Zealand colonies Fairy Prions have an unusual predator, the Tuatara *Sphenodon punctatus* which eats mainly eggs, chicks and fledglings (Harper, 1976; Newman, 1978). The adults seem more capable of defending themselves and a Tuatara was even seen retreating from a burrow with a prion clinging to its tail (Newman, 1978).

Such natural predators and others like Peregrines (*Falco peregrinus*), Short-eared Owls (*Asio flammeus*) and Striated Caracaras (*Phalcoboenus australis*) on *P.belcheri* at the Falklands (Strange, 1980), *Falco novaeseelandiae* on *desolata* at the Auckland Islands, and *F.peregrinus* and *F.longipennis* on *turtur* in Australia (Serventy *et al.*, 1971), probably have little effect on population sizes.

Introduced predators, notably cats and rats, have had big effects, although prion burrows are too small for entry by cats. At Ile aux Cochons, Crozet Islands, prions and diving petrels are still abundant despite the cats that are believed responsible for the demise of larger Procellariiformes (Jouventin *et al.*, 1984). At Macquarie Island perhaps 26 000 Antarctic Prions are killed by cats annually, but here rabbits form a major prey and relieve the pressure on the birds. At the same island the few remaining Fairy Prions nest only on rocks just offshore where cats cannot reach them (Brothers, 1984). Rats appear to be less of a problem for prions, but there are no studies to evaluate their effects.

No species of prion is endangered today and those with small populations are generally protected by the remote and inaccessible nature of their breeding places. Where it is desired to attempt to increase populations locally, alien predators having been removed, the birds can be easily encouraged to use appropriately designed nesting boxes with low roofs that can be slid into rock crevices or buried (pers. obs.).

CHAPTER 6
The Genus Bulweria

Bulwer's Petrel.
Photo: Francis Roux.

Contents

I Introduction

The two living birds in this genus — Bulwer's Petrel, *B.bulwerii* and Jouanin's Petrel, *B.fallax* — are among the least known of Procellariiformes, the precise breeding place of the latter species still being undiscovered.

Although *Bulweria* has been thought to show primitive characters, the fossil record is poor. Olson (1985d) mentioned a humerus of Upper Oligocene age from South Carolina rather similar to that of *B.bulwerii*, but the only described species is *B.bifax*, Olson 1975a, from Pleistocene deposits on St Helena Island. Unfortunately the material comprises only post-cranial elements but Olson concluded that the bird had a mixture of pterodromine and bulwerine characters, viz, short, heavy leg bones like *Bulweria* but slender ulnae like *Pterodroma*, the humerus being intermediate between these two — hence *bifax* (two-faced). He floated the idea that this bird could have been close to the ancestor of those genera.

Bulweria resemble *Pterodroma* in bill structure, and between 1948 and 1965 all gadfly petrels were often placed in *Bulweria*, following Mathews (1948), but distinctive characters of the genus have been recognised. Features that apparently ally it with *Pterodroma* were noted by Pycraft (1899) and Kuroda (1954). More recently Kuroda (1983) has compared many skeletal characters of *B.bulwerii* with those of other Procellariids and concluded that the bird's nearest relationships are with *Pterodroma* (see also Fig. 6.1) rather than with *Pachyptila* and *Procellaria* as sometimes suggested.

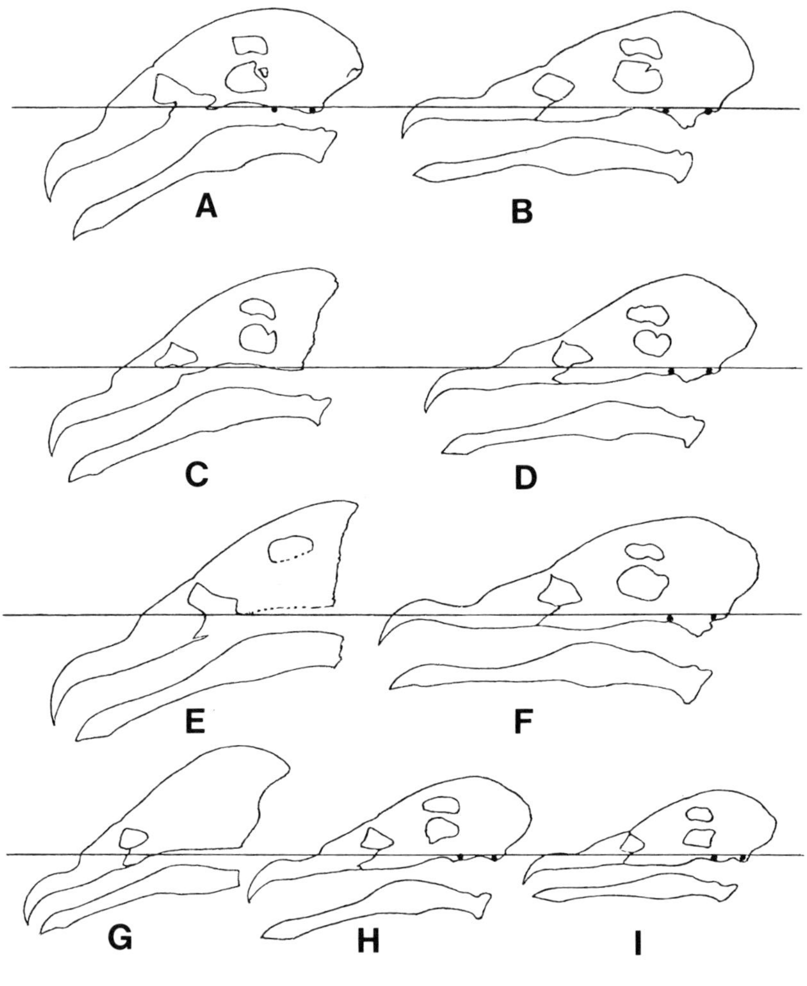

Figure 6.1 *Skull outlines of* Bulweria *and* Pterodroma, *partly from X-ray photographs. Black dots represent lead pellets used to show the basitemporal plate as the basis for comparing the horizontal planes. A.* Pterodroma rupinarum *fossil species from St Helena; B.* P.arminioniana; *C.* P.aterrima; *D.* P.alba; *E.* P.r.rostrata; *F.* P.baraui; *G.* P.r.becki; *H.* Bulweria fallax; *I.* B.bulwerii. *Note the heavy, down-angled bills of the* rostrata *group* [Pseudobulweria *in Imber's classification (1985)]* and the similarities of Bulweria *to the other* Pterodroma. *From Olson (1975a).*

II General characteristics

These are small, long-winged petrels of subtropical and tropical seas. *Bulweria bulwerii* is the smallest Procellariid, not much larger than the biggest storm petrel; *B.fallax* the size of a small *Pterodroma*. Bulwer's Petrel weighs a mere 95 g, wing span 65 cm. Jouanin's Petrel is heavier billed and nearly twice the weight of the other at about 173 g, span about 82 cm. Measurements are compared in Zonfrillo (1988). In both species the plumage is dark sooty brown above and below, the tail rather long and wedge-shaped. The bill is black and heavy-nailed, with a notch at the junction of the latericorn and upper nail (Fig. 6.2) as in some *Pterodroma* (Fig. 4.3). In *B.fallax* the feet are flesh-coloured with blackish edges to webs and toes distally, those of an Italian specimen bi-coloured (Olson, 1985c). The foot colour of *B.bulwerii* is variable. Many of Madeiran birds are greyish-flesh, others bi-coloured grey and pink, and in some the feet are bright pink all over (Zonfrillo, 1988). Bi-coloured feet are found elsewhere among petrels only in *Pterodroma*.

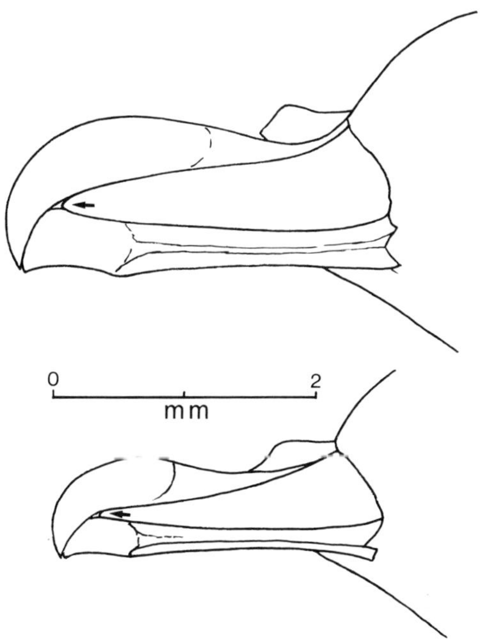

Figure 6.2 *Bills of Jouanin's Petrel (top) and Bulwer's Petrel. Note rhampothecal notch (arrowed).*

The genus was originally defined on the unicolorous dark plumage, but this was later found to be common to several species of *Aestrelata* (=*Pterodroma*). The skeleton of *Bulweria* is so similar to that of *Pterodroma* that Olson (1975b) cast doubt on the validity of the genus. However, a constant feature in the skeleton is that the prominent pre-frontal (lachrymal) bone is free and not fused to the rest of the skull as in fulmars, prions and gadfly petrels (Bourne, 1975). *Bulweria* is also distinct in its combination of tail shape, voice and lack of aerial displays. However, in the absence of any confirmatory biochemical and other evidence from modern techniques, the systematic position of the genus remains somewhat conjectural. Bulwer's Petrel has sometimes been considered primitive on account of its free prefrontal bones, dark 'elementary' plumage, simple voice and displays, and nesting in natural cavities instead of digging burrows for that purpose.

The two species do seem to be distinct, rather than their being large and small forms of one species. For example, each is stated to have its own, unique species of *Haliperus* feather louse (Timmermann, 1960).

Little field data have been collected on these birds; indeed *B.fallax* was only named in 1955 (Jouanin, 1955), although described (in error as *Aestrelata aterrima*) by van Oort (1912), with specimens in the collections of the Leyden Museum for nearly 50 years (Jouanin, 1957). Its precise breeding places are still unknown, though believed to be in Southern Arabia (see below). The most comprehensive account is that of *B.bulwerii* at Madeira and the Salvage Islands (Jouanin *et al.*, 1979).

In hand and field the sexes look alike, but males are the larger in all dimensions except wing length (Robertson & James, 1988). The moulting sequence has not been described, but the birds do not shed feathers while nesting, so that moult evidently takes place on the wintering grounds.

III Life styles

Both species inhabit warm seas. South of Hawaii King (1970, Fig. 34) found Bulwer's Petrel most abundant in waters at 24–27°C, the greatest densities being encountered where the air temperatures were 28–29°C.

Bulwer's Petrels have a very buoyant but erratic flight, lacking the fluttering action of a storm petrel and with wing strokes of greater amplitude than those of a shearwater, being lifted quite high on the upstroke to give a few jerky downbeats followed by a glide or a sideways jinking action, and Harrison (pers. comm.) noted its flight appearance as rather like that of a noddy tern (*Anous* sp.). The birds have also been seen hanging in the wind, using a few lazy beats while facing the breeze, like gulls, a manoeuvre impossible for most other Procellariids, but possible for *B.bulwerii* because of its very low wing loading. Jouanin's Petrels are less buoyant. They fly fast with a swooping flight and may rise high above the waves like a gadfly petrel.

At sea Bulwer's Petrel is generally encountered singly but is sometimes in small dispersed groups and/or in association with Wedge-tailed Shearwaters and other tubenoses. In the Central Pacific only 58 (9.3%) of 624 birds were in flocks (Fefer *et al.*, 1984).

On the ground the birds tend to shuffle along, but their low wing loadings and long, flexible tails should give them good control during landings and take-offs, but these are undescribed.

A peculiarity of Bulwer's Petrel is its tiny vocal repertoire, the sole call being a low, barking 'woof' (c.f. Pl.X in Jouanin *et al.*, 1979), a call frequently repeated by birds on the ground and common to both sexes (Wetmore, in Amerson & Shelton,

1976). It can be heard day and night and birds in burrows readily respond to human imitations particularly during the pre-egg period. The more musical and complex call described by Ogilvie-Grant (1896) was that of a Little Shearwater. Jouanin *et al*. (1979) and Thibault & Holyoak (1978) thought that Bulwer's Petrels do not call from the air, behaviour in contrast with that of *Pterodroma*. However, its simple barking call has been recorded occasionally from flying *bulwerii* in Hawaii (Kepler *et al*., 1984).

Virtually nothing is known of courtship or other sexual behaviour. Lockley (1952) reported a pair without an egg using a quickened version of the barking call and Wetmore (in Amerson & Shelton, 1976) saw birds resting a few cm apart with necks outstretched and throats distended while giving their low barks.

IV Distribution

Bulwerine petrels occur in all the ocean basins. They breed on the North Atlantic islands, including a few on Madeira; on some islands of the Central Pacific; on islands of the Hawaiian Archipelago, notably Nihoa; on the Bonin and Volcano Islands, and on rocks and islets around Taiwan and off the south-east coast of China (Fig. 6.3).

The only *Bulweria* nesting in the Indian Ocean is Jouanin's Petrel. Exactly where is unknown, but three birds, two still with nestling down adhering, were found in the desert 70 km north of Salalah, Oman (Fig. 6.4) on 5 December 1978 (Walker, 1981). Another, probably a young bird, was picked up in December inland to the north-east of the Kuria Muria Islands (Bourne, 1987b), but whether the species nests high in the coastal mountains or on islands offshore, is conjectural (Zonfrillo, 1988).

Bulweria fallax is abundant off this part of the Arabian coast and around the Kuria Muria Islands, and off Masirah from March to August, taking advantage of a flush of food accompanying coastal upwellings generated by the south-western monsoon winds, a situation that may parallel that of *Oceanodroma hornbyi* which seems to nest in the Peruvian desert within range of the plankton-rich upwellings off that coast.

Bailey (1966) found *B.fallax* the commonest and most widespread seabird from June to August off south-east Arabia and up to 320 km from land (Fig. 6.4). *Bulweria bulwerii* is likewise abundant south of and around the Hawaiian islands in the breeding season, May to July (King, 1970), and at least 3000 birds were seen near another probable breeding place, the Sento Shosho Islands (28°48′N; 123°36′E) on 24 May 1972 (Simpson, 1973).

V Feeding and foods

Little information on the feeding style of either species has been reported and it is often stated that they feed mainly at night, but I have found no direct evidence for this. Fefer *et al*. (1984) classified Bulwer's Petrel as a surface seizer, sitting on the water with spread wings and head submerged. However, Jouanin *et al*. (1979) drew

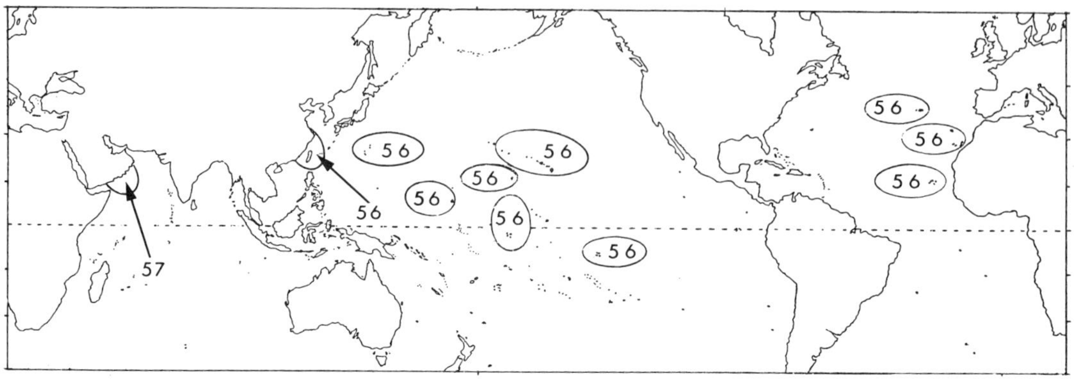

Figure 6.3 *Breeding distribution of Bulwer's Petrel (56) and Jouanin's Petrel (57).*

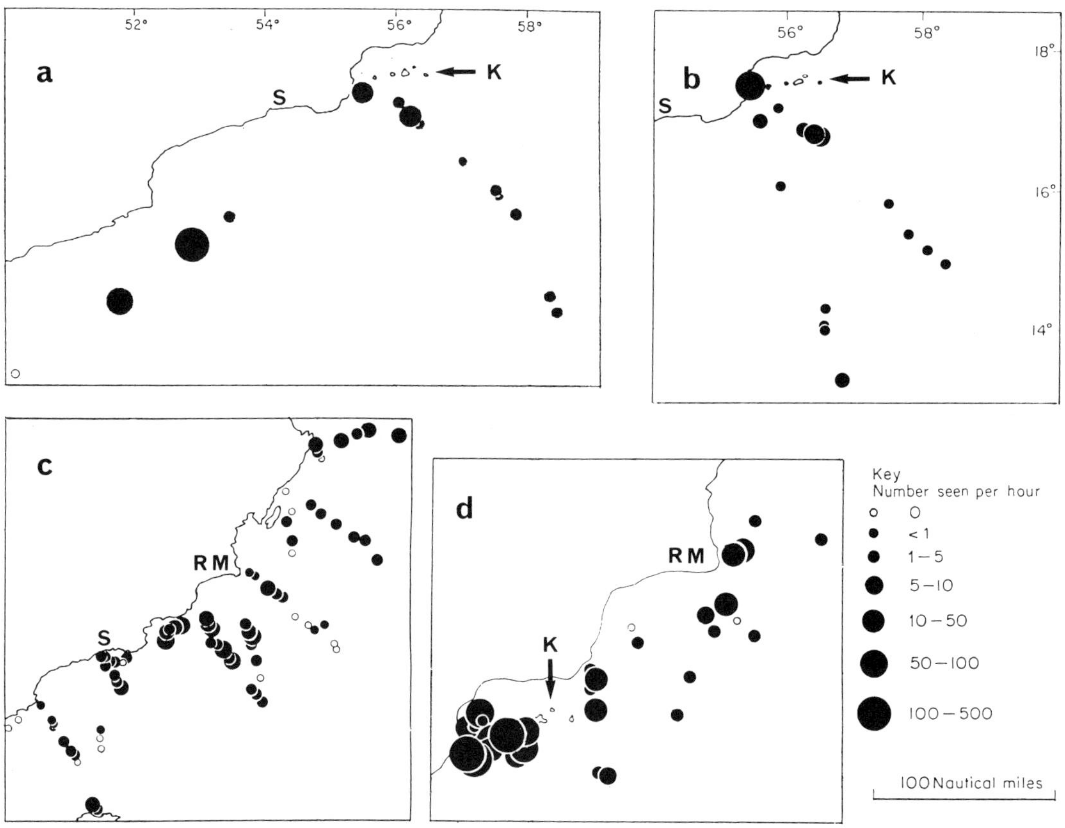

Figure 6.4 *Sightings of Jouanin's Petrel off the south-east coast of Arabia. Based on Bailey (1966). (a) 8–13 March 1964. (b) 25–28 May 1964. (c) 25 June–20 July 1963. (d) 30 July–7 August 1963. K = Kuria Muria Is.; RM = Ras Madraka; S = Salalah.*

attention to the low wing loading of this species which has a higher Hartman's Buoyancy Index than any other tubenose, storm petrels included, so diving seems unlikely. Such buoyancy, together with the long tail for manouevrability, seem to fit the bird for feeding by dipping. Perhaps significantly, both in the North Atlantic and in the Pacific, breeding occurs when fish like bonito and tuna are shoaling nearby: at such times many seabirds converge to feed on fish, including flying fish, driven to or from the surface by these predators. Jouanin's Petrel also associates with whales, dolphins, shearwaters and phalaropes off the South Yemen coast.

Analyses of the foods are given by Fefer *et al.* (1984), Harrison *et al.* (1983) and Zonfrillo (1986). The stomachs of Bulwer's Petrels at Laysan Island contained mainly fish of the families Myctophidae and Sternoptychidae, fish eggs (mostly of Exocoetidae), squid and crustacea, in that order of abundance. Fefer *et al.* (1984) estimated that the north-western Hawaiian population of Bulwer's Petrels would consume about 2520 tonnes of these foods per annum out of some 409 000 tonnes taken by all the seabirds of those islands.

VI Breeding

The principal studies of breeding by *B.bulwerii* are those of Jouanin *et al.* (1979) and Mougin (1989) made at Grande Salvage Island in the North Atlantic and of Amerson and Shelton (1976) at Sand Island, Johnston Atoll, in the North Pacific. The timing is set to take advantage of seasonal upwellings and food flushes in these oceans, but at the Marquesas, where there is little seasonal change in marine conditions, the laying period is extensive (Jouanin *et al.*, 1979).

In general, despite considerable gaps in our knowledge, the breeding patterns seem typical of the Procellariidae, with retention of mate and nest site from year to year. The birds are secretive and strictly nocturnal on land. The age at which breeding starts is not known but one 3-year-old and four 4-year-olds returned to their natal island about 2 months later than the established breeders, some of them developing brood patches and engaging in courtship duets, so evidently these birds must be at least 5 years old before breeding (Amerson & Shelton, 1976). Mougin (1989) deduced that breeding began at about 7 years old.

One difference from many of their family is that, although nesting in concealment, Bulwer's Petrels do not dig their own burrows. They typically lay in natural or man-made cavities, in fissures or holes in rocks, under abandoned building material or the shells of dead marine turtles, or on the surface concealed by grass or other vegetation. When the substrate is soft the egg is laid in a small depression made by the bird. Some that lay in burrows appear to have taken over those of other species like *P.hypoleuca*, *P.lherminieri* or *P.marina*.

In the northern hemisphere nesting occurs in the spring and the laying period is rather short, Amerson & Shelton noting that at Johnston Atoll Bulwer's Petrel was the only seabird to have a restricted season. There they laid over 6 or 7 weeks but hatched only over about 3 weeks, later-laid eggs not surviving. Further south, at the Marquesas, nesting is in the southern summer and possibly laying there is more extended for Holyoak & Thibault (1984) found eggs from September to December, in which month there were not only eggs but also chicks in all stages of development. At the Cape Verde Islands laying may be a little earlier than around Madeira, as de Naurois (1969) found a fresh egg in late January and several at various stages of incubation on 6 March. Bourne (1955) pointed out that seasonal upwellings reached a maximum off the Cape Verdes early in the year but near the northern Atlantic islands did so around midsummer.

Holyoak & Thibault noted that the nests had a strong, distinctive and nauseating odour different from the usual musky smell of petrels: this facilitated their finding of occupied sites.

Nest densities may be quite high, e.g. at the Desertas Islands within 20 cm of one another among rock falls (Jouanin *et al.*, 1979), but elsewhere nests are often widely scattered and sometimes cheek-by-jowl with those of other species like *O.castro*, *P.pacificus* and *P.nativitatis*.

The egg is about 22.6% of the adult's weight (Jouanin *et al.*, 1979), not out of line for a 96 g petrel (Fig. 13.4) but is a greater ratio than that of any other Procellariid and almost within the storm petrel range.

Incubation and chick rearing is shared by the parents, incubating in shifts of 9.5 days (Mougin, in press). Whether there is a pre-laying exodus and how long incubation takes are not recorded. Nestling periods at Johnston Atoll ranged from 57–67 days, averaging 62 days ($n = 8$) according to Amerson & Shelton (1976), who also stated that the chicks were abandoned for several days before they flew. This is unexpected in a species that appears not to undertake any very extensive and urgent post-nuptial migration and requires confirmation.

Productivity at Johnston Atoll was high during Amerson & Shelton's 6-year study, and they believed that their population was increasing. Hatching success ranged from 50–93% (average 67%) and fledging success from 67–100% (average 85%) over those years; i.e. of 71 eggs laid at least 45 (63%) chicks fledged. The demography of the species at Grand Salvage Island is detailed by Mougin (1989).

Unusual behaviour recorded from Laysan Island is the formation of large 'rafts' of 'at least 1000 birds' sitting around in the open after dark (Ely & Clapp, 1973). This was in June at the height of incubation. The status of the birds was unknown: they could have been pre-breeders.

The dates of fledgling *B.fallax* found in Oman show that these flew in November and December, apparently 2 months later than Atlantic *bulwerii* (Zonfrillo, 1988).

VII Dispersal and migration

Bulwer's Petrels are absent from both their Hawaiian and North Atlantic colonies and adjacent seas for about 5.5 months following breeding. As ringed birds have only been recovered at the nesting places, evidence for dispersal is based on sightings at sea.

Outside the breeding season Bulwer's Petrels have been identified off Brazil where upwelling conditions obtain (van Oordt & Kruijt, 1953; Bourne 1955) and Harris & Hansen (1974) found them abundant near Fernando de Noronha in late October. It seems that many winter off these coasts, travelling south-west from the northern islands, others dispersing far to the south of the Equator. Some remain in these winter quarters during the subsequent breeding season. A southwards movement evidently also takes place from the North Pacific as these petrels are absent from seas around and south of the Hawaiian Archipelago in the winter (King, 1970), but good sightings are rare. Some evidently travel through Indonesian waters into the eastern Indian Ocean. For example, considerable numbers were seen in early November off north-west cape, Western Australia, the Exmouth Rise south of Christmas Island and within 95 km of Java (Harrison, pers. comm.). There is also a specimen record from the Maldive Islands.

The pattern for *B.fallax* seems to be rather similar in that the birds leave the seas off Arabia to move south, with some crossing the Equator. Jouanin's Petrel has provided one of the more spectacular instances of long-distance vagrancy. While Bulwer's Petrels regularly invade the Indian Ocean, a male *B.fallax* reversed the pattern, being collected on 4 September 1967 on Lisianski Island, Hawaii (Clapp, 1971). One was found at the head of the Adriatic in November 1953 (Giol, 1957): this was said to have been accompanied by two others. Bourne (1987) suggested that these *B.fallax* might have overshot their breeding places on their northwards migration and entered the Mediterranean via the Red Sea. There are sight records of the species from that sea.

VIII Populations and mortality

With breeding places unknown, the size of the population of *B.fallax* must also be conjectural. Nevertheless, large numbers of these birds have been seen on a single day off Arabia so that the population would appear to be substantial.

Some colonies of Bulwer's Petrel at the North Atlantic islands are small or very small, e.g. only *c.* 100 pairs on Montana Clara in the Canary Islands, and *c.* 10 pairs on Razo and Cima in the Cape Verdes, although the species is still plentiful at the Desertas and Salvages. Some of the North Pacific populations are also small, e.g. most of those on the Bonin and Volcano Islands, but the world's largest aggregation appears to be on Nihoa Island, Hawaii, with an estimated 75 000–100 000 pairs (Fefer *et al.*, 1984). These authors computed that some 412 000 breeding and non-breeding Bulwer's Petrels were based on the Hawaiian Leeward Chain.

There are few data on natural causes of mortality in *B.bulwerii*. At Johnston Atoll Amerson & Shelton (1976) considered that flooding of nests during high tides was a main cause of nesting failure. Natural predators recorded include *Tyto alba* at the Cape Verde Islands, gulls of various species in various places and probably land crabs on islands near Oahu and Maui (Kepler & Kepler, 1980). At the Marquesas crabs take chicks from surface nests (Holyoak & Thibault, 1984).

As usual, man's introductions have had the greater effects. The population at Midway Atoll has disappeared with the build-up in *Rattus rattus* since 1943, and the large population of Lisianski Island has gone due to predation by rats and perhaps mice, while cats have depleted some colonies off the main Hawaiian Islands.

Direct predation by man has also been important. The Hawaiian Polynesians considered the chicks of the 'Ou' a great delicacy, taking them from islets close to the main islands, and Bulwer's Petrel has also been eaten in China. At Corvo in the Canary Islands they were killed for their oil (Bolle in Bannerman, 1914) and Lockley

(1952) found that the quarrymen in Baixo, off Porto Santo, Madeira ate them extensively. Some such predation evidently still goes on for according to Le Grand *et al.* (1984) the seabirds at the Desertas are still heavily exploited and in July 1981 fisherman milled hundreds of live *B.bulwerii* for fishing bait.

Another potential hazard for Bulwer's Petrel around Hawaii is the excessive harvesting of Opelu or Mackerel Skad *Decapterus* spp. for fishermen's bait. This is an important food for this and other petrels (Fefer *et al.*, 1984), and probably becomes available to them when driven to the surface by tuna. The development of an intensive bait fishery based on Opelu during April to October could have a depressing effect on the ability of breeding *B.bulwerii* to raise their chicks.

Long-lived birds like petrels can withstand quite heavy cropping of their young, but uncontrolled killing, often of adults, as still apparently continues on some of the islands of Macaronesia, may be as fatal to the remnant populations as predation by rats and cats. The big populations on Nihoa and Laysan Islands seem to form the main reserve for the species and under the control of the US Fish and Wildlife Service would seem to have a good future. On some of the low lying sand atolls populations seem to be limited by the availability of suitable nest sites. Bulwer's Petrel appears to prefer to lay under cover, perhaps to escape the tropical heat, and provision of well-designed nest-boxes, perhaps made in concrete giving protection overhead and suitable cells or niches below, could well increase populations locally, e.g. at Johnston Atoll.

CHAPTER 7
The Genera Procellaria *and* Calonectris

Cory's Shearwater.
Photo: author.

Contents

I Introduction

Apart from *Diomedea, Procellaria* was the first petrel genus to be described by Linnaeus and all the then known species, including the storm petrels, were placed in it. Subsequently it remained the genus name for the latter birds and was still in use for *Hydrobates* as recently as Godman (1907–1910). However, Gray in 1840 showed that Linnaeus' *Procellaria* was the Shoemaker *Procellaria aequinoctialis*.

Procellaria and *Calonectris* have been loosely associated in many classifications, primarily on similarities in size, bill structure, and in the general resemblances of plumage between the Grey Petrel *Procellaria cinerea* and *Calonectris*. This last genus has usually been separated from *Puffinus* since the discussion by Alexander *et al.* (1965). Kuroda (1983) analysed the skull structures of *P.aequinoctialis* and *C.leucomelas* and concluded that the genera are indeed related, with the former having some special features of its own and others suggesting a link with the fulmars. He regarded the *Procellaria* skull as being built on distinctly shearwater lines.

II The genus *Procellaria*

The four members of this genus are large petrels with particularly robust, heavily-hooked bills (Fig. 7.1). These are usually white or cream in colour with varying amounts of black along the ridge of the culmen, in the mandibular grooves and on the nails. All are blackish-brown in plumage except for the Grey Petrel or Pediunker *P.cinerea* a bird formerly placed in the monotypic genus *Adamastor*. This bird is basically grey above and white below, its bill pale with little black and side plates a grey-green shade. In all four the nasal tubes are upstanding, open forwards with rounded apertures, their free edge V-shaped in dorsal view. The tubes are only slightly bevelled in contrast to those of *Calonectris* and *Puffinus*.

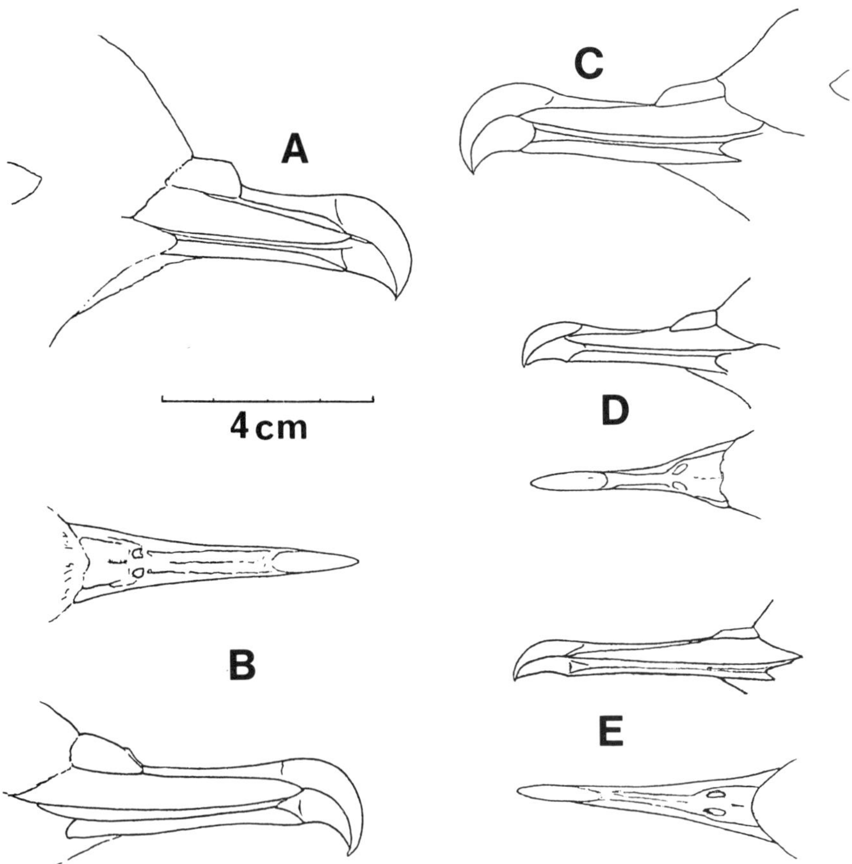

Figure 7.1 *Bills of (A)* Procellaria cinerea; *(B)* Calonectris d.borealis; *(C)* Puffinus gravis; *(D)* P.lherminieri; *(E)* P.huttoni.

With the exception of the giant petrels, these are the largest birds in their family. The Shoemaker or White-chinned Petrel *P.aequinoctialis* is the biggest with a wingspan of some 138 cm, the wings of the smallest, the Grey Petrel, span 120 cm.

The only fossil tentatively assigned to this genus is a part humerus of Pliocene age from South Africa (Olson, 1985b). Bourne (1967b) considered subfossil humeri from the Chatham Islands to be from a *Procellaria* of similar size to *P.a.conspicillata*, and as there were two immature bones, the bird (?*P.cinerea*) could have bred.

A *General characteristics*

The Shoemaker has long been known to sailors in the Southern Ocean and was named by Linnaeus for 'The Great Peteril', described and illustrated by Edwards (1747). *Procellaria* have generally been classified close to *Puffinus*, but are quite distinct in a number of features such as voice (Warham, 1988b) and in their blood plasma proteins and esterases (Harper, 1978) and Fig. 7.2. Harper suggested that their closest relatives were not the shearwaters but the albatrosses. Imber (1985a) related *Procellaria* to '*Pseudobulweria*' (= *Pterodroma rostrata* + *P.aterrima*) because both share the peculiar rhamphothecal notch at the junction of latericorn and unguis (see Mathews, 1912–1913, p.111), by similarities in the septa between the nasal tubes, by both having rather laterally compressed tarsi, partly unfused

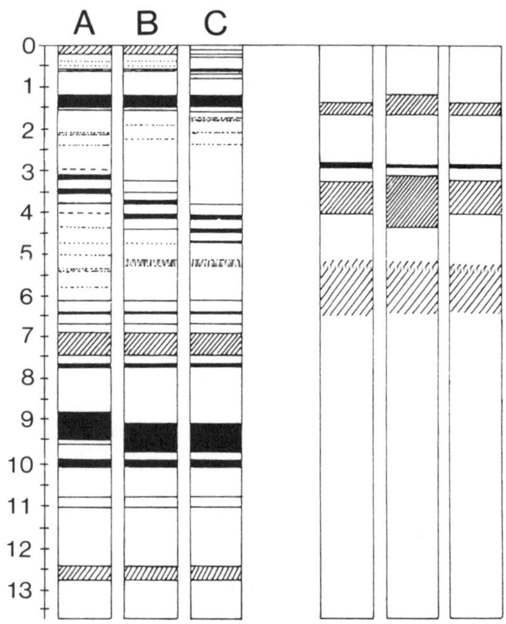

Figure 7.2 *Blood plasma proteins of* Procellaria; *esterase patterns in right-hand columns. (A)* P.westlandica; *(B)* P.parkinsoni; *(C)* P.aequinoctialis. *For key to components see Appendix 2. From Harper (1978), courtesy of* New Zealand Journal of Zoology.

prefrontal bones, and by some other skull features. Imber hypothesised an ancestor of the *Procellaria* type for '*Pseudobulweria*', with *Bulweria* arising from the same stock.

Kuroda (1954) associated *Procellaria* with *Calonectris* and *Puffinus*, noting osteological similarities with both. He commented on the similar plumages of *Calonectris* and *P.cinerea* and pointed out that, whereas *P.aequinoctialis* is an aerial form, gliding well like *Calonectris*, *P.cinerea* differs in diving frequently and in more frequent beating of its wings. He also noted that the tongue is spiny in *Procellaria* and *Puffinus*, smooth in *Calonectris*.

Shoemakers of Inaccessible Island, Tristan da Cunha, have distinctive bands of white feathers on their cheeks which often extend across their crowns (Fig. 7.3).

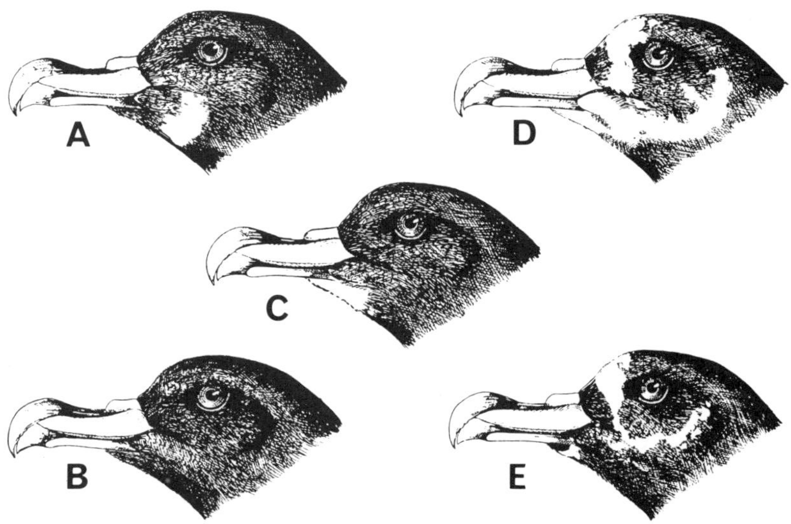

Figure 7.3 *Shoemaker heads showing variation (A) unusual pattern from bird taken at sea, (B) bird lacking white on chin from New Zealand, (C) typical example, (D) and (E)* conspicillata *from Inaccessible Island. Partly from Falla (1937).*

These birds belong to the slightly smaller 'spectacled' form *conspicillata*. Although off-white feathers are also occasionally seen on the heads and bodies of *P.westlandica*, the vocal repertoire and ecology of that bird are quite different from that of the Shoemaker and the two are good species.

Some of these vocalisations have been analysed by Brooke (1986a) and Warham (1988b). Although shearwater-like moaning cries are used, the principal calls tend to be hard, staccato rattles, quacks and clacks which carry far. Shoemakers incorporate bill clicks into their calls — hence that whaler's name for the bird — it sings like a cobbler at his last. The very distinctive call of the Grey Petrel was accurately described by Hutton (1865) as like the bleating of a lamb.

There is some sexual dimorphism in size among *Procellaria* but less than that in

Calonectris, males generally being only slightly bigger than females, see Barrat (1974) and Warham & Bell (1979).

B *Life styles*

These petrels are all powerful fliers and often follow ships to scavenge food. In recent years Westland Petrels have become increasingly dependent in the breeding season for offal discarded from fishing boats within range of the breeding sites (Bartle, 1985). Shoemakers in the Benguela Current also rely heavily on trawler offal (Jackson, 1988). Grey Petrels get much of their food by pursuit and surface plunging, the others also dive but surface seizing is their more usual method. Offal apart, all take many squid, some fish and fewer planktonic crustacea, much at night when their prey are at the surface (Imber, 1976a). Imber believed that this was facilitated by dorsal bioluminescent organs on many of the cephalopod prey species, though species lacking these organs are also eaten.

Both Parkinson's and Westland Petrels tend to crash land through the forest canopy on returning to their nests, doing this without injury, and despite their size they can climb tree trunks to gain height for take-off. On the ground they are rather clumsy and if cornered lunge vigorously and bite hard.

C *Distribution*

Calonectris breeds in warm northern waters and the two New Zealand endemics, *westlandica* and *parkinsoni* also frequent subtropical seas, *westlandica* feeding over continental shelves, *parkinsoni* over deep water (Bartle, pers. comm.). The other *Procellaria* have circumpolar distributions, breeding on widely separated islands in the Southern Ocean (Fig. 7.4).

Both the New Zealand endemics have very restricted breeding ranges, their habitat temperate rain forest. The Westland Petrel burrows in only a few hectares of coastal hills on the north west coast of the South Island, Parkinson's Petrel in similar habitat on Little and Great Barrier Islands off the north east coast of the North Island. Presumably their populations were once much larger and *P.parkinsoni* bred inland in both islands of New Zealand in historic times. It also bred in spurs of the coastal hills of the South Island very close to those used by Westland Petrels today and the two species were then sympatric (Falla, 1946).

In the South Atlantic in the austral spring, Rumboll & Jehl (1977) found Shoemakers widespread in deep water from Cape Horn to about 30°S and 35°W. Grey Petrels were also widely distributed through this region, again over deep water, but more particularly along the subtropical convergence. The largest group (of 13 birds) was associated with a herd of pilot whales.

D *Breeding*

Studies of breeding biology include several on *P.aequinoctialis* (Mougin, 1970b, 1971; Hall, 1987), a few on *westlandica* (Jackson, 1958; Baker & Coleman, 1977), one

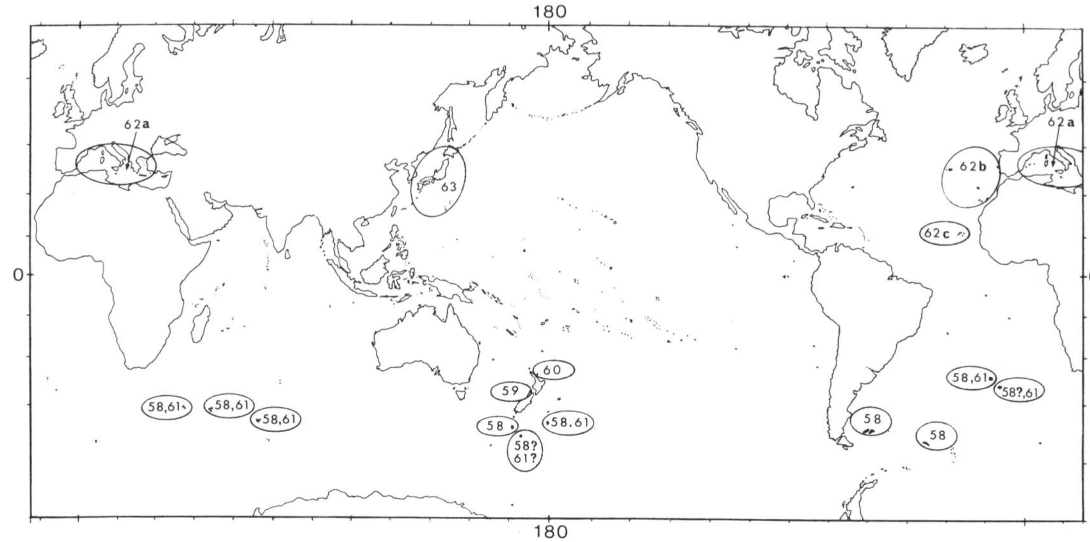

Figure 7.4 *Breeding distribution of* Procellaria *and* Calonectris: *58.* P.aequinoctialis; *59.* P.westlandica; *60.* P.parkinsoni; *61.* P.cinerea; *62a.* C.d.diomedea; *62b.* C.d.borealis; *62c.* C.d.edwardsi.

on *parkinsoni* (Imber, 1987) but very little on *cinerea* (Barrat, 1974). The energetics of chick-rearing by *P.aequinoctialis* were examined by Berruti *et al.* (1985).

Over land all are usually nocturnal visitors with the exception of the Grey Petrel. Paradoxically this, the smallest, weakest and least aggressive of its genus, often comes ashore in daylight despite the presence of predatory skuas. Some Shoemakers may also land before dark. The two New Zealand endemics, however, which, until the recent advent of man-introduced pests appear to have had no enemies on land, are nonetheless strictly night-time visitors to their nesting places.

Breeding seasons vary. Shoemakers and Parkinson's Petrels nest in the summer, Grey and Westland Petrels in the winter. At several places (Fig. 7.4) Shoemakers and Grey Petrels are sympatric, the latter laying before the other's chicks have fledged: the possibilities of burrow-shading and competition between them have not been explored.

The birds are all colonial breeders in burrows, the Shoemaker being the biggest tubenose to nest in this way. Its burrow has a distinctive feature — the egg is laid on a raised saucer-shaped platform within the nest chamber, the surrounding gutter often being half-filled with water (Hutton, 1895). At South Georgia water from melting ice may drain away down the approach tunnel (Murphy, 1936, p.644). On Inaccessible Island the burrows often appear to be deliberately sited on marshy land. The tunnel is usually driven into rising ground so that, while the lower section may hold water, the chamber higher up is relatively dry (Rowan *et al.*, 1951) (Fig. 7.5). Shoemakers are powerful birds and on l'Ile de l'Est, Crozet have been found to take over rabbit burrows, transforming their warrens into Shoemaker colonies (Despin *et al.*, 1972).

Bartle (1987) has found that the sex ratio at the *westlandica* colony favours males

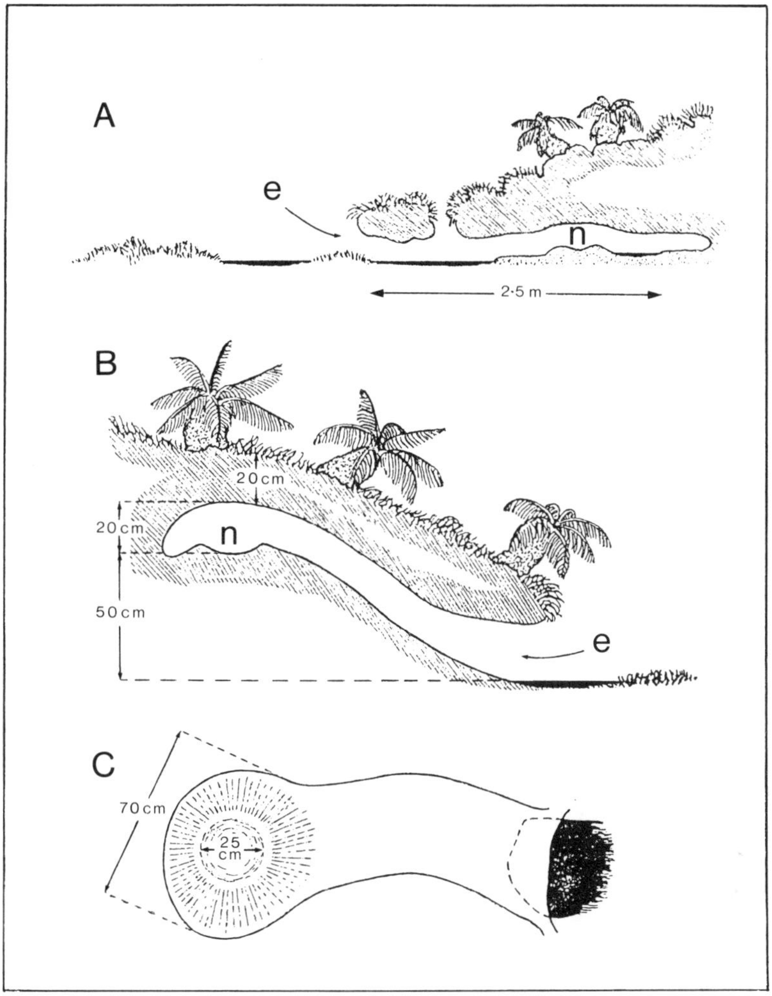

Figure 7.5 *Nests of Shoemakers at Inaccessible Island. (e) entrance; (n) nest cone. (A) section through burrow in bog with standing water; (B) and (C) lateral and vertical sections through nest on slope with pool at entrance. From Hagen (1952).*

by 2:1, and that males have much higher survival rates than females: 35% of males ringed as adults were still alive 17 years later.

Laying dates are very regular in Westland Petrels, eggs appearing over about 3.5 weeks (Bartle, 1985) but are spread over *c.* 4 weeks in Parkinson's Petrel and 8 weeks in Shoemakers (Mougin, 1975). Mougin (1970b) found that hatching occurred over a shorter period than laying because late layings were usually abandoned during incubation.

Other special features of the *Procellaria* breeding system include the long chick-rearing period. This takes 98 days in *aequinoctialis* (Hall, 1987), *c.* 107 days in *parkinsoni* (Imber, 1987), and averages 125 days in *westlandica* (Baker & Coleman,

1977; Bartle, 1985) and, according to the last author, some pairs take substantially longer than this. Such periods are very long even allowing for the large sizes of these birds. Another unusual feature is that *westlandica* chicks tend to be guarded for about 2 weeks, sometimes more (Jackson, 1958; Baker & Coleman, 1977) — again long for any burrowing petrel other then *Pelecanoides*. This does not apply throughout *Procellaria*, however, most Shoemaker chicks, for example, are not brooded after the first 24–36 hours (Mougin, 1970; Warham, unpublished).

That established *Procellaria* may miss a breeding season and have 'sabbatical' years like *Calonectris* (Chapter 11, Section VA), is suggested by the finding by Bartle (1985) that in some years Westland Petrels fail to breed. Some returned without laying, or if they laid, deserted their egg. Likewise Imber (1987) noted that only early-laying Parkinson's Petrels produced an egg the next year but their nestings either failed or their chicks were under weight. Bartle (pers. comm.) has suggested that poor breeding seasons for *westlandica* are partly due to 'El Nino' disturbances that reduced the squid and fish available within range of the colony.

E *Dispersal and migration*

Whereas the migratory classes of *Calonectris* shearwaters move south after breeding, those *Procellaria* that move significantly make longitudinal movements away from the nesting grounds or shift north. Shoemakers leave their oceanic feeding grounds in winter to feed in large concentrations over the continental shelves of South America, South Africa, and, to a lesser extent, of Australia. Earlier reports that *conspicillata* was mainly sedentary have been modified by analyses of winter sightings of this form in the South Atlantic (Enticott & O'Connell, 1985). None was seen near Tristan da Cunha; instead the 16 *conspicillata* identified in June and July were well west of Inaccessible Island, mainly over the South American shelf, the nearest 600 nautical miles from the breeding place.

Jehl (1973) found Shoemakers fairly common in winter within 65 km of the Chilean coast. There were even small flocks, mostly over deep water with surface temperatures of 8–13°C. In the Atlantic off Argentina the birds were widespread in winter but uncommon inshore and, while still not common, they became the dominant petrel beyond the continental shelf (Jehl, 1974b).

There seems to be little evidence for any regular pattern of movements during the southern summer of Grey Petrels, although it turns up in beach wrecks in southern Australia and New Zealand and straggles north along cool currents. Bartle (1990) has examined its occurrence off eastern New Zealand. This is the only *Procellaria* to be seen in the pack-ice, but it is rare there.

Westland Petrels shift eastwards into the Pacific after breeding, where they moult. Banded immatures have been recovered off the Chilean coast, where they presumably spend their first five years before returning to their colony (Bartle, pers. comm.). The position with the rare Parkinson's Petrel is clearer thanks to the specimens that have been taken in the contra-nuptial wintering grounds in the eastern tropical Pacific since 1905 when Rollo H. Beck collected the first of three birds near the Galapagos Islands (Loomis, 1918). Specimens have also been collected off Guatemala near where they have been seen taking bait from fish hooks and chasing Pomarine Skuas from food (Jehl, 1974a). Whether the whole population of Parkinson's Petrel is involved in these movements north is unknown. Those collected were undergoing their annual moult. A banded subadult bird was recovered on the coast of Peru (Imber, pers. comm.).

F *Populations and mortality*

The most abundant *Procellaria* is the Shoemaker, but no accurate census figures are available. Jouventin *et al.* (1984) considered that tens of thousands of pairs nested in the Crozet Group. Populations of similar magnitude were suggested for Marion and Prince Edward Islands (Williams, 1984), with perhaps 2 million pairs on South Georgia (Prince & Croxall, 1983). The Grey Petrel is much less common with 'thousands of pairs' at Marion and Prince Edward and the Crozets (Williams, 1984; Jouventin *et al.*, 1984) and perhaps 10 to 50 thousand pairs on offshore New Zealand islands (Robertson & Bell, 1984). There is also a large population on Gough Island (Elliott, 1970). The Westland Petrel totals about 20 000 birds (Bartle, pers. comm.) and Parkinson's Petrel but 500–1000 breeding pairs, most on Great Barrier Island (Imber, 1987). Imber estimated that there might also be another 2000 non-breeders.

These large petrels are usually strong and aggressive enough to deter alien predators like cats and mustelids, but their chicks are very vulnerable, although young Westland Petrels and Shoemakers have some ability to spit stomach oil if attacked. Predation by Southern Skuas is heavy in some places and at some times, the variation apparently due to the availability of other smaller and easier prey. Skuas apparently have difficulty in killing the large and aggressive Shoemakers but at Antipodes Island they form a considerable proportion of the victims of nesting skuas although, in attacks seen by day, the petrels have managed to deter their aggressors (Warham & Bell, 1979; Moors, 1980). Grey Petrels at Marion Island have survived despite the presence of cats but those formerly breeding at Macquarie Island evidently succumbed to the depredations of these and rats, dogs and probably Wekas *Gallirallus australis*. Being winter breeders, Grey Petrels would be particularly vulnerable then as most other seabirds are absent and invertebrate prey scarce.

The mainland populations of Parkinson's Petrel, now probably all gone, suffered severely from predation by dogs, cats, stoats and even by possums *Trichosurus vulpecula*, which took over their burrows (Imber, 1985b). Between 1971 and 1974 cats on Little Barrier Island killed a large proportion, if not all, chicks that were reared to the stage of emerging from their nests. By 1975 adults were being killed too. Fortunately an intensive cat eradication campaign between 1977 and 1980 was a great success (details in Veitch, 1985) although as yet (1989) that population has not recovered. Predation by local 'mutton-birders' on the Westland Petrels was suspected by Baker & Coleman (1977) but, according to Bartle (1985) in the last 20 years the population has doubled mainly, he believes, due to the ready availability of fish offal from trawlers offshore, offal that is also fed to the chicks. However, after growing at about 5% per annum since at least 1955, Bartle (1987) considers that the population has nearly stabilised.

Parkinson's Petrel was once a favoured 'mutton-bird' of the Maori and, apart from albatrosses and giant petrels, was the biggest petrel they ate. Any such activity as now continues on Great Barrier Island must be very minor. The Grey Petrel was one of the foods of the Tristan islanders but now the main island probably holds only 50 to 100 breeding pairs (Richardson, 1984).

III The genus *Calonectris*

The two species in this genus are large shearwaters, their wings spanning 100–120 cm; *C.leucomelas,* the Streaked or White-fronted Shearwater, breeds around Japan and weighs 500–750 g, *C.diomedea* weighs 600–950 g, and breeds in the North Atlantic islands and the Mediterranean. This latter bird has three well-marked subspecies: the Mediterranean Shearwater *C.d.diomedea,* nesting from the Balearic Islands to the Adriatic and Aegean, Cory's Shearwater *C.d.borealis* breeding off Iberia and on Madeira and the Salvage, Canary and Azores Islands, and the Cape Verde Shearwater *C.d.edwardsi.*

Although no fossil *Calonectris* has been described apart from an earlier Pliocene bone fragment from South Africa (Olson, 1985b) and bones of *C.diomedea* from Pleistocene cave deposits in South Wales (Harrison, 1977a), the Mediterranean Shearwater has been known since antiquity and Aristotle's 'cepphus' may have been this bird. It was also the 'Bird of Diomede' and the colony at the Tremiti or Diomede Islands in the Adriatic was referred to by classical writers and later by St Augustine. This petrel is also figured in colour in the paraphrase of the 'Ornithiaca' of Dionysius (? AD 200) attached to folios of the Dioscurides Herbal. According to Kadar (1978) the illustration is very unusual for this period in showing a bird in flight (Fig. 7.6). Presumably this was one of the earliest 'mutton-birds', and the fishy taste of its flesh was noted by some early writers, but I have found no account of any regular harvesting of the species during classical times.

Figure 7.6 *'Birds of Diomede'* C.d.diomedea. *From Dionysius (? 200 A.D.).*

A *General characteristics*

Mathews and Iredale (1915) erected the genus on a suite of characters including the distinctive plumage colour, the very slight lateral compression of the tarsi (which are as long as, or shorter than, the culmen) and the long, heavy bill (Fig. 7.1B). In all these features *Calonectris* differs from *Puffinus*. Kuroda (1954) noted other structural peculiarities such as the dorsally convex, broad and deep pelvis (Fig. 8.1), short, wide sternum, tibiotarsus with short crista cnemialis cranialis (more albatross-like than puffinine), the humerus rounded in cross-section not somewhat flattened as in *Puffinus*, and the spineless tongue. Shufeldt (1907) also examined the skeleton of *C.diomedea* and compared many features with those of *Puffinus*.

Both *Calonectris* spp. have pale, flesh-coloured feet and pale bluish or horn-coloured bills, often with an orange or reddish cast. Their plumage is greyish or grey-brown above, whitish below and the contour feathers of back and rump have pale edges creating a scalloped pattern on the upper parts.

Another difference from *Puffinus* (and from *Procellaria*) is the strong sexual dimorphism in size; the males being the larger allowed Ogilvie-Grant (1896) to sex members of pairs in the field as also did Ristow & Wink (1980) when working with *C.d.diomedea*. From 50 adult Cory's Shearwaters of either sex collected at the Azores, Murphy (1923) gave mean wing lengths for males (females) as 351.5 (344.3); tail 130.2 (129.3); culmen 55.5 (52.9); tarsus 57.0 (54.6); middle toe and claw 71.6 (69.0). Figures for Mediterranean Shearwaters, e.g. those of Ristow & Wink (1980), show similar degrees of difference, greatest in bill dimensions, those of adult males being 5–7% longer. They used a beak shape index of bill length × height to separate the sexes of birds in the hand. Mayaud (1934) also noted sexual differences in bone dimensions. The dimorphism is even greater in Streaked Shearwaters, bill lengths of males being 13% greater than those of females (Yoshida, 1981).

Moult has been little studied, but breeding Cory's Shearwaters have moulted their flight feathers by the time of their arrival at the nesting places where contour feathers start being shed from about the time of hatching (Jouanin & Roux, 1966).

Throughout their range Streaked Shearwaters appear to be of one form whereas populations of *C.diomedea* show differences in size and bill colour, hence their separation into three geographically separated subspecies. Even within these categories there may be differences in mean sizes between different island populations and, evidently, the birds are strongly philopatric.

B *Life styles*

The flight of both species is strong, with albatross-like banking and gliding without wing beats in strong winds and with rather lazy deep wing strokes interspersed with long glides close to the surface in light airs. Kuroda (1954) correlated the light skeleton and other structural specialisations of *Calonectris* with its aerial flight style and reduced capacity for diving, as contrasted with highly aquatic feeders such as Sooty and Short-tailed Shearwaters whose rather flattened tarsi and humeri and shorter wings facilitate sub-surface fishing. Despite these adaptations to gliding,

C.leucomelas that nest in forests climb trees well to facilitate their departures. On the ground the birds shuffle along with their tarsi but they can blunder off if alarmed, flailing their wings to speed their escape. At sea these petrels are often widely dispersed, but they may form 'rafts' on the water during migrations or when concentrations form around localised food sources.

Sexual dimorphism also extends to the voice, as Lockley (1952) suggested, and Ristow & Wink (1980) have shown for the Mediterranean Shearwater; the female, as with shearwaters, has the lower-pitched voice. The vocalisations are complex and await proper analysis. Those of *leucomelas* include very high-pitched as well as low-pitched phrases, some rather like calls of *C.diomedea*. Both species sing from the air as well as from the ground. In this respect they differ from *Procellaria* which is not very vocal when flying over its breeding grounds. The Portuguese and Spanish 'cagarra' and 'pardela' are said to be onomatopoeic representations of the calls of *C.diomedea*. The name for *C.leucomelas*, 'Aho-tori' (foolish bird), used by Japanese fishermen, presumably refers to the ease with which it becomes enmeshed in their nets.

Wink *et al.* (1982) suggested that *C.d.diomedea* used echo-sounding when approaching its nesting cave on dark nights.

C *Distribution*

Calonectris shearwaters are birds of northern subtropical seas (Fig. 7.4). Although suitable nest-sites are available, Cory's Shearwater is strangely absent from Little Salvage Island only a few km from the main island of that group where many breed (Jouanin & Roux, 1966). In summer Streaked Shearwaters are concentrated around the Japanese islands and the Korean Peninsula, but they then range out to sea as far as 160°E and 20°S. In the northern spring 40% of seabirds seen in the western Pacific north of 30°N were *leucomelas* with *P.griseus* and *P.tenuirostris* making up 30% of the rest (Shuntov, 1974). Cory's Shearwaters are also common within range of their nesting islands during spring and summer — their breeding season.

D *Feeding and foods*

These shearwaters feed primarily by dipping and surface seizing, often associating with tuna and mackerel shoals, and with cetaceans. They take mainly fish, squid and crustacea and can capture fish in mid-air that are forced by predators to leap from the sea. Both species attend fishing boats for discarded offal. Haney & McGillivary (1985) showed how *C.diomedea* concentrated along Gulf Stream fronts off Florida in summer and autumn and they discussed the environmental variables involved. Pulich (1982) showed that they had a similar distribution pattern off Port Aransas in the Gulf of Mexico at those seasons. Despin & Mougin (1988) studied aspects of the energetics of the feeding of *C.d.borealis* and Sara (1983) that of *C.d.diomedea*.

E *Breeding*

Calonectris breeds colonially in the northern summer, the Streaked Shearwater digging burrows on many, mostly forested islands around Japan, the various races of Cory's Shearwater nesting more often in rock crevices and recesses than in burrows. This species tend to form rather smaller colonies than do the North Pacific birds, sometimes nesting in quite widely scattered groups. At Grand Salvage Island many come ashore well before dusk, but elsewhere *Calonectris* seems to be wholly nocturnal on land. 'Rafts' of birds waiting for nightfall form off the nestings islands and Mougin & Stahl (1982) mapped these rafts off Grand Salvage Island on 8 July 1980.

The breeding regime resembles that of other petrels, with nest sites and partners being retained for more than one season, with a pre-laying exodus and a sharing of incubation and chick-rearing between the sexes. Round & Swann (1977) found that *C.d.diomedea* chicks were unfed, on average, for 4.7 days before they took their first flight and *C.leucomelas* is also deserted during its last days in the nest (Yoshida, 1962). Most appear to be annual breeders but Mougin *et al*. (1987c) found that 7% of *borealis* missed a nesting season each year.

The various populations of *C.diomedea* have received increased attention in recent times, particularly the birds at the Salvage Islands. Jouanin & Roux (1966), Zino (1971), Zino *et al*. (1987) and Mougin *et al*. (1984a) dealt with their general reproductive biology, and Mougin *et al*. (1987a) with the demography of the population. Mougin *et al*. (1986a) dealt with pair-bonding and Mougin *et al*. (1987b, 1988a) with factors affecting breeding success. Wink *et al*. (1982) presented results of work at a colony in the Aegean and other studies include those done near Crete (Round & Swann, 1977), off Malta, and on Corsica by Thibault (1985). Many aspects of the breeding and ecology of the Streaked Shearwater have also been examined with at least three books being devoted to the species (Okamoto, 1972; Tan, 1977; Yoshida, 1981).

F *Dispersal and migration*

Both species are migratory, many birds making trans-equatorial movements. Streaked Shearwaters winter in tropical waters of the East and South China Seas and around Indonesia and the Philippines. According to Yoshida (1962) the parents leave the nesting islands to start their migration, leaving their young to follow later. Many reach New Guinea and flocks are seen in the Coral Sea and small ones as far south as Sydney. Shuntov (1974) reported them as common in the Arafura and Timor seas from December to March, some penetrating to the Indian Ocean and even to Sri Lanka with flocks reported as far south as the eastern Indian Ocean as the Abrohlos Islands. Shuntov outlined their movements in relation to food and other factors. He found that Streaked Shearwaters were the most numerous seabirds in the East China Sea in the northern winter, feeding along the fronts forming between cold water flowing from the north and warm currents from the south. Here the birds also benefited from the offal discarded by an intensive fishery. The northward movement of flocks heading back towards Japan in the spring brought birds from further south into this area and over 90% of the seabirds at this season were *C.leucomelas*. By June the numbers were greatly reduced, as was the food supply (though this shearwater remained the most numerous species) and later migrants heading north avoided the area by passing to the east of it (Shuntov, 1974).

The migrations of *C.diomedea* have been analysed by Mougin *et al*. (1988b), from sightings at sea and recoveries of ringed birds. The subspecies *borealis* mostly winters off South America, with large numbers occurring off the eastern United States where many non-breeders feed over the continental shelf in summer and autumn. Yesou (1982) analysed environmental conditions associated with big summer influxes of *borealis* into inshore waters of the Bay of Biscay and as far west as Ireland. Some of this subspecies also reach the southern Indian Ocean, but to the north *C.d.diomedea* have been identified off Eilat. Some of this race are seen off the nesting islands in winter but most migrate out of the Mediterranean and travel south to fish the Benguela Current off South-west Africa in the winter. Telleria (1980) saw a passage through the Straits of Gibraltar involving well over 100 000 birds. *C.d.borealis* has straggled as far as New Zealand, the Streaked Shearwater to California. The Cape Verde Islands' *edwardsi* also disappear from local waters and have been identified at 25°S, perhaps shifting less extensively than *borealis* (Bourne, 1986; Mougin *et al*., 1988b).

G *Populations and mortality*

The largest *Calonectris* populations are those of the Streaked Shearwater, particularly at the Izu Islands south of Honshu, with an estimated 1.75 to 3.5 million birds (Hasegawa, 1984b) and this is Japan's most abundant seabird. Atlantic colonies of Cory's Shearwater are often quite small — a mere 60 to 80 pairs on the Berlengas, for example, in a 1981 census. At the Salvages the birds were decimated in massacres during 1975 and 1976, but since the stationing of wardens there, numbers have begun to recover, and in 1983 Mougin *et al*. (1984c, 1987a) estimated that the total population of breeders was some 25 000 birds. They established that the age at first breeding averaged 9 years and gave a breakdown of the age structure of the population. At the Azores perhaps half a million pairs nest (Le Grand *et al*., 1984). Many also breed at the Canary Islands, nesting in all suitable situations with several thousand pairs on Allegranza Islet, for example (Le Grand *et al*., 1984). The Cape Verde subspecies *edwardsi* was said to breed 'in great numbers' on Branco and Razo (de Naurois, 1969). Mediterranean Shearwater populations tend to be rather small, though some are probably still undiscovered. There are 5000 to 10 000 pairs in the Balearic Islands (de Juana *et al*., 1984), 10 000 birds on Zembra Island off Tunisia (Jarry, 1969), perhaps 7000 and 5000 pairs on islands off Italy and Greece respectively, with a smaller number, about 500 pairs, off Malta (James, 1984).

Apart from losses due to flooding and suchlike hazards, *Calonectris* shearwaters appear to have few natural causes of mortality on land. In one year all the Cory's Shearwater chicks at Grand Salvage Island died through suffocation by the sirocco, but this seems to be unusual. Eggs and chicks are lost to gulls and an elapid snake eats eggs of *C.leucomelas* at some nesting sites. However, Streaked Shearwaters are fairly regular victims of gales and many fledglings die on Japanese beaches from time to time. These 'wrecks' have been the subject of several studies, e.g. by Yoshida (1973). Hayashi (1983) correlated the mortality with wind velocities.

Unnatural predators like cats and dogs have forced the retreat of some populations into inaccessible holes in steep cliffs (as at Madeira). And, whereas Shoemakers can evidently dominate rabbits (see above), on islands of Marseilles the breeding success of *C.d.diomedea* has fallen due to interference by those mammals (O. Fernandez, pers. comm.). Off Japan, netted anchovies are kept fresh in crawls floated in the sea. Streaked Shearwaters attracted by the milling fish visible in uncovered cages make shallow dives to feed on them or settle on the sea to take them by surface seizing. In the process many birds get trapped in the slack of the crawl itself or, if these are covered, get tangled in the covering nets. Nakamura (1974) described a mass accidental death of this kind (Fig. 7.7).

Figure 7.7 *Streaked Shearwaters trapped in anchovy crawls off Honshu, Japan. From Nakamura (1974).*

Most populations of *Calonectris diomedea* have experienced uncontrolled exploitation for their eggs, chicks and adults, and many still do. At the Salvage Islands up to 30 000 chicks were taken in a good year but the species is now protected there. Zino (1985) has provided a history of this shearwater hunt. On Linosa, Pelagic Isles (between Malta and Tunisia), Vaughan (1980) found that the main colony of Mediterranean Shearwaters was being used as a rubbish tip. Local people still collected eggs there as in other parts of the Mediterranean.

In Japan Streaked Shearwaters, formerly eaten, are protected under the Wildlife Protection Laws and only legally taken for scientific purposes or, on Mikura Island, to lessen damage to the forest due to burrows undermining the trees. Between 1981 and 1985 inclusive 900 birds were taken annually, in 1986 none (N.Oka, pers. comm.).

CHAPTER 8
The Shearwater Genus Puffinus

Short-tailed Shearwaters.
Photo: author.

Contents

I Introduction

This genus contains some 15 species of medium-sized to small petrels — the 'true' shearwaters.

There is a reasonably good, and improving, fossil record. The oldest species (and the oldest petrel fossil), is *Puffinus raemondonki* (Van Beneden, 1871) based on a humerus from Belgium of early Oligocene age. There are several Miocene birds from North America, South Africa and France, and Pliocene ones from St Helena (Olson, 1975a) and North Carolina. Olson (1985d) concluded that most modern species-groups of *Puffinus* had evolved by the middle Miocene and that there has been very little morphological change in the last 15 million years. A new species, *P.holei*, that bred at the Canary Islands 32 000 years ago, has been described recently (Walker *et al.*, 1990).

Birds of this genus have been known for a long time and early in the seventeenth century the name shearwater was given to the Manx Shearwater *Puffinus puffinus* by Sir Thomas Browne of Norwich. He kept two in captivity for 5–6 weeks, force feeding them with fish until 'my Servant grew weary, and gave them over. And they lived fifteen days without any food.' (Willughby, 1678). Willughby knew the bird as breeding in the Scilly Isles and Isle of Man. According to Williamson (1940) the earliest mention of shearwaters in the Scilly Isles dates from *c.* 1337 when they were leased by Edward III for 6s. 8d. or 300 'Puffins'. James Chaloner visited the Isle of Man in 1651–1652 and described the breeding of *P.puffinus*, how it feeds its young, its voice, and some of its human uses (Chaloner, 1656). Records of the 'Puffin' harvests at this place were kept and varied from about 10 000 in 1660 down to 3174 in 1716. The chicks were sold for 9d. per dozen, one foot being cut off each and retained to keep a check on the annual take. However, by 1789 the birds had gone, killed by rats from a wrecked ship (B.R.S.M., 1941).

Elsewhere other species have been used as food. Maoris traditionally took squabs and adults of several shearwaters in the North Island of New Zealand (Reischek, 1886), but the chief 'mutton-bird' of New Zealand is the Sooty Shearwater *P.griseus*, harvested mainly on islands to the south of that country. Traditional methods as described by Richdale (1948) have been modified in recent times. The Tasmanian Mutton-bird or Short-tailed Shearwater *P.tenuirostris* forms the basis for a carefully controlled industry particularly on islands in Bass Strait (Serventy, 1974). The Tasmanian aborigines used this species for food at least 2000 years ago but did not preserve their catch (Bowdler, 1974), and it was not until early in the nineteenth century that shipwrecked sailors developed the harvest. In Hawaii Newell's Shearwater or 'Ao' *P.auricularis newelli* was also eaten and in early times reserved to the chiefs (Munro, 1941). Further north still, in Davis Strait, Greenlanders hunted Greater Shearwaters *P.gravis* from kyaks in August when the birds were in full wing moult and almost flightless.

II General characteristics

Birds of this genus are characterised by their slender bills at the bases of which lie paired flat and wide nasal tubes separated by a thick septum and opening obliquely upwards as their distal edges are bevelled (Fig. 7.1). The skull is long and delicate, with well developed temporal fossae (Fig. 8.1). The lachrymals are free and the skull roof rounded, perhaps giving the bird a relatively large brain as compared to *Calonectris* (Kuroda, 1954). The bill is shorter than the tarsus, the latter being laterally compressed (in some species more than in others) and sharp-edged in front. This was regarded as an adaptation for an aquatic life style by Kuroda as were the rather narrow sternum and pelvis (Fig. 8.2). Likewise this tibia possesses a large crista cnemialis cranialis for the attachment of powerful leg retractors (Fig. 4.9A). The wing is long and pointed with primary 10 the longest. The tongue is pointed and spiny — as in *Procellaria* but not in *Calonectris*. For details of morphology and anatomy the monograph of Kuroda (1954) should be consulted. Mayaud (1932a, 1934) also dealt with the osteology of *Puffinus*.

Shearwaters are mostly soberly coloured, tending to be dark brown or blackish above, brown or white below. Some, like *P.pacificus* and *P.creatopus*, are polymorphic with dark, light, and intermediate-phase plumages. Tails are wedge- or fan-shaped. Bills and feet are usually dark, the latter not bicoloured as in many gadfly petrels. Biggest are *P.carneipes*, the Fleshy-footed Shearwater and its North American representative *P.creatopus*, the Pink-footed Shearwater. These have

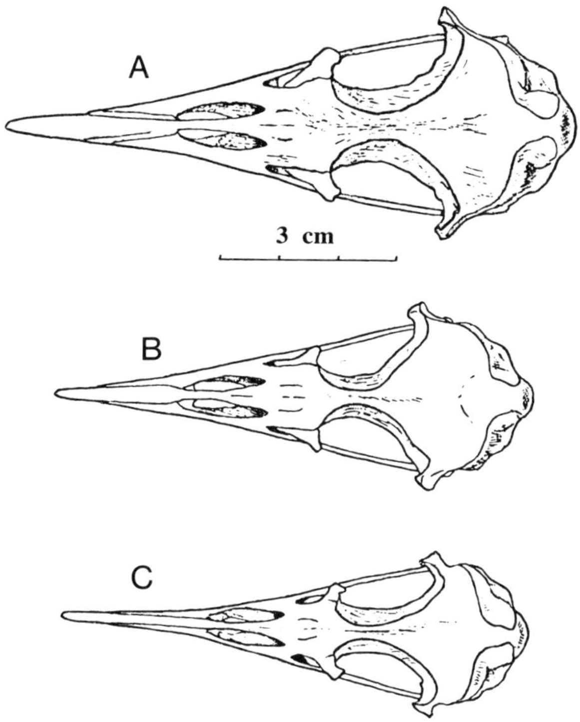

Figure 8.1 *Skulls of shearwaters (A) Fleshy-footed* P.carneipes, *(B) Short-tailed* P.tenuirostris *and (C) Hutton's* P.huttoni. *Note the deep temporal fossae and narrow ridge between the supra-orbital depressions. From Condon (1939).*

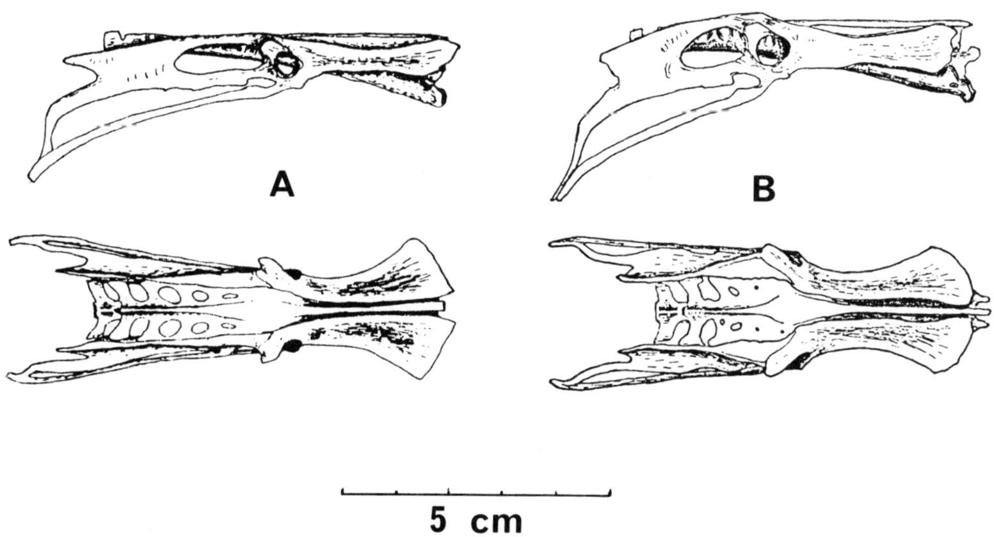

Figure 8.2 *Pelves in lateral and dorsal view of: (A)* P.griseus *and (B)* C.d.diomedea. *After Mayaud (1932a, 1934).*

wings spanning 1.5 m and weigh about 700 g. Smallest are the sibling species — Audubon's and the Little Shearwater *P.lherminieri* and *P.assimilis*, with wing spans of 60 cm and weights of about 195 g.

Sexual dimorphism in body measurements tends to be slight. In *P.tenuirustris* males average very slightly larger than females, have more robust necks, and their bills are slightly longer. Brooke (1978d) found this true also of adult Manx Shearwaters.

Moult of flight feathers after breeding is in ascending order along the primaries and outwards in both directions from the central secondaries (Mayaud, 1931, 1932b). Different age classes moult at different times but good data from known-age birds are lacking. Some renewal of body and contour feathers often occurs during nesting, but those of tail and flight feathers takes place after breeding. Trans-equatorial migrants like *P.tenuirostris* and *P.gravis* delay wing moult until winter quarters are gained and then immediately shed flight feathers (Marshall & Serventy, 1956a; Stresemann & Streseman, 1970; Brown, 1988). This may cause the sea and nearby beaches to become littered with feathers. Chu (1984) found that Sooty Shearwaters off California completed wing moult earlier than they did body moult. The Stresemanns (*loc. cit.*) concluded that *P.gravis* of the year started body moult when 6–7 months old but retained their flight feathers until they were 10–11 months old, moulting 1–2 months before the breeding adults did. On return to their breeding grounds the trans-equatorial migrants are fully winged.

As many as five primaries may be missing at one time, so that, although the birds do not become flightless, their aerial performance is appreciably modified (Voous, 1970; Brown, 1988). Yesou (1985) examined moulting *P.p.mauretanicus* and pointed out that the handicap of reduced flying efficiency might be counterbalanced nowadays by the ready availability of offal released from trawlers.

III The species

The taxonomy is somewhat confused, but Jouanin and Mougin (1979) grouped clearly related species in subgenera (many of these having been originally proposed by Gregory Mathews). Murphy (1952) considered the Manx Shearwater as a bird of world-wide distribution and grouped *puffinus, mauretanicus, newelli, gavia, auricularis, yelkouan* and *opisthomelas* as subspecies. Jouanin and Mougin included only six of these in *P.puffinus*, making *gavia* (and *huttoni*) full species. Other checklists, such as that of the American Ornithologists' Union (1983), have treated *auricularis* and *opisthomelas* as full species, and so on. Various attempts have also been made to group some or all of these birds in superspecies. Many subspecies of the two smallest, the Little and Audubon's Shearwaters *P.assimilis* and *lherminieri*, have been described and some may prove to be valid species, e.g. *P.l.heinrothi*, but more needs to be known about the non-structural aspects of their biology such as their biochemical genetics and vocalisations, for clarification of the taxonomic limits. Harper (1978) sampled blood proteins of six puffinoid shearwaters (Fig. 8.3) finding that *P.bulleri* and *P.assimilis* gave the most distinctive patterns on the gels whereas those of *gavia* and *huttoni* were similar but not identical. From an electrophoretic study of red cell esterases G.Wragg (unpubl.) found that there was a difference in mobility between *gavia* and *huttoni* at two loci, with *P.griseus* being different again.

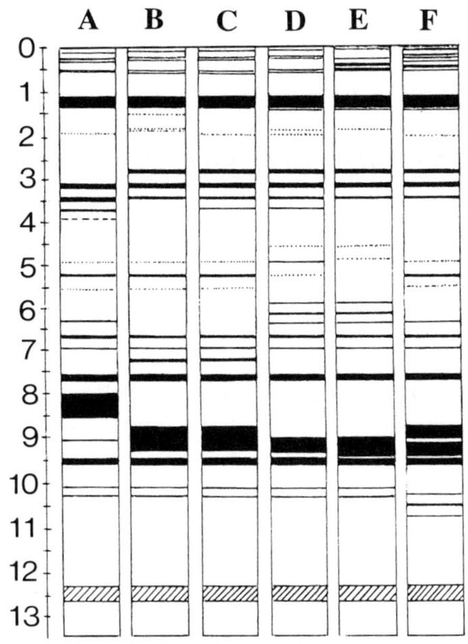

Figure 8.3 *Blood plasma proteins of shearwaters (A)* bulleri, *(B)* carneipes, *(C)* griseus, *(D)* gavia. *(E)* huttoni, *(F)* assimilis. *Note the distinct albumen and pre-albumen pattern of* assimilis. *The banding of* bulleri *suggests that it is not closely related to the rest and all patterns differ considerably from those for* Procellaria *(Fig. 7.2). From Harper (1978), courtesy of* New Zealand Journal of Zoology. *For key to components see Appendix 2.*

IV Life styles

As their name implies, shearwaters tend to skim the sea's surface and in calm weather seldom fly high unless heading inland to montane nesting places. They travel fast on pointed wings and in strong winds skim along the wave furrows and then rise and bank before accelerating towards the surface again. When a short distance above it they reverse their bank and use their newly acquired momentum to skim the sea at an angle to wind and waves before repeating the manoeuvre, following a spiral course. Whole groups often perform these evolutions more or less in unison when the evening sun may catch their underwings as all simultaneously reach the precise position at which to reflect the light from their glossy feathers back to the observer.

Shearwaters are well known for their 'rafting' behaviour in which flocks form offshore, particularly during the breeding season, the birds resting and preening while awaiting the onset of darkness. Similar behaviour is seen among feeding and moulting flocks elsewhere, as with Balearic Shearwaters (*P.p.mauretanicus*) summering off the French Atlantic coast (Yesou, 1986) or *P.griseus* and *P.tenuirostris* on migration off California (Baltz & Morejohn, 1977). Typically the petrels ride facing the wind in strings of birds perhaps 5–10 abreast with newcomers alighting at the heads and tails of the columns. Off the nesting islands a raft of, e.g. Short-tailed Shearwaters, may be 1–2 km long before it breaks up as groups rise at dusk and head for their nesting colonies (Warham, 1960). King (1974) suggested that 'rafting' of *P.pacificus* near Hawaii occurs as a prelude to migration.

Flocks also form when shearwaters feed on fish shoals. Groups of 100 or more birds will abruptly dive from the surface over a restricted area, remain submerged for several seconds when no birds are visible, and then equally abruptly pop up to form a 'raft' before moving on.

Progression under water is efficient. Shearwaters usually submerge quite smoothly head-first and when travelling horizontally at slow speed propel themselves with alternate kicks of their webbed feet. When rising, falling or travelling fast the wings provide the main thrust. These are usually beaten with the primaries folded and the wing only about half extended. In this flipper-like mode the birds are extremely agile and can put on very fast spurts. Brown *et al.* (1978) compared the styles used by several *Puffinus* and *Calonectris* species.

With legs set far back on their bodies shearwaters have difficulty in taking off in calms. At sea they run across the surface with beating wings to rise at a shallow angle. From land under such conditions they climb to some elevated place for take-off, grappling with the hooks of their bills and thrusting with the sharp-clawed feet and depressed tails. In forested situations they may climb trees. Having gained their launching platform they often beat their wings at low amplitude in several spurts; as if warming up their muscle engines, the lift generated straightens out their legs until their bird is on tip-toe: take-off proper, however, is achieved with a leap (Warham, 1960).

The clumsiness of shearwaters on land is often over-emphasised. On open ground a frightened bird can travel fast by half-running, half-flying, and in heavily vegetated habitats can easily elude a human pursuer.

Their wing loadings tend to be high and landings in calms may be difficult. In open country and with some wind, most land quite gently, stalling at touch-down. Where colonies are in trees many birds crash into branches but remarkably few casualties result and even females heavy with shelled eggs seem immune from injury on alighting.

Shearwaters are noisy birds on their breeding grounds after dark, especially immatures and pre-breeders. The smaller kinds like *P.assimilis* and *P.puffinus* are vocal both in flight and on the ground, the larger ones such as *P.tenuirostris* and *P.griseus* are very noisy on the ground but mostly strangely silent on the wing. The calls of female Manx Shearwaters are harsher, 'noisier' and less easily located than those of males (Brooke, 1978e) and this is true of some other species, e.g. *P.griseus* (pers. obs.). The calls are the same for either sex but females pitch theirs lower than the males, differences readily detected by human ears. Male Manx Shearwaters call more from the ground, females more from the wing (Brooke, 1978e). Storey (1984) and James (1985) have discussed functions of the various calls. Sound production both on intake and expiry of breath is particularly noticeable in shearwaters and was described for the Manx Shearwater by Lockley (1942). For example, the main call of the Fleshy-footed Shearwater is a trisyllabic crooning with the middle syllable stressed — 'ku-kooo-ah' — repeated three to six times and tending to become increasingly hysterical in tone, the 'kooo' becoming a scream. The first two syllables are given as the breath is expelled, the final sob on inspiration (Warham, 1958): the muscle contractions accompanying these actions are readily seen in the field.

V Distribution

Shearwaters feed in all the oceans but nest most abundantly in cool temperate seas (Fig. 8.4). New Zealand harbours the greatest diversity, seven of the 15 species breeding there or on its associated islands. The most northerly breeding places are those of the Manx Shearwaters on the Westmann Islands (Iceland) at 63°N, the most southerly the Sooty Shearwaters of Cape Horn at 55°S. Ainley (1977) suggested that the breeding of shearwaters in the North Pacific has been restricted by competition with the many alcids there; these occupy the available nest sites in cavities and burrows. And although three breed in the tropics on either side of the Equator (*P.nativitatis, pacificus* and *lherminieri*) only one, *P.assimilis*, breeds in mid-high latitudes in both hemispheres. This bird's distribution in the North and South Atlantic is similar to that of *P.mollis* (Fig. 4.14.X), except that this is primarily an Indian Ocean species, *assimilis* mainly Australasian. The Wedge-tailed Shearwater *P.pacificus*, so widespread in warm Pacific and Indian Ocean seas, is surprisingly absent from the Atlantic, but the same or a very similar bird lived into the Pleistocene on St Helena and in North Carolina. It apparently disappeared in the face of cooling seas during the glaciation (Olson, 1985d).

VI Feeding and foods

The pincer-like beaks, and tongues and palates with retrose papillae, are well adapted for handling fish, which, with small crustacea and some squid, are typical foods. Shearwaters tend to feed in flocks and to harry school fish, e.g. Sooty Shearwaters off California eat anchovies, young Rockfish (*Sebates* sp.) and the squid *Loligo opalescens* (Briggs & Chu, 1986), Manx Shearwaters follow sardines in the Bay of Biscay (Perrins & Brooke, 1976), and Hutton's Shearwaters take other clupeids off New Zealand (West & Imber, 1985). Great and Sooty Shearwaters off eastern Canada fed opportunistically wherever prey was available, with concentrations of the crustacean *Meganyctiphanes*, spawning capelin *Mallotus villosus* and migrating squid *Illex illecebrosus* being preferred (Brown *et al.*, 1981). These authors noted that *P.gravis* has a stouter, longer, and more hooked bill than *P.griseus*, and so might be adapted to taking larger or tougher-bodied fish. They also confirmed the hypothesis of Kuroda (1954), based on skeletal evidence, that Sooty Shearwaters are better adapted to diving than are the Greater Shearwaters. While both dived, these workers found that *P.gravis* surfaced directly ahead of their submergence points whereas *P.griseus* showed their greater agility by popping up in any direction; to some extent these differences were reflected in the foods taken.

Kuroda (1954) related skeletal features (pelvis width, compression of tarsus and humerus etc.) to the diving ability of shearwaters, concluding that *Calonectris* was primarily a gliding, aerial form, as were *P.pacificus* and *bulleri*, which, like *Calonectris*, seldom submerge fully, whereas *P.carneipes, creatopus* and *gravis* can dive but are less adept than *griseus* and *tenuirostris*. *P.opisthomelas* evidently does not dive as often or as efficiently as members of the *P.puffinus, huttoni, gavia* complex do, but has been seen to submerge for 5–10 s when chased by skuas (Jehl, 1982) — see also Harper (1978). The most efficient divers appear to be the smallest, *assimilis* and *lherminieri*, which are rather short-winged for their body sizes.

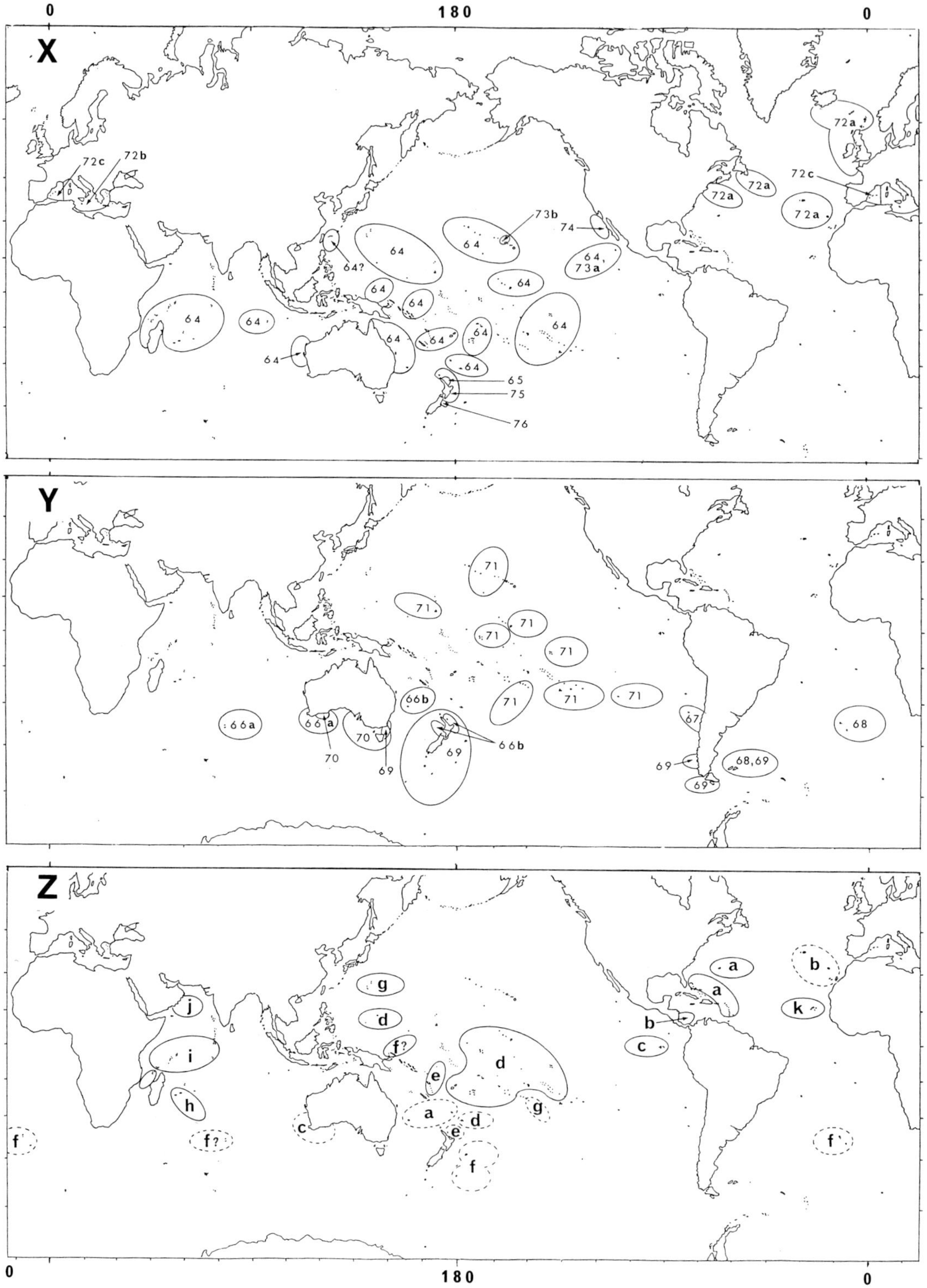
0
180
0
X
72c
72b
64?
64
73b
74
64, 73a
72a
72c
65
75
76
Y
71
66a
66b
70
69
67
68
68,69
Z
j
i
h
f
f?
c
g
d
e
a
b
k
180
0

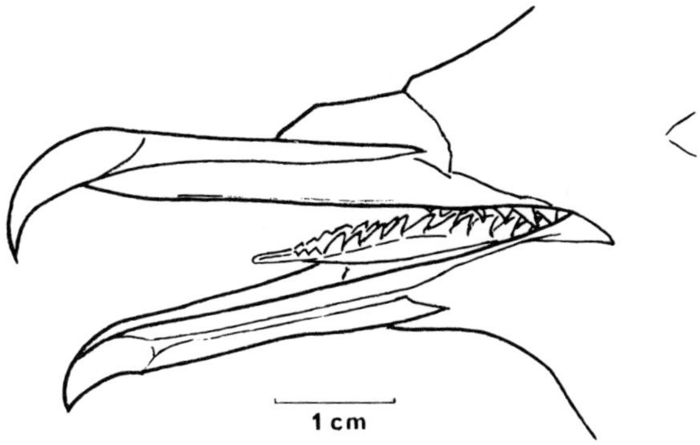

Figure 8.5 *Beak and tongue of Short-tailed Shearwater showing retrose lingual and palatal papillae used to retain food items. After Morgan & Ritz (1982), courtesy of Elsevier Science Publishers, Amsterdam.*

Shearwaters mostly feed by surface-seizing and pursuit-plunging and usually appear not to penetrate very far below water, with much prey taken from the top 2 m. There are accounts of deep diving, however, e.g. Skira (1979) reported a group of 65 Short-tailed Shearwaters feeding among weeds at 12 m depth: their half-opened wings enabled them to dart quickly to and fro. Sooty, Fleshy-footed and Fluttering Shearwaters often put their heads below the surface before submerging, evidently looking for their prey before giving chase. Other species may make clean shallow dives from a few m up or belly-flop to the sea with breasts, wings and feet spread, and at once submerge. Buller's Shearwater has been seen feeding by contact-dipping, flying slowly close to the surface with beating wings held back and feet down while the head was immersed to grab near-surface food (Harper, 1983). Jenkins (1974) reported the same species swimming forward while swinging its partly submerged head from side to side. Sooty Shearwaters may feed similarly and doubtless the style can be varied to match the behaviour of the prey. Ogi (1984a) saw *griseus* grazing on barnacles attached to floating logs.

From a study using the heads of Short-tailed Shearwaters in a flow tank Morgan & Ritz (1982) suggested that as the bird moves under water with open gape the bill architecture produces a low pressure volume into which the prey are sucked. When the bill is open the retrose papillae on tongue and palate are raised forming an interlocking mesh that retains the prey (Fig. 8.5), the water passing out at the corner of the gape.

Figure 8.4 *Breeding distribution of* Puffinus *shearwaters:* (**X**) P.pacificus *(64);* P.bulleri *(65);* P.puffinus *(72),* p.puffinus *(72a),* p.yelkouan *(72b),* p.mauretanicus *(72c);* P.auricularis *(73),* a.auricularis *(73a),* a.newelli *(73b);* P.opisthomelas *(74);* P.gavia *(75);* P.huttoni *(76).*

(**Y**) P.carneipes *(66),* c.carneipes *(66a),* c.hullianus *(66b);* P.creatopus *(67);* P.gravis *(68);* P.griseus *(69);* P.tenuirostris *(70);* P.nativitatis *(71).*

(**Z**) P.assimilis *(dashed outlines) and* P.lherminieri *(continuous outlines).*

P.a.assimilis *(a),* a.baroli *(b),* a.tunneyi *(c),* a.kermadecensis *(d),* a.elegans *(e) and* a.myrtae *(f).* P.l.lherminieri *(a),* l.loyemilleri *(b),* l.subalaris *(c),* l.dichrous *(d),* l.gunax *(e),* l.heinrothi *(f),* l.bannermani *(g),* l.bailloni *(h),* l.nicolai *(i),* l.persicus *(j),* l.boydi *(k),* l.temptator *(l).*

Shearwaters tend to keep in tight-knit flocks when harrying shoaling prey or following whales, the water above fish being continually disturbed by diving birds which follow the shoals moving below, while birds at the fear hastily leap-frog over the rest to attack the van of the moving shoal. Very often more than one species of petrel is involved in these transitory food bonanzas, e.g. *P.auricularis* with *P.pacificus* off Mexico, and *P.bulleri* with *P.griseus*, and particularly *P.carneipes*, off northern New Zealand. Jenkins (1974) reported that in the latter area patches of water 'boiling' with fish were not directly harassed; rather the petrels concentrated down-current of the shoals to feed at the edges on the remains of fish killed by gannets *Morus serrator*.

Many species also follow fishing boats to feed on offal and can be caught on baited lines. Fishermen on the Grand Banks took Greater Shearwaters for use as bait. Off south-western Australia Fleshy-footed Shearwaters may take bait from anglers' lines cast from the shore, and the birds may be glimpsed against the sky as they swim within the breakers. This species has even been known to scavenge dead fish on a beach. Audubon's Shearwater *P.lherminieri* has an unusually long tail and is very manoeuvrable in air and water: it can pluck flying fish from mid-air like a noddy tern *Anous* (Jehl, 1974c) — 'air dipping'. Similar behaviour is known for the Wedge-tailed Shearwater.

Other shearwaters such as *P.tenuirostris* seldom follow ships, as opposed to *P.griseus*, which often does so, while Buller's, Wedge-tailed and Little Shearwaters usually, but not always, ignore fishing vessels.

VII Breeding

The smaller species like *P.puffinus* and *P.assimilis* are comparatively unaffected by judicious handling when breeding and the former has been the subject of experimental studies like that of the resistance to chilling of the embryo (Matthews, 1954), on how burrows are located after dark (James, 1986), on the effects of age, and length of pair bond on breeding success (Brooke, 1978a), and on the growth and survival of artificial broods of two chicks (Perrins *et al.*, 1973). Experimental studies of Hawiian *P.pacificus* have included a detailed examination of its thermoregulatory ability (Whittow *et al.*, 1987). Large southern shearwaters such as *P.tenuirostris*, *carneipes*, and *griseus* make difficult subjects, being very susceptible to disturbance and usually deserting their egg if handled during incubation. The last of these is particularly aggressive and when captured fights to exhaustion. However, even 'tame' species like the Manx Shearwater are affected to some degree by handling.

The breeding system of puffinine petrels is well known, thanks to the pioneer work of Lockley (1930, 1942) on marked Manx Shearwaters, and these birds have continued to attract researchers at the West Wales colonies and elsewhere, e.g. Perrins *et al.* (1973); Storey (1984); Brooke (1990). Southern hemisphere species have been the subjects of particularly long-term studies, e.g. those of *P.tenuirostris* over 30 years (Serventy & Curry, 1984; Bradley *et al.*, 1989) and of *P.griseus* over 17 years (Richdale, 1963). Shorter studies include those of Swanson & Merritt (1974) on *P.pacificus* and of Warham *et al.* (1982) on *P.griseus*.

Breeding procedures generally follow typical petrel patterns with nest-site and mate-tenacity the norm and with fairly equal sharing of incubation and chick-rearing duties by the sexes. Shearwaters are colonial nesters laying under ground except that, in the tropics where aerial predators are absent, *P.pacificus* occasionally and *P.nativitatis* regularly, lay on the surface. Most visit their nests only after nightfall but tropical *P.nativitatis* and *P.pacificus* may come and go by day, and, at the Snares and Nightingale Islands, some *P.griseus* and *P.gravis* alight in broad daylight (Warham *et al.*, 1982; Rowan, 1952). Apart from predation by skuas and gulls, some eggs and small chicks are eaten by *Sphenodon* in New Zealand or by snakes like *Notechis scutatus* in Australia (Warham, 1960). Such reptiles often share burrows with the birds.

Nesting habitats vary considerably, from open coral sand and holes in coral breccia (*P.nativitatis*), to treeless but lightly vegetated ground that allows burrowing (e.g. *P.puffinus, P.assimilis*). Others burrow under heavy fern, bushes and tussock grasses (*P.a.newelli, P.tenuirostris* etc.) or under an almost closed canopy of forest as with some populations of *P.assimilis, P.carneipes* and *P.griseus*. Colonies may be close to the sea or up to 1525 m above sea level (*P.huttoni*). The burrowing of this shearwater into thinly grassed and very steep slopes has led to soil erosion, an effect intensified by the shearwater practise of tearing off nearby vegetation to line nesting chambers and by the grazing damage of goats, deer and chamois.The ground cover is lost, then the soil goes, and the birds perforce must shift (Evans, 1973). This species nests on the 'mainland' of New Zealand where small relict colonies of *P.griseus* also still persist, but usually shearwaters need the safety of small to medium-sized islands free of alien predators for successful breeding.

Particularly noticeable among migratory species is that egg-laying is spread over only a few days and peaks at much the same date every year. It is this regularity that has enabled the 'Tasmanian Mutton-bird' *P.tenuirostris* to form the basis of a small industry, as it has allowed advance planning for harvesting.

VIII Dispersal and migration

The genus includes some notable trans-equatorial migrants. All, except the Manx Shearwater, are southern hemisphere breeders. *P.tenuirostris, carneipes,* Australasian *griseus* and *creatopus* winter in the North Pacific; *P.gravis* and South American *griseus* shift to the North Atlantic. Some of those that undertake very long movements apparently desert their chicks to begin their northwards movements. Fledglings evidently follow later. Manx Shearwaters move in the reverse direction, wintering in the South Atlantic, off South America (Thomson, 1965).

Although large numbers of the birds have been ringed on the nesting islands — *c.* 112 000 *P.tenuirostris,* for example — recovery rates are very low. However, the peripheral figure-of-8 migration route for that bird proposed by Marshall & Serventy (1965b) from sightings, 'wrecks' on northern beaches, and recoveries of marked birds, has been questioned in recent years because flocks have been seen migrating through the central North Pacific. Marked birds are naturally recovered on land and such a bias may mask the movements of major elements of the population moving over deep water. Different age classes may also have different routes. Considerable mortality, mostly of the younger birds, may occur both when the contra-nuptial areas are reached and towards the end of the return movement. Some of the consequential 'wrecks' on Japanese beaches have provided material for studies of the ages and condition of birds at death, e.g. by Oka & Maruyama (1986), while a number of surveys by air and sea have highlighted aspects of the distribution and movements of flocks in relation to weather, food and other variables, e.g. Guzman & Myres, (1983); Ogi (1984a); Ogi *et al.* (1980) and Briggs & Chu (1986). Sampling of birds from the contra-nuptial flocks has given insights into the diets and energy balances, e.g. the work of Chu (1984) on *P.griseus* off California.

Such migrations allow these two 'mutton-birds' to benefit from summer-time food abundances at both ends of their ranges. The Short-tailed Shearwater chooses cooler water in the northern summer than that in which it breeds and many penetrate to the Bering Sea: few Sooty Shearwaters do that.

A distributional puzzle concerns the large, mid-summer flocks of *P.griseus* and *P.tenuirostris* that have been repeatedly reported near the edge of the pack-ice in the south-east Indian ocean. Most have been identified as *griseus*, but the only confirmatory specimens have been *tenuirostris* (Kerry *et al.*, 1983). both of these easily confused species do seem to be involved because in January and February 1984 large flocks of *griseus* with a few *tenuirostris* were encountered in this area (Stahl, 1987). Many came close to the ship for ready identification (J.A. Bartle, pers. comm.). The provenance of these birds so far from their breeding grounds at the height of the nesting season is conjectural. Are they immature and/or failed breeders and do they join their fellows in the North Pacific for the southern winter?

IX Populations and mortality

Shearwaters are among the most numerous of seabirds, with a great capacity for fasting, so that they have been able to breed in large numbers far from their main food supplies. The southern species, *P.tenuirostris*, *griseus* and *gravis*, have huge populations. One colony of the first was estimated to number 250 000 pairs on about 21 ha of burrowable ground (at *c.* 1 m^{-2}) and one estimate for the total population of *tenuirostris* in Tasmania was 5.6 million breeding pairs (Naarding, 1980). Populations of Sooty Shearwaters seem equally large, Warham & Wilson (1982) estimating that there were 2.75 million burrows on 328 ha at the Snares Islands. Rowan (1952) thought that there were 2 million burrows of Greater Shearwaters on Nightingale Island's 200 ha and Elliott (1970b) 600 000 pairs of the same species at Gough Island. All such estimates, however, whether based on line transects or on sample quadrats, are subject to considerable errors mainly due to difficulties in determining burrow occupancy and the status of the occupants.

Such massive numbers are reflected in the sizes of the migratory flocks and in their spectacular movements. Based on a nest density of one burrow per square yard Flinders (1801) calculated that a flock of *P.tenuirostris* that he ran into at sea needed 186 square miles of ground to accommodate the 132 million birds it contained. Off Vancouver Island, Canada, lines of Sooty Shearwaters km long have been seen (Guzman & Myers, 1983) and up to 2.5 million of them occur off

California in May, with one flock of 631 000 estimated to cover 4 km^2 (Briggs & Chu, 1986). Hunt *et al.* (1981) thought that there were perhaps 45 to 65 million shearwaters, mostly *tenuirostris*, in the eastern Bering Sea during the northern summer, when they concentrate along an inner front where intensive vertical mixing of oceanic waters occurs.

Most recent changes in shearwater populations have been decreases due to disturbance, cultivation, over-harvesting (see, e.g. Norman, 1985) and predation by introduced pests like cats, dogs, rats and mustelids. Mortality also arises when birds are trapped and drowned in fishing nets, particularly in the North Pacific, and the few recoveries of banded New Zealand Sooty Shearwaters have come from Japanese fishermen working south of the Aleutian Islands. Short-tailed Shearwaters are particularly prone to die in salmon gill nets, perhaps because they concentrate on swarming euphasids that form the chief diet of the fish (Ogi, 1984b).

However, the typical race of *P.puffinus* has established itself since 1977 in small numbers on an island off Newfoundland (Storey & Lien, 1985). Buller's Shearwater, although still restricted to one small group of islands, has experienced a population boom in recent years with the removal of predation by humans and pigs. Its expansion has been at the expense of the smaller, less aggressive Fluttering Shearwater *P.gavia* (Harper, 1983).

In some years Manx Shearwater fledglings suffer losses from the bacterial disease puffinosis which shows as blisters on their feet (Fig. 8.6), conjunctivitis and paralysis. Harris (1965) found that this was only transmitted by birds in colonies among dense vegetation and Brooke (1990) describes experiments to help determine the pathogen responsible. A mild form of the disease also occurs among Short-tailed Shearwaters, but it is neither fatal nor transmitted to humans. However, in some years a considerable mortality of chicks occurs due to a quite different 'limey-bird disease' in which the rectum is blocked by deposits of sodium urate (Munday, 1966).

As with many other seabirds, influenza is a natural infection, and Type A viruses have been isolated from at least one shearwater — the Wedge-tailed species breeding on Heron Island, Queensland (Downie *et al.*, 1977). The virologists' interest arises from the possibility that new human viruses could arise by recombination of a human strain and a virus from another vertebrate, a combination to which human populations would have no immunity.

The harvesting of Tasmanian Mutton-bird chicks has mostly had little effect on population sizes except at some accessible and over-exploited colonies. Indeed, the species appears to have increased in numbers recently, while annual harvests have tended to decline. In 1985 325 000 chicks were taken under licence: in most years the bulk is exported to New Zealand (Skira *et al.*, 1985).

The conservation of shearwaters usually involves protection against interference and burrow trampling by people, and the control of alien predators. Some enterprising measures by the US Fish and Wildlife Service to help protect Newell's Shearwater involve trying to establish a new colony by cross-fostering onto Wedge-tailed Shearwaters (Byrd *et al.*, 1984), by organising the collection and release of fledglings dazed by light or fog, and by hooding strong lights to reduce their attractiveness to the young birds (Telfer *et al.*, 1987).

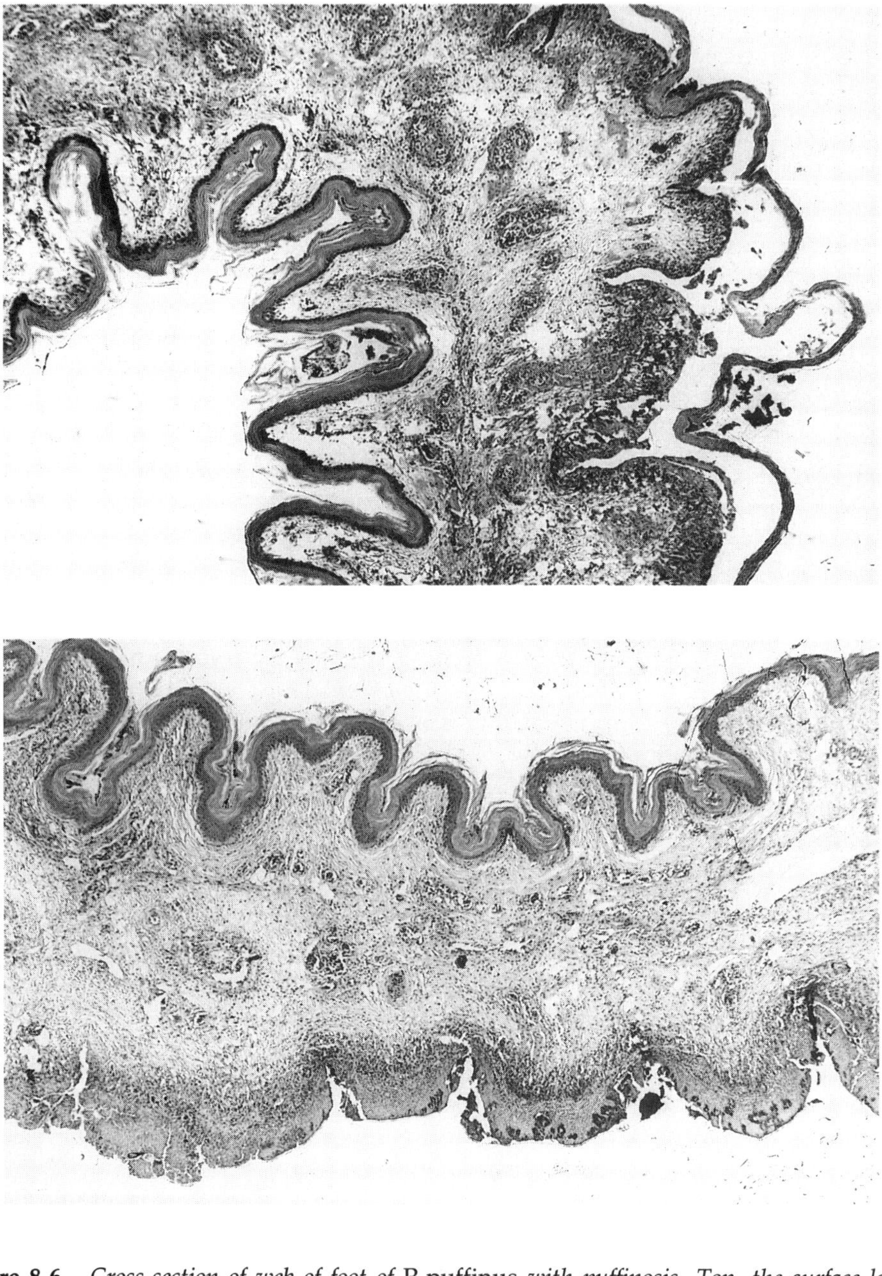

Figure 8.6 *Cross-section of web of foot of* P.puffinus *with puffinosis. Top, the surface layer of keratin on the right is split from the epidermis by a large blister. Bottom, a later stage: the blister has burst on the lower surface and the keratin gone. The epithelium is severely damaged and colonies of bacteria show as dark patches along the lower edge. Photos: G.Holland. From Harris (1965).*

CHAPTER 9
Family Hydrobatidae — Storm Petrels

Black-bellied Storm Petrel.
Photo: author.

Contents

I Introduction

The 'stormy petrels' were well-known to the early mariners like Dampier, their appearance being an augury of bad weather. The British Storm Petrel *Hydrobates pelagicus* had a diversity of vernacular names in the Orkney and Shetland Islands, in Normandy, and in parts of Ireland. This may have been the 'small black and white diver with sharp-pointed bill', the latter 'a little bending at the end, [both Mandibles.]' of Sir Thomas Browne of Norwich, who had seen but two examples (Willughby, 1678). It was one of the five tubenoses included by Linnaeus (1758) in his *Systema Naturae*.

The fossil record is poor. Two species are known from the Upper Miocene of California, *Oceanodroma hubbsi*, L.H. Miller 1951, and a larger, unnamed congener of similar age. The Hubb's Storm Petrel specimen is almost complete. It was entombed in fine sediments and resembled today's Ashy Petrel *O.homochroa* of California, but was heavier in pelvis and hind limb and had a larger bill and shorter skull (Miller, 1951). Olson (1985a) has described early Pliocene *Oceanites zaloscarthmus* from Cape Province, South Africa, of which abundant material of early Pliocene age has been found (Fig. 9.1). It was larger than modern *Oceanites* or *Garrodia*, but similar in size and other features to the White-faced Storm Petrel *Pelagodroma marina* and Olson suggested that *zaloscarthmus* was ancestral to that bird. Subfossil material of the White-throated Storm Petrel *Nesofregetta fuliginosa c.* 800–500 years B.P. has been found in the Cook Islands and on Henderson Island. The species is no longer there and it may have been exterminated by the Polynesians (Steadman & Olson, 1985).

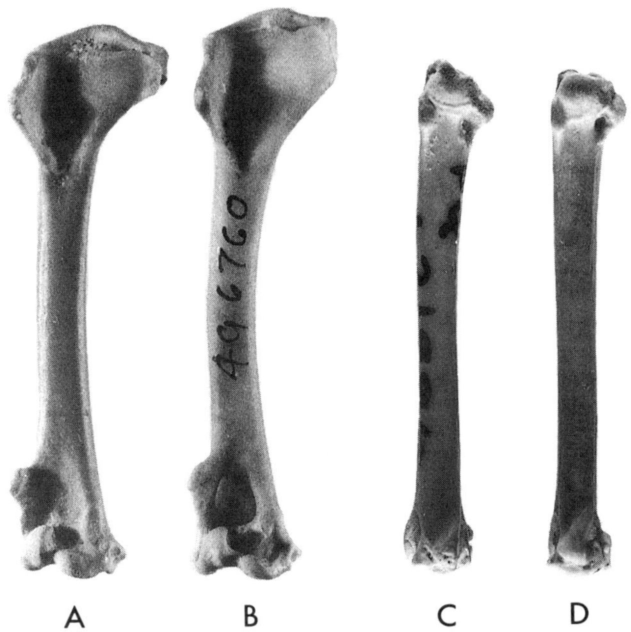

Figure 9.1 *Right humeri and left ulnae in cranial aspect of (A) and (C)* O.zalascarthmus *(Pliocene and (B) and (D)* P.marina *(Recent). From Olson (1985a).*

II General characteristics

Storm petrels are small, delicate birds with rather broad, round-tipped wings due to the large size of the hand compared with the rest of the wing elements and the wing formula in which primary 9, not 10, is the longest (Fig 9.2). The impression is of a rather generalised bird, and Linnaeus (1745) initially though that '*Procellaria pelagica*' was a passerine! The tail is of variable shape, often forked. Bills are black, thin, compressed at the tip and sharply hooked. The fused tubular nostrils are conspicuous, nearly half the length of the culmen, and often uptilted at their ends (Fig. 9.3). Because the anterior edge of the septum is set well back, the nostrils appear to have but a single opening. Storm petrels have long legs and toes, very

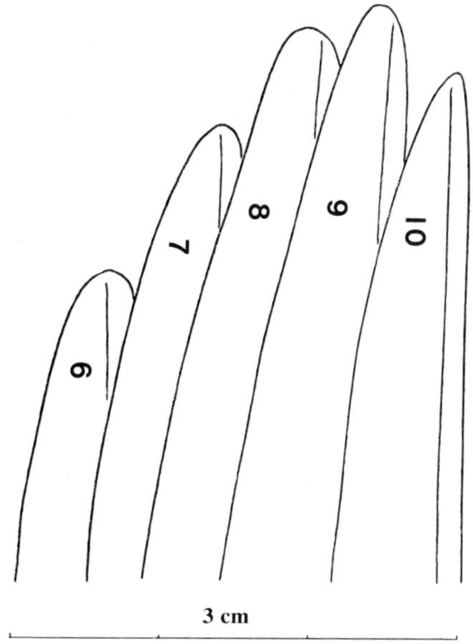

Figure 9.2 *Wing formula of the British Storm Petrel* H.pelagicus. *The outermost functional primary (10) is shorter than nine and subequal to eight, producing the characteristically round-tipped storm petrel wing.*

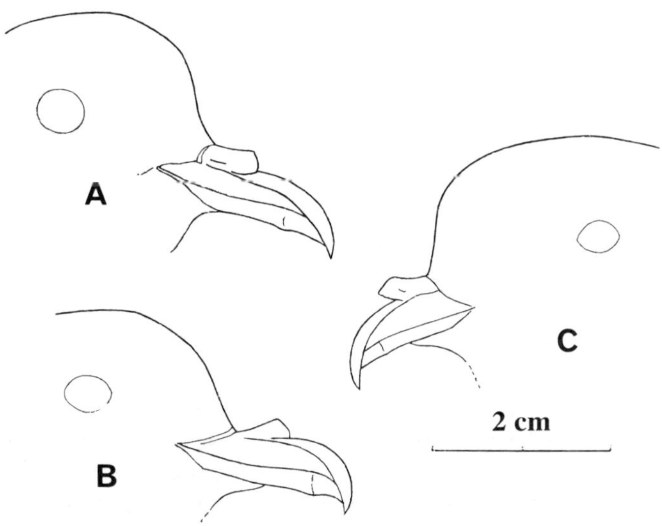

Figure 9.3 *Bills of (A) White-faced Storm Petrel* P.marina; *(B) Leach's Petrel* O.leucorhoa, *and (C) Black-bellied Storm Petrel* F.tropica.

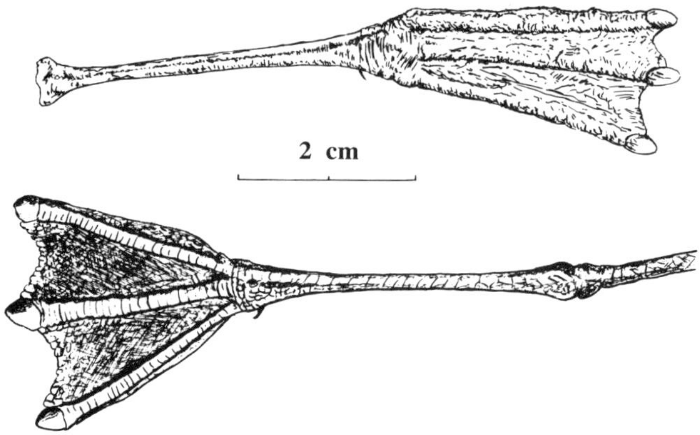

Figure 9.4 *Upper: Foot of* Nesofregetta. *The tarsus is uniquely shaped for a petrel, being anterio-posteriorly flattened and with very long basal phalanges and narrow webs — shown fully extended here. Lower: Foot of* Fregetta *showing spade-like claws and flattened mid-toe. From Murphy (1924) and Mathews (1928).*

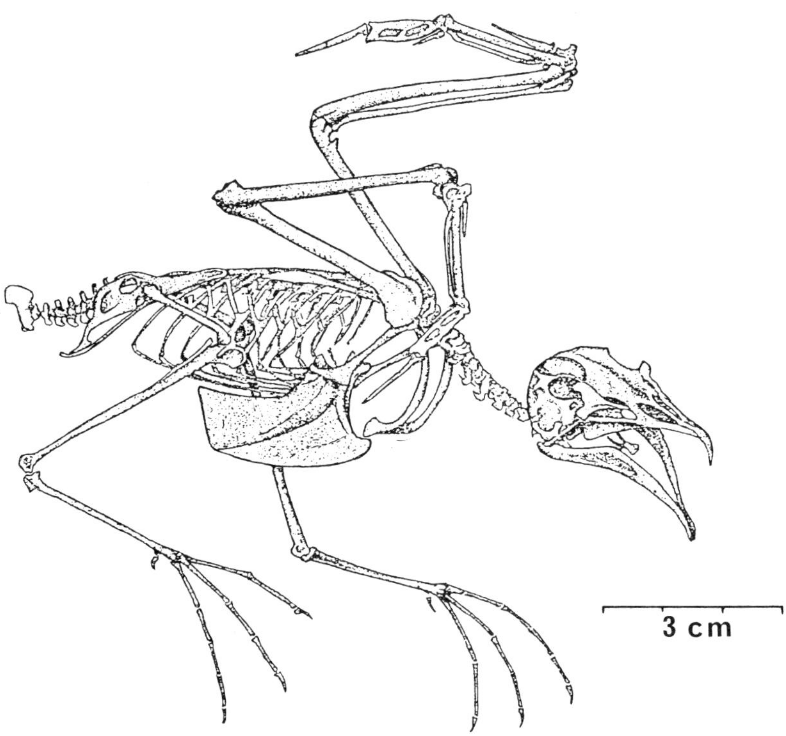

Figure 9.5 *Skeleton of Fork-tailed Petrel* O.furcata, *showing leg proportions, unnotched sternum, and deep pelvis. From Shufeldt (1888).*

long in most southern species, with a minute hind toe (Figs 9.4 and 9.5). The long legs are used to fend off from the water while picking up surface food without alighting.

Murphy & Snyder (1952) drew attention to the variations in foot structure in the long-legged species as evidence of evolutionary progression. For example, in *Garrodia, Oceanites* and *Pelagodroma* the toes are slender, but a trend towards flattening and broadening appears in *Fregetta* and culminates in the strange foot of *Nesofregetta* in which the webs are narrow so that the ventral surface of the toes is of greater area than that of the intervening webs (Fig. 9.4).

Other skeletal features include a sternum with an unnotched hind border and a special arrangement of the quadrato-tympanic region of the skull where the upper tympanic recess lies between the two articular facets for the quadrate, an arrangement differing from that of other Procellariiforms (Lowe, 1925). Osteological peculiarities also include an absence of basipterygoid processes and temporal fossae in the skull and of sesamoid (uncinate) bones at the elbow. Unlike other tubenoses, the clavicles have a downwardly directed hypocleidium at the furcular symphysis.

In size the storm petrels vary from the Least Petrel *Halocyptena microsoma* averaging 20 g, wingspan 31 cm, to the Sooty Storm Petrel *O.tristrami* averaging 83 g, wingspan 57 cm.

Storm petrels are unusual in that, where there are sexual differences in size, the females average the larger. Thus female Black-bellied Storm Petrels *F.tropica* have significantly longer wings and tarsi (Murphy & Snyder, 1952; Beck & Brown, 1971). Female Madeiran Storm Petrels *O.castro*, Swinhoe's Storm Petrels *O.monorhis*, British Storm Petrels *H.pelagicus* and Grey-backed Storm Petrels *Garrodia nereis* all have longer wings and tails than the males and in Wilson's Storm Petrel, bill, tarsus and wing lengths are all greater in females than in males and the females weigh more (Copestake & Croxall, 1985).

The mean size of the birds may vary from one population of a species to another, e.g. *F.tropica* from Signy Island average longer in wing and tail than those from Antipodes Island (Beck & Brown, 1971; Warham & Bell, 1979). However, the greatest size variation is that between the typical race of *F.grallaria* breeding at Mas Atierra Island, Juan Fernandez and the appropriately called *titan*, breeding on Rapa Island, Austral Group. Murphy & Synder (1952) found that 51 male and 12 female *g.grallaria* had wings and tails averaging 155.4, 156.4 and 73.3 and 73.5 mm respectively, whereas for 17 male and 10 female *g.titan* these measurements were 181, 184.1 and 82.4 and 83.8 mm. Rapa is also home for a particularly long-winged and long-tailed form of *N.fuliginosa* (Holyoak & Thibauld, 1984).

Some, mainly southern species, have polymorphic colour patterns. For example, *N.fuliginosa* at the Phoenix Islands breed in a range of plumages from completely dark birds to the 'normal' form with white belly and throat (Crossin, 1974); and Lord Howe Island specimens of the White-bellied Storm Petrel *Fregetta grallaria* show variation in underwing pattern, 'scalloping' due to white edges to the contour feathers, and completely melanistic examples. Also, a proportion of birds in the populations of southern white-breasted storm petrels have dark longitudinal streaks on their underparts. This led some earlier workers to believe that distinct species were involved, and one such variant was named *Pealea lineata*, hence the term 'Pealea' phenomenon. Murphy and Snyder (1952) reviewed the various specimens and concluded that *P.lineata* was merely a mutant of *F.tropica* (Fig. 9.6). These authors considered that other specimens in 'Pealea' plumage were variants of *O.oceanicus* and *F.grallaria*.

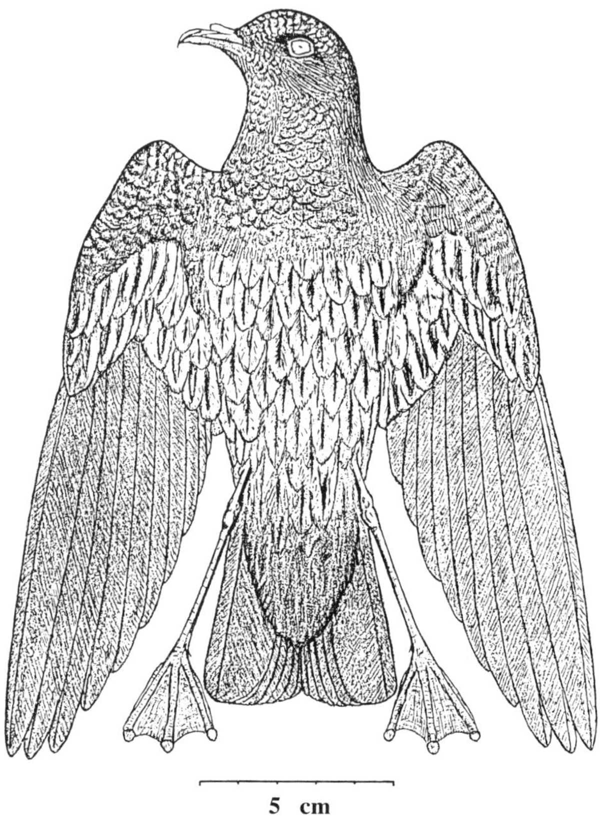

Figure 9.6 'Pealea lineata' *(type) showing the streaked underparts found occasionally in* F.grallaria, P.marina *and* G.nereis. *The bird figured was probably a morph of* F.tropica. *From Mathews (1933).*

Colour polymorphism still raises problems in categorising and naming some populations, notably those of Leach's Petrel *Oceanodroma leucorhoa* breeding along the Pacific coast of North America. These differ in mean size, colour of the rump (varying from white to blackish), and probably in their flight calls. Furthermore, at least at one breeding place Guadelupe Island, Mexico, there are winter-breeding and summer-breeding populations whose rump colours differ. A canonical variates analysis using nine mensural and one colour characters of birds from 13 localities revealed clinal variation and showed that the birds of the two Guadelupe populations differ from the rest and from each other (Power & Ainley, 1986). These authors proposed a tentative arrangement of the species into four subspecies (Appendix 1).

Storm petrels moult wing quills in descendant order (i.e. from the carpal joint outwards) after breeding, although some shedding of the body contour feathers occurs in birds still tending chicks. Migratory species like *H.pelagicus* and *O.leucorhoa* gradually replace most flight feathers while in winter quarters, being full-winged on return to the breeding grounds. Non-breeders, failed breeders and immatures probably moult sooner.

The position is different in the non-migratory Ashy Storm Petrel *O.homochroa*, studied by Ainley *et al*. (1976). The adult moult is spread over 257 days and when one phase reaches its peak the next begins. Thus body moult starts at around hatching, tail moult about 40 days later, 30 days thereafter the primaries start falling, followed almost immediately by the secondaries. Most birds have completed tail moult and lost the first two or three primaries when their chick departs, and the breeders are still not quite full-winged when they return at the start of the next nesting cycle. Leach's Petrels at the same island take longer (*c*. 274 days), and have moulted their tails but not their wings by the time that their fledglings leave. Most adults complete their moult during their contra-nuptial migration and are fully winged at the start of the breeding season (Ainley *et al*., 1976). Other studies of moult are those of Harris (1969b) on *O.castro*, which also appears to need the whole time between breeding episodes to renew its plumage. An early study was that of Murphy (1918) on *O.oceanicus*.

A peculiarity of some storm petrels is that their chicks have only one coat of down. This is true of *P.marina* (Richdale, 1965a) but not of *O.castro* (Allan, 1962), or apparently, of other Hydrobatinae. The adaptive significance of having a single coat and whether this feature is shared with other storm petrels has not been determined.

III The genera

The 21 species of storm petrels are readily separated into two groups — seven species of long-legged, southern hemisphere birds — subfamily Oceanitinae — and the shorter-legged species of more northern seas — subfamily Hydrobatinae. The Oceanitinae are mostly rather short-winged with only 10 secondaries and rather square tails and slender skulls. The Hydrobatinae have relatively longer wings and tails, the latter often forked, 12 or more secondary feathers and shorter skulls with wider gapes. The morphological similarities among the northern forms allow all 14 species to be accomodated in three genera — *Hydrobates, Oceanodroma* and *Halocyptena* — whereas there are greater differences among the southern species, hence their classification into five genera — *Oceanites, Fregetta, Nesofregetta, Garrodia* and *Pelagodroma*.

These genera are characterised on colour patterns, the condition of the nasal tubes, tail shape, structure of the claws and proportions of the leg bones.

Oceanites — Plumage black with white upper- and under-tail coverts and sometimes white on the abdomen; bill short, nasal tubes long (about half length of culmen); tarsus booted (scales fused to a continuous sheath) or with obscure scutes; claws little flattened, webs yellow.

Garrodia — Plumage grey on back, white below; bill short, nasal tubes about half length of culmen and upturned at end; tarsus scutellate in front, claws slightly flattened, webs black.

Pelagodroma — Plumage grey and white; tail slightly forked; bill long, nasal tube <half culmen in length; tarsus booted in front, webs black, basal joint middle toe flattened, claws blunt and flattened.

Fregetta — Plumage black above, white below and white upper tail coverts; nasal tube free at end and upturned, < half culmen in length; tarsus booted in front, webs black, basal joint middle toe flattened, claws blunt and flattened.

Nesofregetta — Large storm petrels; plumage black above, white throat and belly ranging to all dark below; tail strongly forked; tarsus uniquely anterio-posteriorly flattened, unscaled; basal phalanges of toes longer than remainder of toe and claw; webs narrow; claws blunt and flattened.

Hydrobates — Small petrels; plumage black with white rump; tail square; tarsus scutellate in front, slightly longer than middle toe with claw and about half length of femur; claws narrow.

Halocyptena — Very small black petrels; bill weak, compressed and decurved; tail wedge-shaped; wings rather long and pointed; tarsus rather short — slightly longer than mid-toe with claw and scutellate.

Oceanodroma — Medium-sized petrels; plumage dark or greyish, often with pale rumps; tail more or less forked; tarsus short < middle toe with claw and scutellate; claws narrow.

The main specialisations are in leg proportions and foot structure. However, Murphy (1936, p.744) drew attention to the long legs of the northern Black Storm Petrel *O. melania* and the ratios of femur:tarsometatarsus:tibiotarsus. Examples are:

O.leucorhoa	1: 1.48: 2.31	*O.hornbyi*	1: 1.56: 2.34
O.melania	1: 1.85: 2.59	*O.macrodactyla*	1: 1.30: 2.26
O.furcata	1: 1.70: 2.50:		
O.oceanicus	1: 2.33: 3.42	*P.marina*	1: 2.39: 3.51
F.tropica	1: 2.26: 3.13	*F.grallaria*	1: 2.13: 3.02
G.nereis	1: 2.16: 3.34		

Therefore, although *melania* has a slightly longer tarsus than the other Hydrobatinae, none shows the leg development of the Oceanitinae.

From a study of the muscles of the hind limbs of 10 species and of the proportions of their bones, Klemm (1969) supported the separation of the Oceanitinae from the Hydrobatinae. For example, he found *M.fibularis longus* and *M.flexor cruris lateralis accessorius* present in Oceanitinae but absent in Hydrobatinae.

Lowe (1925), ignoring the specialised hind limbs of southern species but using other skeletal characters, concluded that the northern, short-legged birds were the more specialised. His work has never been re-examined: it needs extending to all species for a fuller assessment of relationships among the storm petrels based on their skeletal anatomy.

Some workers group *Pelagodroma, Garrodia* and *Fregetta* in *Oceanites* on the basis of skeletal similarities, but this seems premature without comparative analyses of behaviour and vocalisations, and of biochemical studies (see also Klemm, 1969,

p. 160). *Pelagodroma*, for instance, differs considerably from *Oceanites* in plumage, leg-proportions, voice, behaviour at sea and in its lack of aerial displays. Harper (1978) compared plasma proteins of three southern species — *F.tropica, G.nereis* and *P.marina* (Fig. 9.7). The banding for *P.marina* was the most distinctive and Harper concluded that all three were correctly placed in separate genera.

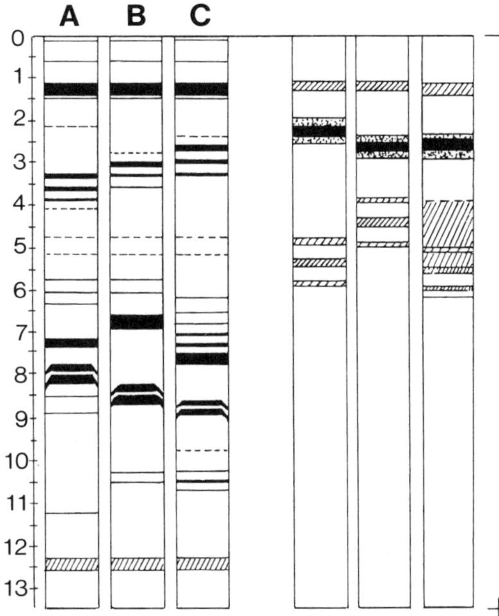

Figure 9.7 *Blood serum proteins of three storm petrels. (A)* F.tropica; *(B)* G.nereis *and (C)* P.marina. *For key to bands see Appendix 2. From Harper (1978), courtesy of* New Zealand Journal of Zoology.

One peculiarity of *Fregetta* and *Nesofregetta* is their unusually blunt, rounded, and spade-like claws (Fig. 9.4), a feature not due to wear as it it seen also in the chicks. The foot of *Nesofregetta* is even more peculiar as the tarsometatarsus is compressed in the anterior-posterior plane, giving a very long foot with broad toes and narrow webs (Fig. 9.4).

The four all-dark *Oceanodroma* of the North Pacific — *markhami, melania, tristrami* and *matsudairae* — differ mainly in mean dimensions and presence or absence of an upperwing patch. They are usually regarded as distinct species. None has been subjected to detailed study so such an arrangement is merely provisional except that, as *O.tristrami* and *O.matsudairae* breed sympatrically on the Bonin Islands without evident interbreeding, their specific status seems clinched. Likewise the tiny all-dark *H.microsoma* and *O.melania* also share nesting places with a dark-rumped form of *O.leucorhoa* at the San Benito Islands, Baja California.

Subspecific differentiation has been considerable in some widespread species with four subspecies recognised for *O.leucorhoa* and four for *P.marina*.

IV Life styles

Storm petrels are small birds that keep close to the surface of the sea from which they pick small items, usually while on the wing and facing the wind. They are very light and buoyant so that as a wave heaves upwards they rise with the displaced air, apparently without effort, and keep much the same distance from the surface. They are adept at taking advantage of air deflected from the windward face of waves or swells, flicking over the crests to the wind shadow behind. Short glides are interspersed with wing fluttering to maintain position at a food source. The long-legged species tend to use their feet to thrust off from the water, the most familiar example being Wilson's Petrel, which bounces along, pats the surface with both feet simultaneously, flutters its wings, pats again, and so on. This bird has been seen feeding and maintaining its position in winds of up to 40 knots. The White-faced Storm Petrel uses its feet similarly but glides more and banks like a prion. *Nesofregetta* is perhaps the most specialised in the structure and use of its feet. A strong kick to the surface with one foot powers an abrupt change of direction and acceleration, then follows a glide for 20–30 s before another kick-off from a wave speeds the bird in a new direction (Crossin, 1974). This author suggested that such a mode of locomotion might be energetically more efficient in equatorial regions with little wind and that the very long, flattened legs and spatulate feet are adaptations permitting this style of progression. Indeed, each kind of storm petrel appears to have its own characteristic mode of flight but one that is modified in accordance with prevailing wind and sea conditions. For example, according to Rollo Beck (in Bent, 1922), *F.grallaria* strikes the water with one foot at a time, using the foot to leeward with a cross wind, while the other is stretched out behind, but either leg is used when flying directly into the wind. *O.melania* flies with steady and deep wing beats either with or without interspersed glides (Brown, 1980). *Garrodia* uses very rapid, almost frenzied dartings to and fro when feeding from the surface of an advancing wave (Serventy *et al.*, 1971), and the flight of *O.leucorhoa* is a bit less erratic, leaping and manoeuvrable. Propulsion is aided by well developed gastrocnemius muscle built for speed.

Storm petrels can travel quite fast. Four Leach's Petrels from Kent Island, New Brunswick, transported by air to Selsey Bill, England, covered the 4796 km direct line distance back to their burrows at mean speeds of at least 392, 392, 187 and 121 km day^{-1} (Billings, 1968).

Storm petrels appear unsuited for diving but Murphy (1918) described *O.oceanicus* going to depths of several times their own length to recover sinking food and Jehl (1974a) saw *O.melania* diving for up to 4 s after suet sinking in the wake.

Wilson's Petrel and *O.gracilis* are well known for their tendency to feed behind ships, but other storm petrels shun them. Mills (1968), who logged many Hornby's Petrels *O.hornbyi* off Peru, noted that they ignored the boat, unlike *O.markhami, tethys* and *O.gracilis*. *O.castro* and *O.leucorhoa* also avoid ships whereas *P.marina* may follow them (Watson *et al.*, 1986). Most species are attracted to lights.

Storm petrels tend to be solitary at sea, but small parties are encountered around appropriate food sources or resting on the water and both Wilson's and Leach's Petrels may form quite tight rafts of single or mixed species, but usually of less than 30 birds, especially late in the day (Finch *et al.*, 1978). Migrants like *O.oceanicus* form large flocks before their departure south, e.g. off Sri Lanka (Phillips, 1955) and off New York (Murphy, 1918). Jehl (1974a) encountered small mixed flocks of *O.tethys*, *O.leucorhoa*, *O.melania* and *O.oceanicus*. He also saw groups of up to 200 *H.microsoma* near the Tres Marias Islands and had as many as 50 *O.melania* following in the wake at one time.

On the ground storm petrels typically rest on their tarsi, but even the very long-legged species can run on their toes for short distances (Fig. 9.8). In climbing obstacles the bill tip is hooked over any irregularity while gripping with the claws. Landings are gentle and controlled, take-offs, with an elastic spring, easy and immediate.

Figure 9.8 *White-faced Storm Petrels normally rest on their extremely long tarsi, but can lift their bodies from the ground to walk or run for short distances. The burrow here is in very soft sand, typical substrate for the species. Binding by plant roots (here by a succulent* Carpobrotus *sp.) is important to reduce nesting failure from burrow collapse. Photo: author.*

Murphy (1918) wrote of Wilson's Petrels becoming hyperactive towards sunset, dashing hither and thither, shooting up almost to mast height and then plunging down at full speed. This was in the Central Atlantic at 4°N on 8 October and Murphy supposed that the display reflected physiological changes as the breeding season approached. Harris (1969b) described rather similar flighting behaviour among thousands of *O.tethys* swooping around over the edge of a cliff in the non-breeding season and where few birds nested later in the year.

The functions of these performances are unclear, for the display repertoires are mostly unstudied. The northern species are well known for their nocturnal flights, with calling, over the nesting grounds. Aerial chases may eventuate or birds may hover over nest sites and then enter, attracted by the calls of birds within — e.g. in *O.castro* (Allan, 1962). Grubb (1973), using taped call playback, showed that the 'purr' calls of *O.leucorhoa* were attractive to flighting birds. Ainley (pers. comm.) has seen *O.oceanicus* and *O.homochroa* circling and calling over their nesting areas and Beck & Brown (1971) described dual flights with calling in *F.tropica* (Beck & Brown, 1971). The White-faced Storm Petrel *P.marina* is mostly silent when flying over its nesting places and aerial displays are not known for that species, apart from hovering performances of uncertain function just above the burrow entrances. Ground and within-burrow actions among storm petrels await examination: only desultory mutual and reciprocal preening have been described, but neither copulation nor territorial behaviour.

Also unexplained are the frequent collisions that occur during aerial chases at the nesting grounds. These are known for *O.tethys, H.pelagicus* and *O.leucorhoa*, and appear to be deliberate rather than accidents of crowded airspace. Williamson (1945) noted that colliding Leach's Petrels which fell to the ground disappeared into burrows, and he suggested that this activity may be the finale to the aerial display and the prelude to duetting in the burrow.

As most storm petrels come ashore only after dark, vocal communication is important. The repertoire of Leach's Petrel has been most studied, including some inter-population differences by Ainley (1980). Common to many northern species are sustained churring or purring sounds given from the ground or from within burrows. Such calls may be sustained for minutes at a time, being broken only by brief, high-pitched squeaks which Hall-Craggs & Sellar (1976) showed in *H.pelagicus* and *O.leucorhoa* are inhalatory 'breath' notes. Similar calls are used by *O.castro, O.tethys* and *O.monorhis* (Harris, 1969b; James & Robertson, 1985), and there are sexual differences in the calls of *H.pelagicus* in which, according to James (1984), only the males 'churr' whereas both sexes use a similar flight call. A series of investigations into the functions of the calls of *O.leucorhoa* has been carried out in Japan — references in Taoka *et al.* (1989). A twitter, reiterated for many minutes, is the main call of *O.melania* and that of *N.fuliginosa* is a soft, guttural 'grrr,grrr,grrr' (Crossin, 1974). Less is known of the vocalisations of southern species but *F.tropica* and *F.grallaria* use high-pitched whistling and piping sounds (Hagen, 1952; Beck & Brown, 1971) and spontaneous low whistles led to the first discovery of the nesting place of Elliot's Storm Petrel *O.gracilis* (Schlatter & Marin, 1983). The whistle of *F.tropica* is very powerful and has a ventriloqual quality (Beck & Brown, 1971; Imber, 1983). Wilson's Petrel, however, uses a harsh disyllabic ground call, monotonously repeated, as well as high-pitched 'peeps' (Beck & Brown, 1972; Roberts, 1940), and when flocking round food at sea these birds use a sparrow-like twitter by day and night (Murphy, 1918; Roberts, 1940). A detailed study of the vocalisations of *O.oceanicus* has recently appeared (Bretagnolle, in press). *Garrodia nereis* also uses a rhythmically repeated but quiet wheezy croak. Perhaps the most distinctive voice is that of *P.marina* whose main call is a regularly repeated low bark, very like that of Bulwer's Petrel, and apparently bearing little resemblance to the call of any other storm petrel. Most of the Hydrobatinae use aerial calls, most of the Oceanitinae are either not recorded as calling in flight, or, at least over the nesting places, call sparingly.

V Distribution

Figure 9.9 shows the known breeding distribution of storm petrels. Precisely where Hornby's, Markham's and Elliot's mainly breed is still unrecorded and more colonies of *Hydrobates* (particularly in the Mediterranean), of *O.oceanicus* and of *Nesofregetta* must still await discovery.

There are few polar storm petrels: in the north only the small colonies of *H.pelagicus* and *O.leucorhoa* at the Lofoten Islands, Norway, and, in the south, many colonies of *O.oceanicus* around Antarctica. As Fig. 9.9 shows, the shorter-legged forms are not entirely restricted to the northern hemisphere as three *Oceanodroma* nest somewhere near the cool Humboldt Current in coastal Peru and Chile (*hornbyi, tethys* and *markhami*), and, in the Atlantic, *O.castro* breeds at St Helena and Ascension Islands. Of the long-legged species only *P.marina* nests north of the Equator — at the Cape Verde, Salvage and Canary Islands, paralleling the distribution of *Pterodroma mollis* and *Puffinus assimilis*. Bourne (1953) suggested that *P.marina* may have reached the North Atlantic along the cool currents off West Africa during the Pleistocene and, if so, the North Atlantic race can only have been isolated since the last Ice Age, *c.* 10 000 years B.P.

In the Southern Ocean three species nest sympatrically at the Auckland, Crozet, and Kerguelen Islands, while a complex of five (plus one recently extinct species) is centred around islands from California to northern Mexico. Least, Leach's and Black Storm Petrels nest at the San Benito Islands and according to Crossin (1974) every possible habitat is occupied and there is competition for nest sites. The concentrations around Baja California may have developed in response to good food availability in local upwellings, complex oceanography and refugial populations during ice age (Ainley, pers. comm.), but diversity in the Southern Ocean may have as much to do with the lack of islands than exceptionally good feeding locally.

Some storm petrels are adapted to cool surface waters, others to warm seas, and occasionally such opposites share a nesting island, as with *P.marina* (warm water) and *F.tropica* (cold water) at the Auckland Islands. Off western USA *O.leucorhoa* prefers warmer waters and is found far offshore unless warm water comes in closer whereas co-nesting *furcata* and *homochroa* occur inshore (Ainley & Boekelheide, 1990).

Distributions at sea are mostly little known, partly due to the problems of seeing and identifying small birds hugging the surface. Crossin (1974) summarised the many sightings and specimens collected in parts of the Northern and Central Pacific during the Pacific Ocean Biological Survey Program of the Smithsonian Institution. Many species, like *O.castro*, range widely, but breeding birds with chicks visit them frequently so must feed locally: many distant sightings in the breeding season must refer to immatures or non-breeders.

Although single storm petrels are encountered at sea apparently very far from others of their kind, at times they may be numerous along the edges of fronts and eddies and other features associated with surface plankton concentrations. Haney (1985) has demonstrated how *O.castro* aggregates along upwellings associated with the western and eastern Gulf Stream boundaries off Florida, Georgia and South Carolina.

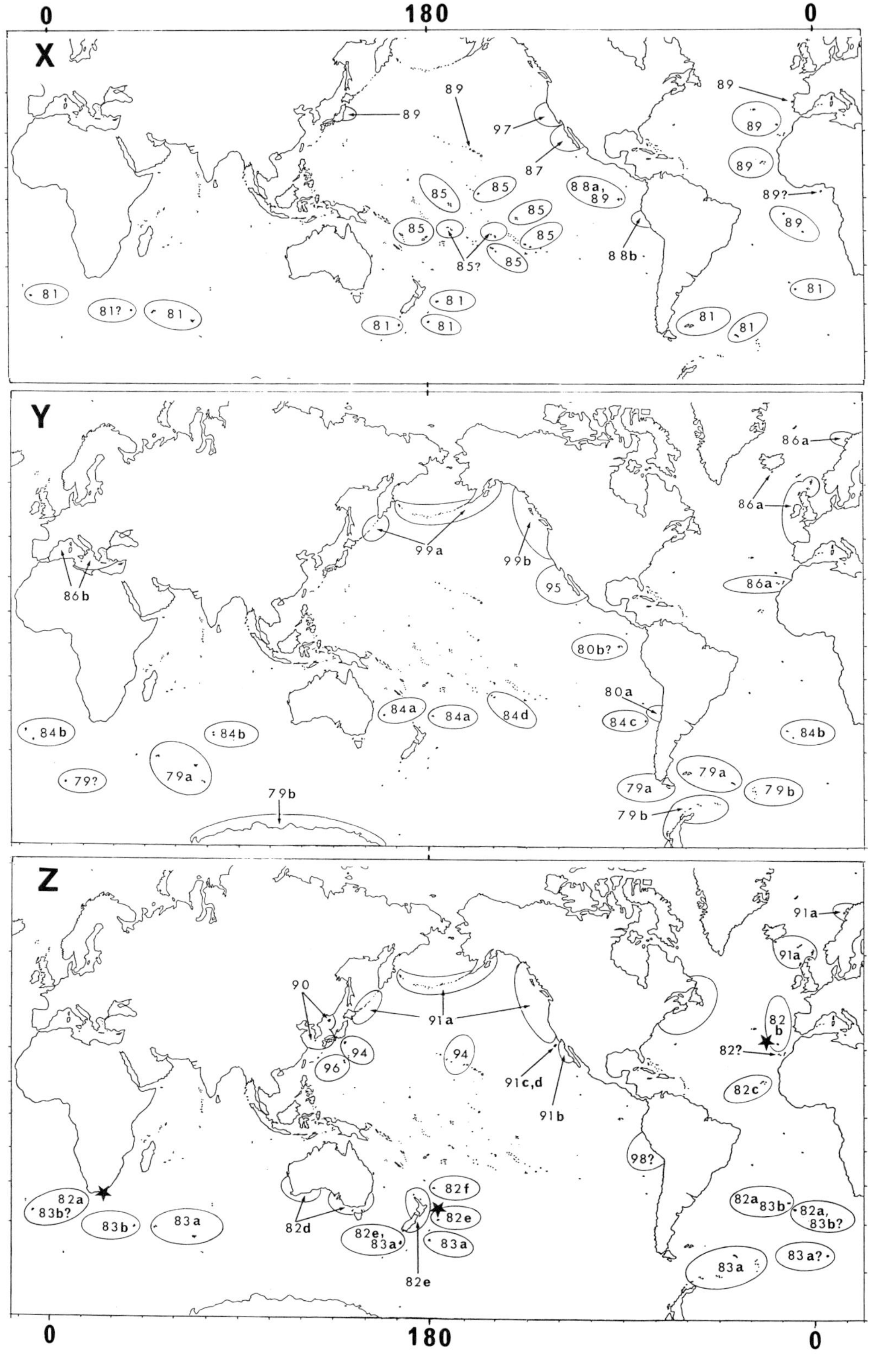
0
180
0
X
89
89
89
97
87
89
89
85
85
88a, 89
89?
85
85
85
85
89
85?
88b
85
81
81?
81
81
81
81
81
81
81
Y
86a
86a
99a
99b
86b
95
86a
80b?
80a
84a
84a
84d
84c
84b
84b
84b
79a
79?
79a
79b
79a
79b
79b
Z
91a
91a
90
91a
94
82 b
96
94
82?
91c,d
91b
82c
98?
82f
82a 83b?
82e
82a 83b
82a, 83b?
83b
83a
82d
82e, 83a
83a
83a?
83a
82e
0
180
0

VI Feeding and foods

The variations in size, relative wing length and tail shape may be reflected in the various flight and feeding techniques, although for most species these await adequate description. Beck & Brown (1971), for example, noted that while Wilson's Petrels flutter and patter with wings held high, Black-bellied Storm Petrels tend to swing from side to side with wings just above the horizontal, bounding along in a series of skips.

All the species seem to get the bulk of their food from the top few cms of the sea, manoeuvring with their flexible wings and tails and using air currents to remain within pecking range of the surface, taking food by pattering and dipping without alighting. Diving and surface-seizing of food seem to be unusual. From a submarine Norris (1965) watched *P.marina* feeding and by dipping the periscope he saw that small planktonic organisms were very close to the surface, despite strong sunlight. Forty-three storm petrels were pattering close to the periscope and the splashes as their feet struck the water were clearly visible. In addition to its normal pattering action with both feet simultaneously, Wilson's Petrel has also been reported as facing the wind, wings held out and both feet immersed as a sort of sheet anchor to hold its position while pecking rapidly at concentrations of small prey.

Storm petrels appear to suffer a high incidence of damage to their feet. Allan (1962) recorded 3.5% of *O.castro* with foot injuries, Harris (1969b) 7% in *O.tethys*, while Threlfall (1974) figured typical examples in *O.leucorhoa*. Birds with a leg missing are sometimes seen and clearly dipped feet are at risk from predatory fish. Presumably some lose both feet, their pattering ability and hence die. Matthews (1951) noted some Wilson's Petrels that seemed to be stirring the sea with their feet and Zink & Eldridge (1980) hypothesised that the yellow-centred webs of *Pelagodroma*, *O.oceanicus* and *O.gracilis* might be dangled in the water to lure (or frighten) krill and make them more conspicuous to the birds.

A good deal of mainly circumstantial evidence implies that some storm petrels feed at night, (e.g. Murphy 1936, p.728; Watanuki, 1985). Murphy noted that fishermen taking Wilson's Petrels by entanglement in weighted threads trailed from the stern, caught more by night than by day. Others have watched storm petrels feeding in ship's lights and while hove to at 8°23'S, 80°45'W, Mills (1968) saw a large number of *O.hornbyi* feeding at the edge of the lighted area around his ship.

Figure 9.9 *Breeding distribution of storm petrels:* (**X**) G.nereis *(81)*; N.fuliginosa *(85)*; H.microsoma *(87)*. O.tethys *(88)*, t.tethys *(88a)*, t.kelsalli *(88b)*; O.castro *(89)*; O.homochroa *(97)*. (**Y**) O.oceanicus *(79)*, o.oceanicus *(79a)*, o.exasperatus *(79b)*; O.gracilis *(80)*; F.grallaria *(84)*, g.grallaria *(84a)*, g.leucogaster *(84b)*, g.segethi *(84c)*, g.titan *(84d)*; H.pelagicus *(86)*, p.pelagicus *(86a)*, p.melitensis *(86b)*; O.melania *(95)*; O.furcata *(99)*, f.furcata *(99a)*, f.plumbea *(99b)*. (**Z**) P.marina *(82)*, m.marina *(82a)*, m.hypoleuca *(82b)*, m.eadesi *(82c)*, m.dulciae *(82d)*, m.maoriana *(82e)*, m.albiclunis *(82f)*; F.tropica *(83)*, t.tropica *(83a)*, t.melanoleuca *(83b)*; O.monorhis *(90)*; O.leucorhoa *(91)*, l.leucorhoa *(91a)*, l.chapmani *(91b)*, l.socorrensis *(91c)*, l.cheimomnestes *(91d)*; O.tristrami *(94)*; O.matsudairae *(96)*; O.hornbyi *(98)*. *Stars mark places where Leach's Storm Petrels have been found in burrows in S.Africa and the Chatham Is., New Zealand and Swinhoe's Storm Petrel in the Salvage Is.*

Few detailed studies of foods have been done but a great array of planktonic invertebrates has been found in storm petrel stomachs or ejecta: amphipods, copepods, decapods, polychaetes, squid and other molluscs. Fish otoliths are also common. Stomachs of *P.marina* taken in the Atlantic contained fish up to 6.5 cm long and Atlantic and Pacific races of this bird take marine water-striders *Halobates* sp.

Watanuki (1985) surveyed the diet of *O.leucorhoa* at a Japanese colony by pumping out 307 stomachs. He believed that this technique did not get the larger items, but, for the rest, squid and *Euphausia pacifica* were the most numerous, together with one species of copepod and a range of amphipods, decapods, fish and plastic debris. There was some seasonal change in the food spectra.

Imber (1981) found an ecological segregation by diet in *P.marina* and *G.nereis* at the Chatham Islands. The former took a great variety of prey 6-25 mm long, but no copepods, and Imber believed that feeding was mainly over the continental shelves and the Subtropical Convergence. The diet of *Garrodia* was different and notable for the abundance (over 8000 in 27 stomachs) of the stalked barnacle *Lepas australis*, both as large cyprids and as attached juveniles. Imber suggested that taking *Lepas* was a dietary specialisation of this small storm petrel, that the young barnacles were picked up by day from mats of floating debris, and that the latitudinally restricted range of *Garrodia* (Fig. 9.9) corresponds with the restricted range of this barnacle. *Lepas australis* has a large cyprid larva and is available at the surface throughout the year. Richardson (1984) watched hundreds of *P.marina* (locally called 'skipjack') feeding in the kelp belt off Tristan da Cunha from mid-November to early January.

Wilson's Petrels do not get all their food while on the wing, and will readily settle to feed on dead birds or to scavenge offal. Large numbers will huddle together in a dense mass over big pieces, behaviour that allowed fisherman seeking bait to kill a score or more with a single flick of a whip (Collins, 1884). There are also many accounts of storm petrels feeding at oil slicks from whale and seal carcasses and some species can be enticed to feed on artificial slicks of vegetable oil or on 'chum' like popcorn! Croxall *et al.* (1988) provide details of the food and feeding of this species at South Georgia, Ainley *et al.* (1984) that for the Ross Sea, Antarctica.

VII Breeding

Accounts of the breeding of storm petrels include a 5-year study of *P.marina* by Richdale (1965a). Other investigations include those of Lockley (1932) and Davis (1957) on *H.pelagicus*, on *O.oceanicus* by Roberts (1940), Lacan (1971), Beck & Brown (1972), and Wasilewski (1986), on *O.tethys* by Harris (1969b), on *O.castro* by Allan (1962) and Harris (1969b), on *F.tropica* by Beck & Brown (1971), on *O.furcata* by Boersma *et al.* (1980) and Simons (1981b) and on *O.leucorhoa* by Gross (1935) and on the latter and *O.homochroa* by Ainley *et al.* (1990).

Many species, e.g. *O.oceanicus* and *O.homochroa* (Beck & Brown, 1972; Ainley *et al.*, 1990) are susceptible to handling or interference, particularly during incubation, and workers like Davis (1957) have reported consequential breeding failure. Boersma *et al.* (1980) found that daily handling of adult *O.furcata* caused nest desertions, with hatching success of those that stayed lower than at sites seldom inspected. Others make no mention of such effects, but they may have been operating nevertheless.

Storm petrels generally breed in the summer months but *O.tristrami* and *O.leucorhoa* of the North Pacific are winter breeders and some tropical species may nest throughout the year. Chicks of *O.markhami* apparently fly from their undiscovered breeding grounds in November, when their numbers increase off northern Peru. One still with down on its head was caught at 09°S 79°W on 10 November 1977 (Brown, 1981).

Site- and mate-tenacity appear to be the norm, e.g. in *P.marina* and *O.furcata* (Richdale 1965a; Boersma *et al.*, 1980) but few data are available for most species. Likewise breeding is an annual event, although Crossin (1974) pointed out that records from two populations of *O.leucorhoa* showing that only half the breeders in one year were seen the following year, suggested either a very high mortality, movement elsewhere, or non-annual nesting. However, that *H.pelagicus* lays over an 85-day period (Davis, 1957) and *P.marina* over 43 days, led Richdale (1965a, p.143) to suggest that such long laying seasons might be due to early layers replacing lost eggs (Chapter 13, Section IIIC).

At their island sites most storm petrels avoid excessive predation by gulls, skuas, and giant petrels by being strictly nocturnal. A few may fly overland in daylight: some *O.oceanicus* do this and in the tropical Pacific *Nesofregetta* may feed its chick in broad daylight and *F.g.titan* also flies over its breeding grounds by day. At the Galapagos *O.castro* is strictly nocturnal whereas *O.tethys* feeds its young during daytime (Harris, 1969b) and Harris believed that breeding *tethys* specialised in feeding at night, as also found by Ainley, Spear & Ribic (pers. comm.).

Bright moon tends to reduce night-time activity over land, e.g. with *O.leucorhoa* and *O.homochroa*, birds that suffer heavy predation from gulls, and their effect on the first species at a Japanese colony has been analysed by Watanuki (1986), *Larus schistisagus* being their main predator.

Storm petrels lay in burrows or in recesses among rocks and talus breakdowns. However, the only nest of *O.gracilis* so far discovered was on the ground under a bush and evidently some *O.tethys* may lay in niches hidden by vegetation, while incubating *Nesofregetta* may be only partly hidden. *Garrodia nereis*, a tiny petrel poorly equipped for digging, tends to lay in natural recesses under plant cover and with a very short or no entrance tunnel (Kidder & Coues, 1876; Despin *et al.*, 1972) or constructs a grassy cup deep in the skirt of a tussock clump at the end of a tunnel made through the leaves. *Fregetta grallaria* may nest similarly (Fig. 9.10).

Burrow densities may be high but those species using natural cavities under stones are restricted by availability of suitable niches and the nests of *G.nereis* at Ile de L'Est, Crozet Islands, are widely scattered for this reason (Despin *et al.*, 1972). At the Tower Island colony of *O.tethys*, sites are also limited and intra-specific competition for them fierce, forming a major cause of nesting failure (Harris, 1969b).

These small petrels lack the reserves to withstand long fasts and incubation spans are usually <3 days unless bad weather intervenes or burrows get blocked with snow — as happens with Wilson's Petrel (Roberts, 1940). Intermittent incubation is common but the eggs resistant to chilling (Chapter 14, Section IV). If the chicks are cold and starved they may become torpid yet revive to fledge if heated and fed in time (Chapter 15, Section V3).

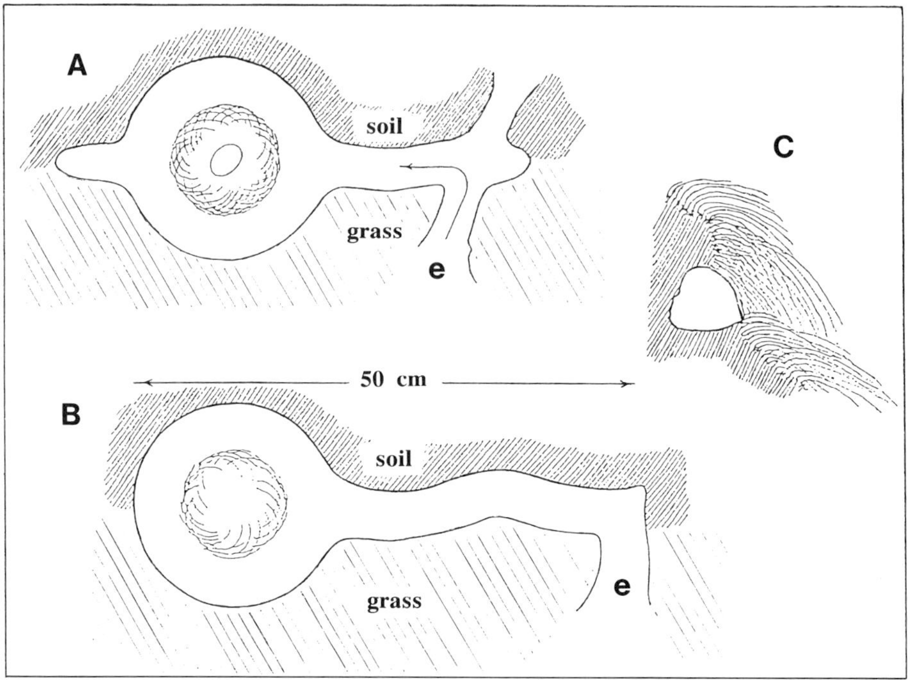

Figure 9.10 *Burrows of* Fregetta grallaria *at Nightingale Island, Tristan da Cunha in cross section: (a) nest with complex entrance, (b) more typical site, (c) entry passage along bank hidden by handing grass. Nests curtained by vegetation or in the skirts of tussocks are common with this species and usual with* G.nereis. *From Hagen (1952).*

VIII Dispersal and migration

Some storm petrels seem to be relatively sedentary. The Galapagos *O.tethys,* for example, may be found at Tower Island in almost any month, and their spectacular flighting behaviour continues when breeders are at sea moulting (Harris, 1969b). Ashy Storm Petrels are also to be found at the Farallones Islands, California, throughout the year. But some younger, pre-breeding storm petrels may wander far during the nesting season. Shetland-ringed *H.pelagicus* in this category have been recaptured in Ireland, St Kilda, the Hebrides, the Faroes, Iceland and the Lofoten Islands (Fowler & Swinfen, 1984).

In general, high-latitude northern species move south after breeding, high-latitude southern ones go north. The most notable migrant storm petrel is *O.oceanicus*. During the Antarctic winter it deserts the Southern Ocean. Its movements have been analysed by Roberts (1940), Serventy (1952), Huber (1971) and others. The birds may travel fast and some go as far as Newfoundland and Scotland in the Atlantic, to the Red Sea in the Indian Ocean, and to Japan in the Pacific Ocean (Nakamura *et al.,* 1983b). Flocks of thousands may be seen in the Arabian Sea during the southern winter when the species is also abundant along the Gulf Stream off eastern North America.

Another southern trans-equatorial migrant after breeding is the White-faced Storm Petrel *P.marina*. Birds from Western Australian colonies winter mainly along the tropical convergences in the Arabian Sea and northern Indian Ocean (Bourne, 1953; Bailey 1968). The movements of the eastern Australia and New Zealand birds are less clear, but Imber (1984a) provided new evidence that the many wintering in the eastern tropical Pacific, particularly west of the Galapagos Islands (Loomis, 1918; Crossin, 1974) are of Australasian origin, and an adult ringed at the Chatham Islands was recovered about 1000 km south of the Galapagos in May 1980. There are also many winter sightings of these petrels in the south-west Pacific from northern New Zealand to the Kermadec and Norfolk Islands and New Caledonia.

Matsudaira's Storm Petrel migrates from its sole breeding place in the Volcano Islands, north-west Pacific, to winter with other storm petrels in the western Indian Ocean, possibly travelling over the shelf water off northern Australia. Some *O.monorhis* also enter the Indian Ocean but that movement is very inadequately documented (Bailey *et al.*, 1968).

Other movements are less spectacular. Some *H.pelagicus* winter off South Africa and British-ringed birds have been recovered, beach-washed, in Durban (J.A. Fowler, pers. comm.). North Atlantic *O.leucorhoa* also migrate south, mostly wintering in the tropics on both sides of the ocean, as do many of this species in the Pacific (Crossin, 1974). Southwards movements of *O.melania* and *H.microsoma* in November occur after breeding, the latter replacing *O.tethys* in the Gulf of Panama, a bird that is abundant there from June to November (Crossin, 1974).

Some wander far and may even investigate possible breeding places well outside their normal breeding range — Leach's Petrel provides two recent examples (Fig. 9.9). In the summer of 1980 two birds were found performing pre-breeding activities in a burrow on an island of the Chathams Group, New Zealand (Imber & Lovegrove, 1982). Unlike North Pacific breeders at that date, these were in fresh plumage and appeared to have switched their normal moult regime by 6 months. Likewise, at St Croix Island in Algoa Bay, South Africa, between 1976 and 1984 a Leach's Petrel called from a hole between late October to January, again with no evidence of breeding (Randall & Randall, 1986). Even more surprising is the Swinhoe's Storm Petrel captured calling from a burrow on Grande Salvage Island in June 1983. This bird breeds on islands around Japan and Korea (Fig. 9.9), some migrating into the Indian Ocean, but none had been recorded in the Atlantic before. Such birds, far from their normal range, may become imprinted on their new location and return year after year with little chance of finding a mate.

IX Populations and mortality

Storm petrel populations range from tiny to enormous. Wilson's Petrel has often been considered one of the world's most abundant birds, but estimating population sizes of small birds, most of which nest under rock piles, is virtually impossible, and most of the figures touted around must be mainly guesswork. 'Several million' may breed in the islands of the Scotia Arc (Croxall *et al.*, 1984). At the other end of the globe the Alaskan populations of *O.furcata* and *O.leucorhoa* seem equally large, estimated at about 5 and 4 million birds respectively and they may be increasing as they have re-established colonies on many islands where they had been exterminated by foxes (Lensink, 1984). And, whereas *O.oceanicus* may have to compete for resources with many other petrel species, the Alaskan petrels mainly share resources with a wide range of small and medium-sized auks, some probably more numerous than the petrels.

In contrast to these huge populations, only a few pairs of *F.grallaria* now breed at St Paul Island (on Roche Quille) where the birds were once abundant, and the large form *titan* on Rapa Island is said to number but a few hundred pairs (Holyoak & Thibault, 1984). Other breeding colonies have been lost in historic times to predation by cats, dogs and rats and some have left no trace, e.g. the *F.tropica* and *G.nereis* that surely bred at Macquarie Island.

On Tower Island, Galapagos, Harris (1969b) found that the most densely populated area (2500 m^2) in the *O.tethys* colony held 15.6 pairs m^{-2}. He estimated a total population of 200 000 pairs (at *c.* 8 m^{-2}).

The age structure of one population of Leach's Petrel has been analysed by Morse & Buchheister (1977). They found that most birds in burrows were 6 or more years old, some reaching 20 years. Mean life expectancy on attainment of breeding status was at least 4.2 years, mean annual survival at least 78.9%.

Storm petrels have plenty of natural predators when on land. Adults and chicks are killed by skuas, sheathbills and gulls. Large gulls like *Larus marinus* swallow adult storm petrels whole, casting up neatly packaged skeletons later. The very long-legged *P.marina*, while quite mobile on open ground, easily becomes entangled in low vegetation or wet grass when its struggles often attract the attention of skuas. In warm climates snakes and other reptiles take chicks and eggs and a large scorpion has been seen carrying off an egg of the Least Petrel (Crossin, 1974). The local Galapagos Owl *Asio galapogoensis* kills large numbers of *O.castro* and *O.tethys* (Harris, 1969b) and River Otters *Lutra canadensis* take some Fork-tailed Petrels on islands in Washington State.

Some storm petrels are notoriously prone to be storm-driven onto land. Leach's Petrel is the best known for this, with many 'wrecks' recorded. A very big episode was in the autumn of 1952 with wrecks throughout Britain, France and Belgium, extending to Germany and even to Switzerland (Wynne-Edwards, 1953). Quite large wrecks of the same species also occur in Japan (Nakamura *et al.*, 1983). In 1913 breeding *O.tristrami* at Laysan Island suffered heavy losses due to flooding of burrows. Shortly afterwards many at the same colony were smothered in their nests during sandstorms.

An unusual cause of accidental mortality has been found among New Zealand and Gough Island populations of *P.marina*. On South-East Island, Chathams Group an estimated 200 000 dead birds were found in 1970 with their feet bound together by a fibrous material that had restricted the birds' movements on the ground so that they became entangled in the vegetation and died. The material proved to be filaments from a larval trematode *Distomum filiferum* (Claugher, 1976). Evidently the birds accidentally picked up the adhesive filaments while trailing their legs in aggregations of these larvae, and the filaments hardened on drying (Fig. 9.11). Imber (1984b) suggested that the storm petrels were the main victims because of the way they use their feet while foraging on the euphausid *Nematoscelis megalops*, itself an intermediate host of the trematode.

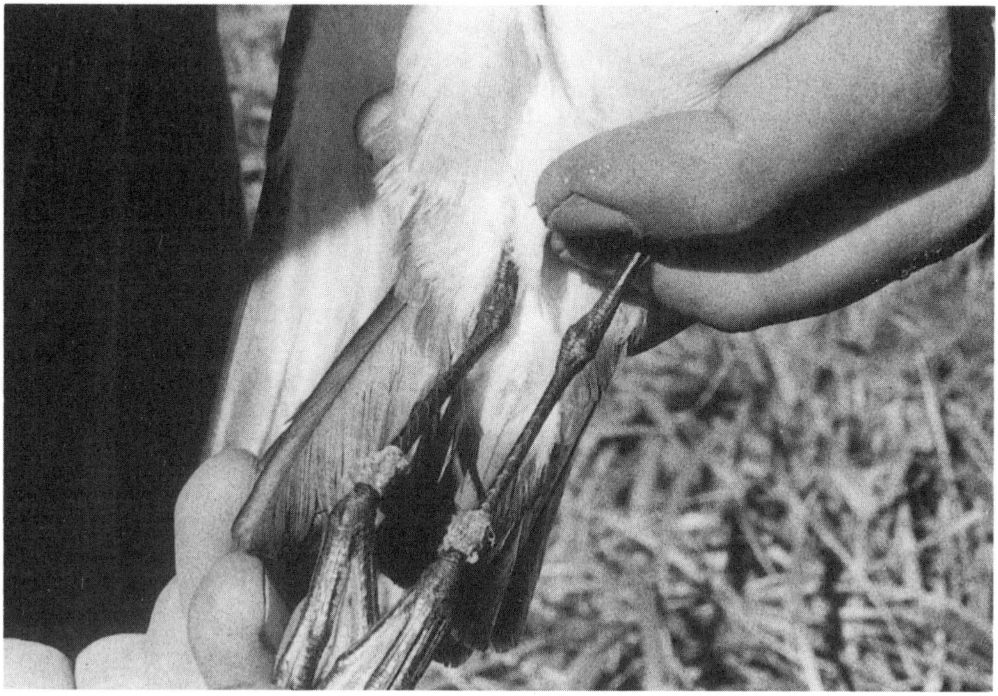

Figure 9.11 *Trematode 'anklets' on tarsi of White-faced Storm Petrel from the Chatham Islands, New Zealand. Photo: T.G. Lovegrove.*

X Human use and conservation

Storm petrels were eaten by the Morioris on the Chatham Islands and by Eskimos and North American Indians in Alaska, British Columbia and Washington State. J.H. Gurney saw a storm petrel (*H.pelagicus*) displayed for sale in Leadenhall Market, London in 1870. Another occasional use was as a candle. Sealers and whalers are said to have threaded a wick through the mouth and out through the cloaca: this drew up the stomach oil which was then lit.

Being small, easily transportable and with some species not too alarmed by handling, storm petrels have been used in several experimental studies. *O.leucorhoa* was the subject of investigations into olfaction by Grubb (1974, 1979); used by Griffin (1940) and Billings (1968) for testing homing ability, by Ricklefs *et al.* (1986) for measuring the energy costs of incubation and Vleck & Kenagy (1980) used *O.furcata* for studying embryonic metabolism at different temperatures.

Man's influence today is mainly indirect — kills due to attraction into lighthouse beams and to the deck lights of fishing boats. Man-made introductions of alien mammals are particularly dangerous and cats, dogs, rats, foxes, mongooses and so on given access to breeding places have wiped out many populations. The Guadalupe Storm Petrel *O.macrodactyla* was exterminated sometime after 1906 by a combination of goat-induced erosion and cat predation (Jehl, 1972). Rats are particularly serious threats to these small birds. The disappearance of Sooty Storm Petrels *O.tristrami* from Midway Atoll has been attributed to the arrival of rats and *Rattus rattus* and cats have completely eliminated the huge numbers that bred on Torishima Island until the late 1960s (Hasegawa, 1978).

Trampling over burrows in soft ground by sheep and cattle has also depleted some colonies, particularly those of *P.marina* in parts of Australia. Trampling by human visitors may also be damaging. At Mud Island, Victoria White-faced Storm Petrels burrow at high densities, from 2–4 m^{-2} (Gillham, 1963) and it is impossible to move over the ground without crushing burrows and occupants. This Mud Island colony was one of the first areas protected by law for the breeding of storm petrels: it was declared a sanctuary in 1903.

As they nest on islands, and often quite small ones, the main practical measures usually needed for protection of the birds is the maintenance of their habitats free from introduced predators and possibly to restrict human access to marked pathways during the nesting season. Local control of gull populations may be necessary where these have increased excessively in response to greater food availability at rubbish tips and the like. Some success has been achieved in re-establishing *O.leucorhoa* on islands off the north-east United States where these birds formerly bred, using artificial burrows and vocal playback lures (Podolsky & Kress, 1989) and such techniques have potential elsewhere.

CHAPTER 10
Family Pelecanoididae — The Diving Petrels

Common Diving Petrel.
Photo: author.

Contents

I Introduction

This southern family consists of four species of small, compact petrels readily recognised by their size, shape, colouring and behaviour. Their written history is short, although the early navigators rounding the Horn must have seen the birds. Georg Forster saw them in the South Atlantic and off Kerguelen during Cook's second voyage. He called them Diving Petrels, *Procellaria tridactyla*, but gave no description (Forster, 1777). This was done by Gmelin when he named the Common Diving Petrel *Procellaria urinatrix* from a drawing made by Forster of a bird seen in 1773 in Queen Charlotte Sound, New Zealand.

Until recently diving petrels did not feature in the fossil record. Two humeri and a coracoid of a bird of early Pliocene age from South Africa described by Olson (1985a) and named *Pelecanoides cymatotrypetes* (Gr. *kyma*, wave; *trypetes*, hover) is the only example to date. The humeri are rather flattened as in modern species and the bird was of similar size to today's Common Diving Petrel. Olson considers that the Pliocene bird would have been readily recognisable as a diving petrel, supporting a widespread belief that the family is an old one.

II General characteristics

These are short-necked petrels weighing 105–130 g and with wings spanning 38–42 cm. Their plumage is very compact, shiny black above and white below. Tail and wings are short, the latter rather paddle-like as part of their specialisation for swimming or 'flying' underwater by wing propulsion. The bill is short (Fig. 10.1) the openings of the nasal tubes being kidney-shaped and opening upwards (Fig. 10.2). Within them are flanges peculiar to the family — the so-called paraseptal processes (Fig. 10.3), named by Murphy & Harper (1921), whose function may be to prevent the ingress of water when flying through sea-spray. The bill is black with a varying amount of blue on the ramicorns. These enclose an area of loose skin forming a distensible pouch for food storage (Fig. 10.4, c.f. *Pachyptila* (Fig. 5.10)). The feet are also rather variable in colour, but usually include some shade of blue, brighter in breeding birds. There is no vestige of a hind toe. In the skeleton the hind border of the sternum is unnotched (Fig. 10.5) with the *Proc. craniolateralis* or 'costal processes' (with which the ribs articulate) very anterior and with outwards turning 'horns'. These arrangements are similar to those in the storm petrels but the sternum of *Pelecanoides* is proportionally narrower.

Latham (1785) noted the similarity of a diving petrel to a Little Auk *Alle alle* and Charles Darwin (in Gould, 1841) was struck by the resemblances of that bird to the Magellanic Diving Petrel *P.magellani*. They have almost identical proportions, plumages, gular pouches and very similar flight and feeding behaviour. Some small auks also lay single white eggs in burrows which they too visit after dark, avoiding moonlit nights. Such similarities extend to many features of the skeleton, e.g. the rather flattened wing bones, free finger joints allowing rotation and extension for underwater propulsion, rearwardly placed feet, the same number and arrangement of vertebrae (Murphy & Harper, 1921) as well as similar wing

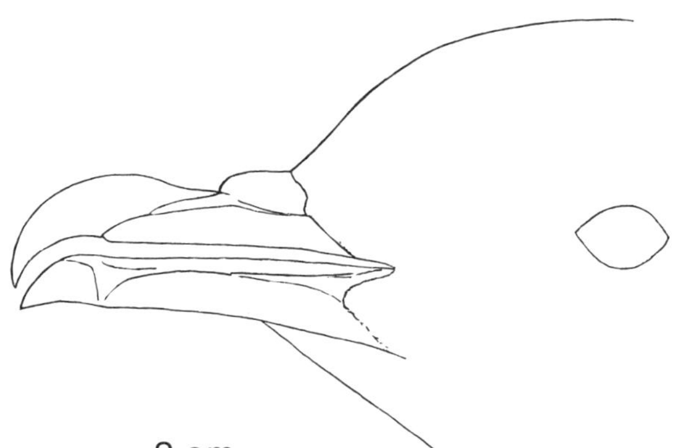

Figure 10.1 *Head of Common Diving Petrel* Pelecanoides urinatrix.

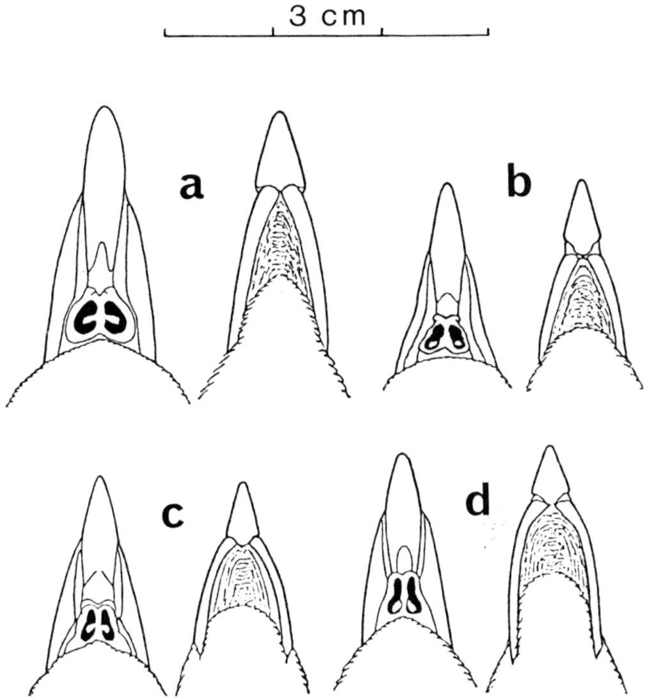

Figure 10.2 *Dorsal (left) and ventral views of the bills of a.* P.garnotii, *b.* P.magellani, *c.* P.georgicus, *and d.* P.urinatrix. *From Murphy & Harper (1921).*

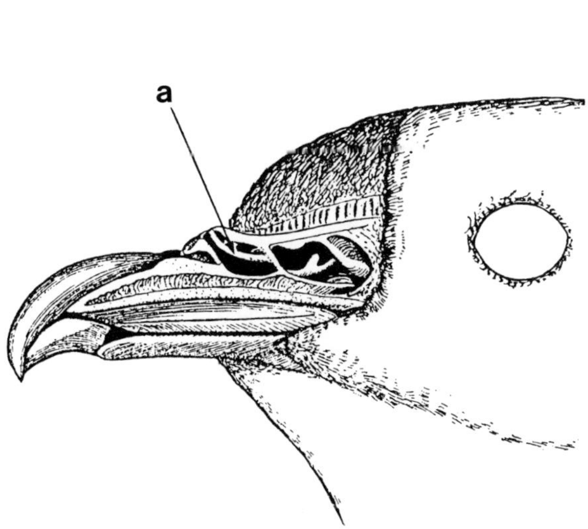

Figure 10.3 *Sagittal section through left nasal passage of the Potoyunco or Peruvian Diving Petrel* P.garnotii, *showing the paraseptal process (a). After Murphy & Harper (1921).*

Figure 10.4 *Inflated gular pouch of the Common Diving Petrel. From Jones (1937).*

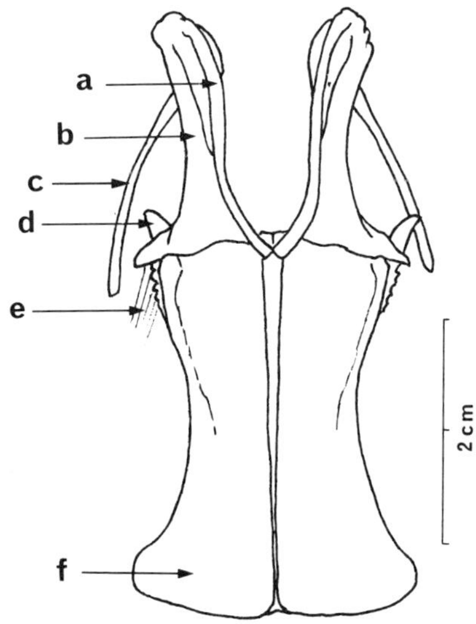

Figure 10.5 *Sternum and pectoral girdle of Common Diving Petrel; a. clavicle; b. coracoid; c. scapula; d. outurned costal process or* Proc. craniolateralis *with which ribs e. articulate, and f. unnotched sternum.*

loadings (Warham, 1977a). In both the sterna and very long ribs with the rib cage elongated caudally, are seen as adaptations for diving and sub-marine progression. Thoresen (1969) listed the comparative habits of *P.urinatrix* and a small auk *Ptychoramphus aleuticus*. The functioning of the pouch of diving petrels has not been examined nor is it known whether it develops only in breeding birds and regresses after the chicks fledge, as with auks like *P.aleuticus* (Speich & Manuwal, 1974).

Thus, were it not for their tubular nostrils, from external examination diving petrels would be considered as alcids, and they have been so considered in the past, Verheyen, as late as 1958, holding that view. In a detailed analysis of the skeletons of *P.urinatrix* and the Ancient Murrelet *Synthliboramphus antiquus* Kuroda (1967) showed, however, that *Pelecanoides* is a true petrel and confirmed the widely held view that similarities to the auks are due to convergent evolution in similar habitats. Harrison (1977b) compared the limb bones of diving petrels and *Alle alle* and found that, although the bones are of similar shape and relative length, fine structural details characteristic of each taxa were recognisable. The main skeletal modification to the petrel humerus and ulna result in stronger limbs for underwater propulsion. In some other features the petrel characters are prominent, e.g. their nostrils are holorhinal not schizorhinal as in the auks.

Thus the picture of *Pelecanoides* is of a petrel highly modified for underwater pursuit of food using its wings, the group forming an ecological analogue of the northern murrelets and auklets.

There are also physiological and behavioural differences from other petrels. For example, *Pelecanoides* seems unique in not storing stomach oil — at least no one has demonstrated its presence (Warham, 1977b). Many of the behavioural ones concern the breeding system — see Section VII below.

Sexual dimorphism in measurements and plumage is either absent or slight. Payne & Prince (1979) confirmed that breeding female South Georgian Diving Petrels *P.georgicus* have standard wings lengths 2.7 mm longer than males, but they detected no such difference with *P.urinatrix*.

The moult of some diving petrels is rather different from that of other procellariiforms. While Payne & Prince (1979) found nesting *P.u.exsul* starting to shed their primaries before their chicks fledged, and thought that moulting of the flight feathers proceeded in the usual petrel style — in descendant order — there are several reports of birds showing synchronous shedding of flight feathers, e.g. by Murphy & Harper (1921) and Watson (1968). This is another convergent feature with some auks. Murphy & Harper reported on some Potoyuncos *P.garnotti* shot at sea that had lost most of their tail quills and all their regimes but despite which there was plenty of food in their stomachs. These birds had become completely aquatic, temporarily. Probably with all the species the different classes of birds time their moulting differently: a clear picture awaits studies on birds of known age and status.

III The species

No group of petrels shows less morphological diversity than do the diving petrels, and most past taxonomists have placed them in their own subfamily or family. Recent classifications follow Murphy (1936) and recognise but the one genus. Harper (1978) analysed the blood plasma proteins of five populations of *P.urinatrix*, including subspecies *chathamensis* and *exsul* as well as the typical form. He found minor differences in their banding patterns but much greater differences between the *Pelecanoides* pattern and those of other petrels. Figure 10.6 is a composite diagram from 83 runs made on this material.

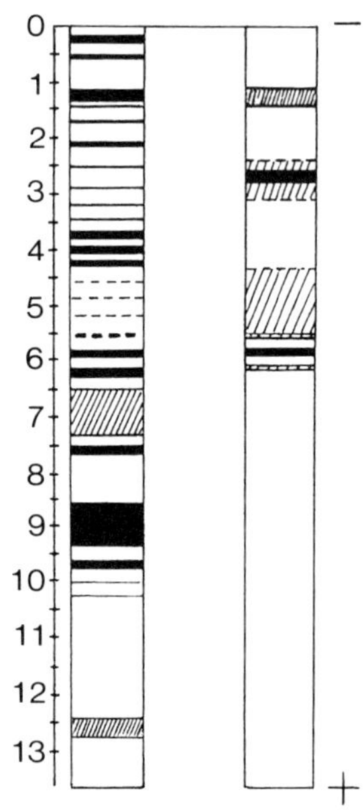

Figure 10.6 *Protein banding patterns from electrophoresed blood plasma of* P.urinatrix; *a composite diagram from analyses of three subspecies. Esterase banding in right-hand column. For key to the patterns see Appendix 2. From Harper (1978), courtesy of* New Zealand Journal of Zoology.

Murphy (1936) regarded the Subantarctic *exsul* as a full species, whereas the differences between this bird and other populations of *urinatrix* have appeared so minor that many, as in the present work, treat *exsul* as a southerly species of *urinatrix*. This circumpolar bird has produced a number of populations with slightly differing dimensions and bill proportions which have been given subspecific names. Their validity is still in doubt and much more data are needed for their evaluation — see discussion in Bourne (1968). Whatever the outcome of such evaluation, the specific integrity of *P.georgicus* and *P.urinatrix* is shown by their nesting on the same islands without interbreeding. They are separate ecologically, breeding at different times and in different habitats, and having different voices, as well as showing slight but clear differences in plumage, leg colours and nostril structure.

The affinities of the diving petrels with the rest of the tubenoses are unclear and different investigators have proposed different classifications. Most have concluded that they have an ancient lineage and Murphy (1936, p.772) presented evidence for a South American origin for the group. Presumably such an ancestor must have been a more aerial animal than the present-day species. Kuroda (1967) cited the presence of *M.pectoralis profundus,* though much reduced, as evidence for such an ancestor, this muscle being adaptive for sailing flight, helping to brace the extended wings against changing air pressures from below.

Several taxonomists have treated *Pelecanoides* as a genus of the Procellariidae. Recent examples are Cracraft (1981) who thought that undue emphasis had been placed on morphological distinctiveness and that their family status should be abandoned. Sibley *et al.* (1988) also places them with the shearwaters, fulmars and gadfly petrels in his subfamily Procellariinae.

IV Life styles

Diving petrels are primarily coastal rather than pelagic birds, usually seen at sea singly or in small groups. When resting they float quite high on the water, their shape and black and white plumage patterns making them look like small penguins and the Tristan islanders called their local diver 'flying pinnamin'. More often the birds are glimpsed whirring along in straight lines just above the surface and diving shallowly through the waves without any perceptible change of beat. This 'whirr' flight is quite different from that of other petrels, their wing stroke frequencies being much higher than those of Little Shearwaters, which also take more wavering courses with longer interspersed glides. When disturbed by a boat diving petrels either hurriedly dive or they burst from the bows like a bundle of flying fish, to travel a short distance before submerging. Under water they travel fast but tend not to use the whole wing as a flipper, penguin style, but scissor their wings open and shut from the carpal joint, with legs and webs stuffly extended (Fig. 10.7). Take-off usually requires a run across the surface to acquire the

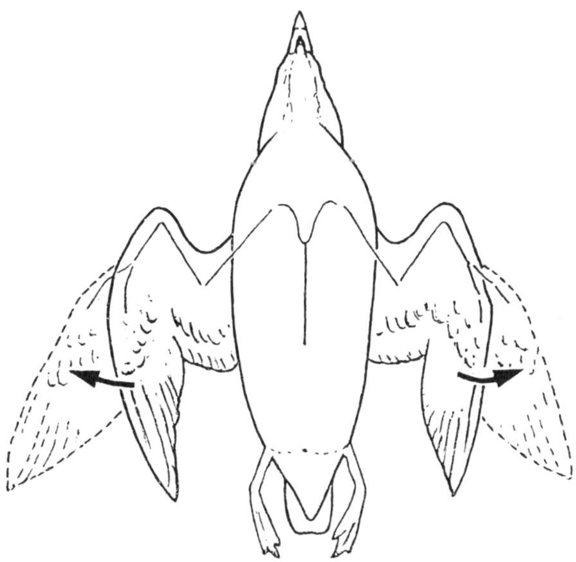

Figure 10.7 *Wing movements of diving petrels during underwater 'flying'. From Kuroda (1967).*

necessary lift unless a good sea is running, their high wing loadings and low buoyancy, while helpful when under water, hinder take off, landing, and aerial manoeuvrability.

When flying to nests diving petrels travel in direct lines a few m above the ground. They do not circle above their nesting areas before alighting but come in directly, to land close to their burrows. Marked birds landed almost at their burrow mouths when Richdale was weighing their chicks (Richdale, 1965a, p.51). And, despite their rather heavy landings, the birds seldom sustain injury and rarely get snagged in trees as shearwaters and other petrels may do. Take-off from level ground (as is usual at many *P.georgicus* colonies) requires a run of several m unless in strong winds, but birds nesting on steep slopes encounter little trouble in getting away. On departure before dawn the birds speed out like high velocity projectiles, the 'zip' of their bodies through the air telling of their unseen passage.

Diving petrels can scuttle quickly along the ground in a rat-like manner. Their folded wings may be held out slightly from their bodies and when cornered they tend to flatten themselves to the earth and acquire a turtle-like shape.

Perhaps even more than other petrels, *Pelecanoides* is attracted to lights, and many museum specimens had crashed after dark into lighthouse lanterns or into ships' lights. The birds can be trapped in a torch beam and may dive straight down it, so that if the ground is rocky or hard, they may be killed on impact. They also tend to fly into flames — hence their name 'Firebird' in the Falkland Islands.

Diving petrels appear to be silent at sea. On land they are quite vocal but the described repertoire is small and little studied. There is a pronounced evening and pre-dawn chorus during the breeding season, mainly from singers on the ground, but also from flying birds. Richdale believed that most singers belonged to the 'unemployed', non-breeding section of the community.

Thoresen (1969) listed a number of variants of the 'kuaka' call of the Common Diving Petrel (for sonagram see Payne & Prince, 1979). This bird also uses a clear, low-pitched moan on a rising note, continuously repeated. The voice of *P.georgicus* is more of a succession of squeaks arranged in phrases of 5–10 syllables (Payne & Prince, 1979). Gwynn (in Downes *et al.*, 1959) heard these birds 'cooing like mewing cats' and Paulian (1953) stated that the calls of *georgicus*, whether used in flight, on the ground or from within the burrow, are the same.

V Distribution

Birds of this family are restricted to the southern hemisphere and although the Peruvian Diving Petrel nests up to 6°S, there are no records of them crossing the Equator even as vagrants.

All live in cold or cool seas with the major breeding concentrations along the Antarctic Convergence and in the Subantarctic zone (Fig. 10.8). For all its northerly range, the Peruvian species nests on islands adjacent to the cool Humboldt Current from just south of Valparaiso to Lobos de Tierra at 6°27'S. As Fig. 10.8 shows, *urinatrix* and *georgicus* have circumpolar ranges, breeding sympatrically at South Georgia, the Prince Edward Islands, at the Crozets, Kerguelen, Heard Island and at Codfish Island in Foveaux Strait, New Zealand.

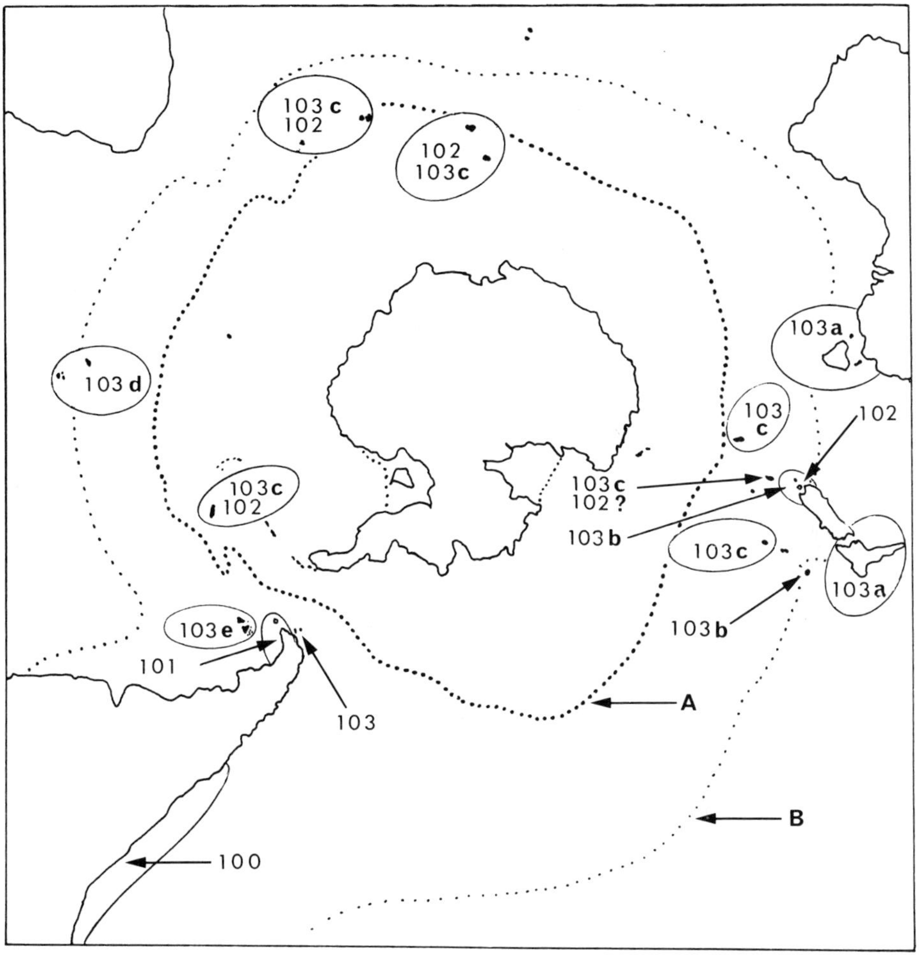

Figure 10.8 *The breeding distribution of diving petrels: 100* P.garnotii; *101* P.magellani; *102* P.georgicus; *103a* P.u.urinatrix; *103b* P.u.chathamensis; *103c* P.u.exsul; *103d* P.u.dacunhae; *103e* P.u.berard. *The recently discovered population breeding at Diego Ramirez near Cape Horn (103) presumably is of the subspecies* P.u.coppingeri. *A. Antarctic Convergence; B. Subtropical Convergence.*

The discovery of the population of *P.georgicus* at Codfish Island (Imber & Nilsson, 1980) was unexpected, this being a very northerly location for the species. Here *georgicus* burrows in unstable coastal dunes whereas *urinatrix* choses more stable vegetated ground nearby. The colony may be a relict of a larger population that flourished when the convergences lay further to the north and Imber & Nilsson noted that the small squid found in their stomachs suggested that these petrels were feeding well offshore along the edge of the continental shelf, whereas *urinatrix* evidently fed closer in. They pointed out that apart from Macquarie Island and two of the Aucklands Group, there is little of the habitat preferred by *georgicus* available on the Subantarctic islands of New Zealand, which might explain the use of dunes as a substitute for scree and scoria.

The overlapping distribution of *georgicus* and *urinatrix exsul* around the Antarctic Convergence may be another example of Murphy's dictum that seabirds can feed where they will but must nest where they can, and Bourne (1968) suggested that *georgicus* might be adapted to feeding south of the Convergence, *urinatrix* to the north of it. Payne & Prince (1979) postulated that their parent stocks evolved in isolated South American and New Zealand populations which came together relatively recently, after which sympatry led to an accentuation of their differences: the two giant petrels may be a parallel case.

As it is usually impossible to identify diving petrels at sea, the distribution away from the nesting islands is unclear. Most birds are seen close inshore. Nonetheless there are numerous sightings of diving petrels, mostly singly or in small groups, far from land; many, for example, in the South Pacific between 50° and 60°S far from any breeding place (Watson *et al.*, 1971), and Ainley *et al.* (1984) recorded 0.16 birds km^{-2} during transects to and from the Ross Sea in December and 4.05 km^{-2} in February. These were in the northern part of the Polar Front Zone south of New Zealand at 60–64°S, the southernmost sightings ending at the 3°C isotherm in those months.

VI Feeding and foods

Close observations of feeding by diving petrels have not been reported but their prey appears to be taken one item at a time and from the top few m of the sea. Common Diving Petrels have been timed submerging for 40 s, surfacing for 2 s and diving for another 40 s. Thoresen (1969) watched many dives of birds feeding in tide rips. They dropped into one end of the rip and, submerging for 15–20 s, gradually drifted with the current until evidently running out of prey when they flew back to repeat the manoeuvre. Flocks feeding in a similar fashion have been seen off Nightingale Island, Tristan da Cunha and by Richdale (1943) in New Zealand.

Precise feeding ranges of nesting birds are unknown, but where changes on egg or chick occur nightly, the parents cannot go far. South Georgian *P.georgicus*, whose incubation stints are longer than those of sympatric *P.urinatrix*, could feed further afield and thereby lessen inter-specific competition. With any such system of frequent reliefs, suitable food supplies must be close at hand and Murphy (1936, p.776) pointed out that diving petrels seem to be confined to regions of cool upwellings. These may not only include major features like the Humboldt Current, but in the southern Indian Ocean, local eddy systems and fronts created by the breeding island and its associated seabed configuration deflecting and disturbing the westerly current drift. Rather oddly, there are no records of diving petrels in the cool Benguela Current off South-west Africa: indeed diving petrels are unknown in Africa, even as beached derelicts.

Stomach contents include juvenile fish and small sardines and stomachs of *P.garnotii* may be distended with anchovies. A range of planktonic crustacea is taken — copepods, isopods and amphipods. The many *Euchaeta antarctica* found in

P.georgicus from Heard Island suggested to Ealey (1954) that the birds had fed at night, as he believed that this crustacean remained in deep water by day.

Payne & Prince (1979) analysed the foods fed to the chicks of the two species at South Georgia. Both ate the same prey but in different proportions, all identified items being small crustacea — in *urinatrix* euphausids predominated, in *georgicus* copepods. However, the food patterns of both changed as the season advanced, with a trend for copepods to predominate in early summer and euphausids later.

The above prey types tend to support Murphy's belief that *Pelecanoides* are the only petrels not to eat squid, but the discovery of the remains of the remains of juveniles of four kinds of cephalopods in the stomachs of two Codfish Island *georgicus* by Imber & Nilsson (1980) form an exception to that generalisation.

VII Breeding

Little is known about the breeding biology of the South American species but Richdale (1943, 1945, 1965a) studied *P.urinatrix* in southern New Zealand over six seasons. Thoresen (1969) made a shorter examination of the behaviour of the same species in northern New Zealand and Payne & Prince (1979) a comparative study of *P.georgicus* and *P.urinatrix* at South Georgia.

Diving petrels have well defined breeding seasons in the southern summer, except that the Potoyuncos of the Humboldt Current nest throughout the year. In Australasia the timing of laying of *P.urinatrix* varies with latitude (Richdale, 1943) the more northerly birds breeding earlier, but the laying of *P.georgicus* does not seem to vary in that way (Derenne & Mougin, 1976). Where *urinatrix* and *georgicus* are sympatric the latter lays about a month later (Payne & Prince, 1979; Imber & Nilsson, 1980). There may also be seasonal effects. At South Georgia both bred about 10 days earlier in 1971/72 than in 1972/73. Richdale found that egg laying in *urinatrix* was spread over 59 days, an unusually long laying period for a small petrel, but he thought one reason for this was that young birds, some of which breed in their second year, lay later than older birds.

Diving petrels nest in burrows, sometimes at high densities, sometimes widely dispersed, occasionally in rock piles, very rarely on the surface hidden in vegetation. At Nightingale Island, Tristan da Cunha, nests are at the ends of burrows lodged between turf and a more or less vertical rock face (Elliott, 1957, p.571). Derenne & Mougin (1976) figured typical cross sections of burrows of *P.georgicus*, this bird being unusual in nesting in level ground, and at Heard Island some began their excavations in the depressions of wheel ruts (Downes *et al.*, 1959, pl.48).

Many of the unusual features of the breeding system of *P.urinatrix* are related to its inshore feeding. For example, each adult tends to return to its burrow nightly during incubation and chick rearing. In consequence, both parents are often found in the nest together. There may also be no pre-laying exodus, although Payne & Prince (1979) suggested that in *georgicus* the females might stay at sea for 9–15 days before the egg appeared. The eggs themselves are unusually small and spherical for petrels of their size (Chapter 13, Section VA1).

The few measurements of incubation and nestling periods are summarised in

Table 15.3. Incubation periods appear to be rather short for 100–130 g birds. Intermittent incubation is also quite common (Chapter 14, Section IVC) and these petrels are easily affected by observer disturbances (Roby & Ricklefs, 1983; Payne & Prince, 1979). Richdale (1965a, p.15) also reported that *urinatrix* chicks never got tame when regularly handled as do most other petrel chicks. Diving petrel chicks are also brooded for far longer than are those of other burrowing petrels and the meals delivered nightly are of less varied size than usual with them (Richdale, 1965a, p.31). A parent is usually with the chick during its first 10–15 days, at least in *P.urinatrix*; by then homeothermy has been achieved (Chapter 15, Section IIIB).

Pelecanoides grow two sets of down but there are bare patches on the throat and sides of the head and neck in *P.georgicus* (Matthews, 1929) and very thin down on the head and back of *urinatrix* (Richdale, 1965a, p.23). Richdale supposed that it was because of the shortness of the down that the chick needed brooding for so long. He also noted that even at 48 days old part of its head is bare and attributed this to wear caused by the parents running their beaks over the top of the young bird's head and behind its neck while orientating to it in the darkness of the burrow before disgorging food. Payne & Prince found that South Georgian *georgicus* chicks had thicker primary down than those of *urinatrix* and attained homeothermy at an earlier age, adapting to a colder climate than that favoured by the latter.

Another feature of the chick's behaviour is that it does not disgorge any oil when alarmed or handled, as do so many small petrel chicks — helping to confirm the absence of stomach oil in the adults.

Richdale (1965a, pp.36, 54) presented data on pair-bond retention. Sixty-five per cent of *urinatrix* bonds only lasted for one season, a high turnover for a petrel, but 58% of broken bonds followed the loss or disappearance of a partner. Such birds soon got another mate and several that had three different partners in three consecutive seasons raised a chick each time. The longer the bond lasted the less likely was a shift of nest site to occur, but in the course of time fewer and fewer retained the same burrow. The overall distance of such shifts in the 6 years of Richdale's study was 2.7 m.

VIII Dispersal and migration

Diving petrels do not perform dramatic large-scale seasonal movements although very few checks seem to have been made on activities at nesting sites out of the breeding season. Birds at the more southerly breeding places seem to disappear for several weeks following the departure of their young, and are presumably at sea moulting. However, even at Heard Island, well into the Antarctic Zone, *P.urinatrix* has been found in burrows as early as 15 August and others, probably of this species, have been attracted to lights and seen offshore in June and July (Downes *et al.*, 1959). Richdale (1965a, p.16) had evidence of their being not far from the nesting places in the non-breeding season and large numbers were coming to land at the Solander Islands (Foveaux Strait) on 25 July (Cooper *et al.*, 1986). The same species winters along the Argentinian coast as far north as 38°S: presumably these are birds from the Falkland Islands. Potoyuncos, nesting at all seasons, are evidently sedentary, although the young birds perhaps travel more.

Probably both *P.urinatrix* and *P.georgicus* shift northwards to moult but otherwise stay within range of their home islands: distant sightings may be of more widely ranging pre-breeders. Specimens of *P.u.exsul* washed up in Western Australia are presumably birds from Indian Ocean colonies driven down wind, but Johnstone (1979), noting that they have atypically long wings and tails, suggested that these birds might have originated from a single, 'uncollected' breeding place.

IX Populations and mortality

Estimating the sizes of diving petrel populations is no easier than with other burrowing species. Some colonies are small, e.g. the 100 or so *P.georgicus* at Codfish Island, and the 20 pairs of *P.urinatrix* at Macquarie Island (Brothers, 1984). Other colonies are large. Croxall & Hunter (1982) estimated that about 4900 pairs of *georgicus* nested on Bird Island, South Georgia at an average density of 0.25 burrows m^{-2} in burrowed areas. Extrapolating to the South Georgian 'mainland' they calculated a breeding population of about 2 million pairs. A similar figure was suggested for that species at Ile de Cochons, Crozet by Derenne & Mougin (1976) and the populations of *P.urinatrix* around New Zealand must also run into the millions.

Working on the survival of birds at least 3 years old, Richdale (1965a, p.47), estimated that an annual mortality of about 18% occurred during the course of his study. He scored 466 of 595 birds as breeders (78%), a very high figure for a petrel population, and mainly explained by the return of young birds to nest in their second year, so that the pre-breeding element was small.

Even on islands free of introduced predators, some diving petrels are killed by skuas, giant petrels, various raptors and so on. Southern Skuas dig out South Georgian Diving Petrels at Heard Island, trenching back from the entrances with their bills to reach shallow nest chambers (Downes *et al.*, 1959). Skuas also kill fledglings at departure, take adults in flight by day, but seem to ignore those resting on the sea.

Such mortality seems easily sustainable but the effects of cats, rats, Wekas and similar introductions have been disastrous. The frequency with which diving petrels visit their nests makes them all the more vulnerable to attack. Richdale (1943) compared the 'millions' of Kuakas on the Herekopare Island (Foveaux Strait) in 1911 with the situation 30 years later when only hundreds were seen, cats having been introduced since the earlier visit. All the diving petrels are now gone. At Possession Island, Crozet, Despin *et al.* (1972) attributed the concentration of *P.georgicus* on unvegetated high ground to the presence of rats at lower levels.

Data for losses at sea are few, mostly derived from casual findings of *Pelecanoides* in the stomachs of Leopard Seals, barracouta and 'snapper'. Birds with leg injuries have been reported. Commenting on Murphy's deduction (from the ability of these petrels to feed despite being almost flightless during the moult) that flight was only necessary for them to reach their burrows, Watson (1968) suggested that the ability to fly from sub-marine predators may have played a part in the evolution of synchronous moult in diving petrels, as this system reduces their flying ability only briefly.

Diving petrels are commonly found 'wrecked' on southern Australian and New Zealand beaches and sometimes inland. In the winter of 1975 a major such influx of *P.urinatrix* occurred in northern New Zealand, mainly on the west coast. After persistent westerly winds 3580 dead birds were collected. Norman & Brown (1987) examined 100 *P.urinatrix* from a 'wreck' in August 1985 in Victoria, Australia. Of these 70 were males and the gonad condition of the females suggested that many were young adults that died around or just after laying, rather than being young birds. Most such mortality appears to be from local colonies.

Other birds die through unnatural accidents: whereas lighter petrels like shearwaters can hit pylons and radio aerials without injury, diving petrels have poor manoeuvring ability and, travelling fast, are often killed on impact.

X Human use and conservation

Being small and breeding in remote places, diving petrels have seldom been used by man. The Potoyunco, a guano bird, is still taken by Peruvian guano workers and fishermen (Duffy *et al.*, 1984) and probably the Moriori in the Chatham Islands and the New Zealand Maori formerly ate *P.urinatrix* (their Kuaka), as it was extremely abundant in some places and easily caught on land.

Only *P.garnotii* appears to be threatened today due to a combination of direct human predation, habitat disturbance, and overfishing of local anchovy *Engraulis ringens*. Duffy *et al.* (1984) estimated that a maximum of some tens of thousands of breeding birds now remains in Peru of this once very abundant species whose nesting substrate was largely removed by the guano miners. Schlatter (1984) also lists rats and guano mining as threats to the reduced and decreasing population on the Chilean islands. However, until the biology of this bird is better known, it seems unlikely that appropriate measures for its conservation can be effective. Duffy and his co-workers discussed the past and present background to this problem.

Elsewhere, for those populations of diving petrels without introduced mammals, the only management required is the maintenance of the breeding habitats from change and interference and, in some places, protection of the burrows from trampling.

CHAPTER 11

Breeding Biology: Introduction

Grey-headed Albatrosses.
Photo: author.

Contents

Much information on the breeding systems of petrels has been gained since the early studies of Lockley (1942) on the Manx Shearwater, of Gross (1935) on Leach's Storm Petrel and of Richdale (1950, 1952) on the Royal Albatross. Key events of a typical cycle are outlined in Fig. 11.1; see also Fig. 1.7.

I Methods of study

A *General*

Investigations are facilitated by the ease with which most petrels can be observed,

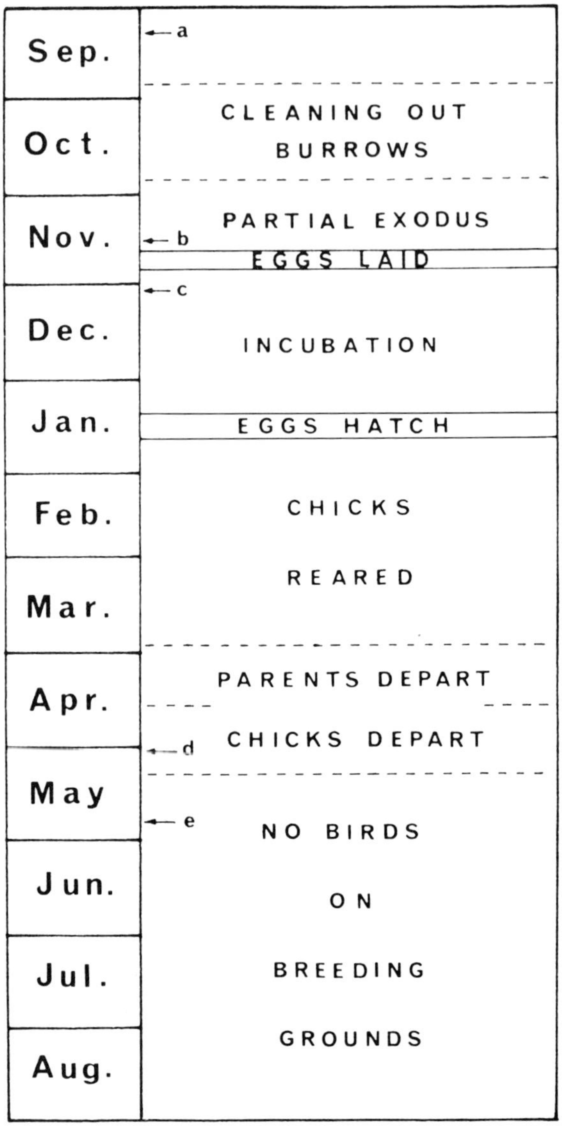

Figure 11.1 *A typical breeding timetable for a southern, trans-equatorial migrant petrel, the Sooty Shearwater* P.griseus. *(a) first birds return (8 Sep.); (b) first egg laid (16 Nov.); (c) last egg laid (4 Dec.); (d) last adult seen (1 May); (e) last chick seen (25 May). From Warham* et al. *(1982).*

caught and marked at their breeding places and by their tendency to return to the same colony, nest and partner during successive breeding episodes. As the colonies are often large, adequate samples can be studied and control groups readily established. Following the fate of a market cohort from birth to death may be less feasible because the larger kinds may outlive both their researchers and identifying rings. The Royal Albatross banded by L.E. Richdale was still laying in

1988 aged 58+ years when Richdale had been dead 5 years. Also, most burrowers are hard to observe, being active ashore only after dark.

Petrels in deep burrows may prove difficult to find and extract, and burrows should not be destroyed in the process. Many birds will grasp a wire pushed down to them and can be drawn out slowly as they latch on with their bills. 'Puffin hooks' have been used to extract prions and storm petrels and even to capture giant petrels. Mist-netting in conjunction with tape playbacks have proved useful for catching samples of storm petrels and dazzling with spotlights can pull a wide range of species from the air particularly on misty nights. Vocal lures can be very effective with gadfly petrels (Warham, 1988a; Tennyson & Taylor, 1990).

Counting nests or burrows in measured plots for census purposes is relatively easy in most situations and a spurt of paint from a spray can on the ground nearby ensures that no nest is counted twice. Determining burrow occupancy may be difficult, even impossible, where the tunnels are deep, tortuous, or serve several nest chambers. Birds may sometimes be felt with a long stout wire, bent if necessary, or with a flexible measuring tape. Often occupants do not call but useful clues are the presence in the tunnel of moulted contour feathers or chick down. There is no practicable device that allows one to see a bird in a burrow beneath 30 cm of peat, soil or granite. One possibility would be to fit radioactive bands as used by Godfrey (1954) for studying the subterranean movements of moles, another approach to monitor breathing or heartbeats with a sensitive microphone inserted down the tunnel.

Leg bands make good markers for albatrosses and giant petrels which stand a lot, but the legs of burrowing species are usually hidden and for identifying individuals with the minimum of disturbance, temporary paint marks on the crown may be necessary. Enamel or fast-drying paints will often last for a breeding cycle. Picric acid and acid (wool) dyes have also been used to mark birds for sighting at sea. Markers with light emitting diodes or Betalights (Hayward, 1987) coded for sex and/or recognition of individual petrels do not appear to have been used for night-time observations. Modern stainless steel rings wear little and often outlast coloured plastic bands.

Idiosyncrasies of plumage are seldom helpful for identifying individual birds, except perhaps with Cape Petrels, but for identifying particular Northern Giant Petrels I used photographs of the patterns of the flecking of their irides, which differed from bird to bird, and from eye to eye of the same bird.

If done expeditiously, many species of petrel will continue breeding activities without apparent ill effects despite handling. However, the ease with which Manx Shearwaters can be taken from the nest led some workers to conclude that this is typical of its genus, which it is not. The larger species like *P.carneipes, tenuirostris* and *bulleri* will not tolerate such treatment. Sooty Shearwaters also usually desert if disturbed in their burrows (by day or by night) during the pre-chick stage. Storm petrels are also sensitive and *O.leucorhoa, H.pelagicus, O.furcata* and *O.oceanicus* desert readily. Some surface nesters too are easily disturbed, and when adult giant petrels have been caught on nests for banding, as has sometimes been done, mass losses of eggs must have ensued. Such losses can be reduced if the eggs are immediately covered with nest material.

Albatrosses and other large petrels are more easily handled than some of the smaller ones, although two people are usually needed to hold them: their heavy bills can be temporarily immobilised with an elastic band or tape.

Ejection of stomach oil is not normally a problem and most albatross chicks usually stop spitting once accustomed to handling. This is not true of young giant petrels whose salvoes are difficult to dodge and oil-covered hands make ringing or measuring impossible. A small shield to deflect the oil and a U-shaped prong to peg the bird to the ground frees the hands for measuring or banding (Fig. 11.2). For measuring the wings of the large albatrosses and giant petrels in the field, a stopped board with scale inserted and hinged so that it can be folded for carrying, is useful.

Ringing of adult birds is best delayed until chicks are hatched. On the other hand, if ringing is done a week or fortnight before laying the females can be

Figure 11.2 *When mass banding giant petrel chicks a lightweight shield can be used to deflect spat oil (top) and (bottom) a U-shaped prong chained to the handle of the shield pegs the chick to the ground freeing both hands to fit the ring.*

identified by the state of the cloaca. If possible, however, the study sample should be marked during a season before that in which the research is to be done.

If eggs have to be examined, these can often be raked out from under a sitting bird with an oval-shaped, long-handled, kitchen ladle. This is slowly positioned beneath the egg and cautiously withdrawn, replacement being done with equal care.

B *Sexing*

In petrels the great enlargement of the oviduct before laying is associated with a distention of the cloacal lips. With *P.tenuirostris* this swelling is visible about a month before laying, reaching its peak just before that event and being still evident 2–3 weeks later (Cole, 1913; Serventy, 1956). In this species the non-breeding females and the males have small, roundish vents but in females approaching laying the cloaca becomes distended transversely, measuring about 22 × 44 mm with the slit 11 mm wide and the epithelium red and raw, even bleeding. This method of sexing has been widely used with tubenoses, the differences quantified by Boersma & Davies (1987) and Copestake *et al.* (1988). Females may also be identified a few days before laying by palpation of the egg. Positions at copulation may also be helpful and, in general, the female is the bird found by day on the newly laid egg. Inspection of the gonads by laparotomy or fibre-optic endoscopy (Jones *et al.*, 1984) has seldom been used, Simons (1981b), working with *O.furcata*, being one exception. The possibility of using fibre optics to examine the cloacal papillae marking the points of discharge of the two ureters and single oviduct (female) or two vas deferens (male) seems not to have been explored, but would enable sex determination at any time of the year. Albatrosses, fulmars and *Calonectris* sp. can often be sexed by the larger size, heavier bills and broader heads of the males when compared with those of their partners or against standard measurements or using a discriminant function (Dunnet & Anderson, 1961).

C *Ageing*

The plumage of most petrels does not change perceptibly with age. Even for albatrosses like *D.albatrus* and *D.exulans*, that take years to acquire fully adult plumage, the data are not yet adequate to age them on such characters, the plumages of different sexes and populations changing at different rates. The degree of skull ossification has been used to separate nestling, fledgling and adult Short-tailed Shearwaters (Sugimori *et al.*, 1985) but most studies of known-age petrels have been based on birds marked as chicks.

D *Observation techniques*

Burrow nesting species pose obvious problems. Their nocturnal behaviour on the surface can be observed using image-intensifiers, infra-red illumination and videocameras (e.g. Simons, 1985), but often observations within the nest chamber are needed. Usually fibre-optics don't help; their lenses soon get clogged with dirt but they have been used in conjunction with a video camera (Dyer & Hill, in press). Inducing breeding in nest-boxes with viewing facilities often helps as most petrels take readily to boxes of appropriate size and design which replace their own nesting chambers — the entrance tunnels being untouched. Groups of linked boxes slotted into rock crevices have been quickly taken over by prions. Fitting nest-boxes is best done during the non-breeding season, otherwise it is done by day during the chick stage — an observation hide or a simple viewing box can be incorporated (Figs 11.3 and 4).

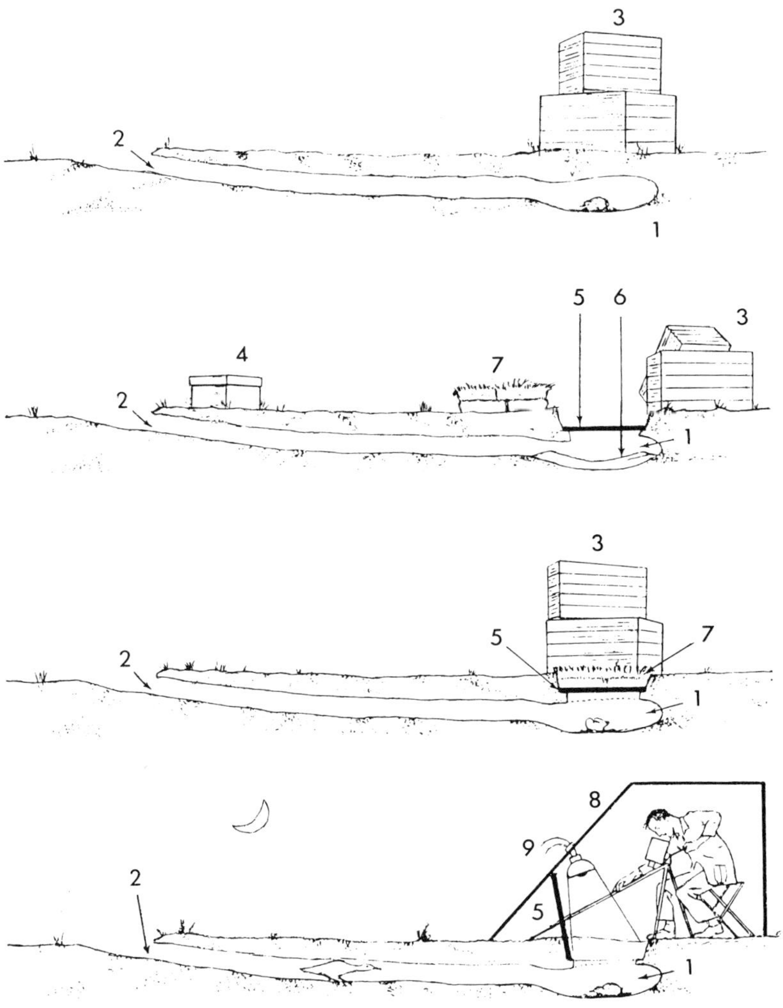

Figure 11.3 *Preparing a 'dark hide' over a petrel's nesting chamber. 1. Nest chamber. 2. Entrance. 3. Boxes to simulate hide. 4. Box holding chick temporarily removed from nest. 5. False roof with polystyrene insulation. 6. Cloth to catch digging debris. 7. Turfs replaced. 8. 'Dark hide' made from hardboard. 9. Infra-red light. From Warham (1966), courtesy of Focal Press, London.*

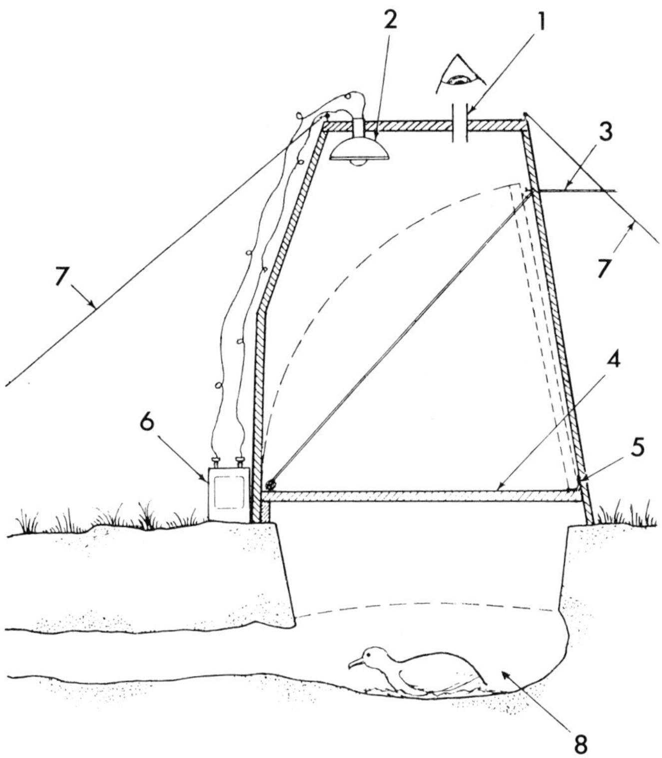

Figure 11.4 *An observation box. 1. Viewing tube, normally corked. 2. Infra-red light. 3. Drawstring to raise false roof. 4. and 5. False roof, insulated with polystyrene sheet and hinged to facilitate lifting. 6. Battery. 7. Guy ropes. 8. Nest chamber. From Warham (1966), courtesy of Focal Press, London.*

Such preparation involves a lot of work if many nests are to be studied, but properly made, easily inspected boxes are preferable to the common practice of cutting removable earth plugs above the chamber. Usually soil falls into the nest at each inspection and the seals are seldom perfect, letting water into the compartment and causing breeding failure. Predators like skuas may also learn to recognise the plugs and to excavate them to reach the chicks. Johnstone & Davis (1990) used soil-filled plastic bags to plug 'peep-holes'.

E *Automatic recording*

The most practicable way to check movements to and from a burrow is usually to fence the entrance with easily movable short wires, toothpicks etc., but parental visits tend to be infrequent and irregular, and obtaining precise data on their activities by direct observation, time consuming. Event-recorders with trips set in access tunnels have been used, e.g. with *P.phaeopygia* and *P.tenuirostris* (Simons, 1981a, 1985; Naarding, 1980; Schramm, 1983b). Something similar is needed for accurate measurement of meals delivered to chicks, to record the parents' passages through the tunnel, and to weigh them and the chick automatically at each transit (Chapter 15, Section VC1). Such equipment must be so designed that soil thrown up by the birds cannot jam the mechanism — Simon's device had trip fingers hanging from the roof of the tunnel to avoid this. A common hazard is the chick that squats at the gate and runs down the battery, or the bird that gives false information by coming to the entrance without leaving. Davis (1957), one of the first to use automatic devices, had them fail when their clocks became clogged with earwigs! Roberts (1940) used a two-bulb thermograph to monitor the visits of *O.oceanicus* by the consequent temperature changes in the burrow. Nest-boxes fitted with treadles and micro-switches were used in work on *O.homochroa* by Ainley *et al.* (1990).

An automatic recorder developed by the British Antarctic Survey to monitor mollymawk activity used a mechanical nest (Prince & Walton, 1984). This consisted of a pedestal topped by a moulded fibreglass nest fixed to a weighing assembly, the whole supported on a steel post driven into the ground. The weighing device comprised an inflated polythene sand with and diaphragm whose internal pressure changed with the changing weight of the nest occupant. A tube connected the diaphragm to a pressure transducer whose signals passed to a multi-channel recorder powered by a battery trickle-charged from solar panels.

Time-lapse photography and/or video recording may also be used but this is expensive and weather- and vandal-proofing may be difficult.

II Breeding habitats

Petrels mostly nest on remote islands or headlands beyond the reach of mammalian predators and usually within sight of the sea. Some nest on mountains and may have to cross intervening ranges before sighting the ocean, as do some Snow and Antarctic Petrels which breed far from the water. One small population of Mottled Petrels nests on an island in a freshwater lake.

Typically too, petrels select open habitats with low or no vegetation and where take-off in calm weather is easy — from cliffs, steep slopes, grassy sidings and the like. About 86% of the 103 species breed in such habitats. The South Georgian Diving Petrel is perhaps the extreme example, using a bare substrate for burrowing, whereas the Common Diving Petrel tends to dig in well-vegetated peaty ground, often below trees.

Some 14% of the species nest below dense vegetation and forest. Newell's Shearwaters do so under an almost closed canopy of fern and many shearwaters, two *Procellaria*, gadfly petrels like *P.hasitata*, and some prions have to dive through forest canopies (and climb out again) in getting to and from their nests. Even some albatrosses — *D.bulleri*, *D.immutabilis* and *D.chlororhynchos* — may nest below trees.

Some 25% of tubenoses, i.e. the larger species and all the fulmars, are surface nesters, laying on the ground or on rock ledges. Medium-sized gadfly petrels of tropical seas also tend to lay on the surface, partly because their islands lack good burrowable ground. A few shearwaters, notably *P.nativitatis* and, less often, *P.pacificus*, also do that. Tropical petrels, however, tend to select sites giving some shade from the sun. A few may lay on the surface in one habitat, below ground in another. Albatrosses build on fairly level ground or on broad ledges, as do giant petrels. Grassland is not essential for Wandering Albatrosses: they will use ground covered only with a sward of moss, and Southern Giant Petrels lay on bare, stony substrates in Antarctica and the Falkland Islands, as does the Galapagos Albatross, while the Black-footed species mostly lays on bare sand.

Rock crevices and talus breakdowns are chosen by small species like some storm petrels, Bulwer's Petrel and some prions. About 4% of the species use such sites.

The majority of petrels burrow. These range from birds the size of the Shoemaker down to smallest storm petrel. Burrowing depends on the availability of suitable substrates, these varying from water-logged and peaty soils, to heavy clay and soft, dry sand.

The most unusual site for a tubenose is that used by many Grey-backed Storm Petrels *Garrodia nereis* which construct a nest of grasses in the skirt of a tussock stool. Nests of *Fregetta grallaria* may also approach this type occasionally (Fig. 9.10). Otherwise *G.nereis* lays either in a burrow with a very short access tunnel or merely under a stone (Despin *et al.*, 1972).

The most enterprising in choice of site is the Northern Fulmar. As its population has expanded so it has come to exploit artificial cliffs like castle and quarry walls, inland bluffs well out of sight and sound of the sea, Rook's nests in trees; it has even laid on the ground amid marram grass with Sandwich Terns. A considerable proportion of fulmars on the Outer Hebridean Monach Islands now nests in sand dunes (Hepburn & Randall, 1975). This is the only procellariiform that habitually breeds on man-made structures. Its oil-spitting ability has presumably enabled it to nest within reach of stoats and cats, but its breeding success at such accessible sites has not been examined.

III Breeding dispersion

Most procellariiforms are social birds breeding colonially, nests being tightly packed where the terrain permits. High densities are mainly achieved by burrowing species, but surface nesters too may crowd together so that sitting birds can almost reach their neighbours, e.g. with *Macronectes giganteus*. In contrast, *M.halli*, while mostly nesting in loose groups, chooses the shelter of rocks and vegetation and inter-nest distances are much greater. Nests made on flat areas or on extensive slopes, whether on the surface or below ground, tend to be over-dispersed. Even so, instances of local clumping are not rare. Among *D.irrorata* Harris (1973) found that such clumping was by birds of similar laying data, but he saw little display to account for the synchronisation.

Burrow densities have been measured for many species since Bishop Montgomery estimated 40 000 per acre of *P.tenuirostris* at Chappel Island at 1 m^{-2} (Montgomery, 1896). The greatest crowding recorded is that of the Galapagos Storm-petrel *O.tethys* at 15 m^{-2} in restricted areas (Harris, 1969b). More typical of medium-sized species are figures of 1.2 m^{-2} for *P.griseus* (Warham & Wilson, 1982). Burrow crowding varies with vegetation type, high densities tending to occur in grassland, the same species having its burrows farther apart in forest where root systems impede digging (Warham & Wilson, 1982). Other studies of nest densities in relation to vegetation type include those of Gillham (1963) on *P.marina* and Maesako (1985) on *C.leucomelas*.

The *Phoebetria* albatrosses provide the best examples of petrels that often nest out of sight of others. This is partly because they are ledge nesters on steep slopes with limited space. Weimerskirch *et al.* (1986) scored 49% of 1908 nests of *P.palpebrata* as being in groups of one, 23% in groups of two, and 12% in groups of three. Their figures for *P.fusca* showed that these preferred to nest in slightly larger groups, at least at The Crozets. The dispersion of ledge-nesting fulmars is also restricted by their sites, but at times the distances between nests are small. The storm petrel *Garrodia nereis* may also nest solitarily, but whether any lone-nesting birds are vocally isolated is unclear. At the Torishima Island colony of *D.albatrus* the nests are in neat files along rows of grass planted across the friable ash terrain favoured by the birds (Hasegawa, 1984a, p.35).

Wandering Albatrosses at the Auckland Islands and Royal Albatrosses at Campbell Island (their major breeding grounds), usually lay many m apart, i.e. at about 20 ha^{-1}; those at Bird Island, South Georgia at about 120 ha^{-1} (Tickell, 1968). At Possession Island, Crozets, densities at three colonies of Wanderers were 10, 8, and 3.5 ha^{-1} (Weimerskirch & Jouventin, 1987). These figures are mainly for habitats with good vegetation: at sites with little ground cover, as at Midway Atoll, densities are higher — for *D.immutabilis* about 250 ha^{-1} (Frings & Frings, 1961). Here few birds will be out of sight of neighbours and even fewer beyond the range of their loud calls. Sometimes highly dispersed nesting may be partly the result of predation — as with *P.phaeopygia* at Maui, Hawaii.

Even the 'great' albatrosses may nest tightly packed. Nest densities of Royal Albatrosses on The Sisters and Forty-Fours, flat-topped stacks in the Chatham Islands, reach 800 ha^{-1} in some areas (Robertson, 1974). At such concentrations levels of aggression are higher than elsewhere and possibly cause nesting failures. Similarly, when Laysan Albatross nests are so close that their rims touch, both may fail to hatch eggs. The relationships between nesting density, mutual stimulation, and nesting success have yet to be investigated in petrels.

Factors limiting excessive crowding seem likely to include the territorality of the nest owners and the increased probability of burrow collapse where their density exceeds the 'holding capacity' of the substrate. Similar factors probably account for the over-dispersed patterns of both surface and underground nests. Despite the crowding on some small islands, most of the bigger ones have plenty of unused ground apparently suitable for nesting.

Where several tubenoses share an island the species may be segregated by habitat, by altitude, by substrate preference, by using different layers of soil, or there may be mixed colonies. On Midway Atoll *D.nigripes* lays on the sand of the beach dunes, areas not used by *D.immutabilis*, but some *nigripes* also nest inland surrounded by the more numerous Laysans. Often several species of dissimilar sizes occur together, a small prion like *P.crassirostris* breeding in crevices among Shy Albatrosses; Mottled Petrels, Common Diving Petrels and Fairy Prion burrows intermingling with those of Sooty Shearwaters, etc. Mixed colonies of Black-browed and Grey-headed Mollymawks are common and at the Crozets these two may share nesting slopes with the Yellow-nosed species. Studies of community structure and ecological segregation at various breeding islands include those of Croxall & Prince (1980), Jouventin *et al.* (1982) and Mougin (1985).

IV Breeding seasons

A *Introduction*

The breeding season can be defined as the period between the return to the nesting place (or, with sedentary species, the time when the adults renew their responses to territorial and sexual stimuli) before the pre-laying exodus, to the departure of the adults, or their entry into the post-breeding phase of moult. Its timing has to be adjusted to the season when all the essential needs of the petrel rearing its chick are adequately met, needs that include a nesting place giving protection from predators, from excesses of heat and cold, from tidal waves and the like, and when food, nutrients and water are within range of the nest for parents, chick and the newly independent fledgling. Kuroda (1960) and Beck (1970), among others, discussed the breeding seasons of petrels.

Most petrels of temperate seas are spring and summer nesters (Fig. 11.5). Those of polar regions, e.g. the fulmars and Wilson's Storm Petrel, are obligate summer nesters, as local seas are ice-bound in winter (Fig. 11.6). In these short summers the birds enjoy continuous or extended daylight, a flush of food at sea with the blooming of the phytoplankton and subsequent zooplankton, together with the shift towards the surface of many organisms that winter at greater depths. Even were the birds physiologically capable of re-laying, failed breeders would not have time to re-nest before the cold set in: as it is, both early and late nesters tend to lose chicks during unseasonal polar storms.

For tropical species with little regular environmental variability, laying may be spread over several months, and the same species may lay at different times at different islands. Most such species are sedentary or their seasonal movements, if any, are unknown.

Biennial breeding albatrosses time laying so that the chicks are fully homeothermic before the onset of winter and fledge when marine foods are at a high level. Even at the Galapagos Islands, where 17 seabirds have loose or non-annual breeding cycles, the Galapagos Albatross has a rigid annual one. It is the only albatross to lay in May. Then the winds are stronger, offshore currents cooler, and food supplies probably better than at other times, starvation of the young albatrosses being almost unknown. By laying early the bird rears its chick during the same period that most of the other Galapagos petrels are raising theirs (Harris, 1969c).

B *Autumn and winter breeders*

The most notable examples of this type are the three North Pacific Albatrosses and their common associate, the Bonin Petrel *P.hypoleuca* (Fig. 11.5).

Why these should breed in the northern winter is unclear: possibly summer heat would kill the albatross chicks and/or the timing has been carried over from that of a southern ancestor. Kuroda (1960) considered that these albatrosses are cold-adapted, breeding during the cooler months and migrating to even colder waters after nesting. However, another petrel that shares their breeding places,

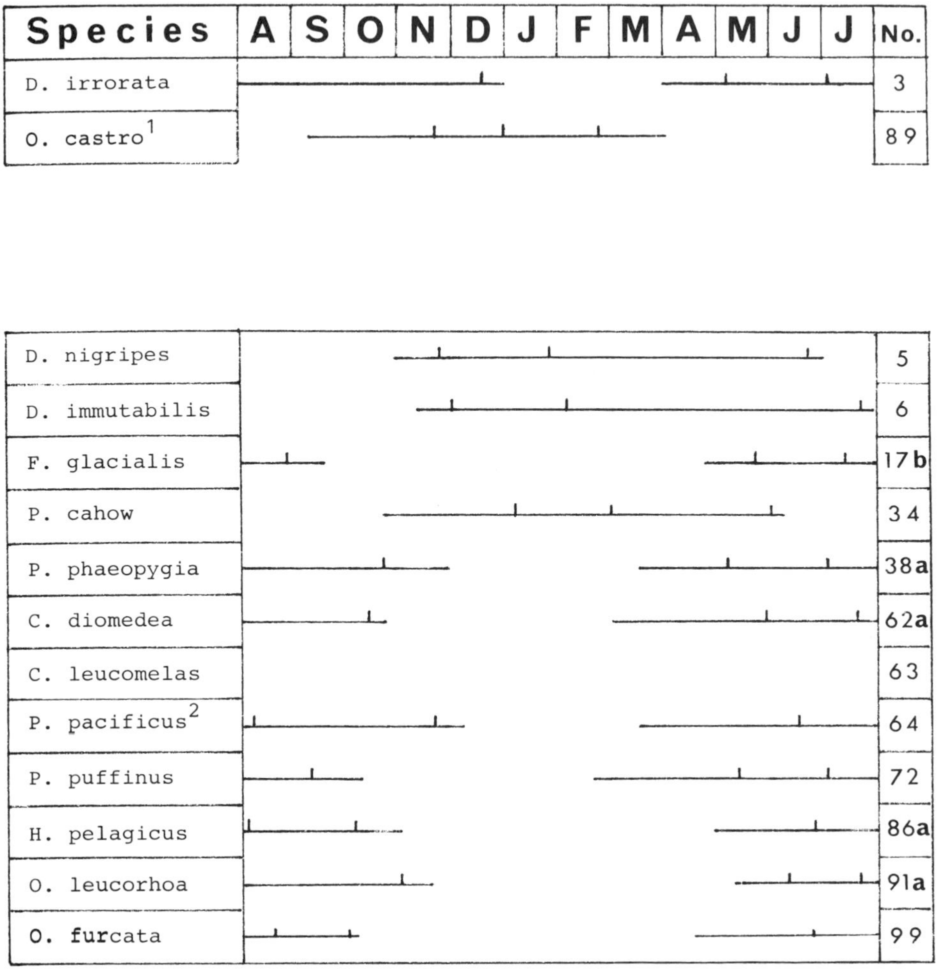

Figure 11.5 *Breeding seasons — from mean arrival to mean departure dates for (top) two tropical petrels and (bottom) 12 northern species. The three vertical marks indicate mean dates of laying, hatching and fledging. 1. Ascension Island, 2. Hawaii.*

P.pacificus, is a summer breeder in both hemispheres (Murphy, 1951). As it is a burrower, summer heat is evidently not limiting, so that seems unlikely to be important with *P.hypoleuca* either. Possibly competition with *P.pacificus* for nests may have forced the smaller *hypoleuca* to lay when *pacificus* is absent: even so the two species overlap, some burrows are shared, and substantial numbers of Bonin chicks are killed or evicted by the shearwaters.

Kuroda (1960) also discussed the North Pacific storm petrels, *O.matsudairae* and *O.tristrami*, which nest sympatrically at the Volcano Islands. Here *matsudairae* lays

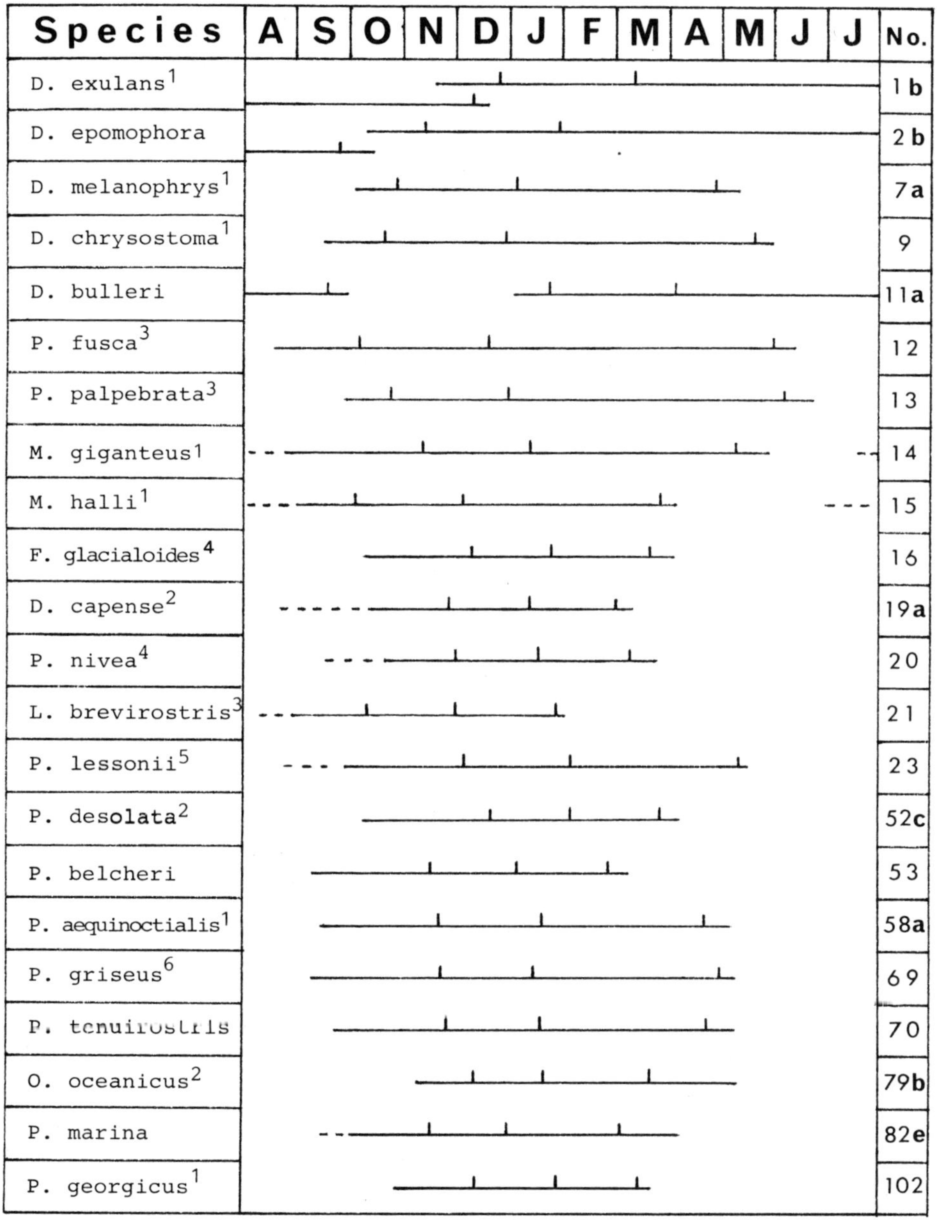

Figure 11.6 *Breeding seasons of 22 southern petrels. Vertical marks as in Fig. 11.5. 1. South Georgia. 2. South Orkney Is. 3. Crozet Is. 4. Terre Adelie. 5. Macquarie I. 6. Snares Is.*

in the spring, *tristrami* in the winter, so ensuring seasonal segregation. Kuroda believed that *matsudairae* is adapted to warm seas, the other preferring cooler waters and, like the albatrosses, migrating northwards after breeding. Recent sightings suggest that at least part of the *matsudairae* population shifts south to warmer seas in the non-breeding season.

The advantages of winter breeding in the south to birds like *Pterodroma macroptera* and *Puffinus assimilis* are also unclear, but they certainly benefit from reduced skua predation. Competition for resources with summer residents at the same sites has often been mooted as a driving force, and perhaps significantly, the populations of winter breeders are smaller than those of most of the summer-time ones at the same places. This suggests that food availability is an important factor, because in winter much plankton is too deep for petrels to reach, photosynthesis is low, and large seabird populations are presumably unsupportable.

Imber (1976a, b) stated that *Pterodroma macroptera* is a nocturnal feeder and that nesting in winter gave it the maximum time for feeding at night, but he thought that inter-specific competition for food was more important. He also considered that most winter-breeding petrels feed in the subtropical zone where seasonal variation in marine productivity is reduced and that this bird's main adaptation to winter nesting was the long pre-laying exodus, allowing the necessary reserves to be laid down despite low food concentrations.

Reduced competition for nest sites and burrows has also been suggested as a factor in winter breeding, but a factor not so far measured. There are cases of more powerful petrels apparently driving out smaller species, e.g. the resurgence of *P.bulleri* and decline of *P.gavia* at the Poor Knights Islands (Harper, 1983). Winter breeding, although lessening competitive interactions, may not completely avoid them, because chicks of similar-sized summer nesters may be still in burrows when the winter ones are seeking territories, e.g. *P.aequinoctialis* and *cinerea* at Antipodes Island (Warham & Bell, 1979) and *P.carneipes* and *P.macroptera* in Western Australia (Warham, 1958). The breeding seasons of such large species are too long for burrow-sharing.

The chicks of winter breeders leave when sea temperatures and local marine productivity are rising and when the newly arrived summer migrants have only to feed themselves. Non-migratory winter breeders may share the summer-time food resources with the migrant species, but having no nesting commitments, can reduce competition for food by ranging more widely than these.

C *Trans-equatorial migrants*

The timing of egg laying by migrants like Sooty and Short-tailed Shearwaters is very constant and there is little inter-colony variation. Thus, while the colonies of the Short-tailed Shearwater extend over 36° of latitude, adjacent to seas differing widely in temperature, chemistry, currents, and zooplankton, laying dates are the same throughout and from year to year (Serventy, 1963). Such precision enables the birds to take advantage of the summer-time peaks of food in the south when rearing chicks and a parallel abundance of food when in Subarctic waters during the northern summer. Truncated laying seasons in these and other such migrants like the Mottled Petrel also help ensure that the adults can get away to time whereas many non-migratory species can have more prolonged laying periods (Falla, 1934).

D *Inter-population variability*

In contrast to the highly synchronised laying of long-distance migrants and of those nesting during short summers, different populations of other species may lay at different times. Thus, while all populations of *P.palpebrata* lay on similar dates, those of *P.fusca* vary from island to island (Mougin, 1970c, p.49). Where such variation occurs the birds tend to be sedentary. Sometimes the differences are large. Most Little Shearwaters of the southern hemisphere are winter breeders like their northern counterparts, but at Gough and Antipodes Islands they lay in the spring (Imber, 1983). Inter-population differences may correlate with latitude. For example, the non-migratory Fairy Prion *P.turtur* in southern New Zealand breeds about a month later than it does in the north of that country (Richdale, 1965b; Harper, 1976). Another example was documented by Derenne & Mougin (1976): at The Crozets and Kerguelen the diving petrel *P.georgicus* lays most of its eggs in November whereas at South Georgia (and also, strangely enough, at Marion Island) these are laid in December.

Perhaps even more unusual is the situation of the storm petrel *O.castro* in the Galapagos Islands, where Harris (1969b) found two separate populations sharing nesting holes, and evidently breeding annually; one in May–June, the other in November–December. Non-breeders and failed breeders remained faithful to their annual cycles over four seasons. How the two populations remain separate without giving rise to a continuous breeding regime (as *P.lherminieri* does at those islands) is unclear.

Summer- and winter-breeding populations of *O.leucorhoa* around Guadalupe Island, Mexico, have been described by Crossin (1974). He suggested that this situation might have arisen from the conjunction of vast areas for feeding with limited land areas for breeding.

E *Year to year variation*

Some petrels are quite flexible in their breeding seasonality. Hutton's Shearwater delays laying when its burrows at 1460 m are buried in late snow, but what effects, if any, this has on fledgling success is unknown. In Alaska, the Fork-tailed Storm Petrel *O.furcata*, like some other storm petrels, has a long laying season, the peak varying from year to year according to snow conditions: evidently for similar reasons laying is later in some years at high elevations than it is for birds nesting near sea level at the same island (Simons, 1981b). In his 5-year study of *P.marina* Richdale (1965a) found a significant 10.6 day difference between mean hatching dates in the earliest and latest seasons. In fact (p.145) he found that, while at Whero Island the timing of *P.griseus* did not vary from year to year, there were consistent differences in hatching dates of the *P.urinatrix*, *P.turtur* and *P.marina*.

The 1941/42 seasons were early and the 1944/45 seasons late for all three. Richdale thought that food availability locally might be the common factor, these small birds having short incubation spans and thus presumably being more dependent on nearby supplies than the shearwater.

Sea conditions may have effects here. For example, when the spring ice breakout from Antarctica is late, birds like Cape Petrels and Antarctic Fulmars may delay their return. However, Cape Petrels at Signy Island did not delay laying when they arrived late (Pinder, 1966).

The greatest variation between seasons arises when a population fails to breed at all. Good examples have been provided during the El Nino Southern Oscillations in the Pacific Ocean. The catastrophic effects of the 1982/83 event led to an almost complete failure of breeding by the five petrel species at Christmas Island (Kiribati). Presumably the missing birds did not lay elsewhere. With other seabird species nesting failures arose through wholesale abandonment of chicks and flooding during heavy rain and doubtless the petrels were also similarly affected. Their numbers remained very low throughout 1986 (Schreiber & Schreiber, 1987).

F *Tropical species*

In tropical seas where seasonal variations in environmental conditions are usually small, the egg-laying period of petrels may be very extended. At Ascension Island *O.castro* lays from mid-October to early January (Allan, 1962) and in the Galapagos the two populations laying in the cold and hot seasons result in eggs being present for 8 months of the year, while those of *P.lherminieri* are found there throughout the year (Snow, 1965) (see Section VC below).

At Christmas Island in the central Pacific Schreiber & Ashmole (1970) gave data on the laying of four petrels (Fig. 11.7). Eggs of *P.alba* and *P.nativitatis* were laid in

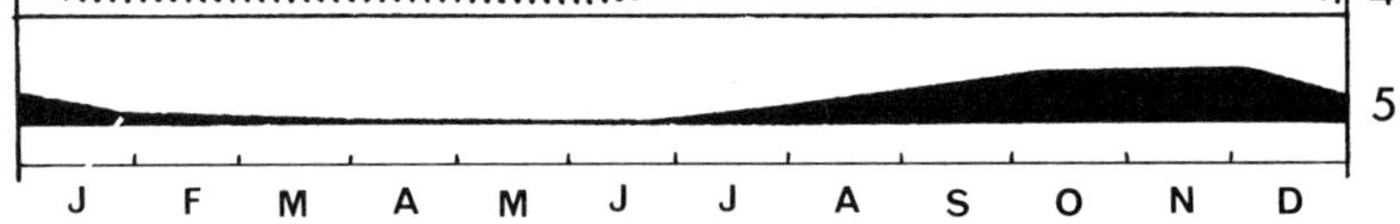

Figure 11.7 *Breeding seasons of petrels at Christmas Island (Kiribati), Pacific Ocean. 1.* Pterodroma alba. *2.* Puffinus pacificus, *3.* P.nativitatis, *4.* P.lherminieri, *5.* Nesofregetta fuliginosa. *Presence of eggs is shown in black, of chicks but not eggs by slanted lines, of adults without eggs or chicks by dashed lines. The pattern for* P.lherminieri *was based on few nests and both eggs and chicks were present between late July and late December. From Schreiber & Ashmole (1970).*

every month and both were annual breeders, but the population of *nativitatis* seemed homogeneous (although most eggs appeared from December to January) whereas that of *P.alba* was thought to be composed of two groups with opposite breeding seasons, as with *O.castro* at the Galapagos. The Christmas Island *N.fuliginosa* laid from August to November inclusive, yet *P.pacificus* there had a short summer laying season, as elsewhere.

Petrels of the subtropics may show similar trends. At the Kermadec Islands (30°S) there are summer and winter laying peaks for *P.neglecta*. At Madeira *O.castro* has summer and autumn breeding populations.

V Breeding frequency

Having obtained a nest site and a partner, most petrels attempt to breed annually. The many non-breeders that come ashore consist of immatures, past breeders that have changed or lost mates, failed breeders and pre-breeders — birds sexually mature but yet to breed.

There are three categories of non-annual breeders among petrels — individual birds that miss a season, species that routinely breed every other year, and those that nest more than once a year.

A *Intermittent breeding*

Probably most species have some mature birds that skip a season, but this is seldom documented, and does not seem to occur commonly even in *F.glacialis*, as suggested by Wynne-Edwards (1939). Fisher (1969) found a considerable annual variation in the numbers of experienced Laysan Albatrosses in his study group, variations not due to their having shifted, but because they arrived too late to breed or had an 'off season' (Fig. 11.8). Mougin *et al.* (1984b) termed this a 'sabbatical year'. At Grand Salvage Island they recorded temporary interruptions of breeding during 4.1% of the Cory's Shearwaters' productive years. In their long-term study of *P.tenuirostris* Wooller *et al.* (1989) found that in any one year 10–11% of breeders were absent and 15–18% present without an egg: both these proportions declined with the birds' increasing ages.

The proportions of experienced Laysan Albatrosses that bred annually varied from 44% to 100% over a 12-year period. In 1964 and 1968, when the proportions were low, Fisher presumed that poor food availability due to anomalous sea conditions was probably a major factor. Harris (1969a) recorded massive egg desertion among Galapagos Albatrosses, these being associated with abnormal heavy rains perhaps due to El Nino events and Beck & Brown (1972) found Wilson's Storm Petrels missing a season when snow blocked their burrows. Imber (1987) suggested that late laying Parkinson's Petrels found difficulty in nesting the following year, early layers performing better. In other seabirds breaks of this kind have been associated with adults arriving in poor condition. In such long-lived birds the loss of a single season's output may be less important than the increased risk of attempting to breed when under-weight.

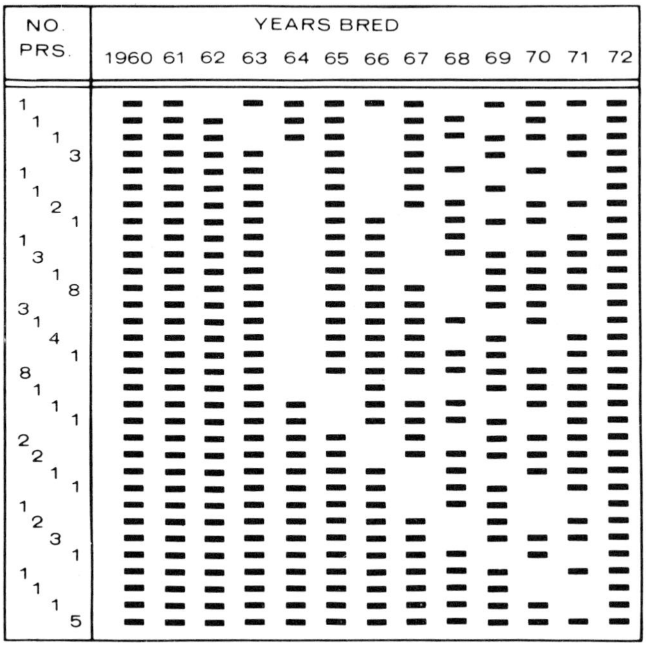

Figure 11.8 *Breeding patterns of 65 constant pairs of Laysan Albatrosses over a 13-year period. Note that all bred in 1960, 1961 and 1972, whereas many did not in 1964 and 1968. From Fisher (1976), courtesy of* The Wilson Bulletin.

Change or loss of partner may also affect breeding frequency. Of Laysan Albatrosses that lost their partner 99% did not attempt to nest the following season (Fisher, 1976). Changing partners also had adverse effects — those that had two different mates over a 10–13 year period bred on 71% of the seasons, those with three mates did so on 64% of them, and those with four mates only 57% of the years (Fisher, 1976).

The situation of *D.irrorata* may be rather special in that one calendar year may not be quite long enough for successful pairs to regain breeding condition — successful females laid slightly later the next year, those that failed laid earlier (Rechten, 1986). Whether consistently successful pairs skipped a season is not recorded. Rechten pointed out that the breeding of this albatross is intermediate between an annual and a biennial system.

B *Biennial breeding*

Less than annual nesting among petrels was suggested by Matthews (1929) who pointed out that the incubation and chick rearing of *D.exulans* took so long that yearly breeding was impossible. Biennial breeding in this and some other albatrosses was discussed in Chapter 2, Section VIIB. The Westland Petrel *P.westlandica* may also nest routinely less than annually (J.A. Bartle, pers. comm.)

and Fisher (1976) stated that about 60% of first-time breeding *D.immutabilis* missed the next year.

C *More than annual breeding*

The sole well-established instance of petrels that lay several times a year (re-laying excepted) is that of Audubon's Shearwater at the Galapagos Islands. Loomis (1918) drew attention to its protracted laying there and Snow (1965) found fresh eggs throughout the year with individual females laying at about 9-monthly intervals. This shearwater is one of eight Galapagos seabirds that breed more often than annually. Harris (1969c) considered that a major factor preventing more rapid breeding when conditions were good, was the need to moult. He also found that over his 21-month study there were two major episodes of food shortage which caused loss of young and stopped laying by *P.lherminieri*. Birds that would normally have laid over several weeks came to synchronise their laying after these events.

Other cases of extended laying periods among tropical petrels may well prove also to involve more-than-annual laying. With Snow's birds at least one of the relaying pair was a previous partner, and it seems doubtful if the new egg came from a new female, but in view of the tendency for more than one female to lay in petrel nests, confirmation that the same females are responsible is desirable in such cases, preferably by palpating eggs in marked birds, or by other means.

Snow & Snow (1966) also suspected a similar pattern applied to Madeiran Storm Petrels at the Galapagos but Harris (1969b), while confirming that there were two cycles of laying, one in the cold and one in the hot season, found that the breeding cycles were too long for successful birds to lay during both episodes. Two populations of morphologically identical birds were involved and, whether successful or not, each evidently remained faithful to its own timetable. According to Riveros-Salcedo & Aparicio (1990) most pairs of *Pelecanoides garnotii* nest twice a year.

VI Factors influencing the timing of breeding

The ultimate factor influencing the timing will be the life-time fitness of the bird in question. Proximate triggers stimulating gonad resurgance such as photoperiod may be effective among sedentary species but are difficult to envisage for the trans-equatorial migrants. These leave northern seas for the south when day lengths are shortening. Winter nesters also attain breeding conditions as daylight contracts.

Short-tailed Shearwaters arrive at their breeding grounds with well developed gonads, so that their seasonal sexual cycles must have developed in the northern hemisphere under decreasing day lengths (Marshall & Serventy, 1959) (Fig. 11.9). These authors could not explain how changes in photoperiod could work to achieve the very precise timing necessary simultaneously to activate birds scattered over a wide area of northern seas. They sought some astronomical factor to be involved, but when they subjected adult birds to captivity under local and artificial photoperiods throughout the southern winter, these came into sexual condition

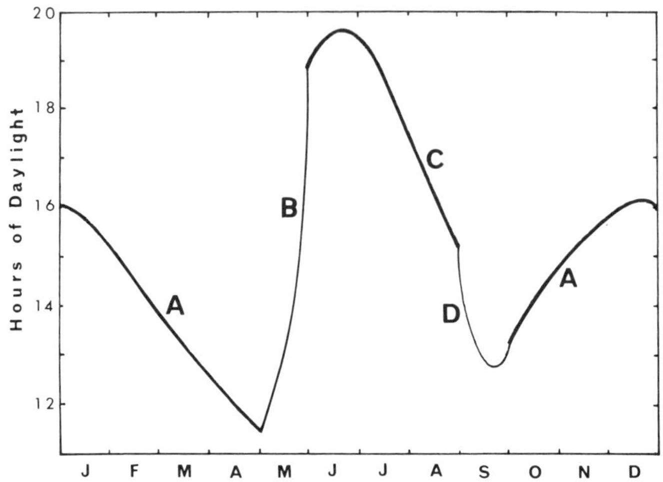

Figure 11.9 *Variation in the hours of daylight during the annual cycle of the Short-tailed Shearwater* P.tenuirostris. *(A) in breeding area, (B) during northward migration, (C) in North Pacific, (D) during southward migration. From Marshall & Serventy (1956b).*

only a little less advanced than that of the wild birds at the end of their return migration. Their breeding rhythms had persisted for at least one year despite having experienced quite atypical photoperiod, temperature, food and habitat conditions. Not only gametogenesis but also wing moult and pre-migratory fat deposition occurred at approximately the usual times: a fixed internal rhythm maintained the gonad cycle on its normal course.

Apart from that work, day length as a proximate factor has not been studied in petrels. Murton & Westwood (1977, p.400) found that for *M.giganteus,* a bird breeding over 21° of latitude, the photoperiod at egg laying was significantly correlated with latitude, whereas with *M.halli,* with a 15° latitudinal range, it was not, and this bird can breed on shorter photoperiods than its sibling species. These authors did not investigate the relationship, if any, between sea or air temperatures and laying dates.

The gonad cycles in relation to the timing of breeding of southern albatrosses have been investigated — references in Hector (1988). Hector found that in their non-breeding year *D.exulans* had high progesterone levels which were thought to block vitellogenesis and thereby inhibit egg production.

Food availability and quality may also exert a direct influence on laying as these could be sampled during the female's pre-laying exodus and deposition delayed or abandoned according to her rate of energy storage in yolk and other lipid deposition. Kuroda (1960) pointed out that all four petrels nesting on Torishima Island (30°N) are winter breeders and evidently choose that season in order to exploit the availability of cooler waters and of the suitably sized squid that they contain.

Brooke (1978a) argued that laying by Manx Shearwaters is timed so that there is strong probability that whatever the food supply in the pre-laying period, there will be adequate food during incubation, so that incubation itself is the ultimate factor involved. Laying is seen as being delayed until there is a high chance that both parents can store sufficient fat to fuel their long incubation fasts. Brooke considered that this is the primary determinant of laying date, not the need for adequate food reserves to produce a large egg (cf. Perrins *et al.*, 1973). Because this is a migratory species, the end of the laying period is presumably determined by the need for both fledglings and parents to lay down reserves before departure and that this is critical is suggested by the adult's desertion of their chick during its last 8.5 days in the nest.

Uncertainty of food supply may have helped the retention of a wide range of laying dates among some tropical species: this lessens the risk that all nesting will fail should conditions deteriorate. With petrels, adjustments to clutch size such as occur with boobies are not possible: they must either lay or not lay.

The role of mutual stimulation through vocalisations in compressing laying is virtually unstudied among petrels (but see Jones, 1986) and may be important, while among sedentary species reduced stimulation could help reduce laying synchrony.

Competition for resources of food and/or nesting sites have been proposed as determinants of breeding timetables. Croxall (1984) gave several possible examples from the southern hemisphere such as that of the Blue Petrel and Fulmar Prion which breed earlier than other prions that share their nesting places and by so doing presumably reduce competition for food while they are raising chicks and their food needs are high.

Unavailability of burrows due to snow accumulation was one factor considered by Payne & Prince (1979) to account for the 40-day difference in hatching dates between *P.georgicus* and *P.urinatrix* at South Georgia. The steep sites used by the latter accumulate less snow and clear earlier than do the nearby burrows of *georgicus* which are dug in bare, stony soil with little or no slope.

VII Season lengths

The time that elapses between the arrival of the adults and their departure after having bred varies roughly according to body size (Figs 11.5 and 6). The period is a long one, even for small species, and for those of high latitudes allows little flexibility in the timing of breeding. Precise data are rather scarce partly due to the difficulty of determining mean dates of return and departure, particularly for sedentary petrels, and the many that visit their nests, apparently erratically, during the non-breeding season.

Rough comparisons are possible. For *D.exulans* the season extends over about 380 days, significantly more than the 352 days for *D.epomophora sanfordi* determined by Richdale (1950, 1952). Likewise for *D.melanophrys* the figure is about 195 days whereas the similar-sized *D.chrysostoma* takes 235 days. Other approximations for albatrosses are *c.* 270 days for *P.palpebrata*, 290 for *D.immutabilis*, 275 for *D.nigripes*, 270 for *D.bulleri*, and about 210 days for *D.chlororhynchos*.

Some smaller petrels may take as long as these albatrosses, e.g. *C.leucomelas*, *P.pacificus*, *P.cinerea* and *P.lessonii* all take about 250 days, *P.puffinus* about 240 days.

Highly migratory species do better — for *P.tenuirostris* the season extends over some 195 days, and for *P.griseus* and *P.inexpectata* over 205 and 210 days respectively. Among storm petrels about 195 days sees the breeding of *O.leucorhoa* completed whereas *H.pelagicus* takes only 150 days.

Antarctic tubenoses have compressed breeding seasons (Chapter 15, Section VIC), many taking 140–150 days, the Antarctic Prion (at Signy Island) about 165 days. Snow Petrels at Durmont D'Urville take only 125 days — 15 days pre-exodus, 18 days exodus, 45 days incubation, and 47 days to rear their chick (Isenmann, 1970). Presumably continuous daylight is a factor here, but even so Southern Giant Petrels only just succeed in getting their chicks away before the Antarctic winter sets in.

VIII Philopatry

The tendency for young colonial animals, from scallops to penguins, to return to breed near their birthplaces, is well developed among petrels. In long-term studies such returns are readily recorded, but birds that shift are often not discovered or are presumed to have died.

Philopatry depends on the young bird's ability to navigate back to its breeding grounds and to remember characteristics of the natal colony. Translocation experiments by Serventy (1967) of chicks of *P.tenuirostris* suggested that they became imprinted on the environs of their nests during the period towards the end of rearing when they emerge from their burrows after dark. A further analysis by Serventy *et al.* (1989) confirms that imprinting must be very precise and raises the possibility that some locality clues may be available to the chick even when it is in the burrow.

Quantitative data on the philopatry of Laysan Albatrosses were given by Fisher (1976). Of 1078 old enough to breed that fledged from within a marked plot, only two males and one female nested outside it, the males 6 and 13 m and the female 7 m away. Within the plot 59 males were incubating at a mean distance of 19 m from their birth nests, 49 females 26 m away. This high degree of philopatry and apparent inbreeding was suggested as one possible explanation of the existence of 'subcolonies' within the main colony — areas where synchronisation of laying and chick fledgling was greater than elsewhere (Fisher, 1976). Local synchronisation of breeding rhythms is also reported of *Calonectris diomedea* by Mougin *et al.* (1988).

At Possession Island in The Crozets most Wandering Albatrosses whose birthplaces were known were found nesting in their home colonies, but at one colony with scattered nests (35 km^{-2}) 36.7% bred elsewhere on the island, i.e. faithfulness was greater to denser colonies (Weimerskirch & Jouventin, 1987).

Richdale (1963, p.63) measured the distances from their natal burrows at which he found adult Sooty Shearwaters marked as chicks in his small island colony. The mean was 17.9 $\pm$ 12.8 m, a much greater distance than that of any other category of movement in the colony. He suspected that some of those he did not find might have moved to other islands. He gave very similar figures for the Common Diving

Petrel (Richdale, 1965a) — 24 former fledglings were caught 18.1 ± 15.6 m from their natal burrow.

Of 17 Northern Fulmars recovered on their home island, the distances between their birth nests and their breeding sites as adults were < 100 m for five birds, between 100 and 500 m for seven and between 700 and 1000 m for the remaining five birds (Dunnet *et al.*, 1979). However, these birds nested on cliffs, so that fewer sites were available near to the birthplaces than with the species already discussed. Only 3% of fledglings actually returned to their colony and these authors estimated that 89–94% of those surviving must have bred elsewhere, as would be expected of a species that is expanding its range.

There are many other instances of birds shifting to places far from their natal sites. Fisher & Fisher (1969) found 232 Laysan Albatrosses (mostly juveniles) on Midway Atoll that had been ringed on Kure Atoll. A much lower proportion of Midway birds was found on Kure, more than half of breeding age, and at least two born on Midway nested on Kure. Beck (1969) found that 25% of 28 breeding known-age Cape Petrels had been born elsewhere, the furthest 4.8 km from where it was then nesting.

Harris (1972) warned against assuming that a petrel has changed its colony just because a bird from one has been found in another. He reported three sexually mature Manx Shearwaters from Skokholm Island that were re-trapped on colonies several hundred km distant but were later back on Skokholm. Despite large-scale banding he found only one bird breeding away from its birthplace and this was but 3 km away on Skomer Island. Furthermore, the Fishers also found that juvenile Laysan Albatrosses often visited other islands briefly on their way back to their home base. In later work at Skokholm, Brooke (1978b) established that permanent emigration of young birds did indeed occur, but that this was mainly confined to the females. Perhaps as many as half of these emigrated and, if they survived, bred elsewhere, probably mostly on Skomer. Greater fidelity of the male to its known area is likely to increase its chances of obtaining a territory, the female avoids inbreeding — advantages that have been claimed for similar behaviour in other seabirds. However, it is difficult to account for persistence in returning to a particular colony when competition for space there results in considerable nesting failure and when alternative sites are readily available nearby, as applies with *O.tethys* on the Galapagos (Harris, 1969b).

Other shifts of single birds are a Royal Albatross of the typical form (from Campbell or the Auckland Islands) that has moved to the Taiaroa Heads colony of *D.e.sanfordi* and bred with a bird of that race. A Wandering Albatross from Macquarie Island nested on Heard Island, about 5000 km upwind (Johnstone, 1982) and another shifted from The Crozets to Marion Island (Mougin, 1977). In December 1984 two Antarctic Petrels were caught at Ardery Island, East Antarctica (Van Franeker & Montague, 1987) which had been ringed as chicks at Mirny Station in early 1979. Both had brood patches and may have attempted breeding and as they were found almost side by side, the authors speculated that fledglings may keep to the same flocks for long periods, as also suggested for *D.capense* by Weimerskirch *et al.* (1985).

Serventy & Curry (1984) analysed the sightings of Short-tailed Shearwaters at night on islands near Fisher Island. Sixteen thousand were picked up from the surface, about 1% of them having been born on Fisher Island. Conversely, a few born on nearby islands were recruited to the Fisher Island population. The shifts seem to have been mainly away from there, however, where the population has declined over the study years. Serventy & Curry suggested that a major factor has been disturbance by field workers, particularly to 'prospecting' birds early in the breeding season, for more than two-thirds of the emigrants had been handled previously on Fisher Island. Apparently that experience led them to shift.

IX Breeding age

Delayed breeding is a feature of tubenose reproduction, none nesting at less than 2 years old, most taking 5 or 6 years to reach that status, while great albatrosses do not usually produce viable eggs until they are about 11 years old. When breeding does start most pairs have already been formed for several years, the pre-breeding rituals can be reduced and the cycle shortened.

However, such pre-breeding birds are not necessarily sexually immature (Fisher, 1975c). They acquire brood patches and display vigorously for several years before they try to breed and Hector (1988) suggested that male Wandering Albatrosses were sexually adult after their fourth year despite their not breeding until at least 7 years old. Fisher established that *D.immutabilis* which bred when younger than the 8–9 years normal for this species, did not experience greater mortality, as might have been expected. Higher mortality did not occur until they had bred 3–6 years, and then without sexual difference. Overall, however, those *D.immutabilis* that started breeding early also tended to die young.

Lack (1954, 1966) saw retarded breeding as having evolved because the care and feeding of the young was too great a strain on inexperienced parents, whereas Wynne-Edwards (1962) considered it part of a self-regulatory system of population control. Correct synchronisation of behaviour is essential for successful breeding and has to be maintained over long periods — for over a year with great albatrosses — so that the idea of younger birds gradually attaining the necessary skills on land and at sea during a pre-breeding phase of their lives seems reasonable. Such birds gradually come back earlier to their nesting grounds and stay longer as they become older (c.f. Fig. 12.8). During this time they find suitable partners and nest sites and acquire the breeding rhythms of established pairs. That they become adept at predator avoidance is suggested by the way that birds that have been banded and recaptured become more difficult to catch, apparently learning by their experience.

At sea their ability to forage, to develop specialised feeding techniques and, in some species to migrate extensively, may improve with practice during the run-up to first breeding. For example, most of the birds in 'wrecks' on beaches prove to be from the younger age classes: the more experienced older birds either die elsewhere or, more likely, die less often. Once breeding has started those with more breeding experience produce more young but their success may decline in later years (Wooller *et al.*, 1989).

That older birds feed better is supported by the finding that they tend to be heavier than young ones, a trend well shown for the Manx Shearwater by Brooke

(1978d). He found that mean weights increased appreciably between 3 and 10+ years, the most likely factor being increasing foraging efficiency, and postulated by Lack.

The mean age at which breeding begins is partly related to size, small birds breeding earlier. Croxall & Gaston (1988) gave the relationship as:

$$y = 2.0\log_e x - 5.85;\ n = 13,\ r^2 = 0.84,\ P < 0.01$$

where y = mean age at first breeding (years) and x = adult weight (g).

However, the fastest maturing petrel is not the smallest but the Common Diving Petrel. This starts nesting when but 2 years old (Richdale, 1965a, p.43). Perhaps food-finding skills are more easily acquired by near-shore feeders than those that forage pelagically.

At the other end of the size range most male *D.exulans* do not breed until 12.1 years (range 9–16 years, n = 24), most females at 11.2 years (7–16 years, n = 64) (Weimerskirch & Jouventin, 1987). The much smaller *Phoebetria* albatrosses take just as long. Jouventin & Weimerskirch (1988) gave the mean age at first breeding of *palpebrata* as 12.0 years and of *fusca* as 12.2 years. Their cumulative curves showing the proportions breeding for the first time for 5- to 16-year-olds of five albatross species emphasised this slow attainment of maturity by these sooty albatrosses compared with that of the other similar-sized *Diomedea*. The sexual differences in the ages at which breeding status is reached may be size-related: the differences seem to be less in smaller birds with little sexual dimorphism, e.g. with *P.tenuirostris* 7.3 years for males, 7.0 years for females (Wooller *et al.*, 1988). The data are few for other species. Other general data on ages at first breeding are given by Jouventin & Mougin (1981), Mougin *et al.* (1986b) and Ollason & Dunnet (1978, 1988).

X Fidelity to site and mate

A *Introduction*

Once a petrel has bred in a particular colony it seldom shifts to another, particularly if it has reared a chick. Colony sites persist over many years: even with giant petrels, which are prone to shifting *en masse* if disturbed, some colonies have existed for at least 46 years. From radiocarbon dating of stomach oil deposits, Hiller *et al.* (1988) suggested that some colonies of Snow and Antarctic Petrels might have been established for up to 30 000 years. The habit among some species like *Macronectes* of occupying nest sites out of the breeding season — protogamic behaviour — will also tend to stabilise colony locations.

Like penguins and many long-lived seabirds, petrels also tend to remain faithful to the same partners and to return to the same nest during successive breeding episodes. Lockley (1942) was one of the first to show this during his study of the Manx Shearwater. Among 240 pairs of this bird caught in successive seasons only two divorces occurred (Perrins *et al.*, 1973). However, many petrels do not pair for life. The bond seems to be stronger among large species like albatrosses, less so among storm and diving petrels. Presumably the benefits of keeping a mate with whom a bird has bred successfully and of the knowledge of the nest surroundings (including that of neighbouring birds) are greater in long-lived large species, which must make many breeding attempts in order to replace themselves, than obtain with smaller, and shorter-lived ones. However, if they survive long enough, most petrels do shift their sites, though over rather short distances. Many also change partners when one of them dies, or they divorce, particularly when nesting fails — apparently incompatibility of some kind is involved.

Retaining the same nest and partner saves time at the start of the season. The nest provides a focal point for partners to meet over many seasons, their arrival times tending to differ. Such a meeting place is particularly important for birds of high lattitudes or for long distance migrants. Experienced pairs also tend to lay their eggs and fledge their chicks earlier than do recently formed pairs and early chicks may have a better chance of survival.

B *Site tenacity*

Available data on retention of nest sites are limited because few studies have followed marked birds for more than 5 or 6 years. Site tenacity is linked to mate tenacity and, in general, males hold onto their nest when they lose a partner. Nevertheless shifts occur without the pair-bond being broken.

The great albatrosses often do not re-use their old nests, partly because these tend to disappear by being broken down and overgrown during their off-duty years, but Richdale (1865b, p.152) found that over 18 years all his birds returned to their particular areas. The pedestals of mollymawks are usually re-occupied in subsequent seasons, but some get washed away and have to be re-made. Among 318 Buller's Mollymarks banded as breeders up to 40 years ago, none has been found at any other colony than that at which it was banded and one pair was still at the same site 19 years later (Richdale & Warham, 1973). This bird breeds annually like its relative *D.chlororhynchos*, which also shows a high degree of site fidelity (Jouventin *et al.*, 1983).

Even among the great albatrosses shifts do occur. At The Crozets, of 236 Wandering Albatrosses, 34% bred at or very near their previous site 1 year later (i.e. being failed breeders), 17% bred there again 2 years later, 12% in the third year and none at the same nest 4 years later. Fidelity was greater among birds that came back to try again a year later, less if they had a year off (Du Bost & Segonzac, 1976). These authors do not state what proportion of old nests remained to be re-used. Shifts between nestings of *D.exulans* averaged about 7 m at South Georgia (Tickell, 1968).

In the unusual case of *D.irrorata* the males return to the same general territories as before but the sites themselves change because the birds move with their eggs tucked into their brood 'pouches' (Harris, 1973). Among 1288 established breeding *D.immutabilis* $<1\%$ moved nests significantly in later years and then only when there was a change of partner. Three widowed males moved only short distances but seven females moved further to mate with new territorial males (Fisher, 1976).

There are reasonable data for only a few Procellariidae. Mougin (1975) summarised findings on some southern birds, recording that of a sample of 39 adult *M.giganteus*, 49% used the same nest 1 year later, 31% did so 2 years later and

6% 3 years later. For other fulmars site retention was higher. Of 75 *F.glacialoides*, 91% of those still alive re-used the nest the next year, 73% doing so 2 years later, 43% 3 years later, 34% 4 years later, 36% 5 years later and 40% of the 15 survivors did so 6 years later. Rather similar figures were given for *P.nivea* and *D.capense* — see also Pinder (1966) and Guillotin & Jouventin (1980). The greater site tenacity of these smaller fulmars compared with that of *Macronectes* may be an artifact of the timidity of giant petrels and of shifts triggered by the disturbance involved in identifying banded birds. Among Northern Fulmars 10% of the males in 456 nestings changed site and/or partner in the following year, as did 9% of the females in 443 nestings (Ollason & Dunnet, 1978).

The site tenacity of shearwaters appears to vary considerably. Of 1205 Cory's Shearwaters 91% used the same site the next year, 84% of 680 stayed there 2 years later and 79% of 306 birds did so 3 years in succession. The mean distance shifted was 2.2 ± 3.8 m in 28 cases (Mougin *et al.*, 1984a). In contrast, Richdale (1963, p.60) found that Sooty Shearwaters did not keep to the same burrow year after year. Of 274 birds that nested 2 years in a row, the mean distance moved was 1.8 ± 2.6 m (range 0–19 m). Of 55 pairs that bred for 2 seasons but shifted, the mean distance moved was only 0.8 ± 1.8 m (range 0–9.8 m), whereas of 219 that changed partners the mean shift was 1.8 ± 2.6 m (range 0–19 m), the difference between the last two means being significant, implying that birds maintaining a bond tended to shift little between years.

Smaller petrels at the same island showed similar retentiveness, with repeated use of the burrow not usual and as time went on more and more shifted to other nests. Data for Common Diving Petrels and Fairy Prions were rather similar and 87% of 511 of the latter were in the same burrow the next season, 74% of 237 recovered 2 years later had not moved nor had 52% of 99 prions seen in the third season, but only 26% of the 23 found during the fourth season had not shifted. Among mated pairs, as with *P.griseus*, there were fewer shifts, 93% of 211 pairs remaining true to their burrow for at least 2 seasons. Nevertheless, in time all would have changed their sites.

Distances moved were generally small. For example, with 86 Fairy Prions that moved the mean shift was only 1.9 m, for 22 intact pairs it was 1.6 m, whereas the survivors of broken pairs moved on average 2.4 m (n = 37). In all cases the birds spend their time ashore in a very restricted area (Richdale, 1965b, p.113).

This researcher found that shifts were caused by collapse of burrows and the swamping of their entrances by new vegetation during the contra-nuptial season, effects that would be less important at colonies with better consolidated soils. Furthermore, Richdale spent much time on the colonies after dark inspecting burrows and weighing chicks, and the effects of his disturbance may have induced some of the shifts, divorces and even emigration of immature birds, more particularly among the Sooty Shearwaters.

The precise locations of burrows change from year to year due to their destruction. Richdale (1965a, b) analysed losses of *P.turtur*, *P.marina* and *P.urinatrix* burrows. On average 39% of the prion nests were lost each year, the others losing 33% and 35% respectively. However here, as in other stable colonies, the total number of burrows changed little, lost nests being replaced by new ones.

Apart from season-to-season faithfulness to nest sites, breeding birds may be remarkably tenacious once the egg is laid. Fisher (1971, p.54) reported a Laysan Albatross floating above an egg in a flooded nest and how birds blown off their eggs in gales or washed away in floods, persisted in trying to return.

In field experiments Rice & Kenyon (1962) tried to get Laysan Albatrosses to accept shifts to new sites, but if their nest was moved for more than about 1 m most shifts were not accepted: displaced chicks tended to be ignored by their parents which returned to the place where their nests had been.

C *Mate tenacity*

Fisher (1976) found that of 1968 *D.immutabilis* whose pair-bond ages were known, 62% bred with the same partner for one or two seasons and 90% had four or fewer partners over a 9-year period. For albatrosses of this species that kept the same partner for 10–13 years the breeding frequency/pair was 0.82–0.86 times per year, for those with two mates in that time the frequency fell to 0.71, for those with three mates to 0.64, and with four mates to 0.57 times/year. Loss of mate in a breeding season meant the loss of that year's egg or chick plus the loss of one or two subsequent breeding attempts by the survivor while it formed a new bond.

Some long-term pair-bonds are recorded. Three pairs of Buller's Mollymawks were together at least 21, 22 and 23 years (Richdale & Warham, 1973) and 12 pairs of Galapagos Albatrosses kept the same mates from 1961 to 1971 (Harris, 1973). At the Taiaroa Heads colony of Royal Albatrosses Richdale (1952) had two pairs whose bond lasted at least 15 years and today some bonds are known to be of much greater age.

Of 223 pair-bonds between Fairy Prions, 43% remained intact during the next season, in 17% of instances only one partner re-appeared and in 39% neither did so but in only 2% were birds divorced, each partner taking a new mate. In this study the mean bond length was 1.6 years, although some persisted for 8 years (Richdale, 1965b, p.123). Among Northern Fulmars fidelity to mate from year to year is about 90% (Ollason & Dunnet, 1978). Most Cape Petrels at Signy Island retained their partners over successive seasons, at least 73% of 41 pairs doing so, with a further 17% in which only one partner was sighted (Pinder, 1966).

Retention among Cory's Shearwaters is also high, 93% of breeding birds still present re-mating one year later, 89% doing so again 2 years later, and 81% still being together 3 years later, the rest taking new partners (Mougin *et al.*, 1984a, 1987). Richdale's data for the similar-sized Sooty Shearwater did not cover many seasons but he distinguished the breeding birds from those simply found sharing a nest — 'keeping company'. His figures are minima as he knew that some birds had been missed. Of 79 breeding over two seasons 33% re-nested with the same partner and 15% were found in burrows with that partner but without breeding. He considered that most of these latter were failed breeders whose eggs he had not seen, if so nearly 49% of the sample bred with the same mate in successive seasons. Despite the relatively short lives of the bonds of his birds, 18 pairs were still together 4 years later, eight together 5 years later, four together 6 years later and one each 8, 10 and 11 years later (Richdale, 1963, p.108). Again, some pair-bonds could have been broken through his activities.

In general, tenacity to nest site is stronger than that to the partner, for example, among *O.leucorhoa* fidelity to mate at 68.4% per annum was significantly lower than fidelity to site (95%), suggesting that the primary attachment was to the site not to the partner (Morse & Buchheister, 1979).

D *Divorce*

The changing of partners while both are alive probably occurs among all petrels, but is evidently quite rare in some species. Conroy (1972) found no divorces among 150 pairs of *M.giganteus* studied over three summers and in a colony of *C.diomedea* only 6% were divorced (Mougin *et al.*, 1987). Among Snow Petrels followed over 13 years the divorce rate from year to year varied between 7.5% and 26%, mean 17.3% (Guillotin & Jouventin, 1980). Sixty-one percent of these divorces had failed in their breeding attempts the previous year whereas only 46% of those retaining their bonds had failed.

Richdale (1963, p.109) stated that of 156 Sooty Shearwaters that had changed partners only 37% bred with their new partner while 31% got a mate but evidently did not breed. He concluded that, in this colony at least, finding a new partner after divorce was not easy. With Northern Fulmars Ollason & Dunnet (1978) noted that the success of divorced birds in rearing chicks rose compared with that before — 75% as against 64% in their small samples.

E *Pair formation*

Bonds may be formed between birds that have never bred and between breeders that have lost their mates or been divorced and who find new partners among other past or failed breeders or from young, inexperienced birds.

The best information comes from observations on albatrosses — their activities are easily followed. Here pair formation is a slow, gradual process. Young, pre-breeding birds come to the nesting grounds progressively earlier with increasing age to display in small groups or 'gams' and gradually pair off to establish territories where they indulge in a variety of mutual behaviours, nest building etc. which apparently help to cement the bond. Older pre-breeders arrive early and go to their previously held territories. Initial nestings mostly fail (few Laysans breeding for the first time even complete incubation), but on their return the next season the birds are already paired. Like the established breeders, what mutual activity now occurs before laying helps consolidate the bond. However, most of the display seen at the colonies primarily involves the non-breeding sections of the community. Most Laysan Albatrosses need 2 or 3 years together before the necessary behavioural and physiological coordination is achieved for success (Fisher, 1975b). Laysan and Royal Albatrosses that desert and lose their egg may return to the site repeatedly, and the partners may sit side by side on or near the nest, perhaps preening for hours; again, presumably this strengthens their bond (Fisher, 1971; Richdale, 1950).

In a sample of 46 Sooty Shearwaters found together with no egg, 48% were still together the following year, 35% of them actually breeding, 11% had acquired a new mate without breeding and 28% were seen but without a partner. Richdale considered that this is how pair-formation occurs: the birds sharing a site for a season or more before breeding proper is attempted (Richdale, 1963, p.112). Pickering (1989) has recently analysed the attendance patterns and behaviour of Wandering Albatrosses in relation to experience and pair-bond formation.

Richdale (1950) cited circumstantial evidence that pair-bonding among albatrosses began in the open ocean outside the breeding season, and that even when nesting, they could find their partners at sea. Certainly albatrosses are well known to form displaying groups at sea far from the nesting places and there are many reports of petrels being seen in twos over the ocean, but even with sedentary species there is little direct evidence that bonds are forged there. (Mougin, 1975, p.80) pointed out some of the problems with that idea and Voisin (1988) reviewed this topic briefly.

CHAPTER 12

The Pre-egg Stage

Copulation — Buller's Albatross.
Photo: author.

Contents

The period between the return of the breeding birds and egg deposition varies considerably between species and groups according to body size and migratory status. With great albatrosses it is about 7% of the length of the whole reproductive cycle, 8–14% for *Phoebetria* and medium-sized albatrosses, and 30–38% for shearwaters, gadfly petrels, prions and storm petrels. Despite the shortness of polar summers, the fulmars do not appear to have reduced the relative length of the pre-egg stage, this being about 35–37% of the breeding cycle (Mougin, 1975).

Unusually long pre-egg periods with particular birds often arise due to delayed returns of their partners.

I The return to the breeding grounds

A *Sedentary species*

For petrels that make contra-nuptial visits to their nests and indulge in photogamic behaviour, the start of the breeding season is poorly defined. Such sedentary species occur in all families except the Diomedeidae, out-of-season visits being known for most of the fulmars and prions, the gadfly petrels *L.brevirostris* and *P.macroptera,* for *H.caerulea,* the shearwater *P.assimilis,* and the storm petrel *O.tethys,* among others.

This part of the annual cycle has been little studied, the best data being on *F.glacialis,* whose presence ashore almost all the year round has been recorded since the seventeenth century (Martin, 1698). This bird even visits its Arctic colonies in the mid-winter darkness providing that there are ice-free leads where it can feed. At Jan Mayen (71°N) fulmars arrive plump, with large breast muscles, but after a fortnight become very thin (Lovenskiold, 1963). Coulson & Horobin (1972) found that after chick departure in August and September, adults at Marsden Rock, England, reappeared singly and briefly in early November. By late November, with the days shortening, many birds returned just after dawn, their numbers increasing during the morning and falling off in the afternoon. Some stayed after sunset, but none overnight. Night-time occupation began just before egg laying — the whole pre-egg period being made up of daily occupations and vacations of the colony. Many birds were non-breeders but over 50% of the sites had two birds, confirmed as established pairs by Macdonald (1980). Numbers ashore correlated with wind speeds: the stronger the wind the fewer the birds. Coulson and Horobin thought that the synchronised departures with freshening breezes were for wind-assisted feeding trips. The fulmars were not driven from the cliffs by the winds and they left after spells of fine weather during which the colonies were crowded, with more birds ashore than at any time after egg laying. Slater (1990) has shown that the first fulmars to arrive at a 36-year-old colony did so progressively earlier as the colony aged, a change presumably related to an increase in the mean age of the birds.

Presumably these winter-visiting fulmars engage in pair-bonding displays and nest-making as do the giant fulmars of the south. The relationships of the latter to the weather are unrecorded but *Macronectes* seem undeterred by snow or rain. Autumn nest-building is a prominent facet of the protogamic activity of giant petrels, with quite substantial structures being produced (Warham, 1962; Warham & Bell, 1979), sometimes in places with no nests during the previous breeding episode. The status of such birds is unknown but they are in adult plumage and appear to be of breeding age. Conroy (1972, p.18) identified some autumn-nesting *M.giganteus* as failed breeders. Clearly they are efficient at food-finding even in winter or could not afford to spend so much time ashore. Conroy found that the winter numbers on land were far greater by day, but as the time for laying approached over-night attendance increased considerably. Other reports of winter-visiting are those of Holdgate (1963) and Mougin (1968) for Antarctic *M.giganteus*, of Pinder (1966) for *D.capense*, and of Prevost (1964) and Brown (1966) for *P.nivea*.

Out-of-season occupation by Northern Fulmars may allow maintenance of their hold on a nest site — a dwindling resource in an expanding population — or acquisition of an empty one. No such limitation applies with *Macronectes* which are ground nesters: there are vast areas apparently suitable for their colonies at Macquarie Island and elsewhere so that there is no competition for space as such.

Jouanin (1964) found that protogamic activity in *Puffinus assimilis*, which included burrow visiting, display and much calling, was from birds in full wing and tail moult, and he suggested that this was when true courtship and pair-bonding took place in preparation for breeding later in the year.

B *Non-sedentary species*

Whether they have been dispersed at sea or on distinct migrations, the return of petrels to their colonies may be a gradual process, fairly abrupt with long-distance migrants, or involve a mass influx. There is some evidence that the earliest birds tend to fly over the sites, only landing a few days later. In general, apart from some albatrosses and fulmars, precise information on the activities of birds of different ages and status, sex by sex, is scarce.

During the pre-egg stage petrel colonies have a predominantly male complexion. Breeding males usually return first and spend more time ashore than their partners do. For both to arrive at their nest on the same day seems quite exceptional.

The greatest difference in arrival times has been recorded with *Phoebetria fusca*. Jouventin & Weimerskirch (1984) gave mean arrival dates as 16 ± 5.3 September for males and 21 ± 5.6 October for the females. More usually with albatrosses the mean differences are less — 10, 11 and 8 days for *D.epomophora*, *D.exulans* and *D.immutabilis* (Richdale, 1950; Tickell, 1968; Fisher & Fisher, 1969). For some smaller petrels the time lag is shorter — with *F.glacialoides*, *T.antarctica*, *P.nivea* and *D.capense* the sexes arrived 2 to 5 days apart (Mougin, 1975). Precise data for burrowing petrels are lacking and not easily obtained even with marked birds.

Females may arrive first; two out of 20 marked Laysan Albatrosses did that (Fisher & Fisher, 1969), but only two females from 82 pairs of Wandering Albatrosses landed before their mates (Tickell, 1968). Richdale (1950) gave similar instances and noted a consistency in the order of arrival of this Royal Albatrosses with two particular males always the first ashore.

While awaiting their partners the males inspect their nests or clean out their burrows and stand guard nearby. They also wander a good deal and may display and sing with other birds. S.W. Harris (1974) found that Leach's Storm Petrels mist-netted at this time were mostly re-captured in the same areas, suggesting that even species which display in the air do not wander far. Male Royal Albatrosses may build new nests but these are generally ignored by the females on their return. Lone male albatrosses and fulmars will sit around, pluck vegetation with their bills and drop it to one flank; lone females may do the same. Burrowing birds like *P.phaeopygia* spend much time cleaning out debris from the nest chambers, which may contain beetles and other inquiline species (Simons, 1985), and carry in new material.

The behaviour of the male now seems to have two main objects — the retention of the nest site and to be on hand for copulation once the female appears. Fisher & Fisher (1969) found that coition between Laysan Albatrosses followed within 24 h of their reunion; within 36 h both had left. The male's readiness for coition helps explain the frequent instances of rape during the pre-egg period (see Section VI below).

Among burrowing species the first signs of return are on dark nights when occasional calling may be heard and newly thrown spoil seen at the entrances to a few burrows. The visits tend to be irregular and most early visitors depart by dawn. Gradually the occasions when two birds are together at the nest by night increase and some pairs remain there by day. Harris (1969b) found some *O.castro* spending as long as 5 days in their burrows at this time. Beck & Brown (1971) noted that with *F.tropica* both partners tended to remain ashore by day whereas they seldom saw *O.oceanicus* in the nest before the egg appeared.

The build-up in numbers of birds on land is fairly gradual at most colonies and even among migratory species like *P.griseus* it probably takes about 20 days for all the breeders to arrive. Pryor (1968) watched the return of four fulmars in Antarctica. The bulk of the Cape Petrels arrived over 6 days. However, the Antarctic Petrels came in large groups and within 36 h the whole population had landed, individual birds going to their nests, courting and pairing off. The Southern Fulmars also returned *en masse* whereas the Snow Petrel's return was less dramatic, and although many went straight to their nest sites, courtship did not begin immediately. Some sat around on deep snow outside their nests; others dug down through the snow to reach them. For these and other high-latitude species, the re-occupation of burrows or nest crevices obviously depends on snow and ice conditions. Wilson's Storm Petrel, a bird too weak to excavate heavy snow and too small to withstand very low temperatures, does not return to Antarctica until 2–4 weeks after the other petrels, by which time air temperatures have begun to climb and winds to drop. Lacan (1971) found that this bird's return on 1 November 1967 coincided nicely with a sudden rise in temperature from -12.6°C in late October to -1.5°C on 1 November. Despite this late return, the bird lays 3 weeks earlier here than at South Georgia, perhaps as a consequence of the shorter Antarctic summer (Copestake & Croxall, 1985).

The most detailed examination of the re-occupation of a colony is that of Laysan Albatrosses by Fisher & Fisher (1969, p.180) at Midway Atoll, where many were marked and of known age. There, former breeders tended to arrive *en masse* during two periods — mid-November and in late December. About 45% of the birds appeared in each period. These authors were unable to account for the separate waves of arrival.

II The nest

Petrels make their nests from material to hand, some build substantial structures where vegetation is plentiful, and may continue adding to it during incubation; elsewhere the same species may lay in a shallow scrape on bare gravel with but a few stones to mark the site. Royal Albatrosses sometimes build around the egg after laying. Some petrels are quite adaptable. Some Wedge-tailed Shearwaters will lay in self-dug burrows, others will nest in cavities beneath concrete slabs on disused airfields, in 'crawl' space beneath buildings, and even on the surface hidden only by a mat of vegetation.

Nests are sited for ease of access and departure, and where shelter is given from excesses of wind and heat.

At high densities, particularly with those that burrow, petrels may change the character of their habitats. Digging may undermine trees and tussock pedestals, while the strikes of falling bodies below pre-dawn launching sites may cut swathes through the vegetation and hasten wind and water erosion. Conversely, the birds fertilise the soil with faeces, dead eggs and chicks, leaving surface soils with high nitrogen and phosphorus levels which often support special plant associations (Gillham, 1961).

A *Surface nests*

Surface-nesting petrels are either too big to burrow or are smaller forms mainly of the tropics, without land predators. About 20% of petrel species nest on the surface.

The largest nests are those of the Royal and Wandering Albatrosses which begin as shallow depressions where the birds have squatted in loose material raked together by their beaks. On hard ground the structures are quite low, but all except the Galapagos Albatross usually nest on soft soils, often of a peaty nature, and substantial structures result, as the birds tear off local plant materials and lay them down to one flank, rake in soil and peat, and press the resulting heap down with webbed feet while rotating on their breasts. The resulting truncated cone has a shallow crater and often a moat around the base which helps drainage. In wet weather the birds tend to plaster more soil onto the sides: their breasts become heavily stained. The structure is just a heap of earth and vegetation moulded to shape, with no inter-weaving of materials. Several starts may be made and abandoned, and most of the effective building seems to be done by the female, the male bringing some of the material.

Laysan Albatrosses form their nests by squatting in a depression and laying on one side, scraping the sand backwards with feet and claws while rotating their bodies in the depression. Twigs, leaves, sand and old bones are dragged back to form a rim which is then trowelled down with pressure from the side of the bill, creating a smooth surface that resists erosion by wind or water. Fisher (1971, p.48) noted that when the incubating male rakes material onto the nest some soil falls into the centre which then tends to get built up below the egg as this rests on his webbed feet. Giant petrels also make nests of this general type, those of *M.halli* often being untidy but substantial structures of loose tussock grass. Those of *M.giganteus* are often more exposed and may be mere rimmed depressions in the peat with peripheral moats. Conroy (1972) described sites where these birds laid on a bed of pebbles through which melt-water drained without cooling the egg.

The voluminous pillars of mollymawks are built with the same plastering actions, body rotations and foot thrustings, and they are refurbished and added to during successive seasons, an old mound often forming the foundation for a new nest

(Fig. 12.1). Surplus nests as made by the great albatrosses seem not to be built. The nest cup is a tight fit so that by slightly extending its folded wings the sitting bird can divert rain over the sides leaving the central depression dry. Murphy (1936, p.529) thought that the bird's faeces helped cement the structure. Certainly it is well designed to withstand the heavy and persistent rainstorms of the subantarctic.

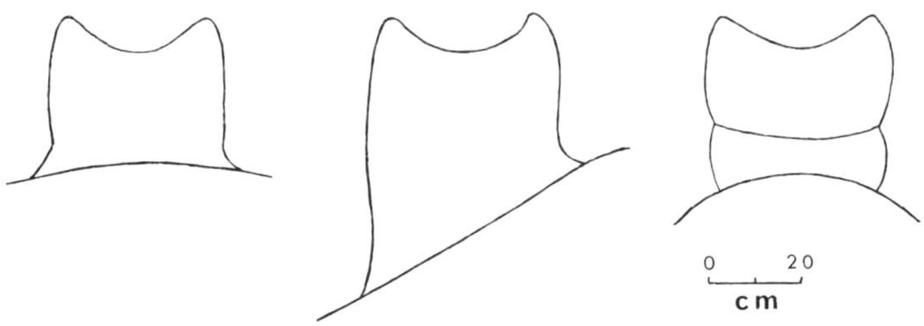

Figure 12.1 *Pedestal nests of mollymawks often last several years. Examples of Grey-headed Albatross nests, the right hand one built on a base of an old one. From Despin* et al. *(1972).*

Surface nests of gadfly petrels like *P.alba* and *P.neglecta* on atolls are mere scraps and often seem to result from lack of burrowable ground. However, the shearwater *P.nativitatis* lays on the surface or in fissures and cavities among coral rock, even though ground suitable for burrowing is available nearby; e.g. at Laysan Island (Ely & Clapp, 1973), where *P.pacificus* does nest underground. Competition with that larger bird may be a factor here.

Surface nests are not placed at random but sited according to local conditions, particularly as regards wind speed and direction. Royal and Wandering Albatrosses nests tend to be in shallow depressions where some shelter is given from the wind, although Mougin (1970a) found that the chill factors in the latters' nests were 82–85% of the value for the climate in general, and so the birds were still facing quite severe conditions.

Phoebetria albatrosses raise their nests on the sides of gullies (where take-off is facilitated even in calms) often facing the prevailing westerlies. Nonetheless, air temperatures were higher and wind speeds lower in a gully where *P.palpebrata* nested than at the station base, with a chill factor of 460 cal $m^{-2}\ h^{-1}$ instead of 960 cal $m^{-2}\ h^{-1}$ at the base (Mougin, 1970c). On Marion Island most nests of *P.palpebrata* were backed by a wall of rock whereas only 20% of *P.fusca* nests were so sited (Berruti, 1979). The Northern Giant Petrel at Macquarie Island lays in the shelter of rocks and large tussocks, its southern sibling there more often choosing very exposed slopes (Bourne & Warham, 1966). Voisin (1976) found similar patterns at the Crozets and compared the differences using an 'index of protection'.

B *Ledge nests*

All the fulmars apart from *Macronectes*, some 4% of all petrel species, are primarly ledge nesters, laying on fairly level surfaces in coigns of cliffs and rocky breakdowns, usually on coasts but sometimes well inland. With little material to hand, nests are bare, mere collections of stones and debris, or are shallow scrapes where there is an earthy substrate. Snow Petrels' nests are often rimmed with frozen stomach oil and guano.

Ledge nesters breed in high latitudes with long days but short summers. The style has the advantage that time has not to be spent on building an elaborate structure, although snow clearing may be needed. The open situations are possible despite the presence of skuas and giant petrels because of the efficiency of the fulmars' oil spitting artillery, A disadvantage is that the exposed chicks must be guarded longer on a ledge than in a burrow.

As demonstrated by Mougin (1975) for birds at Pointe Geologie, Antarctica, these ledge nesters have species-specific preferences in choice of site (Fig. 12.2). Cape Petrels generally select quite wide, level platforms often little sheltered from the wind but enjoying slightly milder temperatures than in the open. Snow Petrels choose snow-free outcrops or cliffs of bedded, strongly jointed rock, often with a jumble of talus below. They avoid volcanic rocks as these lack the necessary crevices and slots for nesting and they tend to lay deep in recesses where they may

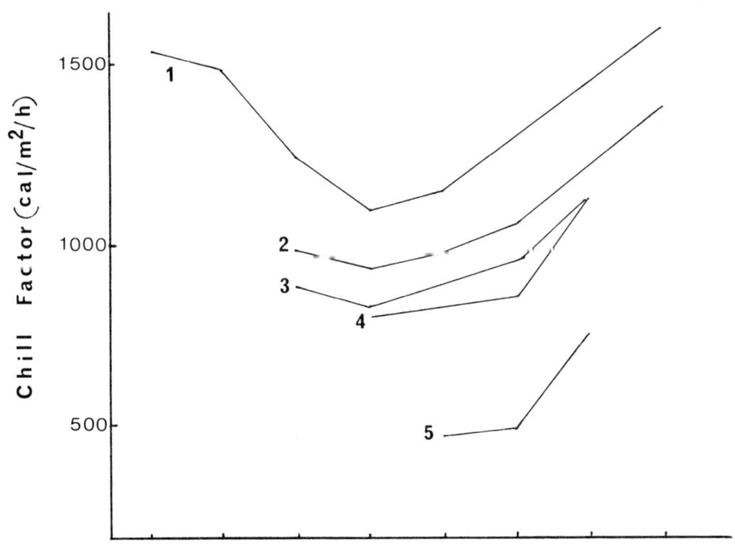

Figure 12.2 *Chill factors measured at fulmar colonies and the Base Station at Pointe Geologie, E. Antarctica from September to April. 1. Climate at Base Station. 2.* M.giganteus. *3.* F.glacialoides. *4.* D.capense. *5.* P.nivea. *Nests of the largest species,* M.giganteus *are the most exposed, those of the smallest* P.nivea, *the most protected. From Mougin (1975).*

get no direct sunlight but greatly reduced winds. However, Somme (1977) found Snow Petrels nesting inland on the northern sides of mountains facing the sun and at Pt. Geologie the Antarctic Fulmars select spots in the lee of the prevailing south-west wind on north-facing slopes where they too get the sun's heat in clear weather.

C *Crevice nests*

About 4% of petrels lay in crevices in cliff faces, under rocks, in caves, in dry stone walls and among talus debris. Most are small birds like the storm petrels *O.tethys* and *O.oceanicus*, prions like *P.crassirostris* and shearwaters like *P.lherminieri*. Even larger burrowing forms such as *C.diomedea* and *P.griseus* may lay in niches beneath rocks, often using both burrows and crevices on the same island. *P.turtur* and *O.leucorhoa* do this, so that a strict separation of ledge, crevice and burrow nesting is not possible. Crevice sites are also often shared by several species, e.g. on Grand Salvage Island where *O.castro, B.bulwerii* and a few *C.diomedea* may be found in the same stone wall.

Eggs may be laid on the bare rock or soil but some crevice nesters carry in moss or debris to form simple linings to the shallow scrape they try to form for the egg. Beck & Brown (1972) described nests of *O.oceanicus* substantially lined with scraps of paper, wood-wool and celluloid film dragged in from a rubbish dump near the base huts.

Crevice nesting gives protection from harsh climates and predators — both of which bear heavily on small species — and Harper (1976) pointed out that a satisfactory crevice site for *P.turtur* was bounded on all sides by rock to prevent its enlargement and take-over by larger petrels.

D *Burrow nests*

This is the predominant mode of nesting among tubenoses, the preferred style for about 65% of them. Burrow-nesting protects the occupants from wind and climatic extremes, the chambers being cooler in the heat and warmer in the cold. Incubating birds can leave to break their fasts with their egg protected from predatory skuas or gulls. Burrowing also helps make the most use of the land available, as this may be in short supply: albatrosses and giant petrels may nest above ground and a range of smaller petrels below them. Although energetically expensive to build, most burrows can be used for several, perhaps many, years, with only minor annual adjustments. Burrows that lose their owners are often quickly taken over by younger birds. However, sites subject to sand, soil, and snow blockage may need considerable work to re-open; for example, those of *P.georgicus* often do not survive from one year to the next. Underground nests tend to be subject to flooding and well-drained soils or raised egg cups (Fig. 7.5) may be important to reduce losses. Nests of *P.georgicus* in loose scree, being well drained, are more successful than those under a mossy cover (Croxall & Hunter, 1982).

How birds incarcerated in the nest chambers for days at a time cope with increased levels of NH_3 and CO_2, particularly in deep burrows, is conjectural. Unlike many cavity nesters such as pardalotes and kingfishers, visits to burrows are infrequent, so that the piston effect of movement in the tunnel in clearing the air must be greatly reduced and circulation by winds very slight. If the birds 'come up for air' this is clearly unusual, at least by day.

Burrows are easier to defend than are open nests, but once another has gained access it may be impossible to dislodge. Burrow nesters are also vulnerable to natural predators like snakes and very vulnerable to alien species like rats, mustelids and cats, burrowing petrels lacking the ability to spit oil effectively in defence. They may also experience interference from burrowing mammals, as *P.lessonii* does from rabbits (Warham, 1967). Elsewhere rabbits themselves provide burrows for petrels, e.g. for *P.aequinoctialis* and *P.puffinus*. Although birds will walk for quite long distances to reach good take-off rocks, Bartle (1968) thought that burrows of Pycroft's Petrel were sited close to trees that provided easily climbed launching pads.

The length of the tunnel and depth of the nest depend largely on the ease of digging. Burrows in peat of shearwaters like *P.griseus* may be 2 m long and 1 m below the surface, those of *P.opisthomelas* in soft guano may be longer, honeycombing the substrate (Anthony, 1900); other burrows in hard ground may be shallow, their occupants even visible from without. Whether one particular sex is mainly responsible for building the burrow and lining the chamber is unclear, but as the males arrive first, presumably they do most of the initial cleaning out. With *P.urinatrix* both sexes dig (Richdale, 1965a, p.17) but seven Leach's Petrels collected from unfinished burrows were all males (Gross, 1935).

The nests proper are terminal enlargements of the tunnels and big enough to accommodate the pair in comfort. The tunnels fit the maker's profile, being round above and flat along the floor with the entrance portal usually bigger. Where caves are available many burrowers will nest on the surface instead of burrowing.

Tunnels may be straight but very often they are crooked as the digger works around rocks and roots. In shallow soils on slopes, diving petrels, prions and others dig down and then turn to one side to run parallel with the slope. The chambers may have more than one entrance, probably due to one bird 'sapping' into an existing tunnel. On level ground the tunnels tend to slope down at rather shallow angles. Examples of typical burrow profiles are shown in Fig. 12.3 and Mougin (1985) gave precise dimensions of samples of burrows for four southern species.

The floors of the nest chambers are usually dry and often lined with leaves, moss, twigs etc. carried from outside (Fig. 12.4). Some birds make voluminous nests, others of the same species only sparse ones. Coloured threads laid down at the entrances are soon incorporated into nest linings as are researchers' toothpick entrance 'palisades', but it is often unclear whether adult or chick is responsible (Warham, 1960). Nests of Bonin Petrels were lined 36 days before laying and this was done mainly during a single night (Grant *et al.*, 1983). With *P.pacificus* both sexes carry in nest material (Shallenberger, 1973). If dry, well-lined nests presumably help heat retention in the egg.

In tunnelling the bill picks the soil loose which is then raked back with the toes and expanded webs of the feet. The bird lies to one side resting on its wrist and

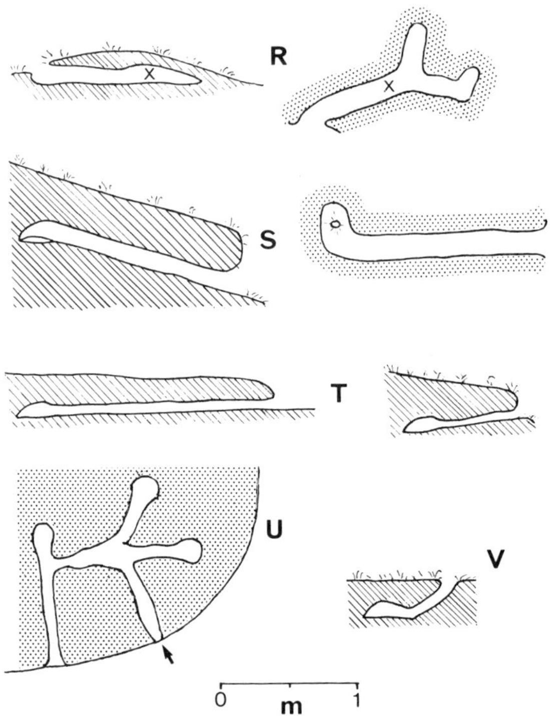

Figure 12.3 *Petrel burrows (diagrammatic).*
(**R**) *Burrow of Kerguelen Petrel: left, in longitudinal vertical section, right, in horizontal section, the latter showing deeper extensions from the nest chamber (x).* (**S**) *Burrow of Kerguelen Petrel showing raised cup to take the egg — left in longitudinal vertical section, right in horizontal section.* (**T**) *Longitudinal vertical sections of South Georgian Diving Petrel burrows.* (**U**) *Complex burrow system of Antarctic Prions from above with three chambers served by one tunnel, that arrowed having probably been excavated from within and abandoned when the bird broke through to the outside.* (**V**) *Burrow of Salvin's Prion seen in longitudinal vertical section. From Mougin (1969); Derenne & Mougin (1976) and Downes* et al. *(1959). [Although vertical sections are drawn as if in one plane, burrows are seldom perfectly straight as shown.]*

unused foot while making a series of rapid thrusts with the freed limb, then rolls to the other side to engage the second foot. In light earth the plumes of spoil gushing up into the air are a feature of the pre-laying activity after dark and arcuate mounds of earth often mark the entrances to nests. Digging *P.hypoleuca* used 3.3 kicks s^{-1} to eject sand up to 1 m from the entrance (Grant *et al.*, 1983).

Small species like storm petrels are too feeble to burrow into anything but soft substrates and *G.nereis* and *F.grallaria* may simply delve into the fibrous debris around the bases of tussock clumps. It has been alleged that storm petrels do not use their bills for digging but there are reports of their having soil on their beaks and direct observations of them picking away to enlarge the nest chamber.

There is little information on the time taken to make a burrow. Crockett (1975)

Figure 12.4 *Sooty Shearwaters carry considerable quantities of leaves and sticks into their burrows. In forest sites the ground may be bared of surface debris. Photo: author.*

saw a Wedge-tailed Shearwater complete its nest on the sixth night after it had first begun to scrape the surface: a further week elapsed before the chamber was lined. Gross (1935) found that Leach's Petrel took about 3 days to complete a nest.

Burrows made in unconsolidated ground or dug too close to one another are liable to cave-in. A nesting chamber so exposed is normally abandoned, although some birds may extend the tunnel and form a deeper chamber. Roots may impede progress but they tend to stabilise the soil and when colonies are on dry sand a complex root matrix may be essential to reduce erosion (Fig. 9.9). Campbell (1933) described how during gales dry sand dribbled into burrows of *P.marina* with every vibration of the wind, and mortality of Bonin Petrels through sand burial occurs at Lisianski Island (Clapp & Wirtz, 1975; Ely & Clapp, 1973).

Occasionally burrowers destroy their own habitat. Hutton's Shearwaters digging on steep mountain slopes in pockets of shallow loessal soils kill the blanketing tussock grasses and undermine the soil which is washed away in heavy rains (Evans, 1973). Under entirely natural conditions petrel-induced erosion is unusual; more often it arises from the combined effects of the birds and some introduced animal. Heavy erosion of some steep slopes on Macquarie Island appears to have been the result of undermining of the soil by burrowing *P.desolata* combined with removal of holding vegetation by rabbits. The resultant bare screes are useless for both.

Colonies of burrowing petrels may consist of a single species, but often smaller kinds intermingle with larger ones and sometimes two petrel species will share a tunnel giving access to separate chambers. Associations of Tuataras *Sphenodon punctatus* with the shearwaters *P.carneipes* and *P.gavia* are well known, and accidental burrow sharing of *P.marina* with the parrot *Neophema petrophila* and of *O.melania* and bats have been recorded.

1 *Burrow microclimates*

Numerous studies of conditions within burrows have been made since those of Roberts (1940) on Wilson's Storm Petrel and Paulian (1953) on the Blue Petrel. Such work shows the amelioration of conditions that burrowing confers. Relative humidities and wind speeds have seldom been measured, but Boersma (1986) found that the mean humidity in *O.furcata* nests was higher than that outside and Mougin (1969), showed for *L.brevirostris* that when the mean wind speed over land averaged 10.5 m^{-s} and was 3.5 m^{-s} just outside the burrow it was 0 m^{-s} within the nest chamber. Similar figures were obtained at burrows of *P.salvini* by Derenne & Mougin (1976). Perfaur (1974) repeated Roberts' experiment, confirming the higher nest temperatures in the burrows and his graphs (Fig. 12.5) show the rapid rise in temperature when an adult is present and its abrupt fall when this leaves. The chill factor, however, seems to be but one influence, as at Adelaide Island, Antarctica, Beck & Brown (1972, p.50) found many Wilson's Storm Petrel nests merely in the lee of boulders with attendant birds visible. They were evidently at little risk from skua attack or snow burial and many chicks were being reared.

For hot climates Simons (1985) found that within *P.phaeopygia* nests the daily temperature variation averaged 2.39°C as against a range of 25°C outside, the moderating effect varying with the depth of the chamber.

2 *Burrow blocking*

This is an unexplained behaviour of some shearwaters and gadfly petrels. Some burrows of *P.tenuirostris* have their entrances stuffed with dry grass or other dead

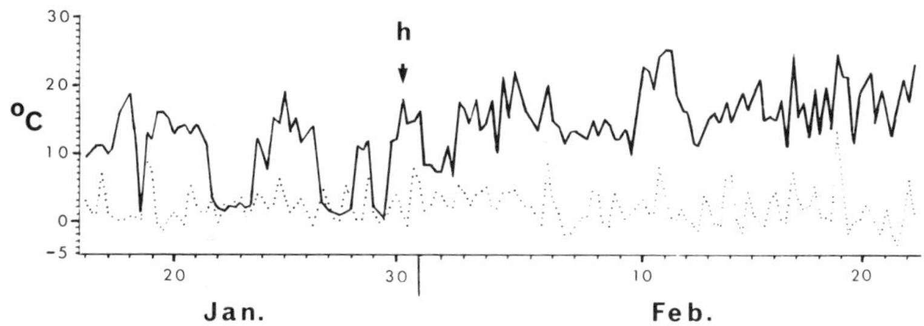

Figure 12.5 *Temperatures during January and February inside (upper graph) and outside a nest of Wilson's Storm Petrel at the South Shetland Islands. Arrow marks the hatching date. The four troughs in January mark periods when the egg was left uncovered. From Perfaur (1974), courtesy of* The Wilson Bulletin.

vegetation, a feature commented on by Montgomery (1896). He thought that this was done by the adult when it left but I found that, as reported by mutton-birders, most such burrows had occupants. The habit has since been noted of *P.carneipes*, *P.pacificus*, *P.macroptera*, *P.lessonii*, *P.inexpectata*, *P.pycrofti* and *P.hypoleuca* (Crockett, 1975; Warham, 1956, 1958, 1967; Warham *et al.*, 1977; Bartle, 1968 and Grant *et al.*, 1983).

Warham (1960) found that only some *P.tenuirostris* blocked burrows. He never saw the birds actually placing the vegetation, but one nest entrance was stuffed with *Bromus* grass whose nearest clump was 5 m away. He related the behaviour to the bird's practice of tearing up grass and dropping it over a shoulder during courtship and nest-building. At a colony of *P.carneipes* where blocking was common, the chambers were often only sparsely lined despite the amount of material jammed into their tunnels (Warham, 1958). With Bonin Petrels the plugs often had a central hole about 10 × 7 cm through which the birds squeezed when entering or leaving and Grant *et al.* (1983) suggested that the plug was a way of adjusting large burrows of *P.pacificus* to match the profiles of the smaller birds.

Bartle (1968) saw adult Pycroft's Petrels collecting and kicking leaves into burrow entrances. As such obstructions are often present during the pre-egg and incubation stages adult petrels must be responsible for them. Warham (1958, 1960) suggested that burrow-blocking resulted from the petrels' antipathy to light when ashore, pointed out that such burrows looked unused, but that they offered no barrier to potential reptilian predators like snakes. With *P.inexpectata* blocked nests tended to have short tunnels or be poorly concealed (Warham *et al.*, 1977). Possibly the behaviour reflects an excess of drive to line the nest, which declines once the egg appears and the birds have other responsibilities.

III Patterns of activity at the colonies

A *'Rafting' offshore*

Shearwaters often assemble in flocks or 'rafts' on the sea off their breeding grounds to await nightfall. Strings of resting and preening birds develop which grow in size as newcomers alight at front and rear as described for *P.tenuirostris* (Warham, 1960). Similar behaviour is known for *P.puffinus, P.griseus, P.lherminieri, C.diomedea,* and *C.leucomelas*, only rarely for *Pterodroma* and not for storm petrels or diving petrels. During their long pre-egg stage Buller's Mollymawks also assemble in rafts close inshore, as do Northern Fulmars before their winter visits to the cliffs (Macdonald, 1980).

Rafting occurs mainly during pre-egg to chick rearing, its conspicuousness dependent on colony size. The birds mostly appear not to feed unless schooling fish are present. Development of these rafts during the course of a day is described in *P.tenuirostris, C.diomedea* and *C.leucomelas* by Warham (1960), Mougin & Stahl (1982) and Okamoto (1972). Presumably rafting occurs because birds away for several days at distant food sources cannot time their returns precisely, so assemble to wait until it is safe to land.

B *Landfall*

Single birds are usually the first overhead, but as the light fades the rafts become increasingly restless, many groups taking to the wing and re-alighting elsewhere. Presently one such group heads for the land; others soon follow. As groups circle overhead, individual birds peel off to alight, the group formations tending to be lost in the general melee, although some stratification and flock cohesion may be perceptible for a while.

Whether they are diurnal birds like albatrosses or nocturnal visitors like shearwaters, incoming petrels usually make a series of circuits above the general area of their nest, hesitating as they pass above it. They eventually pitch down at that point of hesitation and tame species can then often be caught by hand (Warham, 1959). The main exceptions are diving petrels: these appear to alight directly near their nests with little or no preliminary circling.

C *Distance of touch-down from the nest*

Diurnally active petrels generally land at their nests although albatrosses with sites in heavy vegetation may come down on more open ground. Nocturnally active kinds with low wing loadings like storm petrels and small *Pterodroma* may also alight close to their nests. Shearwaters and *Procellaria* petrels have high wing loadings and the larger ones have difficulty making controlled landings, particularly during calms, so that precise touch-down is less common. Those with nests in forests appear to select particular gaps in the canopy through which to drop to the ground, others crash through the branches and are usually quite unharmed.

Richdale (1963, p.21) found that the direct distance from landing to burrow of *P.griseus* in a rather open habitat averaged 5.1 ± 3.5 m (n = 35). Comparable figures for this bird alighting in a tussock meadow were 6.5 ± 4.1 m (n = 24), while the actual distances walked to the burrow entrance averaged 9.1 ± 6.7 m (Warham, unpubl.). Birds that crashed through the forest canopy tended to travel further — for 13.8 m in four instances — most travelled still more but could not be followed in the dim light without frightening them. And, although birds with LED or similar identifying markers have not been followed at night, there is some evidence that particular birds use the same gaps in the canopy and the same route to their nests on successive occasions. Olfactory clues may help in burrow location (Grubb, 1974) although visual ones seem very important (Warham, 1955; Serventy *et al.*, 1971, p.28; Brooke 1978c; James, 1986) and hovering near the nest site may form part of the orientation process. Richdale (1965b, p.122) saw a Fairy Prion drop through the trees and scuttle to a burrow 2 m away and wondered at its ability to pick the precise opening in a very uniform canopy.

D *Arrival times and ambient light levels*

A landfall after dark is essential for small petrels whose habitats are shared by giant petrels and gulls. Even darkness does not confer total immunity. For example, Southern Skuas take adult *P.vittata* and *P.marina* under a forest canopy even on dark nights (Young *et al.*, 1988). When predators are present, even small species like *N.fuliginosa* can come and go by day. Ambient light levels affect both the numbers of birds coming ashore and their times of arrival.

The numbers flying in on clear moonlit nights are generally fewer than those appearing under overcast skies. This has been found, among others, for *O.leucorhoa* by Gross (1935), for *H.pelagicus* by Davis (1957), *P.marina, P.turtur* and *P.vittata* by Richdale (1965a,b) and for *P.puffinus* and *P.tenuirostris* by Lockley (1942) and Warham (1960). Gross found that under a full moon Leach's Petrels flew swiftly and silently to their nests but some still fell victims to the resident gulls. Lockley (1942) and Harris (1966) found that Manx Shearwater chicks were fed less often on moonlight nights. The reduction in numbers ashore on light nights is generally ascribed to a temporary decline of non-breeding birds, which, having no burrow for sanctuary, would be more vulnerable to predators, and with some species a reduction in feeding visits by the breeders is not apparent.

Imber (1975) suggested that petrels feeding at the surface when the moon is bright get little food because the diurnal migration of their luminescent prey is suppressed and remains beyond their reach. Those that are ill-fed therefore cannot afford to visit land.

Night-time landing times also vary with ambient light levels and, in general, with the time of sunset. However, at colonies of diurnally active species like albatrosses and fulmars, numbers ashore also tend to increase in the late afternoon with influxes of younger, non-breeding birds. And even Wandering Albatrosses will fly over their colonies in the dark.

Most burrowing species land early, sometimes hours earlier, when there is low, heavy cloud. For example, Boersma *et al.* (1980) found that Fork-tailed Petrels arrived 1–2 h later under a clear moon and Simons (1981b) showed how arrival times of that bird change during the season with changing times of sunset. At the Snares Islands many Sooty Shearwaters nesting beneath *Olearia* forest land in broad daylight, sometimes sitting on the surface in the sun. This is evidently the only place where that species is partly diurnal. Greater Shearwaters at Nightingale

Island are even more active by day, although seemingly in the pre-egg stage they are crepuscular (Rowan, 1952, p.102).

Hamer & Read (1987) found that breeding *C.diomedea* returned earlier at night than non-breeders did, with individual birds tending to land about the same time from one night to the next.

Once ashore petrels are noisy, the non-breeders very active in vocal and other displays. These peak in the first hour or two after their arrival, then decline for much of the night when many birds sleep on the surface. Some departing birds leave silently throughout the night, hugging the contours of the land as they drop towards the sea, but the activity and din peak again with the main outward movement which usually ends in the half-light of dawn. Bretagnolle (1988a) has analysed some seasonal and circadian activity rhythms for five Antarctic species.

Activity at large colonies is at a high level through the pre-egg stage to chick rearing with aerial calling a feature of those of storm and gadfly petrels and of some shearwaters like *Calonectris diomedea, P.assimilis* and *P.puffinus* as they circle around before descending. However, at the huge colonies of *P.tenuirostris* and *P.griseus* the silence of the flying masses at dusk (Montgomery, 1892; Warham, 1960) is uncanny and in great contrast to the din of *P.puffinus* over their much smaller colonies; but once grounded the song of the southern birds may be deafening.

The hypothesis that the level of activity at a petrel colony may be governed by the activities of predators was tested by Watanuki (1986) working with *O.leucorhoa* where the annual loss of adults to *Larus schistisagus* was about 13%. Both arrival time and the amount of aerial activity were inversely correlated with light intensity early in the breeding cycle, but later, when the predation risk declined, moonlight avoidance also declined, presumably because by then the non-breeders had left. At a Newfoundland colony the main effects of bright nights on non-breeding Manx Shearwaters were on their behaviour rather than on their attendance. They made themselves inconspicuous by calling less and spending more time under ground, evidently to lessen losses from the big gulls nesting nearby (Storey & Grimmer, 1986).

As well as the lunar cyclic patterns, other periodic patterns of returns have been reported. At the Crozet Islands Jouventin *et al.* (1985) found *P.salvini, F.tropica* and *G.nereis* abundantly present in burrows and active on land during two peaks separated by 7.4, 15.5 and 16.5 day intervals. These peaks did not appear to be correlated with the state of the weather or any identifiable factor. Periodic variations in colony attendance were found by Bretagnolle (1988a) for *M.giganteus* (Fig. 12.6) again without any clear link to weather conditions. Jouventin *et al.* (1985) suggested that the periodicities were involved with the establishment of the necessary rhythms for incubation stints, but the timing of the two types of rhythm seem quite different and Bretagnolle suggested that, at least for *P.nivea* and *M.giganteus,* they formed part of the mutual stimulatory system to ensure synchrony of laying — important for species breeding at high latitudes.

E *Non-breeders*

Petrel communities include an important element of non-breeding birds. Many are pre-breeders, birds too young and inexperienced to nest successfully, many

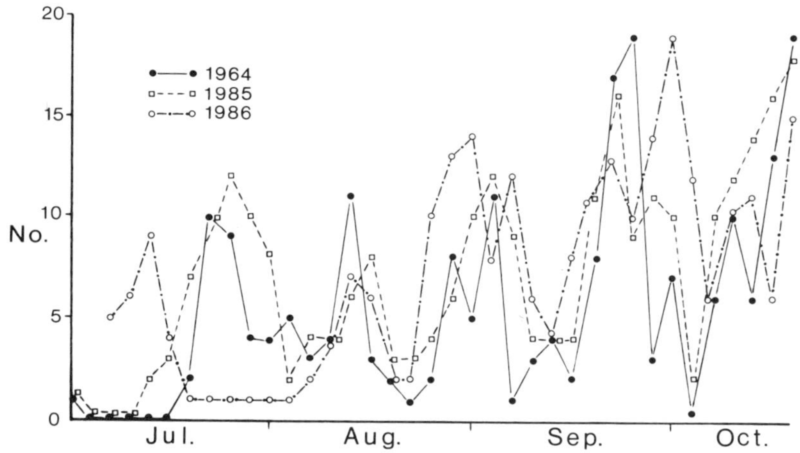

Figure 12.6 *Attendance rhythms of Southern Giant Petrels during the pre-laying period at Pointe Geologie, E. Antarctica, over 3 years. From Bretagnolle (1988a).*

lacking nests or partners. Others are birds that formerly bred but for various reasons are not currently doing so — they may have lost their nest or their partner. Failed breeders also join the non-breeding category later in the season.

Unless the population has been intensively marked and followed for many years, these different classes of birds may be unidentifiable in the field. Richdale lumped them together as 'unemployed' birds, although all are employed in some way — in acquiring a partner or territory — for example. In general, the older the bird the earlier it returns to the colony, former breeders and birds old enough to breed arriving with the established pairs. Younger birds arrive later in a stepped sequence according to age. The pattern was demonstrated by Fisher & Fisher (1969) and Serventy (1967) (Figs 12.7 and 12.8).

The non-breeders are the birds most active in the air and on the ground during the pre-egg and incubation stages, whereas established birds alight and, after a short pause apparently to get their bearings, promptly disperse to their nests. It is mainly non-breeders of the nocturnal species that stay away on bright nights, but Storey & Grimmer (1986) believed that many non-breeding *P. puffinus* returned on bright nights, but did so later than on dark ones. Non-breeders probably comprise the bulk of the birds that wander through the colonies prospecting nest sites and getting into scuffles with established owners, and they are probably also those that take over abandoned eggs without persisting in incubation. As chick-rearing progresses and younger birds and failed breeders leave, non-breeding activity ashore declines until, when the last chicks are being fed the colonies are quiet and appear deserted.

F *Nest repair*

Nests are improved and added to throughout the pre-laying and incubation stages.

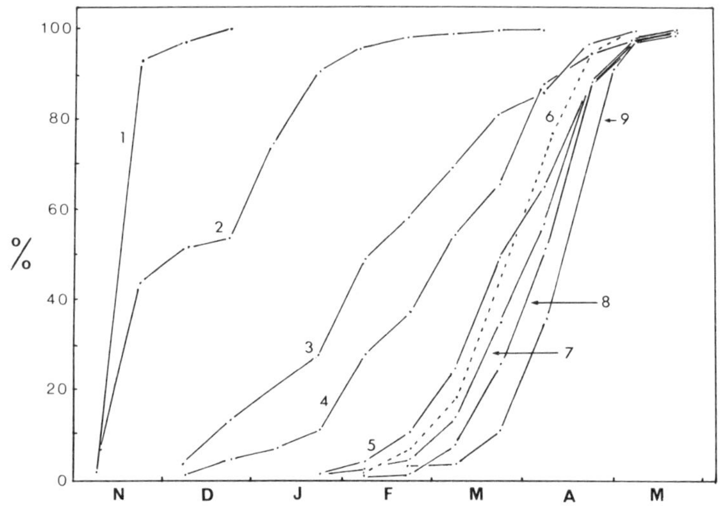

Figure 12.7 *Cumulative arrivals of Laysan Albatrosses of various ages and categories at Midway Atoll in 1964–1965, computed as a per cent of total recaptures in each group. 1: 911 experienced breeders; 2: 607 former breeders; 3: 273 8-year-olds; 4: 505 7-year-olds; 5: 878 5-year-olds; 6: 53 6-year-olds; 7: 984 4-year-olds; 8: 1032 3-year-olds; 9: 181 2-year-olds. From Fisher & Fisher (1969).*

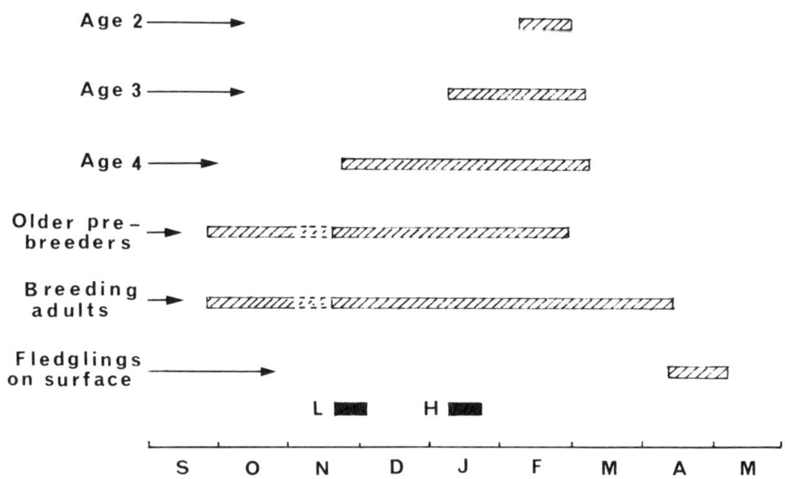

Figure 12.8 *Seasonal events and composition of Short-tailed Shearwaters by age and status. The break in bars of the older pre-breeders and breeders marks the pre-laying exodus. From Serventy (1967).*

Sitting mollymawks reach down to scrape up peat, soil or leaves exposing the coloured skin at their gapes as they do so. The 'load' is put on the rim of the nest and tapped into place, even distribution being achieved by switching attention from side to side and by occasional body rotation. The sitting bird also deepens the nest bowl by scraping backwards with its feet. These activities may result in the pillars getting taller. The mean height of 142 nests of *D.bulleri* at the start of the season, with few birds ashore, averaged 23.5 ± 6.8 cm compared with 25.0 ± 4.8 cm (n = 42) by the time incubation was well advanced (C.J. Fitzsimons, unpubl.)

IV Nest defence

Petrels defend an area around the immediate vicinity of their nest, albatrosses and giant petrels mostly that within reach of the sitting bird's bill; burrowing forms the area around the entrance and the nest tunnel itself. Defence against conspecifics may be vigorous and lead to severe wounding and tends to occur more at high nest densities. Albatrosses like *D.exulans* and *D.chlororhynchos*, whose nests are well spaced out, are seldom aggressive, and an unwanted bird coming too close is usually deterred by bill clappering. Ledge nesters fend off aggressors with barrages of stomach oil and so probably fight little. Pryor (1968) found newly arrived Snow and Cape Petrels ready to spit at any trespasser and all Antarctic fulmarine species do that. Petrels seem to be immune but contamination may be very serious for other recipients (Warham, 1977b). Burrowing species grapple each others' heads and napes and make raucous cries.

Prions packing together in crevices and/or sharing entrances may fight furiously when they infringe personal space (Tickell, 1962, p.18; Harper, 1976; Strange, 1980). In *O.tethys* crowding and competition and the resultant fighting were the main causes of egg and chick loss (Harris, 1969b). Mougin (1969) found that Kerguelen Petrels did not explore nest sites other than their own and did not fight, whereas Shoemakers often explored and thus became embroiled in conflicts. Cory's Shearwaters also fight ferociously and badly wounded birds have been seen (Zino, pers. comm.).

Probably much of the conflict is between older non-breeders and first-time breeders rather than between established pairs, for these are ashore early when competition for space is less severe. Disputes arise when a bird mistakes another's nest for its own and when trios have been formed. Birds trying to evict others may be extremely persistent: in most cases the bird in possession wins unless it has entered the burrow of an established pair in their absence.

Some burrows seem particularly attractive and have many visitors and sometimes several pairs will attempt to use the same burrow and lay there, perhaps evicting early tenants and throwing out egg, chick and even adult birds (Jouventin *et al.*, 1985).

When the conflict over burrow rights is between a larger and a smaller species the former usually wins. Sometimes a similar-sized bird may take over from a petrel. Thus Coulson & Horobin (1972) followed the gradual usurpation of Northern Fulmars by an expanding colony of Kittiwakes *Rissa tridactyla*. Although the fulmars will spit at the gulls (Warham, pers. obs.) the Kittiwakes succeeded because they took over the sites during the petrel's pre-laying exodus, fought tenaciously, and could not then be dislodged. Yet Macdonald (1980) found that fulmars always won in competition with Herring Gulls *Larus argentatus*.

Among albatrosses and giant petrels the (larger) male is the more aggressive sex but with burrowing species probably both sexes take part in nest defence. Fisher (1971, p.52) described pre-laying female *D.immutabilis* attacking any other females within reach; they slashed at each other with partly opened beaks which often interlocked in tugs-of-war that could leave faces, throats and mouths bleeding.

V Copulation

Copulation between petrels takes place on the ground and in mated birds usually at their nest, whether on the surface or below it. Coition of burrow nesting species often seen on the surface is probably of birds lacking burrows: any resulting eggs tend to be abandoned on the surface (Chapter 13, Section VIII).

Pre-copulatory behaviour may be complex as with albatrosses, or simple as with diving petrels, but almost always involves mutual allopreening of heads, napes, cheeks and throats. The actions involved have been best described for diurnally active species, e.g. *D.epomophora* (Richdale, 1950, p.37); *D.exulans* (Tickell, 1968); *D.melanophrys* and *D.chrysostoma* (Tickell, 1984); *D.immutabilis* (Fisher, 1971, p.41); *Macronectes* (Warham, 1962, Conroy, 1972); *F.glacialis* (Hatch, 1987); *P.gravis* (Rowan, 1952). Routines with nocturnally active species include descriptions of *P.pacificus* (Shallenberger, 1973) and of *O.furcata* by Simons (1981b).

Coition may climax a long series of courtship rituals, but between established pairs there is often little pre-copulatory activity apart from mutual preening. Sometimes there may be no visible preliminaries at all; the male may simply respond to some invitatory posture and walk onto his partner's back up her depressed tail. Copulation has been seen repeatedly with some species in the days before the exodus, and Hatch (1987) estimated that female *F.glacialis* were inseminated at least 34 times in a season, this activity peaking 24 days before laying and only a few days before departure on the 'honeymoon' flight. Yet Brown (1966), who witnessed many instances with *P.nivea*, saw only one copulation per pair. Coition of *D.melanophrys* and *D.chrysostoma* was seen almost entirely during 1–2 day visits of the females, by which time yolk deposition was well advanced (Astheimer *et al.*, 1985 and Fig. 12.9).

VI Rape

As already noted, males at their nests during the pre-egg stage are very ready for copulation. Wilkins (1923) found Wandering Albatrosses promiscuous, with groups harassing single birds and Tickell (1968, p.19) described males running from their nests to attempt coition with passing females. Richdale (1949) saw 'intruder' male *D.bulleri*, sometimes those already with partners, rushing at females to attempt copulation. These tended to resist and extricate themselves to join their mates. Fisher (1971, p.46) investigated this behaviour among Laysan Albatrosses. A bird of either sex that, due to accident or other causes, simulated a female's precopulatory signal, might be rushed by a mob of males attempting to mount it. Sometimes another male would mount the first and so on so that a stack,

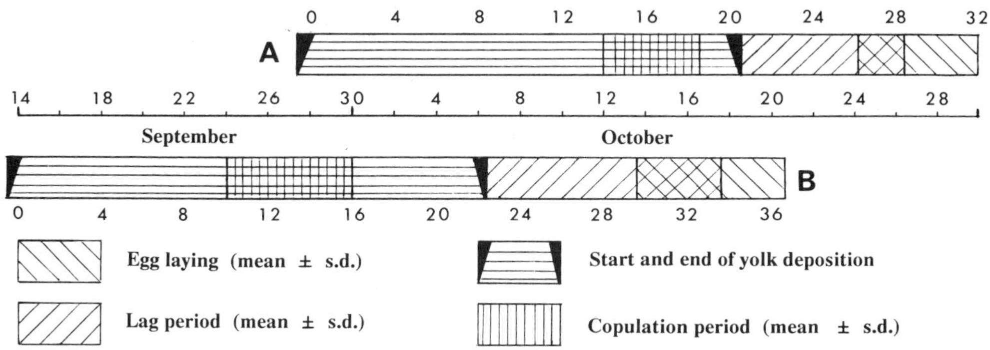

Figure 12.9 *Timing of the main events to egg laying for (A) Black-browed and (B) Grey-headed Albatrosses at South Georgia. Although of similar size,* D.melanophrys *has the shorter pre-egg stage, partly because times from arrival to laying and from copulation to laying are longer in* D.chrysostoma. *From Astheimer* et al. *(1985).*

four birds high could result, which then collapsed allowing the escape of the badly ruffled or injured lowest bird. Similar stacks of Buller's Mollymawks have been seen, the victim sometimes sitting on its nest pedestal: the risk of damage to the egg during such incidents seems very high.

Fisher believed that such assaults did not result in insemination. He inspected 'hundreds' of birds after attacks using cloacal smears and gross examination and found no evidence of cloacal contact. He could see no advantage in this persistent behaviour. Fisher thought that most rape attempts were not deliberately initiated by the victim, but similar 'extra-pair copulations' of *F.glacialis* studied by Hatch (1987) apparently resulted from females soliciting coition. Hatch considered that 20% of such attempts may have led to insemination. He also discussed the natural selective implications of insemination by several competing males.

VII The pre-laying exodus

A *Introduction*

After copulation some or all of the established breeders (and often non-breeders too) go to sea while the egg is made. When the females return they lay almost immediately. Such behaviour is seen among small and large species, among those of warm, cold and temperate seas, with winter and summer nesters and among both migratory and sedentary forms. It is possible because, in at least two species, *F.glacialis* and *O.leucorhoa*, and presumably in all, sperm is stored in special uterovaginal structures and released once the yolk has been laid down (Hatch, 1983). Yolk deposition itself takes about 14 days in *P.marina*, 18 days in *P.bulleri* and about 30 days in *D.epomophora* (Grau, 1984).

The exodus of the Short-tailed Shearwater is complete; all the birds ashore, non-breeders included, desert their busy nesting grounds starting on the night of 3/4 November. Within 5 days all have gone. Their return about 3 weeks later is equally dramatic and the first eggs are laid on the night they re-appear (Fig. 12.8). Although details of these 'honeymoon' flights were not worked out until Marshall & Serventy's study was published in 1956, the behaviour had long been known in Tasmania (Davies, 1846; Elwes, 1859).

Such a total exodus is not typical. With many species, while both sexes may take part, it is the females that stay away longest and in some species only the females leave. The time spent at sea also varies considerably from bird to bird. Exodus data for 48 species are summarised in Table 12.1.

Table 12.1 The pre-laying exodus

Species		Both (B) ♂ or ♀ Total (T)	Length (days)	Authority
DIOMEDEIDAE				
1.	*D.c.chionoptera*	♀	5–14	Tickell (1968)
2b.	*D.e.sanfordi*	♀	2–12	Richdale (1950)
		♀	6(1–17)	Robertson (pers. comm.)
3.	*D.irrorata*	♀	10+	Harris (1973)
6.	*D.immutabilis*	B	10(2–17)	Fisher & Fisher (1969)
7a.	*D.m.melanophrys*	B	15♀♀:16.4±5.2	Astheimer *et al.* (1985)
9.	*D.chrysostoma*	B	15♀♀:10.3±3.3	Astheimer *et al.* (1985)
11a.	*D.b.bulleri*	♀	*c.*6	Richdale (1949)
13.	*P.palpebrata*	?	0–6	Mougin (1970c)
		♀	2–3	Kerry (pers. comm.)
PROCELLARIIDAE				
14.	*M.giganteus*	?	*c.*12	Conroy (1972)
		B	3–18	Mougin (1975)
16.	*F.glacialoides*	?	4+	Mougin (1967)
17b.	*F.g.auduboni*	B	♂14; ♀16	Dunnet *et al.* (1963)
			♂9; ♀20	Macdonald (1977)
			♂16.9; ♀18.7	Hatch (1987)
18.	*T.antarctica*	T	*c.*23	Pryor (1968)
			*c.*20	Luders (1977)
19.	*D.capense*	B	*c.*10	Valette (1906)
		B	*c.*10	Eagle Clarke (1906)
		B	♀♀14–21	Pinder (1966)
20.	*P.nivea*	B	18(14–23)	Brown (1966)
		B	♂♂12	Isenmann (1970)
			♀♀18(11–25)	Isenmann (1970)
		B	♂♂14(7–19)	Mougin (1975)
			♀♀16(7–29)	Mougin (1975)
21.	*L.brevirostris*	B	*c.*30	Jouventin *et al.* (1985)
22b.	*P.m.gouldi*	B	♂*c.*50; ♀*c.*60 (53–80)	Imber (1976b)
31.	*P.mollis*	B	*c.*30	Jouventin *et al.* (1985)

Continued

Table 12.1 The pre-laying exodus (continued)

Species		Both (B) ♂ or ♀ Total (T)	Length (days)	Authority
34.	*P.cahow*	B	40	Wingate (1977)
35a.	*P.h.hasitata*	B	*c.*50	Labat (1722)
38b.	*P.p.sandwichensis*		*c.*40	Simons (1985)
39.	*P.n.neglecta*	B?	14–21	Oliver (1955)
40.	*P.hypoleuca*	B	*c.*15 (2–24)	Grant *et al.* (1983)
42.	*P.inexpectata*	?	9–16	Warham *et al.* (1977)
46.	*P.cookii*	?	*c.*30	Imber (1976b)
50.	*H.caerulea*	B?	*c.*14	Paulian (1953)
52c.	*P.d.banksi*	♀	7–14	Tickell (1962)
53.	*P.belcheri*	♀?	*c.* 17	Strange (1980)
56.	*B.bulwerii*	?	?	Amerson & Shelton (1976)
58a.	*P.a.aequinoctialis*	B	14–21	Mougin (1975)
59.	*P.westlandica*	B?	*c.*15	Baker & Coleman (1977)
62a.	*C.d.diomedea*	?	10–15	Gaultier in Thibault (1985)
62b.	*C.d.borealis*	B?	7+	Zino (1971)
64.	*P.pacificus*	B	14+	Munro (1940)
		B	7–14	Oliver (1955)
		B	6–21	Shallenberger (1973)
		B	*c.*14	Swanson & Merritt (1974)
		B	*c.*14	Crockett (1975)
65.	*P.bulleri*	T	*c.*30	Harper (1983)
66a.	*P.c.carneipes*	?	*c.*20	Thoresen (1967)
68.	*P.gravis*	B	7–14	Shaughnessy & Fairall (1976), Voisin (1979)
69.	*P.griseus*	♀?	*c.*14	Warham *et al.* (1982)
70.	*P.tenuirostris*	T	*c.*28–35	Montgomery (1892)
		T	*c.*21	Marshall & Serventy (1956b)
		T	20–22	Naarding (1981)
71.	*P.nativitatis*	B?	14?	Amerson & Shelton (1976)
72a.	*P.puffinus*	B	14–21	Harris (1966)
		♀	*c.*14	Perrins & Brooke (1976)
75.	*P.gavia*	B?	7+	Thoresen (1967)
76.	*P.huttoni*	B?	14–21	Harrow (1976)
HYDROBATIDAE				
79a.	*O.oceanicus*	♀	11(6–18)	Beck & Brown (1972)
82d.	*P.m.dulciae*	B?	7?	Littler (1910)
83.	*F.tropica*	♀	8–15	Beck & Brown (1971)
86.	*H.pelagicus*	?	?	Davis (1957)
89.	*O.castro*	B	0–20	Harris (1969b)
PELECANOIDIDAE				
102.	*P.georgicus*	?	♀♀9–15?	Payne & Prince (1979)
103c.	*P.u.exsul*	?	*c.*5	Jouventin *et al.* (1985)

B = both breeders; ♀ or ♂ = one breeder only, of sex stated; T = total, i.e., all birds ashore, whether breeding or not; ? = possibly, details unclear.

Little is known of the whereabouts of any petrel during the exodus. From the recoveries of birds aged 6 years and more, Perrins & Brooke (1976) deduced that females of *P.puffinus* from the Irish Sea colonies spent a fortnight or so in the Bay of Biscay feeding on schooling sardines while forming their egg. Up to the peak of laying (May 10) 112 of these birds were captured (in fishing nets), but only 26 thereafter to the end of August. These authors believed that the males also took part in the movement, feeding up intensively in anticipation of their 6.3 day stints on the egg.

Pinder (1966) caught male *D.capense* feeding close inshore during the exodus, but the only female he got did not lay until 17 days later. He surmised that the females might have gone far out to sea whereas the males staying inshore could visit their nests from time to time. Dunnet *et al.* (1963) reported the recovery of a female Northern Fulmar 370 km from its nest at the height of the exodus. Voisin (1979) found *P.gravis* abundant in the seas about the South Atlantic breeding grounds during its exodus migration. Also instructive is the specimen of *P.mollis* with a fully developed yolk in its oviduct, collected on 18 November 1971 about 2100 km north of the nearest possible breeding place at Antipodes Island (Hellyer *et al.*, 1973). This was most likely a breeding bird from that place on its exodus migration, as Imber (1983) found recently prepared burrows of this species there, mostly without occupants, from early November to early December, and supposed that the rest were away on their exodus.

Neither departures nor returns are spontaneous. Probably, as Serventy (1967) suggested (and showed in one instance), the first to leave are the first to return. And, as the partners come back on different days, it seems unlikely that they stay together at sea. Only Fisher (1971, p.42) has postulated this — for *D.immutabilis*.

The time taken for the exodus flight may vary from bird to bird and from year to year by the same bird. Beck & Brown (1972) noted that for two pairs of Wilson's Storm Petrels that bred together during three consecutive seasons, the exoduses lasted from 8 to 17 days and the dates of laying ranged from 31 December to 26 January. They suggested that such variability implied annual variations in the time taken to form the egg and/or variations in the dates of departure after copulation. Such differences could be accounted for by differences in food availability between years: for example, in changes in the distances the birds had to fly across the ice sheets and/or in the state of the pack-ice at the feeding grounds.

B *Diomedeidae*

Recorded exodus periods for albatrosses are short. For example, Robertson (pers. comm.) found that male *D.e.sanfordi* did not leave, whereas the females were away for about 6 days (2–27 days), returning to spend 2–6 days putting the final touches to their nests before laying (Fig. 12.10). Although the female spends little time ashore until the 2–6 days before the egg deposition, she and her partner are in nearby waters and his visits to the site increase in frequency as the time for laying approaches (Richdale, 1950, Tables 3 and 4).

Fisher & Fisher (1969) gave the average absence for 34 *D.immutabilis* as 10 days (2–17 days). In 85% of these instances the females were back at least 12 h before

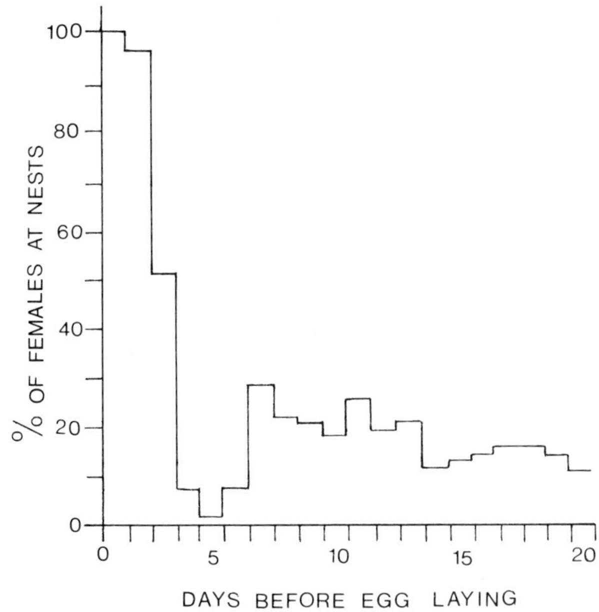

Figure 12.10 *Attendances of female Royal Albatrosses* D.e.sanfordi *at their nests in the 20 days before laying their egg. Note the short pre-laying exodus. From C.J.R. Robertson (pers. comm.).*

their partners and 50% were still alone when their egg appeared. An unusual feature was that both partners tended to begin their exodus on the same day, sometimes on the very day that the female had appeared after her contra-nuptial absence, and the delay before she left never exceeded 3 days (n = 11).

C *Procellariidae*

An exodus, at least by the breeding females, seems universal among the birds so far studied. Like *P.tenuirostris*, Buller's Shearwaters completely desert their nesting islands and for about 30 days are neither seen nor heard on or offshore (Harper, 1983). Immediately they re-appear laying begins. Total abandonment of nesting ledges was also reported for Antarctic Petrels by Pryor (1968) and Luders (1977) calculated a mean absence of 20 days for them.

Otherwise, although both sexes may leave, the exodus is usually not total, the females staying away much longer than their partners. The Northern Fulmar may be typical and, as with other fulmars, both sexes go. Dunnet *et al.* (1963) saw a clear fall-off in numbers at their Orkney colony before laying to 3–10% of the breeding population (Fig. 12.11). Macdonald (1977b) obtained more precise data; the first females left about a month before the first eggs appeared, followed about a week later by the males. Most of these were back before their partners. The females always laid within 2 days of their return, the absences averaging 19.6 days, those of

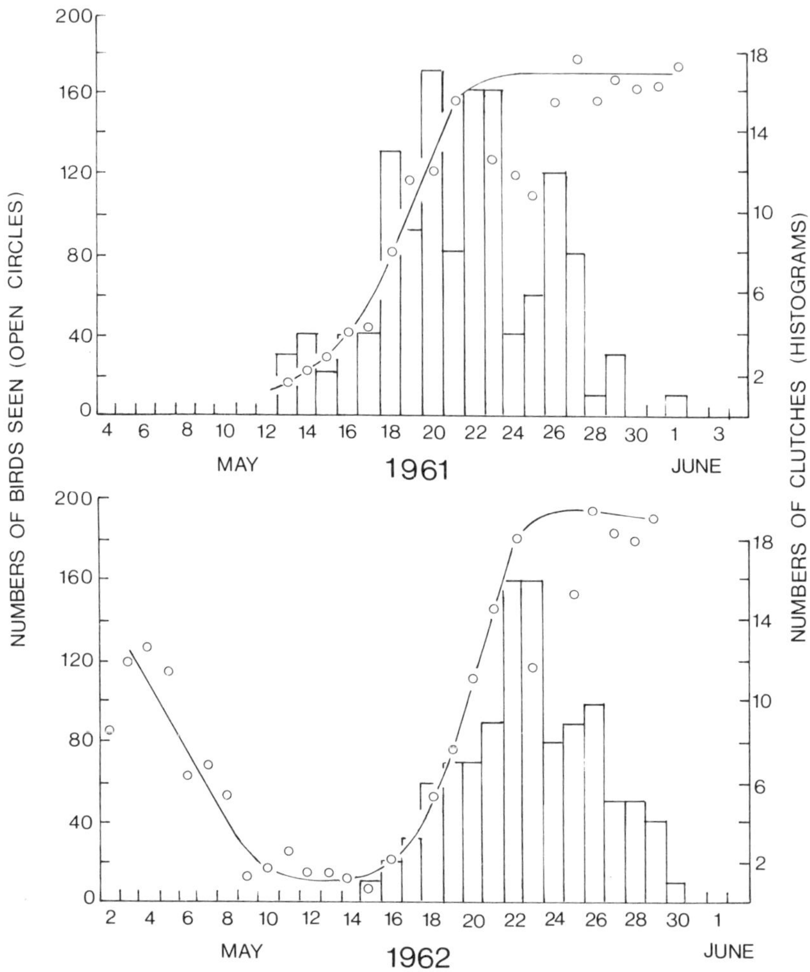

Figure 12.11 *Attendances of Northern Fulmars at their colony before and during egg laying. From Dunnet* et al. *(1963).*

the males about 9.3 days, although some males never left. The exodus of *F.glacialoides* is also not total, most birds going for about 3 weeks (Pryor, 1968). Absence patterns seem similar for other fulmars. The 'honeymoon' flight of *D.capense*, first reported in 1906 by Valette, is less obvious due to a lack of synchrony in the timing of departures. The longest pre-laying absence recorded is that of *P.macroptera*, which Imber (1976b) estimated to last for 2 months for the females. The males were absent about 7 weeks, returning to take over incubation several days after the lay rather than just before that event.

Among Antarctic Prions Tickell (1962) found the exodus restricted to the females — as also, evidently with *P.belcheri* (Strange, 1980). Tickell's males were often in their burrows at night until egg laying whereas their partners were usually absent for 7–14 days before then. Occasionally some females did return after dark, so cannot have been feeding far away. Many non-breeders also came back and, as laying extended over 3 to 4 weeks, the disappearance of part of the population was not obvious.

Little is known of the exodus, if any, of *Bulweria*. The numbers of *B.bulwerii* at sea 280 km south-west of Johnstone Atoll showed peaks in May and September in 1963–1967 (Amerson & Shelton, 1976). As laying begins in May, that month's peak could reflect the presence of birds on their honeymoon flight, while the September rise may mark the population increase after fledging.

Mougin (1970b) found that male Shoemakers gradually stopped coming ashore during the 25 days before the lay. Those that did appear stayed for only 1 or 2 days. The females were even less attentive, and in the last 15 days before the lay none was seen ashore. They were back up to 4 days before depositing their egg, their partners preceding them by a day. For Cory's Shearwaters *C.diomedea*, Zino (1971) calculated an exodus of at least 7 days and thought that both sexes were involved as the increase in activity on their return were dramatic.

Sooty Shearwaters have a major exodus, probably extending over about 14 days (Warham *et al.*, 1982). Not all birds go and that some do not go far is suggested by a sudden and temporary influx on a single night about a week after the general decline in numbers began. Among *P.tenuirostris* Marshall & Serventy (1956b) found that the peak of spermatogenesis was reached at the end of October and they believed that coition took place in the burrow at that time. Laying began on 19 to 21 November and, as in *P.griseus*, is highly synchronised (Serventy, 1963). The constancy of the timing of the exodus is shown by the records of Naarding (1981). His *tenuirostris* left on 3 November 1978, 5 November 1979 and on 3 November 1980 and returned after absences of 21, 20 and 22 days respectively.

A sea sojourn also occurs with smaller shearwaters. In the tropical *P.nativitatis* Amerson & Shelton (1976) found a fall-off in burrow visits 2 weeks before laying. In the Manx Shearwater only the females go, the male breeders and significant numbers of 3- to 5-year-olds remain ashore.

D *Hydrobatidae and Pelecanoididae*

In Wilson's Storm Petrel *O.oceanicus* the females of 31 pairs studied by Beck & Brown (1972) left for 10.7 days (16–18 days) the males being ashore regularly during that time, perhaps, as with *P.desolata*, to help keep the nest snow-free. These authors examined the burrows by night and day so that there is less chance that brief overnight visits were missed, as could have happened in some other studies. Beck & Brown (1971) found that male *F.tropica* were more often in their nests by day than their partners before the lay; two females being away for 8 and 5 days respectively. At the Galapagos Harris (1969b) found a partial exodus occurred mainly by the female *O.castro* whereas the males tended to roost in their nests by day from time to time.

Circumstantial evidence for an exodus among *P.georgicus* was provided by Payne & Prince (1979). In the 5 days before the lay they never saw more than a single bird in any burrow, suspected that these were males, and if so calculated that the females undertook an exodus of from 9 to 15 days. Jouventin *et al*. (1985) reported a rather similar situation in *P.urinatrix*.

E *Timing of coition in relation to the exodus*

Copulation occurs among established pairs throughout the pre-egg period but tends to increase in frequency before the start of the exodus, e.g. in *P.pacificus*, *P.desolata*, and *P.belcheri* (Shallenberger, 1973; Tickell, 1962; Strange, 1980). This may well be the typical pattern.

Copulation has, however, been recorded as late as a day or two before laying in mollymawks (Richdale, 1950; Tickell & Pinder, 1975), and Sorensen (1950) collected a pair of *P.palpebrata* immediately after seeing them copulate, to find the female with a 'fully developed' (? shelled) egg in her oviduct. This episode does not appear to have been an instance of rape by a rogue male. Such activity, and coition between non-breeders, continue after egg laying, and can be misinterpreted unless the animals are identifiable.

It is often difficult to tell which copulations are effective in transferring sperm, but in albatrosses and fulmars the apposition of the cloacae involves awkward manoeuvring and contact lasts several seconds.

The time elapsing between coition and laying by marked birds, or between peaks of copulation and peaks of laying, are given in Table 12.2. The figures suggest that these periods are fairly short in albatrosses (Fig. 12.9) and extend over 3 to 4 weeks among prions, fulmars, shearwaters and storm petrels: this is roughly in accord with the exodus periods of Table 12.1, in that these are short among albatrosses and longer among the Procellariids.

F *Sperm storage*

Marshall & Serventy (1956b) and Serventy (1967) followed gonad development of *P.tenuirostris* from first landfall to desertion of the young. By the time of the exodus the 10 males examined had reached or passed the peak of spermatogenesis with free spermatozoa in the tubules while in 10 females the largest oocyte then averaged 7.1 mm in diameter, as opposed to a mean of 3.4 mm on first arrival. On their return the seven males sampled had greatly changed testes showing no spermatogenesis and much post-nuptial lipid in the tubules. The females now contained shelled eggs developed while at sea and, after laying, the largest oocytes in five ovaries were only 3.0 mm across, i.e. smaller than the largest on arrival at the colony. Less complete data collected by Tickell (1962) for *P.desolata* showed similar patterns.

Sperm is stored by the hen and some other domesticated birds in utero-vaginal 'glands'. These release packets of sperm just before ovulation. Fertilisation occurs in the infundibulum (Fig. 12.12) where sperm may be held temporarily for release in synchrony with ovulation. Sperm may remain viable for up to 40 days in the utero-vaginal glands of the turkey, but with most birds studied viability lasts for only about 10 days.

Table 12.2 Records of copulation among petrels of breeding age

Species		n	Mean copulation date	Mean laying date	Days before lay	Authority
1b.	*D.e.chionoptera*	47	14 Dec.	24. Dec.	10	Tickell (1968)
2a.	*D.e.sanfordi*	10			11–29	Richdale (1950)
3.	*D.irrorata*	3			8,13,28	Harris (1973)
6.	*D.immutablis*	?			8–10	Fisher & Fisher (1969)
7.	*D.m.melanophrys*	15	14.7 Oct.	27.0 Oct.	12.3	Astheimer *et al.* (1985)
9.	*D.chrysostoma*	15	28.3 Sep.	17.4 Oct.	19.1	Astheimer *et al.* (1985)
11.	*D.b.bulleri*	?			1–23	Richdale (1949)
13.	*P.palpebrata*	?			*c*.14	Mougin (1970c)
14.	*M.giganteus*	13	27 Oct.	13 Nov.	17	Conroy (1972)
16.	*F.glacialoides*	2			17,21	Prevost (1953)
		?	1st week	14 Nov.	*c*.37	Pryor (1968)
17a.	*F.g.glacialis*	?			*c*.24	Hatch (1987)
18,	*T.antarctica*	?	17 Oct.	23 Nov.	*c*.36	Pryor (1968)
19a.	*D.c.capense*	4			23,27, 37, 38	Pinder (1966)
		?	5 Nov.	1 Dec.	*c*.26	Pryor (1968)
20.	*P.nivea*	7			26.5	Brown (1966)
		?	16 Nov.	*c*.5 Dec.	*c*.19	Pryor (1968)
40.	*P.hypoleuca*	?			*c*.21	Grant *et al.* (1983)
49.	*P.d.banksi*	?	2 Dec.	20 Dec.	*c*.18	Tickell (1962)
50.	*P.belcheri*	?	25 Oct.	16.5 Nov.	*c*.21	Strange (1980)
69.	*P.p.puffinus*	1			23	Lockley (1942)
84.	*O.castro*	1			33	Harris (1969b)

The necessity for sperm storage has been correlated with the special needs for fecundity in birds imposed by the sequential nature of ovulation, allowing a succession of ova to be fertilised without further intervention of a male.

Among petrels no such need arises. Only one egg is produced. However, it is clear from the small oocyte of *P.tenuirostris* on the eve of the exodus, that the rest of the yolk must be laid down and ovulation occur while the female is at sea. Fertilisation cannot be effected before ovulation, so some form of sperm storage must operate in these birds. Copulation at sea has not been recorded of tubenoses and seems impossible: apposition of the cloacae clearly requires a firm support for the participants.

Hatch (1983) found 'sperm storage glands' in the utero-vaginal region of *F.glacialis* and *O.leucorhoa*. The structures were simple unbranched tubules averaging 225 μ in length and extending into the lamina propria of the vaginal folds near the utero-vaginal junction (Fig. 12.12). In cross-section these tubules were seen formed by rings of columnar cells, mean inside diameter 19 μ (Fig. 12.13A). With *Fulmarus* some tubules contained spermatozoa bunched at the blind ends, their heads pointing away from the openings (Fig. 12.13B). In a bird taken 7 weeks after the last eggs had been laid in the colony no such glands were visible,

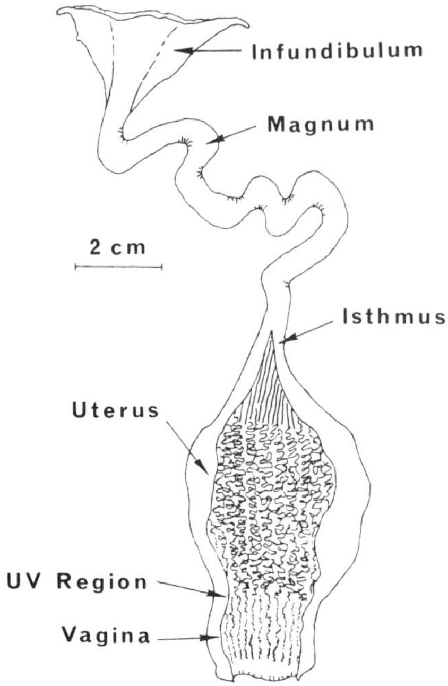

Figure 12.12 *Oviduct of a Northern Fulmar as it appears around the time of oviposition. From Hatch (1983).*

suggesting that there is a seasonal cycle of development and regeneration. Neither in these petrels, nor apparently, in any other bird, has it been determined how the sperm are released from the storage tubules nor how they reach the infundibulum.

Sperm storage is a mechanism for delaying fertilisation and, as Hatch pointed out, may be a major pre-adaptation in petrels, making the pre-laying exodus possible. This in turn permits the production of large eggs from food resources that may be patchily distributed and/or distant from the nesting sites.

Yolk deposition requires 14–30 days according to species (Grau, 1984; Astheimer & Grau, 1990).

In the domestic fowl the average time for transit down the oviduct and for shell formation is about 26 h, but there are no direct data for petrels.

Considerable amounts of calcium and phosphate are mobilised during shell formation. Shell thickness is not simply correlated with egg size, but some groups, including *Pterodroma*, tend to have thinner shells than, for example, fulmar eggs of similar size (Chapter 13, Section VIB). However, fulmars seem not to have longer exodus flights than gadfly petrels, as might be expected if calcium, phosphate or other nutrient levels were limiting in egg synthesis.

The very long exodus of *P.macroptera* reported by Imber (1976b) implies a considerable sperm-storage ability. Viability of sperm for 60 days seems to be greater than that of any of the domesticated birds referred to earlier. Imber also reported a good deal of competition for burrows and of interference with established pairs by intruders, ascribing much of this to the long pre-laying

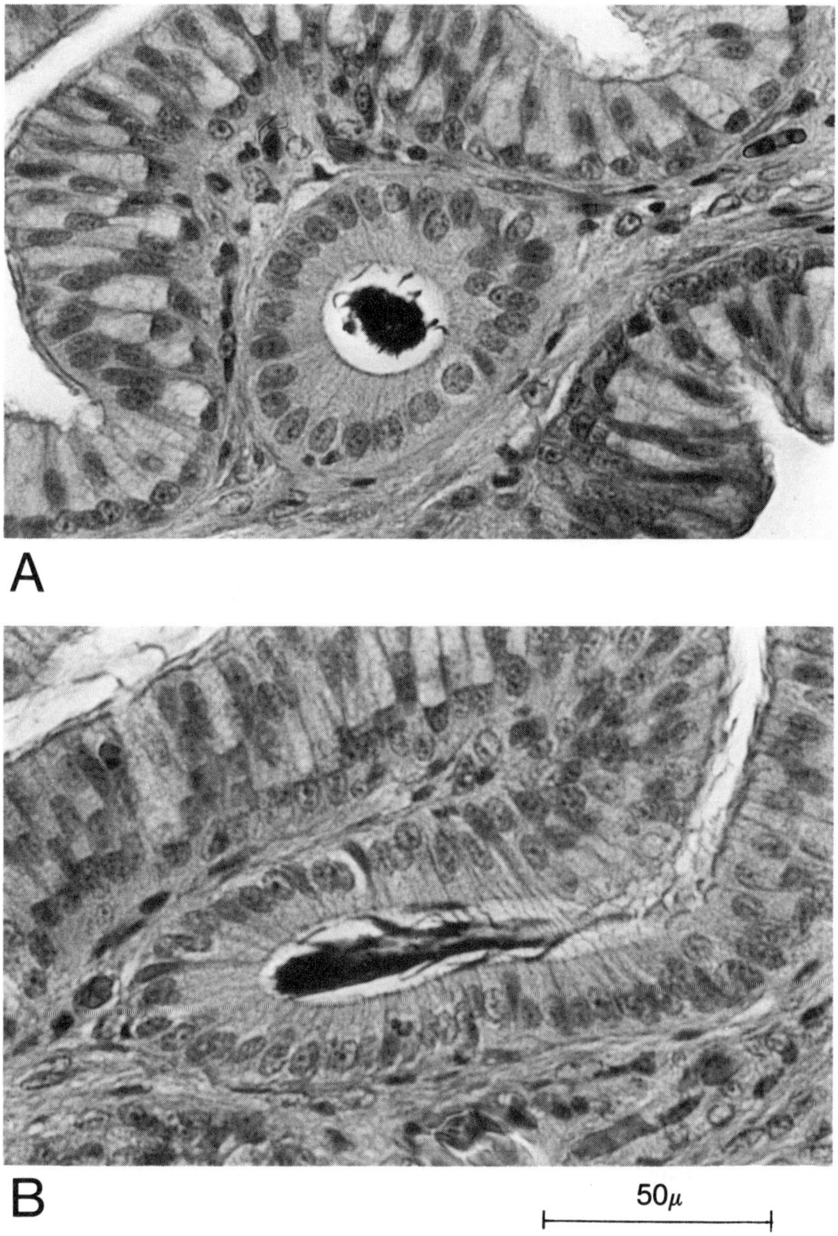

Figure 12.13 *(A) cross-section near the distal end of the sperm-storage gland of* F.glacialis *containing spermatozoa, and (B) magnified longitudinal section showing spermatozoa with heads directed towards blind end of the tubule. From Hatch (1983).*

absence. Some of these intruders may have been non-breeders, although this was not clearly established.

Pterodroma macroptera is a non-migratory petrel breeding in the southern winter when the much larger populations of shearwaters and other potential competitors for food and space are in the northern hemisphere. Where the birds feed while making their eggs is not precisely known and they may forage widely. With the winter descent of the plankton, food near the surface may not support a rapid build-up of reserves. Such factors may necessitate the slow development of the ovarian follicle that is suggested by the very long exodus period.

G *Factors affecting the exodus flights*

A major factor will be the distance of the main feeding grounds used during the exodus. Inshore feeders would be expected either to have no exodus or a brief one. The suggestion from Payne & Prince (1979) that *P.georgicus* may leave for 1 or 2 weeks before the lay runs counter to this idea, diving petrels being the more coastal of Procellariiformes. So are many storm petrels, and feeding within range of their colonies during egg formation should lessen their energy needs.

The short exodus of female *F.glacialoides* (Table 12.1) may reflect a rich food supply close at hand. If so, the long absence of *Thalassoica* petrels on the same cliffs suggests that these may be exploiting quite different and more distant foods.

Distance from the food supply may be a key factor in determining whether an exodus is total or partial. Warham *et al.* (1982) compared two trans-equatorial migrant shearwaters, *P.tenuirostris* and *P.griseus*, whose breeding ecologies and timetables are very similar. The exodus of *tenuirostris* is total, for *griseus* only partial, perhaps being restricted to females. Both have huge populations and must depend on correspondingly large food stocks. The locations of the main feeding grounds for the breeding birds are unknown, but if the smaller *tenuirostris* has to go further than the larger *griseus*, then the former would be expected to be away longer. Indirect evidence that this is so is shown by the lengths of their incubation spells, averaging 12.6 days for *tenuirostris* and 9.5 for *griseus*. Specimens of *tenuirostris* have been collected in Antarctic seas during the breeding season, but whether nesting birds feed that far away is not established.

A breeding system involving long fasts at the nest, e.g. during incubation, may require the male's departure as well as that of his partner because he takes the first spell on the egg. A fast of 12.6 days such as the average male *tenuirostris* undergoes during the first stint could not be supported without adequate fat reserves. Mougin (1975) pointed out that the female *P.aequinoctialis*, who has a prolonged stay at sea after laying, obligates the male to take part in the exodus himself.

H *Functions of the exodus flight*

Tickell (1962) related the pre-laying departure in the female Antarctic Prion to the energy needed in producing the egg (20–25% of her body weight), and this functional interpretation of the exodus has been widely followed (Warham, 1964; Lack, 1966) as all petrels lay eggs that are large in proportion to body sizes (Fig. 13.4). This view is supported by the females having palpable eggs when they return, their immediate laying, and their prompt return to sea, leaving the males to start incubation. This is what would be expected of birds drained by the production of a large egg and needing to replace metabolised energy, nutrients and calcium. Marshall & Serventy (1956b) found that female *tenuirostris* carried only moderate fat deposits at the time of laying, the males abundant ones. One source of energy for the female after laying is in the resorption of her oviduct In *F.glacialis* this organ grows from about 2 g at the start of the season to 20 g at the time of laying (Hatch, 1983). The ovary and oviduct of *P.tenuirostris* comprised 3.2% of the adult body weight at egg laying and had a calorific value of 91.96 ± 7.28 kJ, an increase of 1500% from its calorific content at exodus departure (Fitzherbert, unpubl.). As the petrel oviduct regresses rapidly after the lay, its resorption may help to fuel her outwards flight.

Other seabirds also need extra energy for egg production and feeding of the female by her partner may provide an important part of this, e.g. in *Larus novaehollandiae* (Tasker & Mills, 1981). This does not happen with tubenoses. Nor are they like penguins and arrive at the breeding places with enough stored energy to produce their eggs.

If the main purpose of the exodus is to enable the female to get extra food for egg production, then the smaller petrels with higher metabolic requirements and relatively giant eggs, might need longer to collect the necessary food. Added to which, the small forms like storm petrels appear to consume more energy when feeding, having a bouncing and erratic flight and picking single items from the surface. Even allowing for the support of the air pushed upwards by the heave of a wave, such birds seem much less efficient on the wing than are the larger species that use dynamic and slope soaring so extensively. However the data in Table 12.1 hardly support the idea of the length of the exodus being correlated with relative egg size.

The behaviour of non-breeding birds during the exodus varies. With *P.tenuirostris*, *P.bulleri*, and probably *Thalassoica* and *Halobaena*, all categories of birds desert the colonies. Some non-breeding *Pagodroma* also go (Brown, 1966). But with other species, e.g. *P.desolata*, many non-breeders return after dark. Shallenberger (1973) found that *P.pacificus* which visited the colonies at night during the exodus were unattached to burrows.

Why should non-breeders take part in these honeymoon flights when they have no eggs to make and no incubation duties to perform? It might have been expected that older males would stay to guard or obtain a potential nesting site. Accompanying the breeders may form part of the younger bird's 'apprenticeship', enabling them to learn the routine and the best feeding places; this implies that such experience is selected for in preference to staying back, feeding locally, and trying to steal a burrow or site while the owners are away. This could be futile as most sites are taken up by the time of laying and conspecific intruders seem to be easily displaced by established pairs. Pairs of Southern Fulmars that invaded the ledges of Antarctic Petrels during the latter's exodus were quickly displaced when the rightful owners returned (Luders, 1977). It may also be easier to identify unused sites during the pre-exodus period. Even for those species with only a partial exodus, it is possible that older non-breeding females leave while the younger females and the males stay behind, but I know of no evidence for that.

In summary, possible advantages of having a pre-laying exodus are:

1. It allows the female to go to distant, food-rich areas to acquire the energy and nutrients needed to make a large egg, to fuel her return to the nest and still have enough energy to get back to the feeding area for recuperation.
2. It permits the male to acquire enough reserves to return to the nest, take on the first incubation stint and in turn still retain enough to get back to a feeding area for recharging his reserves.
3. To the non-breeders the value of going on the exodus may be to learn this phase of the breeding cycle. Such birds develop brood patches and presumably experience changes in hormone levels and gonad development similar to those of the breeders. Some birds too inexperienced or perhaps of low hierarchical status to win nest sites develop eggs and have to lay these on the surface (Chapter 13, Section VIII).
4. Where food is patchy and far from the nesting grounds, the pre-egg exodus may allow breeding without a dependable food supply close by, making possible the use of temporary food aggregations that could otherwise not support a breeding population.
5. This system has the potential for allowing the female to assess oceanic resources before committing herself to breeding (Astheimer, pers. comm.).

One disadvantage is that, because of the long distances and times often involved, the exodus provides another opportunity for the partners to miss one another and for their activities to get out of phase, when breeding fails even if a viable egg is produced.

In some situations there are advantages in having the male staying behind so that he can visit the nest at intervals. Its usurpation is then less likely and he may be able to keep the nest in readiness for the female — by clearing blockages of snow or sand from burrows, for example. Without immediate access to the burrow the returning female may have to lay on the surface and lose her egg to predators.

CHAPTER 13

The Petrel Egg

Surface-nesting Wedge-tailed Shearwater.
Photo: author.

Contents

I Introduction

Basic data on petrel eggs, e.g. dimensions, shell weights and thicknesses are listed by Schonwetter (1960) for most species. Some eggs still await description. These include those of *Pterodroma macgillivrayi* and *Bulweria fallax* and of the storm petrels *O.markhami* and *O.hornbyi*. On the other hand there are good samples of the eggs of the Guadalupe Storm Petrel *O.macrodactyla* in the collections although the bird itself is extinct.

II Egg formation

The development of petrel eggs before the lay is similar to, but slower than, that of most seabirds (Grau, 1984). He identified a period of rapid yolk deposition in a single follicle of the developing ovary and a 'lag' period of up to 10 days between yolk completion and laying. After ovulation the albumen, membranes, shell and pigments are deposited. There is a final period when the shelled egg is held in readiness for laying.

The yolk is added to the outside of the oocyte in layers, that deposited by day being of a different structure to that laid down overnight, thus permitting estimation of days of yolk formation. The various components of each layer can also be sampled and related to daily energy and protein assimilation. By feeding the female with oil-soluble dyes during the process these layers can be marked, yolk development followed, and lag time determined (Riddle, 1910; Grau, 1976) (Fig. 13.1). However, as most female petrels spend much time at sea during egg formation, dosing at intervals is usually impossible.

Grau (1984; Astheimer & Grau, 1990) present curves for yolk growth in eggs of nine species of petrels and Astheimer *et al.* (1985) data on circadian patterns of deposition for *D.melanophrys* and *D.chrysostoma* (Fig. 12.9). The yolk of a 440 g Royal Albatross egg was estimated to have taken 30 days to form, with a lag period of another 10 days, whereas in a 13 g egg of the storm petrel the yolk was complete after 14 days.

Astheimer *et al.* (1985) found that the 12-day lag period for Grey-headed Albatrosses was significantly longer than the 9 days taken by Black-browed Albatrosses breeding at the same place. In both, rapid yolk deposition began at sea one or two weeks before the females arrived at the colonies so that it seems unlikely that this process is triggered off by interactions with the males.

Nothing appears to be known precisely of events in the oviduct during the lag period (apart from the sperm storage during the exodus) and Grau (1984) noted that while at laying a hen's egg consists of about 20 000 cells, no data are available on the cleavage processes in seabird eggs.

III Egg deposition

A *The act of laying*

There are few descriptions of laying by petrels. Rice & Kenyon (1962) saw a Laysan Albatross standing with drooping wings, cloaca contracting and dilating so that the egg became visible on dilation. After 9 min she leant back, lifting her breast and lowering the rear of her body to the nest, and almost immediately dropped the egg. She then bent to inspect it and then settled down upon it without touching it with her bill or 'talking' to it. Similar actions accompany laying by mollymawks. With *D.bulleri* the drooped wing posture lasts about 2 min with the bird breathing heavily, the head is bowed forwards as a bulge develops around the cloaca and she makes grunting sounds. The back becomes concave — a sign that laying is imminent (Fig. 13.2). Actual release of the egg takes 1–2 s following intense

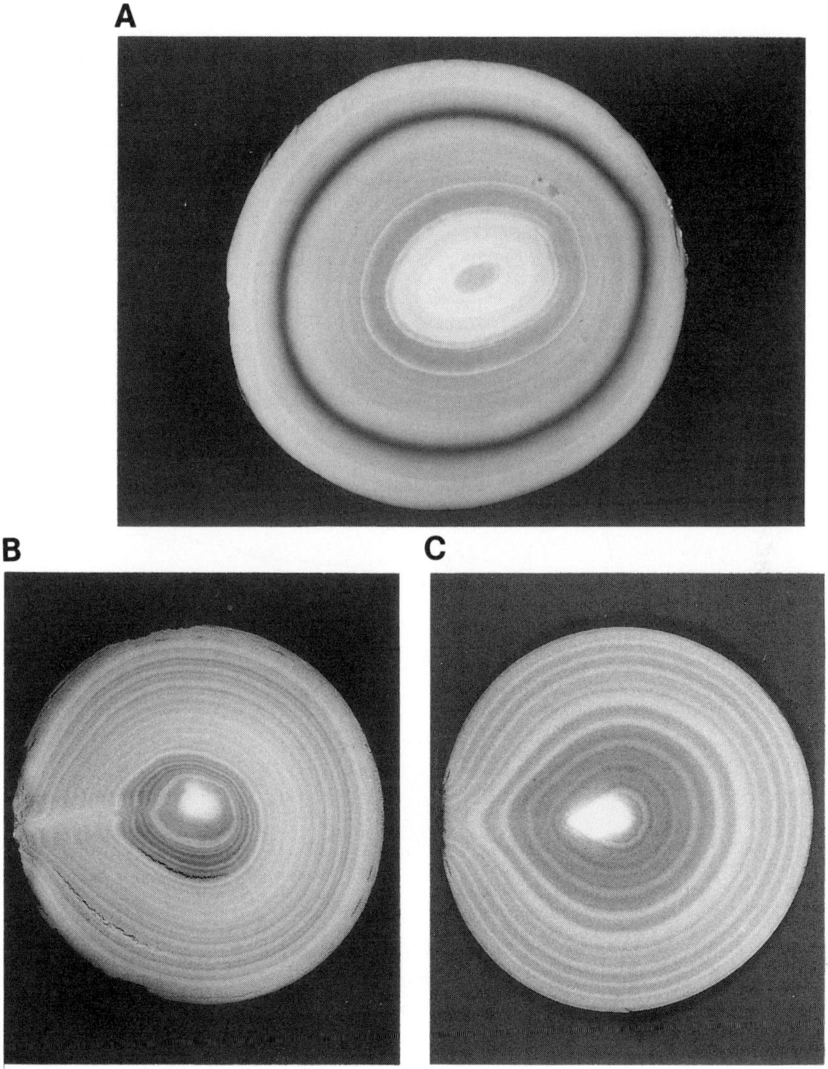

Figure 13.1 **A**. *central section of an egg of* Diomedea chrysostoma *before staining with potassium dichromate. The dark ring resulted from a dose of Sudan Black and served as a date marker. From Astheimer* et al. *(1985)* **B**. *Yolk of* Pelagodroma marina *egg after slicing and staining, showing 12 dark and 12 light rings marking 12 days of rapid yolk deposition. Photo: C.R. Grau.* **C**. *Yolk of* Pachyptila turtur *egg: staining has shown 17 rings formed during rapid yolk deposition. Photo: C.R. Grau.*

straining. After laying the female touches the egg with her bill for about 10 s before settling to incubate. Simons (1981b) described *O.furcata* assuming 'an unusual flattened posture' during egg extrusion.

Judging by observations on *D.immutabilis* and *D.bulleri*, the albatross egg is extruded pointed end first, presumably travelling down the oviduct in this

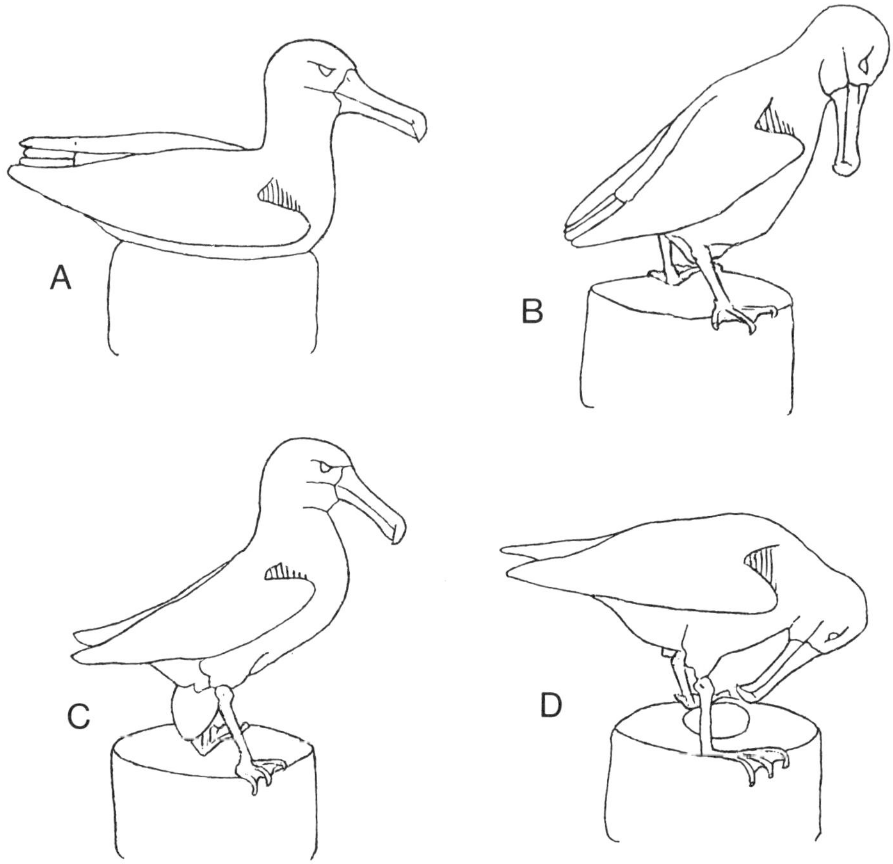

Figure 13.2 *Egg release in* Diomedea bulleri: **A**. *Pre-laying attitude with folded wings held out from the body.* **B**. *About to lay — fleshy sliver exposed at gape and feathers bulge as egg is extruded.* **C**. *Egg is released narrow end first: note characteristic hollowing of back giving a 'swayback' appearance.* **D**. *Inspecting egg before settling upon it. From photographs by C.H. Fitzsimons.*

orientation, there being no evidence that even the relatively small albatross egg is turned through 180° before being laid blunt end first as with some domestic hens (Bradfield, 1951). And with relatively much larger eggs such as those of storm petrels, such rotation would seem impossible (Warham, 1968; Fig. 13.9).

B *The time of laying*

1 *The time of day*

Rice & Kenyon (1962) reported that of 115 Laysan and 10 Black-footed Albatrosses checked, 65 of the former and five of the latter laid during daylight hours, whereas Fisher (1969) found that most Laysans (75%) laid by day and wondered if Rice and

Keyon's birds had been affected by the station lights. Tickell (1968, p.24) concluded that most of his Wandering Albatrosses laid by day. Many burrowing petrels lay on the night of their return from the exodus, though an unknown proportion of eggs appears by day. According to Harris (1969b) the egg of *O.castro* was always laid at night.

The possibility that petrels can delay the laying of their fully shelled egg if baulked from reaching the nest by contrary winds, snow blockage etc;. has not been investigated. Cole (1913) found that female *P.tenuirostris* confined when with palpable eggs could delay laying for at least 2 days.

2 *The laying period*

Egg laying at a colony may be spread over a few days as with highly migratory forms and those in high latitudes or over many days as with sedentary species like *P.alba* (Figs 11.5 and 11.6). Table 13.1 summarises data on the spread of laying of species for which there is adequate information, those with ill-defined or double laying seasons being excluded. Very often the distribution of laying dates is skewed, most eggs being laid in the first half of the period, as with *P.aequinoctialis* (Hall, 1987).

These data are rather heterogeneous, being gathered in different ways and liable to distortion by the inclusion of aberrant early or late eggs. Asynchronous laying is common among storm petrels and in Table 13.1 the most extended laying is of the two sedentary storm petrels of subtropical seas — *O.castro* and *O.homochroa*. The British and Fork-tailed Storm Petrels also lay over many weeks and are not highly migratory, while part of the spread with *O.furcata* and *O.oceanicus* may be due to delays from randomly occurring snow blockage of burrows, so that a wide spread of egg dates may ensure better survival (Beck & Brown, 1972). Wasilewski (1986) suggested that the dates of laying at each colony of *O.oceanicus* had a characteristic pattern independent of environmental conditions, perhaps genetically determined.

The other species with a noticeable spread to its egg laying is the Galapagos Albatross, a bird known to make only short post-nuptial shifts towards the coast of South America (Chapter 11, Section VA).

The high latitude fulmars — *F.glacialoides, D.capense* and *P.nivea* compress laying into no more than 15 days, the larger *M.giganteus*, perhaps less affected by the short summer, taking 21 days (at South Georgia). The Kerguelen Petrel also lays over a very short period, similar to that of the fulmars. The shearwaters, *P.griseus* and *P.tenuirostris*, provide examples of compressed laying associated with tight breeding schedules needed to fit in vast trans-equatorial migrations between seasons, but *P.inexpectata*, a bird performing very similar migrations, spreads laying over nearly 3 weeks. Although both *P.puffinus* and *P.pacificus* make extensive migrations after breeding, their laying is spread over nearly three times the period taken by *griseus* and *tenuirostris*. Other notably short laying periods are those of the Blue Petrel (9 days) and the Sooty Albatross (12 days): the former is a rather sedentary species and the latter not known to make extensive migrations, so that compressed laying seems anomalous.

Some of the differences in these data doubtless arise through a series of

Table 13.1 Laying periods of petrels[a]

Species	*n*	Days	Authority
DIOMEDEIDAE			
1b. *D.e. chionoptera*	260	38	Tickell (1968)
2b. *D.e. sanfordi*	57	36	Richdale (1950)
3. *D. irrorata*	495	77	Harris (1973)
5. *D. nigripes*	118	19	Rice & Kenyon (1962)
6. *D. immutabilis*	3540	25	Fisher (1969)
7a. *D.m. melanophrys*	302	23	Tickell & Pinder (1975)
9. *D. chrysostoma*	278	19	Tickell & Pinder (1975)
10. *D.c. bassi*	147	17	Jouventin *et al.* (1983)
11a. *D.b. bulleri*	50	40	Warham unpubl.
12. *P. fusca*	40	12	Jouventin & Weimerskirch (1984)
13. *P. palpebrata*	?	24	Mougin (1970)
PROCELLARIIDAE			
14. *M. giganteus*	95	21	Hunter (1984b)
15. *M. halli*	?	20	Voisin (1968)
16. *F. glacialoides*	?	15	Mougin (1967)
17b. *F.g. auduboni*	?	21	Mougin (1967)
19a. *D.c. capense*	74	15	Pinder (1966)
20. *P. nivea*	31	15	Isenmann (1970)
21. *L. brevirostris*	26	16	Jouventin *et al.* (1985)
22b. *P.m. gouldi*	200	37	Imber (1976b)
31b. *P.m. fusca*	17	19	Jouventin *et al.* (1985)
42. *P. inexpectata*	71	20	Warham *et al.* (1977)
50. *H. caerulea*	23	9	Jouventin *et al.* (1985)
51c. *P.v. salvini*	62	22	Jouventin *et al.* (1985)
52c. *P.d. banski*	56	26	Tickell (1962)
58a. *P.a. aequinoctialis*	31	42	Mougin (1970)
	41	27	Hall (1987)
62b. *C.d. borealis*	49	11	Zino (1971)
64. *P. pacificus*	28	33	Shallenberger (1973)
69. *P.griseus*	30	12	Warham *et al.* (1982)
70. *P. tenuirostris*	403	14	Serventy (1963)
72a. *P.p. puffinus*	128	37	Harris (1966)
HYDROBATIDAE			
79a. *O.o. oceanicus*	27	43	Copestake & Croxall (1985)
79b. *O.o. exasperatus*	20	27	Jouventin *et al.* (1985)
	68	35	Beck & Brown (1972)
86a *H.p. pelagicus*	57	61	Davis (1957)
89. *O. castro*	115	110	Allan (1962)
97. *O. homochroa*	127	102	Ainley *et al.* (1990)
99b *O.f. plumbea*	94	57	Boersma *et al.* (1980)
PELECANOIDIDAE			
102. *P. georgicus*	18	41	Jouventin *et al.* (1985)
103c. *P.u. exsul*	33	50	Jouventin *et al.* (1985)

[a] With burrowing species only burrow-laid eggs are included.

unstudied factors like the age composition of the population examined and the timing of the food available during the exodus, while the larger the sample size the greater the chance of including abnormally late or early eggs. The data do not allow extensive comparisons of, say, the time over which a given proportion of eggs is produced.

3 *Laying periods of individual birds*

Richdale (1952, p.20) and Tickell (1968, p.23) found that individual Royal and Wandering Albatrosses tended to lay over much more restricted periods than did their populations as a whole. The divergence from season to season of particular birds average 3.8 ± 0.4 and 4.2 ± 1.2 days respectively. Consequently the Royal Albatrosses tended to lay in the same order each breeding season, there being early-, medium- and late-layers; even when the season as a whole was delayed the sequence was maintained. Serventy (1963) also found that, despite the truncated laying of *P.tenuirostris*, most individual females laid within 1 or 2 days of the same date each year. Similar trends have been noted with *P.puffinus* and *O.oceanicus* (Harris, 1966; Wasilewski, 1986).

4 *Effects of age and change of partner*

First-time breeders tend to arrive later at the nesting colony than experienced ones and also tend to lay late. Thus Fisher (1969) found that female Laysan Albatrosses produced their first egg significantly later than did the established nesters (Fig. 13.3) but that with increasing age all eventually became early layers. A female's second egg was laid 2–4 days earlier than her first and her third about 4 days before her second, providing that the pair-bond remained intact. First-time laying Northern Fulmars and Galapagos Albatrosses also lay late (Ollason & Dunnet, 1978; Harris, 1973).

Individual Manx Shearwaters that retained their partners also tended to lay on very similar dates in subsequent years (Harris, 1966) and Brooke (1978a) found that newly formed pairs laid 5.6 days earlier the next season. His data suggested that the age of their pair-bond had more influence on the time of laying than either their age or breeding experience. Perrins (1966) showed that early-hatched chicks of this species (i.e. from early-laid eggs) survived better than later-hatched ones, and Ollason & Dunnet (1978) that Northern Fulmars laying early produced more fledged chicks than those that laid late. In the Arctic, however, these birds may lay when the ledges are still iced over (Lovenskiold, 1963) and at temperatures of −25° to −30°C (Belopol'ski, 1961, p.256).

There is little information on the effect that changing partners has on the laying dates of petrels. Richdale detected no changes of laying date with change of partner among Royal Albatrosses but Fisher (1969) found that older Laysan Albatrosses that remated tended to lay later than before — hence the second peak in Fig. 13.3B — whereas younger females with older partners laid early; and yet birds paired with mates of the same age tended to lay later. Northern Fulmars breeding together for at least the second time laid significantly earlier by 1.5 days than did newly formed pairs (Ollason & Dunnet, 1978).

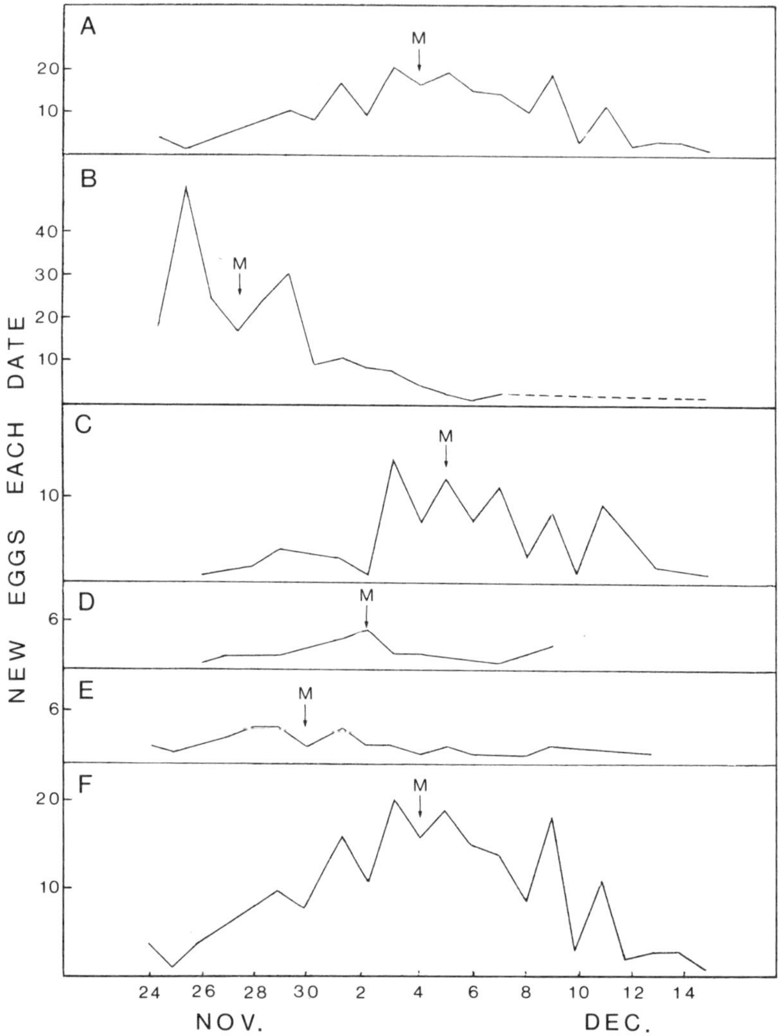

Figure 13.3 *Egg laying dates of Laysan Albatrosses:* **A**, **B**. *Effects of age and experience of the female.* **A** = *females laying for the first time,* **B** = *birds having bred for at least six seasons.* **C** *to* **F**. *Effects of age of mate.* **C** = *mates of the same age (85 pairs);* **D** = *female the younger (26 pairs);* **E** = *male younger (35 pairs);* **F** = *all females (205 pairs). M = median date. From Fisher (1969).*

C *Re-laying*

It is rare for a petrel to lay a second egg in the same season following the loss of her first. The most conclusive data are perhaps those for the Laysan Albatross because, during control attempts at Midway Atoll, large numbers of eggs were taken both early and late in the incubation stage without any replacements (Rice & Kenyon, 1962b; Fisher, 1968).

Even when both members of a pair are identifiable, relaying is difficult to verify as the finding of a known female with a second egg does not prove that she laid it unless continuous observation has shown that no other female was implicated or no other bird has entered the nest.

Lack (1968) pointed out that because of the need for an exodus, it was difficult to envisage how repeat eggs could be produced and it is the more surprising that the most reliable data for re-laying come from the storm petrels (their eggs being nearly a third of the female's body weight), rather than from bigger species like albatrosses with relatively small eggs.

Re-laying seems to be best authenticated among Fork-tailed Petrels *O.furcata*, a species whose egg averages 21% of the adult body mass. Boersma *et al.* (1980) removed 36 eggs on the day laid, 29 of which (81%) were replaced about 21 days later. Although it was not proved that the same females were always involved, most seem to have been genuine re-layings, presumably after a second sojourn at sea. The replacement eggs were significantly smaller. Whether such re-layings resulted in fledged chicks was not established, but chicks hatched late never attained the peak weights of those hatched early and second-egg chicks would have been lighter at birth, anyway.

Other petrels for which there is circumstantial evidence of occasional re-laying include *P.urinatrix* (Roby & Ricklefs, 1983); *H.pelagicus* (Davis 1957 p.98); *O.castro* (Allan, 1962; Snow & Snow, 1966; Harris, 1969b, p.121); *O.leucorhoa* (Wilbur, 1969; Morse & Buchheister, 1979) and even *O.oceanicus* (Beck & Brown, 1972, p.21). Ainley *et al.* (1990), in a 12-year study of *O.homochroa*, also attribute some late eggs to replacement clutches. Morse & Buchheister suggested that re-laying of *O.leucorhoa* might be more frequent in years when the birds' nutritional condition was good.

All the species mentioned above except *O.oceanicus* are likely to be found ashore over many months. With the exception of one possible record for *P.puffinus* by Harris (1966), re-laying not been authenticated for any long distance migrant. Furthermore, the rapid regression of the gonads after fertilisation (Marshall & Serventy, 1956b) makes a second effective copulation highly unlikely: the ability of storm petrels to re-lay suggests that their gonadal cycles are different.

D *Two-egg 'clutches'*

Although laying but one egg is characteristic of petrels, almost all detailed studies turn up instances of nests with two eggs.

Few data on their incidence are available. Fisher (1952, p. 94) quoted a figure of 10–15% of such clutches with Northern Fulmars from various observers and samples of unknown sizes at the Westermann Islands, Iceland. Such figures seem remarkably high, whereas a figure from 1944 of about 1.7% based on 4150 nests is more compatible with data for other species. Warham (1962) found only 0.14–0.18% of some 3000 nests of *M.giganteus* with two eggs of two chicks, H.I. Fisher (1968) about 1% of those of 3440 nests of *D.immutabilis*, and Tickell & Pinder (1966) 1–2% of nests of *D.melanophrys* and *D.chrysostoma*, sample sizes unstated.

That the various accounts of two or more eggs in nests are unlikely to relate to genuine two-egg clutches was noted by Gain (1914), emphasised by Richdale (1949), and confirmed by Tickell & Pinder (1966).

Petrels only lay a single egg and second eggs are due to the intervention of a second female who probably lays when the first is away. Various situations can facilitate double layings: a male may tolerate another female and a trio form at a nest, or two pairs may try to use the same nest, or a bird deserts and exposes the egg allowing a second, probably a younger and later-laying female, to add hers. No pairs are known to have raised two chicks naturally and in most cases complete failure occurs.

Because the size and shape of the brood patch matches the egg it is hard to see how two could be heated and hatched successfully and Fisher (1952) reported that the Northern Fulmars sat on only one of their eggs. The sole instance of a petrel that apparently did hatch two was the Southern Giant Petrel found guarding two chicks of similar size (Warham, 1962, Fig. 2.) All three were of the white-phase in a population with about 8% of such birds. Thus the possibility that a white-phase male had fertilised two brown-phase females and allowed each to lay, as suggested by Tickell & Pinder (1966) seems low, and Conroy (1972, p.63) pointed out that genetic probabilities did not support this view. Whatever the truth concerning this episode, hatching two petrel chicks in one nest must be excessively rare.

IV Egg size

A *Mass and volume indices*

In comparison with their body size petrels lay large, heavy eggs. They range from a mass of about 477 g for eggs of *D.e.chionoptera* to about 4.9 g for those of the smallest species, *H.microsoma*.

Because for most purposes eggs must be weighed on the day they are laid, good figures are not easily obtained, and many of those published are clearly of part-incubated eggs. Some may have been judged new-laid because their shells were very clean. Good data for burrow-laid eggs are particularly scarce because of the difficulty in extracting them without disturbing the adult. The eggs of 18 species lost a mean of 15.7% of their initial weight as water up to starring, when a sharp drop in weight occurs coinciding with direct contact with the ambient air (Rahn & Whittow, 1988).

Figures that compare well with fresh eggs sampled in the field can be obtained from measurements of their capacities by running diluted ethanol into their shells from a burette. A fine wire whisk is used to break occluded air bubbles. Warham (1968) measured the capacities of samples of 38 species of petrel in this way finding that:

$$\text{Egg Capacity (ml)} = 0.017 \text{ Egg Mass (g)} + 0.090$$

Otherwise the air cell of an incubated egg can be filled with water and the egg weighed (Grant *et al.*, 1982c; Rahn *et al.*, 1985). Egg mass figures can also be checked from egg capacities on the assumption that the specific gravity of the contents is the same as that in those of the fowl, i.e. 1.035m using

$$\text{Egg Mass (g)} = 1.035 \text{ Egg Capacity (ml)} + \text{Shell Mass (g)}.$$

The shell is about twice as dense as the egg contents, Rahn *et al.* (1984) finding that for a 100 g petrel egg the density was 2.04 g cm^{-3}.

For many purposes an index of egg volume or mass based on the equation for the volume of an ellipsoid is adequate. This is the product of the square of the maximum breadth (B) and the maximum length (L) since Mass = kLB^2. There are good data for egg dimensions for most petrels including many whose initial egg weights have not been determined. For tubenoses the value of 'k' was determined as 0.551 ± 0.02 L and B measured in cm. Data for 40 species from all four families were used, the mean sample sizes for egg masses and dimensions being at least 10 and 15 respectively (Warham, 1968).

B *Relative egg size*

An early investigation into the relationship between egg mass and body mass in birds was by Heinroth (1922) whose data on 432 species weighing from 2 to 90 000 g were re-examined by Huxley (1927). Heinroth noted the relatively large egg of *F.glacialis* and in recent years, because of their large relative size, a good deal of attention has been paid to this aspect of procellariiform biology (Fig. 13.4). Their fresh egg weights range from about 6% to 29% of the female's own weight. These are larger egg masses than other birds that lay but a single egg. Warham (1968)

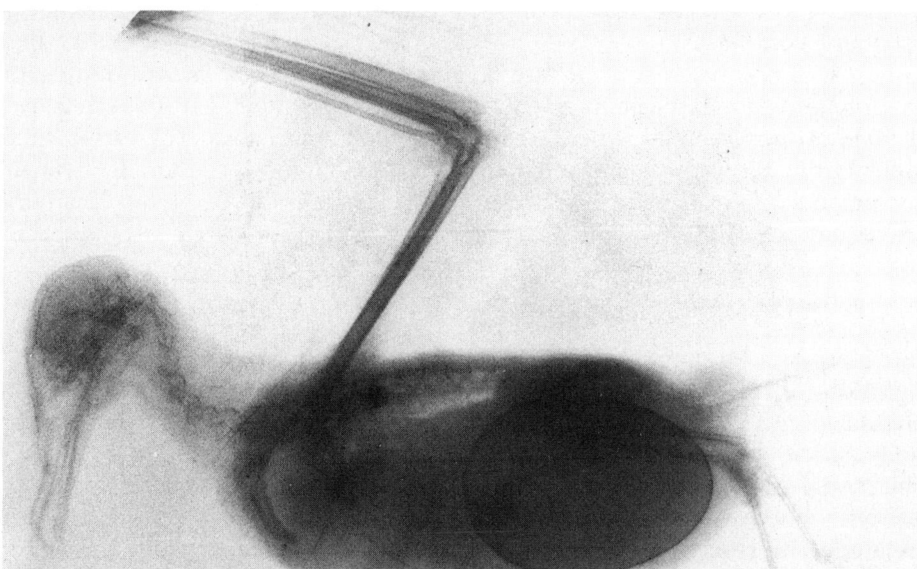

Figure 13.4 *Radiograph of* Pterodroma hypoleuca *with shelled egg in pelvic cavity. Note that pointed end is caudal. Photo: G.S. Grant.*

determined that the ratio of $\log_{10}$ female body mass to $\log_{10}$ egg mass was 0.680 ± 0.052, using data for 19 petrels or:

$$\log_{10} \text{Egg Mass (g)} = 0.859 \log_{10} \text{Female Body Mass (g)}^{0.708}$$

Subsequently much new data have been collected and a number of investigators have presented plots of egg mass *v.* body mass (Jouventin & Mougin, 1981; Croxall, 1984; Rahn *et al.*, 1984; Rahn & Whittow, 1988), mostly in connection with studies of metabolism or ecological energetics.

There are not only the problems in getting accurate initial egg weights, but more in determining reliable figures for body masses. Most investigators have not used birds of known sex or of known status. Where sexual dimorphism exists, males usually being the heavier, the masses will be higher than if females alone were used. Even more of a problem is that fat-free weights are not available, nor likely to be. Because of the large stores of fat that are laid down and expended during the long pre-egg fasts of males and of both during incubation when, for example, a Manx Shearwater will drop during a 12-day stint from an initial weight of about 477 g to 358 g at 10 g day^{-1} (Harris, 1966), body masses vary greatly over time (Fig. 14.3). Usually unknown also are whether the female's weight includes that of an egg, food or stomach oil.

The most practical solution to lessen distortion due to these causes is to use the largest samples available and take masses of birds of a given sex where sexual dimorphism occurs on, if necessary, of mixed sex for shearwaters and other groups without marked sexual dimorphism in size and to draw the mass data from the same populations. Published tables lack such information and often sample sizes are small. More precise data will be given in tables of petrel statistics in a subsequent volume.

In Fig. 13.5 egg mass and LB^2 are regressed against body mass on logarithmic axes, only female masses being used except for shearwaters, gadfly petrels, prions and some storm petrels, where sexual dimorphism is low. All variables were sampled from the same subspecies and none contained less than 10 determinations, mean sample sizes for egg mass, body mass and LB^2 being 39, 72 and 56.

The equations of Fig. 13.5 are:

$$\text{(a) } Y_1 = 0.745X^{0.732};\ r = 0.991;\ P = <0.001;\ n = 34 \text{ species}$$
$$\text{(b) } Y_2 = 1.275X^{0.743};\ r = 0.991;\ P = <0.001;\ n = 52 \text{ species}$$

where Y_1 = Egg Mass (g); $Y_2 = LB^2$; X = Female Body Mass (g)

Equation (a) is close to the regression of Rahn *et al.* (1984) of Egg Mass = 0.674 Body $\text{Mass}^{0.742}$ for 36 Procellariiformes. However, their data were from samples of mixed sex and some weights appear to have been estimated.

The straight line relationships of Fig. 13.5 show that the rate of change of egg mass with body size is constant within the order, the variables being highly correlated. The regression has predictive value. For example, there are very few figures for the mass of the albatross *D.cauta eremita* but good samples of eggs measured in the field (museum collections often include many atypical examples). Robertson & van Tets (1982) gave mean dimensions of 107 eggs of this species as 102 ± 4 × 67 ± 2 mm. With an LB^2 index of 458 a female body mass of about 2500 g is indicated.

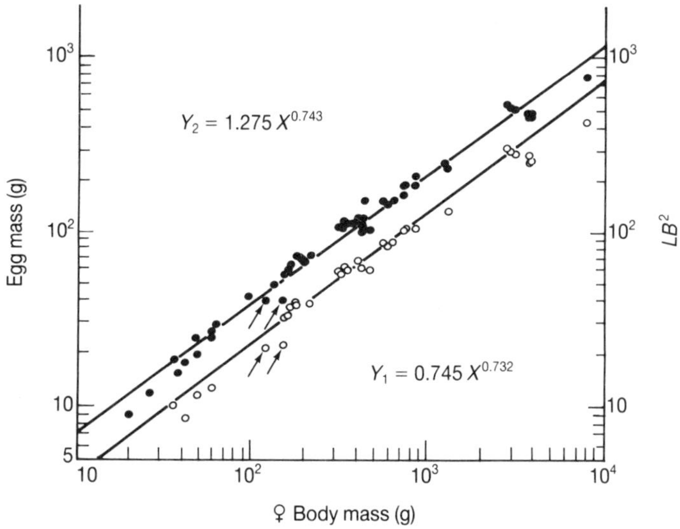

Figure 13.5 *Relationships between female body mass X and egg mass Y_1 (lower graph) and volume index Y_2. L and B in cm. Points for* Pelecanoides *arrowed. From Warham (1968).*

Because egg mass is correlated with body mass, egg mass is also correlated with other variables tied to body size. For example, as mean standard wing lengths of petrels are highly correlated with mean body masses (Warham, 1968):

$$\text{Mean Body Mass (g)} = 0.00587 \text{ Mean Wing Length (cm)}^{3.36}$$

egg masses are also correlated with standard wing lengths, e.g.

$$\text{Mean Egg Mass (g)}^{1/3} = 0.40 \text{ Mean Female Wing Length (mm)}^{1/2} - 2.51$$

(Warham, 1968 using data for 27 and 22 species respectively and excluding diving petrels).

The figures for the two diving petrels plotted in Fig. 13.4 show that these lay surprisingly small eggs for petrels of their sizes: being inshore feeders which raise their chicks quite rapidly, they might have been expected to lay typically large eggs. A similar trend is suggested by the plots for the storm petrels not all of which, however, are inshore feeders. There are few good data for these birds and the divergence may be partly due to the distortions of the extremes in logarithmic plots, but Huxley (1927) also drew attention to a steepening of the slope of his curve at about a body weight of 100 g. More figures are needed, e.g. for the smallest petrel *H.microsoma* (body mass 19.8 g) and of larger ones like the 93 g *O.tristrami*, calculated egg mass 17.5 g.

C *Factors affecting egg size*

It might be expected that egg size would be affected by the feeding conditions encountered by the female during her pre-laying foray at sea but in the population of Manx Shearwaters studied by Brooke (1978a) mean egg volumes (measured by LB^2 values) did not change over 4 years although the food supply is thought to have varied during that time.

Age does affect egg size in petrels as with many other seabirds. In the storm petrel *O.leucorhoa* older birds lay larger eggs (Grimmer in Montevecchi *et al.*, 1983). Manx Shearwaters gradually increased the sizes of their eggs until for 10 to 16-year-olds the LB^2 index was 105.5 but in a sample of birds 15 years old or more had declined to 101.8 (Brooke, 1978a). He also found that individual shearwaters tended to produce similar-sized eggs in different years. Boersma *et al.* (1980) found that the mean sizes of eggs of *O.furcata* laid during three seasons by birds of unknown ages were virtually identical but that in two of these years, early-laid eggs were significantly bigger than those that were laid later in those seasons, but the possible reasons were not established.

Within a species large females would be expected to lay large eggs, but few figures are available apart from those of Imber (1976b). He found that with *P.macroptera* a high correlation existed between the female's weight and that of her egg, with $n = 60$; $r = 0.47$; $P < 0.001$.

V The shapes of petrel eggs

Petrel eggs are characteristically oval in outline but having one end more rounded than the other. Of a sample of 100 surface-laid eggs of *Puffinus griseus*, collected at random, 33 were oval, 25 long oval, 19 long sub-elliptical, 13 sub-elliptical, 8 long pyriform, 1 pyriform and 1 short sub-elliptical (terminology of Palmer, 1962).

Complex mathematical formulae have been used to represent the profiles of eggs, but a simple Shape Index (S.I.) — Breadth/Length × 100 — provides a convenient measure for intra-ordinal comparisons.

A *Inter-specific comparisons*

1 *Egg shape and body size*

Using standard wing lengths as a measure of body size, Fig. 13.6 shows how the shapes of petrel eggs vary with the sizes of the females that produced them. Plots of shape indices against female body masses give a similar pattern but there are fewer species with adequate data and wing lengths vary much less with time than do body weights — if moulting birds are excluded.

In general, the larger the species the longer the egg (lower S.I.), the longest eggs being those of the great albatrosses, the most spherical ones those of the smallest storm petrels and the diving petrels. The position of these latter is not an artifact of their abnormally short wings; both still retain their extreme positions when body masses are plotted against shape indices. Otherwise the most rounded eggs are laid by *L.brevirostris*. Although the smallest petrel, *Halocyptena*, lays the smallest egg, this is not, on average, the most spherical of petrel eggs.

Birds of particular genera tend to plot together and in *Puffinus* and *Diomedea* the

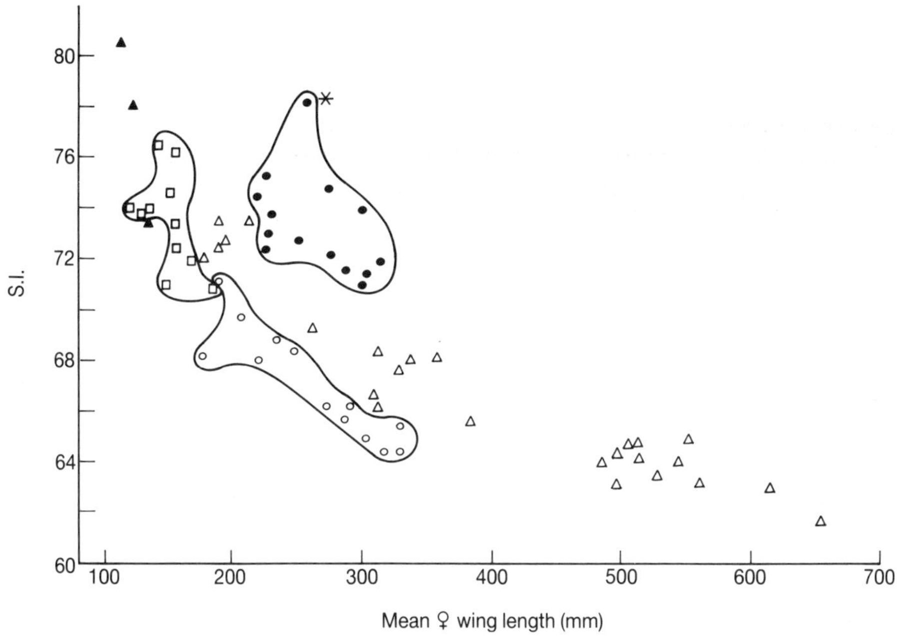

Figure 13.6 *Egg shape indices* v. *standard wing lengths for 66 species of petrel. Solid triangles =* Pelecanoides *(3 spp.); open squares = storm petrels (11 spp.); filled circles =* Pterodroma *(14 spp.) with* L. (P.) brevirostris *starred; open circles =* Puffinus *(13 spp.); open triangles = all other genera. From Warham (1968).*

tendency for egg shape to vary with body size still holds. The greatest scatter occurs among birds with wing lengths between 180–320 mm, due mainly to a marked difference between eggs of *Pterodroma* and *Puffinus*. All 13 of the *Puffinus* plotted lay more elongate eggs (S.I. 64.4–71.1) than do the 14 similar-sized *Pterodroma* (S.I. 71.0–78.2). *Halobaena* plots closer to *Pterodroma* than *Bulweria* does, while the smaller fulmars plot in a graded sequence, their shape indices intermediate between those of the gadfly petrels and the shearwaters. The four *Calonectris* and *Procellaria* also have more spherical eggs than similar-sized shearwaters.

The most marked deviation is the very rounded egg of *Lugensa brevirostris*: with a mean S.I. of between 78.2 and 79.0 based on samples of 27, 19 and 28 eggs measured by different people at different colonies, this bird plots atypically for a *Pterodroma*: nor does it fit the fulmarine shape pattern.

2 *Egg shape and egg size*

Because the larger the petrel the larger its egg, in inter-specific comparisons large eggs tend to be longer than small ones. Plotting S.I.s for 73 species against LB^2 gave a similar pattern to that of Fig. 13.6 with *L.brevirostris* again falling away from the fulmars or the gadfly petrels. These in turn plotted away from shearwaters of similar sizes, while among the 13 species of storm petrels which were adequate data, there was no correlation between size and S.I., all laying rounded eggs (S.I. 71–77).

B *Intra-specific comparisons of egg shape*

1 *Intra-specific variation*

The trends in changes of egg shape between species might also be expected to apply to comparisons between populations of species like *P.puffinus* and *P.assimilis* with many subspecies, but data to test this are poor and any correlations appear weak.

2 *Intra-population variation*

While there is much variation in egg size and shape in general, the amount of variation is about the same in different species. Among large samples of eggs of 12 species, coefficients of variation of the mean shape indices ranged from 2.9 to 6.2, mean 4.79 ± 0.93.

Data to examine whether, within a population, large eggs are longer than small, are few and summarised in Table 13.2. All eggs were measured by the same person at the same colony during the same breeding season. The figures suggest that, within a population, there may be a tendency for large eggs to be more elongate than small ones, but more and better data are needed to confirm this.

Table 13.2 Intra-population relationships between egg size and shape

Species	Mean egg size (mm)	*n*	Mean shape index	Mean LB^2	*r*	*P*
D.b. bulleri	103.16±3.19 X 64.24±1.75	75	62.35	425.9	0.268	0.02
P. inexpectata	60.40±2.16 X 44.09±1.42	42	73.08	117.6	0.166	n.s.
P. carneipes	70.03±2.80 X 46.09±1.80	51	65.88	148.7	0.336	<0.02
P. griseus[a]	76.16±2.83 X 49.26±1.87	100	64.71	185.9	0.671	<0.001

[a] Abandoned surface-laid eggs.

C *Variations in egg lengths and breadths*

Changes of egg shape with changing body size can be examined in terms of egg dimensions. The relationship between egg length or breadth and body size as measured by standard wing length is curvilinear but is converted into a linear one when square roots of the wing lengths are plotted against egg dimensions or by using a log: log transformation. Using data for 53 petrel species, but omitting short-winged *Pelecanoides*, which again fall away from the regression lines, I found that:

$$Y_1 = 6.66\ X^{1/2} - 47.02\ (r = 0.986;\ P<0.001)$$
$$Y_2 = 3.97\ X^{1/2} - 22.45\ (r = 0.987;\ P<0.001)$$

where Y_1 = Mean Egg Length (mm); Y_2 = Mean Egg Breadth (mm); X = Mean Female Wing Length (mm).

In the corresponding logarithmic plots data for 68 species were used for Fig. 13.7 and the least squares regressions, again omitting the diving petrels were:

$$Y_1 = 0.340X^{0.926};\ (r = 0.976;\ P<0.001),\ \text{and}$$
$$Y_2 = 0.458X^{0.807};\ (r = 0.978;\ P<0.001)$$

The slopes of the regression lines differ, that for lengths being the steeper and the greater the distance between the lines at any particular wing length the smaller the S.I. and the more elongate the average egg. (Where the two lines intercept when produced and S.I., = 100, i.e. a sphere, the hypothetical petrel would have a wing about 15 mm long.) Part of the scatter around the regression lines arises from differences in the eggs of similar-sized gadfly petrels and shearwaters already referred to.

This figure gives generalised curves based on data from all genera except *Pelecanoides*. The trend towards a greater increase in length than breadth with

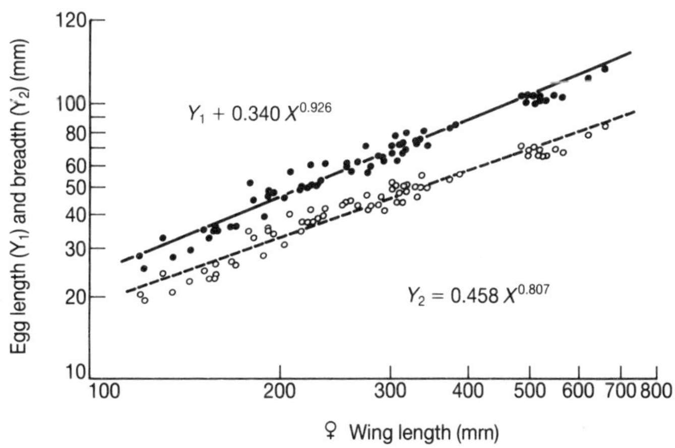

Figure 13.7 *Log: log plots of standard wing lengths against egg lengths (filled circles) and breadths (open circles) for 68 petrel species. Sample sizes 10 or more.*

increasing size is not apparent among the storm petrels which have similar shape indices whatever their size, their lengths tending to change proportionately with their breadths. As with many of these allometric relationships, with better data it may prove that several distinct curves are needed.

D *Factors influencing egg shape*

1 *Age*

Some of the variability in the shapes of eggs within a population will be due to the varying ages or breeding experiences of the layers. Richdale (1952, p.125) showed that the first egg of *D.e.sanfordi* tended to be narrower and lighter than the second which, in turn, was narrower and lighter than the third, but his data were few. Serventy (1967) found that mean egg breadths of *P.tenuirostris* increased up to the fifth year of laying. Young Manx Shearwaters produced smaller eggs but gradually increased their size until they are 10 to 16 years old when their egg volumes declined — sample based on birds 15 years old or more (Brooke, 1978a). He also found that young Manx Shearwaters laid narrower eggs than did older birds, this tendency stabilising at about 10 years. Among a small number of eggs of *D.bulleri* laid 13–22 years apart by the same females, egg lengths decreased from 13 years onwards, breadths staying the same (Warham & Richdale, 1973). The breadths too declined with birds that had laid for 21 years or more, a trend noted in Yellow-eyed Penguins *Megadyptes antipodes* more than 14 years old (Richdale, 1957, p.112) and in domestic hens more than 4 years old (Romanoff & Romanoff, 1949).

Because older birds tend to come back to colonies before the younger ones, and lay earlier, samples of early-laid eggs average broader and with higher S.I.s than later-laid ones, so that the time when the eggs are measured may affect the mean dimensions recorded. For example, Harris (1973) found that 100 eggs of *D.irrorata* from the start of the season had values for LB^2 between 506 and 522, S.I.s 65.7–66.2, whereas the 28 late-laid eggs were significantly smaller (LB^2 of 443–447) and more elongated (S.I. 58.7–62.1). Harris ascribed these effects to age, and the layings of 15 known 5-year-olds supported this, their eggs being significantly narrower ($P<0.001$) and longer ($P<0.05$) than those of older birds. Boersma *et al.* (1980) found that the mean sizes of eggs of *O.furcata* were the same over three seasons but that in two of these seasons early laid ones were significantly longer and wider than those appearing later.

Age effects also appear to be involved in differences between eggs of burrowing petrels laid under ground and on the surface — Section VIII below. Other factors that may link body size with egg shape include anatomical and physiological restrictions on making the egg and anatomical adaptations for specialised modes of locomotion, particularly in the pelvic region.

2 *Egg shape and the mechanics of egg formation*

The oviduct has an inner circular muscle and an outer longitudinal one and their influences on the plastic egg before secretion of the shell could explain most shape variations, but exactly which part of the oviduct is in primary control of final shape is not clear, even in the domestic fowl. Bradfield (1951) suggested that the sphincter-like muscles at the end of the isthmus adjacent to the shell gland would tend to squeeze the egg to a point at its caudal end during the secretion of the egg membranes.

As has often been pointed out, a small egg passing down a loose oviduct might retain the spherical shape of the yolk core, whereas one moving along a tight oviduct would tend to be elongated by the pressure. This may help to explain why eggs of birds breeding for the first time are often narrower than those laid subsequently, as with the Kittiwake *Rissa tridactyla* (Coulson, 1963). If the oviducts of petrels change in a negatively allometric manner with body size, as do many other parameters, then these organs may be relatively smaller and tighter with increasing body size and thus influence egg shape.

3 *Egg shape and pelvis shape*

Nicholsky (1893) suggested that birds laying rounded eggs usually have vertical postures because the weight of their egg counteracts the pressures of the oviduct walls, whereas those that lay oval eggs tend to have a horizontal carriage with the weight of the egg assisting oviducal pressures. Rensch (1947) pointed out that these trends were also associated with the shapes of the pelves, birds that maintain vertical postures and lay spherical eggs also having deep pelves, those that lay elongate eggs having shallower pelves and a more horizontal carriage.

The connection between pelvis shape and posture among petrels is almost the reverse of that suggested by Rensch, for those with the most upright stances, the albatrosses and giant petrels, have rather elongated pelves, many that have a horizontal carriage often having deep pelves that are also wide open caudally. However, the connection between pelvis shape and egg shape among petrels agrees with Rensch's postulate: those petrels with the most spherical eggs also have the deepest pelves.

The pelvic condition of 27 petrels drawn from all four families and mainly from skeletal material is given in Table 13.3 and Fig. 13.8 shows the method of measuring depth and width. Despite much variation and the small samples, pelvis depth and width were well correlated with shape index (r = 0.731 and 0.820 respectively; $P < 0.001$ in both). That is, species laying the most spherical eggs generally have deeper pelves which are more widely spread caudally, those laying elongate eggs shallower ones which are narrower caudally.

Such variations in pelvic anatomy could directly affect the eggs if these are shaped while in the ventral part of the pelvic girdle before the shell is laid down. The hen's egg remains in the shell gland for about 20 h and before that undergoes a 25% increase in volume probably due to uptake of water by the albumen. It is conceivable that final shaping of the petrel egg occurs during or after this swelling when it is pressed up against the curved ventral part of the sacrum and between the pubes. Figure 13.9 shows how the egg of an albatross may fit into the pelvic cavity, being embraced for much of its girth by the flexible arms of the pubis as it rotates about its major axis.

In contrast, for small petrels like that in Fig. 13.9 A, B, D, the egg is so large that it

Table 13.3 Pelvic angles of petrels

Species	*n*	Depth (b/a)	Lateral spread (c/a)	Egg shape index
D.e. epomophora	4	0.46	0.35	62.0
D.c. cauta	4	0.42	0.37	63.3
D. immutabilis	1	0.41	0.33	63.5
M. giganteus	5	0.47	0.41	63.3
F.g. rodgersi	1	0.44	0.41	68.1
P. lessonii	1	0.60	0.56	71.1
P. solandri	1	0.59	0.70	74.2
P. inexpectata	3	0.66	0.46	71.2
L. brevirostris	1	0.73	0.58	79.0
P. turtur	4	0.67	0.60	72.8
P. aequinoctialis	1	0.43	—	65.8
P. westlandica	2	0.46	0.43	66.7
C.d. borealis	1	0.59	0.39	68.2
C. leucomelas	2	0.48	0.45	64.5
P. gravis	1	0.36	0.31	64.4
P. carneipes	2	0.43	0.35	66.2
P.p. chlororhynchos	1	0.43	0.38	66.2
P. griseus	6	0.37	0.30	64.4
P. tenuirostris	6	0.38	0.37	66.2
P.p. puffinus	1	0.37	—	68.8
P.p. mauretanicus	1	0.43	—	69.8
P.a. baroli	1	0.53	0.41	69.4
P.m. maoriana	4	0.75	0.52	72.4
H. pelagicus	1	0.82	0.59	76.3
P.l. leucorhoa	1	0.87	0.50	73.4
P.c. castro	1	0.50	0.42	73.0
P.u. chathamensis	3	0.53	0.58	78.0

is difficult to envisage the pelvic girdle influencing its shape, even given that the posterior parts of the pubic arms are quite flexible and could conceivably exert some pressure on the shaping egg, particularly if the inter-pubic distance increases before laying as it does in the fowl (C.R.Grau, pers. comm.). Resolution of this matter appears to need examination of live birds during the formation of the egg, e.g. by a technique like acoustic tomography, but, as before, a major difficulty remains: the females form their eggs at sea so are unavailable to the researcher.

Shearwaters like *P.gravis*, *tenuirostris*, *griseus* and *puffinus* lay particularly elongate eggs for petrels of their sizes. Kuroda (1954) related the form of the pelvis in shearwaters with their modes of life. Those with long, narrow pelves actively swim and dive for food and have a long post-acetabular ilium for attachment of large muscles needed to power the strong back strokes of the feet. He included all the above except *P.gravis* in this aquatic category. Other shearwaters like *Calonectris* have relatively shorter and broader pelves and short post-acetabular ilia, are highly

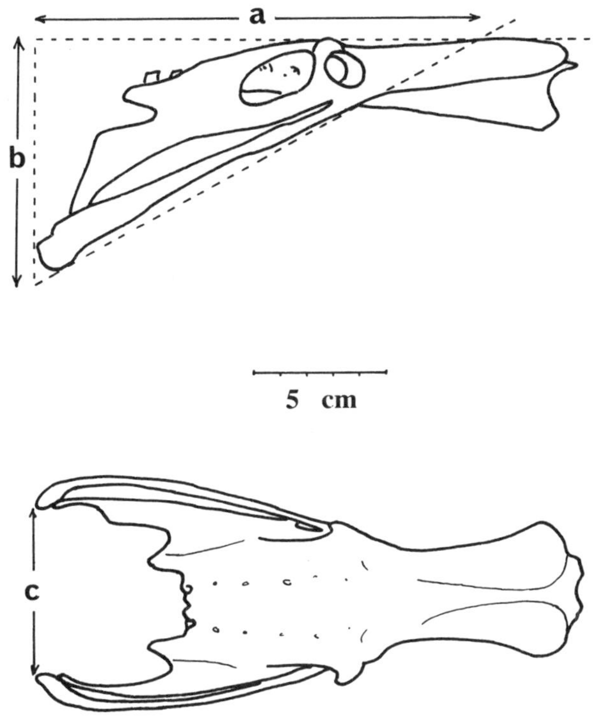

Figure 13.8 *The pelvis of* Diomedea exulans *in lateral and dorsal view showing methods of measuring pelvic depth (b/a) and lateral spread (c/a). From Warham (1968).*

aerial and seldom dive. Kuroda had two intermediate categories, (a) the partially aquatic shearwaters *pacificus* and *gravis,* with long pelves and post-acetabular ilia but better adapted for gliding than the *griseus-tenuirostris* group and (b) a group represented by *P.carneipes* whose pelvis is fairly narrow but whose post-acetabular ilium has proportions intermediate to the aquatic and aerial condition.

Kuroda's classification fit quite well with the egg shape-pelvis shape situation. The most aquatic species, *griseus, tenuirostris* and *puffinus* all have low egg S.I.s in comparison with wing length and pelvis depth and, to a lesser extent, pelvis breadth. The partially aquatic and intermediate species have slightly higher S.I.s and *Calonectris* spp. still higher indices relative to those characters.

Shearwaters apart, relationships between egg shape, pelvis shape and life style are obscure. For example, the diving petrels, efficient underwater swimmers, albeit wing-propelled, have broad, deep pelves and rounded eggs, the albatrosses, birds that rarely submerge, shallow narrow pelves and elongated eggs.

For storm petrels and prions the deep, open pelvis may be associated with the need for wide attachments of the leg and tail muscles, particularly in long-legged forms which fend off from the water with their feet and twist their tails during their erratic dancing actions while snatching food from the restless surface. The deep

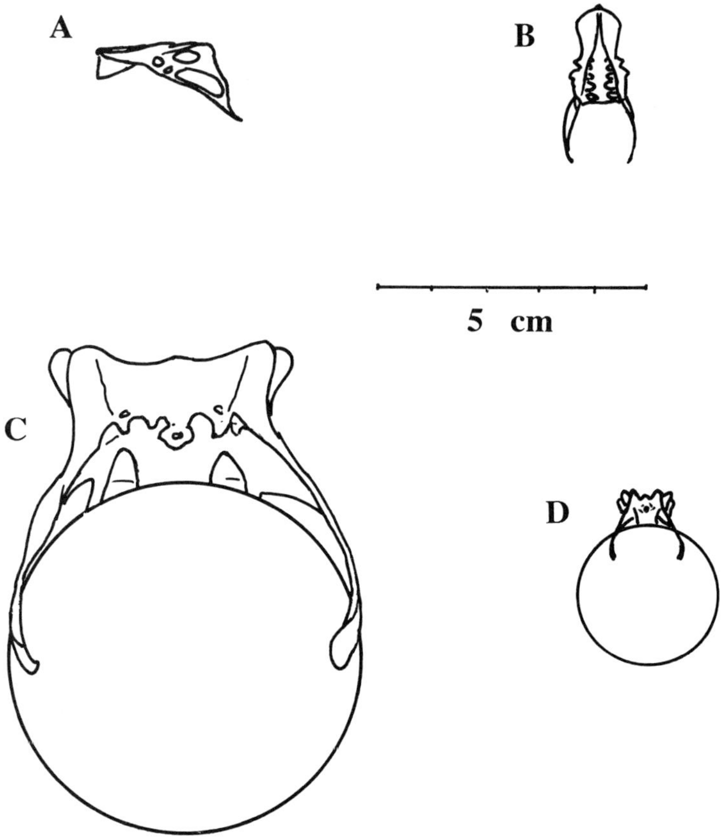

Figure 13.9 *Upper: pelvis of* Pelagodroma marina *in* (**A**) *lateral and* (**B**) *dorsal view. Lower: pelves of* Diomedea cauta *and* P.marina *in caudal view in relation to the breadths of their eggs. From Warham (1968).*

pelvis may also provide a wide attachment for the ventral body wall to help support the female's huge egg.

The position of the Kerguelen Petrel *L.brevirostris* is peculiar. Its S.I. is very high, the pelvis deep and wide spread laterally, and not a consequence of distortion in the dry skeleton as radiographs of recently dead birds show. In these respects it is an extreme example of the gadfly petrel condition associated with surface dipping when feeding, which is also its feeding style (Harper, 1987).

VI The egg shell

A *Structure*

The shells of all petrel eggs, even those of surface nesters, are basically white or cream in colour, so perhaps the ancestral petrel was a burrower. Their texture is usually dull, but in some small species may have a distinct lustre and many albatross eggs have rusty brown, pinkish or purplish marks usually towards the blunt end and these may form a solid band or cap. Those of storm petrels are also usually lightly freckled with brownish spots, again at their broad ends. Such marks can often be removed by washing. In *D.immutabilis*, and probably in other species, they arise from blood spotting from capillary breakdown during laying (Fisher, 1969) or perhaps from the bites of fleas. They often carry the musty smell so typical of tubenoses.

Little research has been done on the structure of petrel eggs. Schmidt (1967a,b) described that of *D.exulans* and Board *et al.* (1977), in their classification of the pore systems of avian egg shells, noted that those of *P.puffinus* had the outer openings of their pores occluded, those of *P.nivea* opening as a network of tubules. Silyn-Roberts & Sharp (1985) found the crystal structure of the shell of the gadfly petrel *P.macroptera* was quite typical of carinate birds.

Tyler (1969) tested the snapping strengths of petrel eggs (species unstated) when snapped inwards and outwards. Snapped inwards eggs of tubenosed birds were the weakest among the non-passerine eggs he tested, but when snapped outwards were as strong as those of loons and geese. His eggs may have included some of fulmars, which, being thick, would be expected to be stronger than those of other petrels.

Considerable attention has been given to the pore geometry of petrel egg shells mostly in the course of work on water balance and gas exchange during incubation — see Whittow (1983), Mehlum *et al.* (1986) and Rahn & Whittow (1988).

B *Shell thickness*

In Fig. 13.10 data for 47 species are plotted against mean standard wing lengths (as indices of body size) and in Fig. 13.11 shell thicknesses are regressed against mean egg lengths and breadths. Thickness data here and in Table 13.4 mostly come from Schonwetter (1967).

Omitting the short-winged *Pelecanoides*, the linear relationship between shell thickness and wing length for the 44 species in Fig. 13.11 is

$$Y = 0.009X_1 + 0.021 \ (r = 0.961; P<0.001)$$

and, for the 47 species of Fig. 13.12 (now including three *Pelecanoides* spp.):

$$Y = 0.005X_2 - 0.019; \ (r = 0.979; P < 0.001)$$
$$Y = 0.008X_3 - 0.071; \ (r = 0.973; P < 0.001)$$

where Y = Shell Thickness (mm); X_1 = Female Wing Length (cm); X_2 = Mean Egg Length (mm) and X_3 = Mean Egg Breadth (mm).

Shell thickness increases with increasing body and egg size, that of the 28 g storm petrel *Hydrobates pelagicus* being 0.12 mm, of the 406 g *P.puffinus* 0.28 mm and of the 7.2 kg albatross *D.exulans* about 0.58 mm. The increases are not uniform, however:

1. Seven of the *Puffinus* plot above the trend lines in both graphs whereas six of the eight *Pterodroma* lie below it. That is, the shearwaters tend to have rather

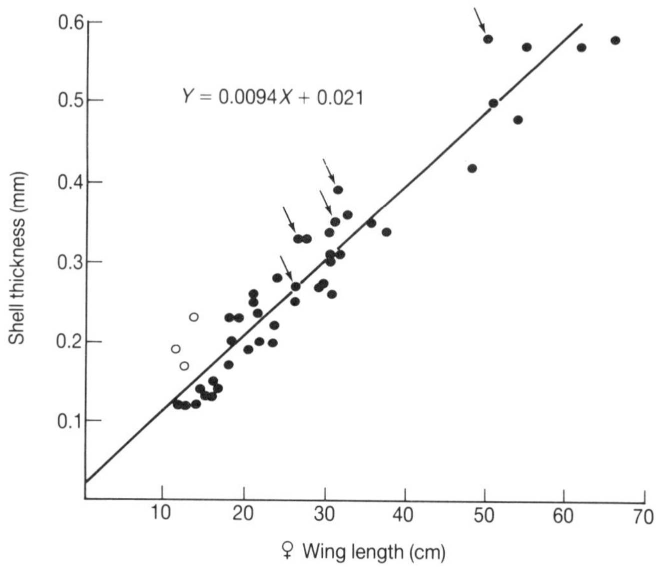

Figure 13.10 *Shell thicknesses and standard wing lengths for 47 spp. of petrels. The three* Pelecanoides *(open circles) were excluded when calculating regression equation. Plots for fulmars are arrowed. From Warham (1968).*

thicker shells than other petrels of similar wing lengths whilst the gadfly petrels have slightly thinner ones. *Puffinus* egg shells average 7.3% of the initial weights of their eggs (6.9–7.6%), for those of *Pterodroma* the mean was 6.2% (range 5.9–7.1%).

Whittow *et al.* (1984) suggested that the shells of *P.phaeopygia* eggs were abnormally thin as a consequence of the altitude of the nesting site — 3000 m. However, at 0.26 mm these would fall in line with similar-sized *Pterodroma* in Fig. 13.10, so that any altitudinal effect may be small. At 0.23 mm the shell of *L.brevirostris* is also rather thin for an egg of its dimensions: it plots well below the regression line of Fig. 13.10 rather than aligning with the fulmars.

Thicker shells may well be stronger but as both groups tend to lay on soil, shearwaters mainly underground, it is difficult to see how the differences are adaptive. Perhaps shearwater eggs experience more pressure in the shell gland and so stimulate greater deposition of calcareous materials. Bourne (in *litt.*) pointed out that a developing egg might be influenced by increased pressure generated when shearwaters swim, the gadfly petrels being more aerial.

2. Three of the fulmars — *M.giganteus, F.glacialis* and *D.capense* have unusually thick shells plotting above the regression line. This is borne out by their shell weights which, for five species, averages 8.2% of their mean egg weights (7.2–9.3%). The one available figure for *Thalassoica* is similar but *Pagodroma* is different, not being thick-shelled. *Macronectes* has the thickest of all petrel shells. As the strength of an egg shell is proportional to thickness^2 (Ar *et al.*, 1979) this could represent an adaptation for surface and ledge nesting on the hard subtrates typically chosen by fulmars. Data for *Pagodroma* need checking, this bird laying on

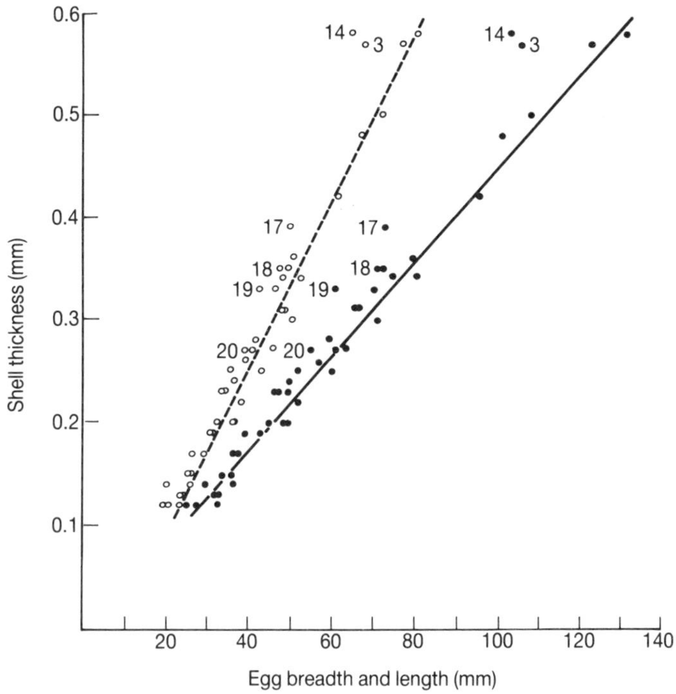

Figure 13.11 *Shell thicknesses and egg breadths (upper graph) and egg lengths. Points 14, 17, 18 & 20 are fulmars, point 3* D.irrorata. *From Warham (1968).*

a rocky bed, and more information for *Thalassoica* would also be useful. Significantly, the other petrel laying a particularly thick-shelled egg is *D.irrorata*, shell about 8.9% of egg mass, and this again is a bird that not only lays on hard laval ground but shifts its egg around, and breaks some in the process.

Breakage also occurs with other petrels, e.g. with albatrosses: those that somehow rest a foot on the egg and then transfer their weight to that foot, break it. Such accidents seem to occur during disputes and probably involve mainly young, naive birds. Thicker shells should also give better insulation against heat loss and Belopol'ski (1961) considered that the thick calcium layer in the shell of the Northern Fulmar was an adaptation to breeding in sub-zero temperatures.

3. The points for three *Pelecanoides* in Fig. 13.10 fall away from the regression line due to their short wings, but prove quite typical of petrels of their sizes when scaled against equatorial egg dimensions in Fig. 13.11.

Paganelli *et al.* (1974) found that for seabirds as a whole, shell mass increased more rapidly than egg mass with increasing size, a trend masked in Procellariiformes by the relatively thicker and heavier shells of the fulmars. Using Schonwetter's figures Rahn *et al.* (1984) calculated for 88 tubenoses that:-

$$Y = 0.055X^{0.40}; \ (r = 0.96)$$

where Y = Shell Thickness (mm) and X = Egg Mass (g). My measurements of 34

Table 13.4 The composition of petrel eggs

Species	Shell thickness (mm)	*n*	Fresh wt	Shell		White		Yolk	
			g	g	%	g	%	g	%
DIOMEDEIDAE									
Diomedea exulans	0.57	1	410.0	35.0	8.5	274.5	67.0	100.5	24.5
D. epomophora	0.58	1	440.5	31.9	7.2	301.5	68.5	107.1	24.3
D. nigripes	0.53	8	299.0 ± 14.64	24.6 ± 2.65	8.2	185.5 ± 10.72	62.1	88.9 ± 5.47	29.7
D. immutabilis	0.49	16	279.9 ± 12.50	22.6 ± 1.21	8.1	172.4 ± 10.56	61.6	84.9 ± 4.76	30.3
D. melanophrys	0.50	9–15	256.0 ± 12.91	20.7 ± 1.32	8.1	167.1 ± 8.7	65.2	68.4 ± 4.9	26.7
D. chrysostoma	0.52	13–15	288.9 ± 17.20	22.5 ± 1.85	7.7	188.3 ± 13.3	64.8	79.6 ± 4.0	27.4
Phoebetria fusca	0.48	3	241.9	22.0	9.1	152.4	63.0	67.5	27.9
P. palpebrata	0.48	1	254.3	25.2	9.9	151.8	59.7	77.3	30.4
D. bulleri	0.47	3	247.5	18.2	7.4	160.8	64.9	68.5	27.7
PROCELLARIIDAE									
Macronectes giganteus	0.58	11	253.2	30.7	12.1	140.7	55.6	81.6	32.2
M. halli	0.55	3	253.3	30.9	12.2	136.0	53.7	86.4	34.1
Fulmarus glacialoides		9	103.4	11.2	10.9	61.6	59.6	30.5	29.5
F. glacialis	0.39	4	97.0 ± 4.33	9.6 ± 0.29	9.9	58.7 ± 3.40	60.6	28.7 ± 2.69	29.5

Daption capense	0.33	10	67.3	7.6	11.2	38.5	57.0	21.4	31.8
Pagodroma nivea	0.27	6	56.8	5.5	9.7	31.9	56.2	19.4	34.1
Pterodroma macroptera	0.31	2	78.9	7.3	9.3	41.5	52.6	30.1	38.1
P. solandri	0.31	2	78.3	5.4	6.9	47.4	60.5	25.5	32.6
P. mollis	0.20	1	49.3	5.6	11.4	26.6	54.0	17.1	34.7
P. neglecta	0.27	3	68.7	5.2	7.5	38.5	56.2	25.0	36.3
P. hypoleuca	0.20	4	41.7± 3.09	2.6±0.23	6.3	23.4±1.93	56.1	15.7±1.00	37.6
P. nigripennis	0.20	1	42.6	2.4	5.6	23.6	55.4	16.6	39.0
Pachyptila vittata	0.24	1	32.3	3.2	9.9	16.7	51.7	12.4	38.4
P. turtur	0.20	10	25.1	1.7	6.8	14.0	55.8	9.4	37.4
Procellaria aequinoctialis	0.34	1	125.2	12.4	9.9	63.4	50.6	49.5	39.5
Puffinus carneipes	0.34	12	73.9± 7.83	6.0±0.58	8.1	43.4±4.76	58.7	24.5±2.55	33.2
P. pacificus	0.27	3	76.9	6.1	7.9	45.7	59.5	25.1	32.6
P. bulleri		6	66.4± 2.55	5.2±0.60	7.8	37.8±2.14	57.0	23.4±0.72	35.2
P. gravis	0.36	2	107.2	7.8	7.3	61.3	57.2	38.1	35.5
P. griseus	0.34	15	105.0± 7.27	9.2±0.73	8.7	62.5±5.95	59.5	33.4±1.94	31.8
P. tenuirostris	0.33	1	80.0	8.5	10.6	43.5	54.5	28.0	35.0
P. lherminieri	0.25	6	25.7± 2.02	2.0±0.32	7.8	14.8± 1.60	57.5	8.9±0.74	34.7
HYDROBATIDAE									
Pelagodroma marina	0.15	2	13.4	0.8	6.0	7.7	54.7	4.9	36.6
Oceanodroma leucorhoa	0.13	20	10.5± 0.7	0.84±0.09	8.2	5.45±0.54	53.1	3.9±0.4	37.4

data sets for petrels whose egg weights came from field measurements of samples of at least 10 eggs and matched precisely for subspecies, gave virtually the same relationship between shell thickness and shell mass viz:

$$Y = 0.048X^{0.424};\ (r = 0.98;\ P<0.001)$$

In general, both relative egg mass and relative thickness of petrel shells are lower than those of other non-passerine eggs (Rahn *et al.*, 1984).

VII Egg composition

The proportions of albumen, yolk and shell in petrel eggs vary in a regular manner with their size, large eggs having proportionately more albumen and less yolk than small ones, while the proportion of mass due to shell increases only slightly with increasing egg size (Fig. 13.12, Table 13.4). For at least three species similar relationships obtain intraspecifically. Figure 13.13 shows these for an albatross and a shearwater: the pattern with a sample of eggs of *P.carneipes* was similar (Warham, unpubl.).

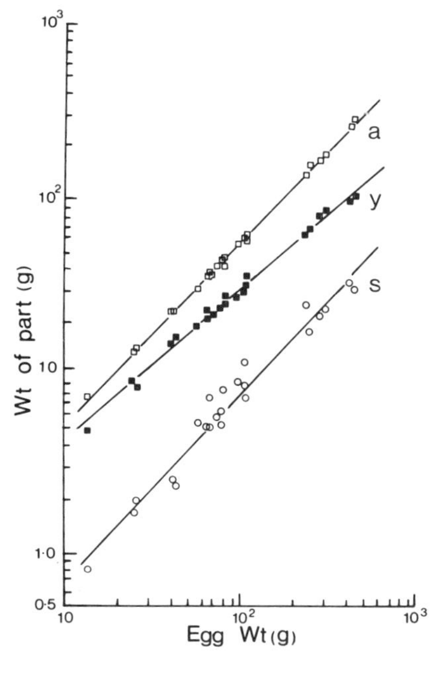

Figure 13.12 *Logarithmic plots for the constituents of eggs of 23 petrel species from data in Table 13.4. a = albumen, y = yolk, s = shell. Least square regressions: albumen mass = 0.493 egg mass$^{1.043}$; r = 0.999; yolk mass = 0.747 egg mass$^{0.878}$ r = 0.996; shell mass = 0.058 egg mass$^{1.075}$; r = 0.983; P<0.001 in all cases. From Warham (1983).*

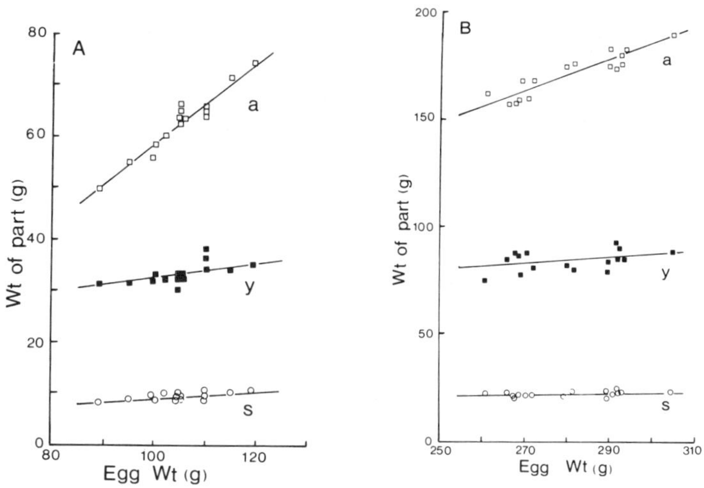

Figure 13.13 *Components of 15 eggs of Sooty Shearwaters* (**A**) *and 16 of Laysan Albatrosses* (**B**). *Symbols as in Figure 13.12. Equations for regression lines in* (**A**):
albumen weight = *0.79 egg weight* − *20.39;* r = *0.964;* P <*0.001*
yolk weight = *0.15 egg weight* + *17.83;* r = *0.553;* P<*0.05*
shell weight = *0.06 egg weight* + *2.56;* r = *0.628;* P<*0.01*
and in (**B**):
albumen weight = *0.78 egg weight* − *45.49;* r = *0.918;* P <*0.001*
yolk weight = *0.18 egg weight* + *35.55;* r = *0.462;* P <*0.1*
shell weight = *0.05 egg weight* + *9.49;* r = *0.485;* P <*0.1. From Warham (1983).*

The proportions of the three egg components of Fig. 13.12 varied consistently. For example, the four fulmars included did not have abnormally small yolks despite having the fastest development rates both as embryos and as chicks (Warham, 1968, 1983), so that their short incubation periods do not result from small-yolked eggs which might hatch sooner. The 'quality' or energy content of their yolks may, of course, be different.

Nice (1962, p.26) speculated that the big yolks of penguin and fulmar eggs might be adaptations to extreme cold if they were not typical of these groups throughout their range. They are indeed typical of petrel eggs, with no suggestion that the component variations are correlated with climate or latitude (Warham, 1983; Williams *et al.*, 1982) and Table 13.4

There have been several recent discussions on the place of tubenosed birds in the altricial-precocial neonatal continuum. Partly because of their yolky eggs, Nice (1962) and Ricklefs *et al.* (1980) suggested that petrels should be classified as semi-precocial. Boersma (1982) maintained that they were truly precocial. Williams *et al.* (1982) however, listed Procellariiformes as semi-altricial. Boersma pointed out that *O.furcata* is totally down-covered at birth, thermoregulates early and is only irregularly and briefly attended by parents. This is typical of burrowing petrels with the exception of *Pelecanoides*.

When compared with those for altricial birds of similar sizes presented by Ar &

Yom-Tov (1978) the data suggest that (1) petrel eggs have larger yolks and smaller whites than eggs of altricial birds of similar sizes, (2) petrel eggs weighing less than about 25 g have less albumen than similar-sized precocial eggs, whereas those weighing more than 25 g have more albumen than the precocials of Ar & Yom-Tov's curves, and (3) petrel eggs weighing under 55g have larger yolks than similar sized eggs of precocial birds, so that, as far as yolk goes, small petrels are highly precocial. Thus on the basis of egg constituents, petrels neither fall conveniently into any of Ar & Yom-Tov's categories, nor into the more finely graded ones of Nice. They seem to form a special category of their own (Warham, 1983). Southerland & Rahn (1987) also take this position; in petrels there is no simple relationship between relative yolk content and hatching maturity.

Rahn *et al.* (1984) plotted yolk as a % of egg content (not yolk mass) for 32 species of Procellariiformes against egg mass:

$$\% \text{ Yolk} = 40.6 - 0.028 \text{ Egg Mass (g)} \ (r = 0.80)$$

and commented that despite their yolky eggs, the precocity of petrel hatchlings was hardly comparable to that of the precocial species of Nice.

Analyses of lipid, water and energy contents of petrel eggs have been made by Pettit *et al.* (1984), Williams *et al.* (1982), Rahn *et al.* (1984) and Sotherland & Rahn (1987). The latter also summarised available data on the water content of the albumen (0.85–0.88%), of the egg contents (71–78%) and of lipid in egg contents (7–12%), for 16 species of petrels. They reasoned that as the water in the fresh egg is similar to that in the hatchling — water lost through the shell being balanced by that generated by the embryo's metabolism — the initial figure for water content of the egg may be better guide for assessing maturity than relative yolk contents.

These authors also pointed out the value of large yolks to pelagic or offshore-feeding birds, and regarded all petrels as belonging to that category, although some such as *O.homochroa* and *P.urinatrix* are mainly inshore feeders. No comparable data are available for these.

Because both intra- and inter-specifically, the proportions of their constituents depend on their sizes, it is necessary when comparing the egg components of different species either to use eggs of the same mass or correct for differences in their masses.

VIII 'Surface' eggs

These are eggs laid in the open by burrowing petrels (as opposed to eggs on the surface that have been raked out from burrows during ownership contests). Such laying is widespread, and particularly noticeable where petrels nest at high densities. Its occurrence in *P.tenuirostris* was noted long ago (Davies, 1846) and these eggs were once collected in Australia for local consumption and cake making, as were those of *P.gravis* by the Tristan Islanders (Rowan, 1965).

Richdale (1963, p.18) suggested that the large numbers of Sooty Shearwater eggs on the ground might have been the work of young birds not yet ready to breed and that a test of this hypothesis would be to compare their measurements with those of burrow-laid eggs. Serventy (1967) did that for Short-tailed Shearwaters, finding that they averaged significantly narrower (46.5 mm) than those laid in burrows (47.1 mm). As he had already established that egg width increased with the female's age, he concluded that the surface eggs were the product of young birds, though not solely by those in their first breeding year. Warham *et al.* (1982) compared samples of Sooty Shearwater surface and burrow eggs over two seasons, finding that in one the mean sizes did not differ but in the next season surface eggs were narrower (48.8 ± 1.7 mm; n = 81) than burrow-laid ones (50.0 ± 1.3 mm; n = 77; P = <0.001). They were also significantly smaller. The difference between seasons could have been due to more older birds laying on the surface and/or that food supplies were better during the earlier season. Similarly, surface eggs of *P.bulleri* and *P.externa* were significantly narrower than burrow-laid ones (Harper, 1983; Brooke, 1987).

Harvesters of 'mutton-birds' in the last century believed that eggs were laid on the ground because there were not enough burrows to go round or that the females could not reach them in time. Richdale (1963) dismissed these ideas as there were plenty of disused burrows available at his study colony of *P.griseus* where surface eggs nonetheless appeared. This may have been true of that small colony, but at other flourishing sites burrow occupancy seems high and very probably some surface eggs are produced by females lacking burrows, e.g. in *P.pacificus* (Swanson & Merritt, 1974). Others may be forced to lay willy-nilly when their nests are unavailable, e.g. Antarctic Prions excluded by snow blockage (Tickell, 1962). At Nightingale Island, where *P.gravis* also lays on the surface, there is said to be severe competition for burrows (Rowan, 1952) but at Inaccesible Island where nests are more dispersed surface eggs are evidently not seen (Rowan, 1965; Elliott, 1970b). However, at a Bruny Island, Tasmania colony, surface eggs of *P.tenuirostris* were laid in 1979 despite only about half the burrows being occupied (Naarding, 1980).

Warham *et al.* (1982) found a highly significant difference in mean dates of laying of *P.griseus* surface and burrow eggs — the latter peaking about 3 days earlier. Byrd *et al.* (1983) also noted that most surface eggs of *P.pacificus* appeared after burrow breeders had laid. These findings again suggest that surface egg-layers are mostly young birds. On the other hand Naarding (1980) reported that the number of surface eggs of *P.tenuirostris* peaked at the same time as did those laid in burrows.

The extent of surface-laying has seldom been measured. Warham (1960) collected those laid on an area of 2500 m^2 estimated to hold 2200 burrows of *P.tenuirostris* and calculated that 1150 such eggs were laid over the colony as a whole. At the Snares Islands the number of *P.griseus* surface eggs did not appear to depend on nest densities in the immediate vicinity — 1.84 per 100 m^2 in a low density area and 1.92 per 100 m^2 in a high density one; but none was found where the soil was waterlogged or shallow and thus unburrowed. Some 27 000 surface eggs were estimated to have been laid on 147 ha of forested ground (Warham *et al.*, 1982).

Burrowing petrels have 'good' and 'bad' seasons, often from unknown causes, and the incidence of surface eggs varies from year to year. In some seasons large numbers of these eggs, locally known as 'gluts', are laid on the islands of Bass Strait.

Jones (1936b), noted that many surface eggs of *P.tenuirostris* were addled and Rowan (1952) that none of those of *P.gravis* showed any signs of incubation. As they all would have been laid overnight, earlier ones devoured by gulls and skuas, they would not be expected to show overt signs of development and certainly were not addled.

Naarding (1980) kept 48 Short-tailed Shearwater surface eggs in an incubator for 21 days and switched 12 surface eggs under birds in nests, transferring the burrow-laid ones to the incubator. These latter showed signs of development and eventually hatched after being returned to their owners, but none of the surface ones ever developed under natural or artificial incubation. This finding is surprising because some earlier trials had suggested that some surface *P.griseus* eggs were fertile. Further work seems to be needed.

Fertile or not, surface eggs are not hatched but abandoned on the morning of the night when they are laid. Even *P.pacificus*, which often lays on the surface, does not usually persist in covering eggs in the open for more than a few days. One Sooty Shearwater that covered a surface-laid egg under a forest canopy in daylight soon desisted.

The white eggs are highly visible in the open and are eagerly sought after by a range of birds — gulls, skuas, crows, and even by a parrot: the Kea *Nestor notabilis* eats the eggs of Hutton's Shearwater (Harrow, 1976).

IX Egg fertility

There is little information on the fertility rates of petrel eggs that are not abandoned and Fisher (1971, p.32) pointed out some of the difficulties in assessing the effects on productivity. He found that the proportion of infertile eggs of *D.immutabilis* varied widely between years. In 1964–1965 4.5% of 72 eggs laid by young birds and 9.8% of 135 laid by experienced females were infertile. In another year 3.4% of 75 examined were infertile. Fisher regarded these as minimal figures as some of the eggs scored as lost through other causes were infertile; also such eggs broke more easily than fertile ones and so were missed. In one season 11% of 64 eggs broken by the birds were infertile. Others scored as infertile might have had early embryos that died — a class that Fisher believed amounted to less than 1% of incubated eggs.

Other figures for losses attributed to infertility include 9.8% of 61 laid by *P.fusca* (Berruti, 1979), 3.6% of 166 laid by *P.turtur* (Richdale, 1965b), 1.8% of 56 laid by *P.marina* (Richdale, 1965a) and 13% of 89 laid by *O.furcata* (Boersma *et al.*, 1980).

Breeding failure due to egg infertility usually only becomes apparent when eggs fail to hatch and are found to be addled. These may or may not have been infertile: their embryos could have died due to various unrecorded causes. Mougin (1975, p.135) gave figures for Procellariidae of from 0–20% for egg 'mortality' as due to putrefaction, but others that he scored as 'abandoned' could also have been infertile.

CHAPTER 14
Incubation

Incubating Great-winged Petrel.
Photo: author.

Contents

I Introduction

Development of the petrel chick within and without the egg is slow and for incubation to succeed proper coordination of the actions of each parent is essential. Both develop brood patches and both incubate, and their general patterns of behaviour are similar throughout the order.

Because of the long time taken and the single-egg clutch, the energetics of incubation have been looked at in some detail — for reviews see Croxall (1982b); Grant (1984); Pettit *et al.* (1984a) and Ricklefs *et al.* (1986).

II The brood patch

All petrels of breeding age develop brood patches during the nesting season whether they breed or not, but with few species has the development and feathering been followed closely, the most complete being the account of Fisher (1971) on Laysan Albatrosses of known age. That of Beck & Brown (1972) of Wilson's Storm Petrel is also quite detailed but their non-breeders were of unknown ages. Other accounts refer to *O.leucorhoa* (S.W. Harris, 1974; Ainley *et al.*, 1975), *O.furcata* (Boersma *et al.*, 1980), *O.castro* (Harris, 1969b) and *P.l.pycrofti* (Bartle, 1968; Dunnet, 1985).

The brood patch extends from the rear of the sternum towards the cloaca, its shape and size quite closely matching those of the egg and with a tendency for the caudal boundary to be the rounder. The patch is down-covered and invested by thick contour feathers most of the year, the down being gradually shed, a narrow midline being the last to disappear. This may persist in some birds on eggs so that at first sight they appear to have two brood patches (Warham *et al.*, 1977). The egg is invested among the feathers encircling the brood patch and very well insulated from ambient conditions. In albatrosses and fulmars the egg is tucked so firmly into place that it may be lifted when the bird rises. The ability of Galapagos Albatrosses to shift around with the egg held in their pouch has already been noted and when ledge-nesting fulmars are suddenly startled they may carry their egg into the air and lose it on the rocks below.

The skin of a de-feathered brood patch is smooth and pink, its vascularisation often not apparent and lacking knotted or visible blood vessels. It forms a very well insulated container for the egg.

Development of the patch depends on the age and status of the bird, de-feathering taking place progressively earlier until the breeding condition is attained. The Laysan Albatross appears to be typical, bearing in mind that this usually starts to breed when 8 or 9 years old. Birds less than 4 years old did not develop brood patches. Some, but not all of the 4- and 5-year-olds showed signs of de-feathering from the time when chicks were hatching in late February. Among 6-year-old birds maximum development was in late January or in February, re-feathering in March and April. Paired 7- and 8-year-olds (potential breeders the next year) has well developed patches by December, those of the males being the more advanced. Re-feathering began quite soon for this group, few retaining brood patches after February.

In some breeding albatrosses down had begun to be shed before their return and in all was completed by the start of laying around 20 November, in the males about a week before their partners — in readiness for the first long incubation stint. The timing of re-feathering varied: some nesting birds showed this in mid-January while still on the egg, and most had begun re-growing down in February, only a few having bare patches after 1 March. Birds that lost eggs followed the same re-feathering sequence as successful ones. Seventy per cent of the males and 45% of the females that had formerly bred, but which were not currently doing so, developed naked patches.

Laysan Albatrosses did not have completely de-feathered patches until 7–10 days before laying, but in several small species down loss may begin weeks before the egg appears — in *O.oceanicus* about 3 weeks before, in *O.leucorhoa* about 4 weeks before, and some *O.castro* had bare patches 6 weeks before the lay (Harris, 1969b), but the timing was very variable in that species.

Thoresen (1969) found that re-feathering with *P.urinatrix* began at hatching time and was complete in 35–40 days. In *O.leucorhoa* it began 2–10 days before the hatch (Ainley *et al.*, 1975) and was completed in 3 to 4 weeks. This seems to be the usual pattern among small petrels, non-breeders and those that have lost their egg starting down renewal about a month after the peak of laying, maybe a month before the successful birds do so.

Beck & Brown (1972) found no sexual differences in the brood patch cycle of *O.oceanicus* which also had bare patches 3 weeks before the egg appeared, re-growth of the down being quite slow and starting with a midline, some birds still having naked patches 60 days after laying and even when their chicks were 3 weeks from fledgling. As these authors pointed out, it seems strange that birds that normally only brood the chick for its first 3 days of life should retain a bare patch representing a considerable proportion of the body surface, the more so in a very cold climate creating special thermoregulatory problems for this smallest of Antarctic birds. Similarly, in Alaska, *O.furcata* also retains a de-feathered brood patch longer than *O.leucorhoa* does in that cold climate (Boersma, pers. comm).

Wilson's Petrels that lost eggs started re-growth of down after hatching time. Non-breeders, mostly with feathered patches on arrival, took about 30 days to shed down, but with much variation, so that Beck & Brown considered the state of the brood patch of little use as a guide to age or status.

Hormonal control of brood patch development has not been studied in tubenoses, but Hector & Goldsmith (1985) found experimentally that prolactin was not secreted in response to tactile stimulation of the brood patch by the egg, at least in *D.melanophrys* and *D.chrysostoma*. For vascular control and anatomy of the brood patch in *M.giganteus* see Tirrell & Murrish (1979a,b).

III Incubation behaviour

A *General*

Petrels sit tightly and, except in the event of intermittent incubation and ventilating actions (see below), the egg is uncovered only during change-overs and then only briefly. Counters at five nests showed that incubating *P.tenuirostris* never emerged during their 14-day stints (Naarding, 1981) and Simons (1985) also found that incubating *P.phaeopygia* seldom left to go to the nest entrance. Bonin Petrels in their burrows appear to uncover their egg twice each night, not at all by day (Grant *et al.*, 1982a). Most surface nesters cannot leave eggs uncovered without loss to aerial predators.

North Pacific albatrosses sitting in the open on hot tropical atolls rise to their feet from time to time as if ventilating their egg, but how much this is a response to possible overheating and how much to the usual comfort movements seen among all surface-nesting petrels which typically rise, ruffle their feathers and shake their wings, bend down to touch the eggs and 'address' it with a call, before quickly resuming incubation, is unclear. Among albatrosses such actions often appear to be nervous responses to the proximity of man. Whether surface-nesting petrels of the tropics show the tightness of sit typical of cold-climate species is uncertain, but surface-nesting *P.pacificus* left their eggs when exposed to the sun and black-bulb temperatures above 45°C (Howell & Bartholomew, 1961b).

In *D.nigripes* and *D.immutabilis* nest-ventilating behaviour occurred 0.77 and 1.30 h^{-1} respectively and these breaks showed up as dips in graphs of brood patch temperatures (Grant *et al.*, 1982b). There was no tendency to ventilate more by day than by night but these authors thought that the air seal between the bird and its nest bowl was tighter in *immutabilis* which might have been a factor in its higher ventilation rate.

Although the brood patch is shaped to match the egg, birds will sit with either pointed or blunt end caudal. Sometimes the egg lies crosswise but then the bird is restless and eventually re-settles with the egg's long axis in line with the body. A Black-footed Albatross on two eggs had one with pointed end forwards, the other backwards. Petrels do not appear to turn their eggs deliberately even during the early stages of incubation when premature adhesion of the extra-embryonic membranes might occur, but this aspect needs further research and Lockley (1930) stated that a Manx Shearwater turned its egg and altered its position at least once a day.

Tubenoses sitting in burrows may face away from or towards the entrance tunnel — chambers tend to be oval so crosswise sitting is uncommon. Wheelwright & Boersma (1979) provided statistical evidence for a preferred orientation: most *O.furcata* sat with their heads away from the entrance (Fig. 14.1) which these authors considered might be a means of reducing heat loss. In contrast, *F.tropica*, also a petrel of cold climates, invariably faced the entrance (Beck & Brown, 1971). Yet again, in a warm climate *P.pacificus*, faced inward by day, outward by night (Shallenberger, 1973). In my experience burrowing petrels after dark normally face the entrance, perhaps stimulated by the night-time chorus and movements overhead, being ready to deter an intruder or to greet a partner. Burrowing species appear to experience less disturbance than do surface nesters because these tend to be alerted by passers-by by day and by night. Surface nesters turn to face strong winds, rain, snow, sleet, hail or driving sand.

Incubating petrels spend much time sleeping. This may be quite deep but is generally shallow and even with the quietest approach it is difficult to get up to a sleeping albatross without awakening it. Simon's nest recorder showed that *P.phaeopygia* spend 95% of the time sleeping. The respiration rate of alert resting birds was 23.7 ± 1.14 breaths min^{-1} (n = 29) but fell to only 12.3 ± 1.20 min^{-1} (n = 60) in sleeping birds, $P < 0.0001$ (Simons, 1985). The sleeper's head may be tucked into the scapulars, its feet into the breast feathers or the head may be erect and the eyes covered by the nictitating membranes. Sleeping mollymawks often droop their wings over the sides of their nest columns.

The drive to incubate may be strong in non-breeders which may adopt uncovered eggs. Such birds may be adult non-breeders, immatures and occasionally even mobile chicks, and they seldom persist for long. Fisher (1971, p.60) placed a Laysan Albatross egg in a nest which he 'built': in 5 hours six different birds sat upon it. Adopted eggs of Galapagos Albatrosses may be hatched

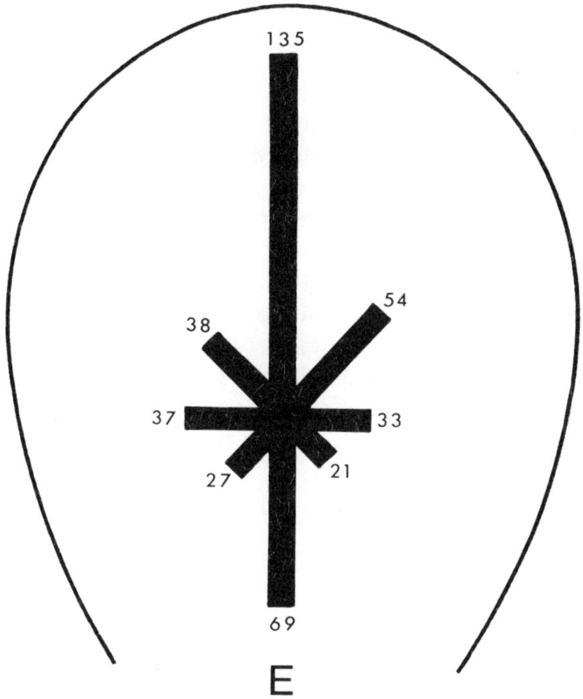

Figure 14.1 *Orientation of Fork-tailed Storm Petrel in respect to burrow entrance (E). Bar lengths are proportional to the number of times the bird was found facing in each direction. From Wheelwright & Boersma (1979). © 1979 by the University of Chicago.*

(Harris, 1973) and he pointed out that egg adoption could explain some of the hybrid Laysan × Black-footed Albatrosses, which as chicks, could become imprinted on one of the foster parents.

B *Change-over on the egg*

Petrels very seldom return to an incubating partner without relieving it. Diurnally active birds like fulmars and albatrosses seem to recognise their mates at some distance as they may suddenly call as the newcomer approaches. The primary focus of the returning bird, however, appears to be the nest, and to this is goes whether the nest be empty, occupied by a stranger or by its own partner, but for *D.irrorata* mate recognition seems to be essential as mate may have shifted and change-over must occur wherever the egg may be. If the partner is missing the relieving bird may take over an abandoned egg.

At nest relief the sitting bird may rise to reveal the egg, and, with mollymawks, either takes over at once, or after some circling of the nest and mutual preening. Straddling the nest, the newcomer looks at the egg and usually utters a series of 'eh -eh' or similar calls. Relieved petrels generally soon leave and many change-overs take place without any overt ceremony, e.g. in giant petrels (Warham, 1962). Brown (1966) described nest relief among Snow Petrels involving mutual calling with the sitting bird quietly relinquishing its place to the other accompanied by low clucking noises. Nest relief of the *F.glacialis* is also accompanied by low sounds (Fisher, 1952).

Change-overs are often infrequent and rarely seen, particularly among gadfly petrels and other burrowers, but the smaller species change more often (Table 14.1) and *O.furcata* spent on average 32.8 ± 38.4 min together (5–235 min, n = 43), resting with occasional bouts of mutual preening and calling (Simons, 1981b), longer than most petrels spend together at nest relief.

There is no courtship feeding among petrels. Some early reports of food being passed between Northern Fulmars may have been occasioned by the dripping of stomach oil or salt gland excretion from the beak; or well-feathered chicks have been mistaken for adults.

C *The shift system*

1 *The shares of the sexes*

As Lockley (1942) put it, during incubation one bird fasts while the other feasts. After laying the female usually departs within a day or two. With nocturnally active species she may leave on the night of the lay or stay until the following night and then go in the dark. Delayed departures by females are usually a consequence of the absence of their mates. Such short stints apart, the male is responsible for the first long spell on the egg (perhaps partly to defend the site at the start of the breeding effort), and a sequence of turn and turn about continues to hatching (Fig. 14.2). In total the two sexes spend about the same amount of time incubating, their shift lengths tend to be similar and to decrease towards hatching so that the on-duty bird has food for the chick. Available information is summarised in Table 14.1.

With some species the males incubate more. Male Laysan Albatrosses averaged a total of 36 days on the egg, the females 29 days (Fisher, 1971, p.68). Fisher also found that the major variations in span lengths occurred in the first and second spans, the birds tending to adjust the later ones so that the total time spent by each sex was close to the average. Other species in which the males perform more than half the incubation are *P.aequinoctialis*, *H.caerulea* and *P.salvini* (Jouventin *et al.*, 1985), *F.glacialoides* and *F.glacialis* (Mougin, 1967).

Variations in on-duty spells also occur between populations. Wandering Albatrosses at South Georgia undertook a mean of 9.7 (range 5–14) incubation stints, those at Crozet 7.4 (4–10) stints, those of South Georgian and Crozet males averaging 8.9 and 11.4 days respectively, of the females 7.7 and 9.3 days. However, the proportion of the total time spent on the egg was 52% for males and 48% for females at both places (Du Bost & Segonzac, 1976).

2 *Variations in shift lengths*

a. Body size. A major determinant of the possible lengths of shifts is body size, large

Table 14.1 Incubation shifts of petrels

	Species	Mean no. shifts	Mean length (days)	Authority
1b.	Wandering Albatross	10	8.1	Tickell (1968)
2b.	Royal Albatross	10	7.2	Richdale (1962)
3.	Galapagos Albatross	(3)[a]	19–22	Harris (1973)
5.	Black-footed Albatross	4	15	Rice & Kenyon (1962)
6.	Laysan Albatross	4	17.5	Rice & Kenyon (1962)
7a.	Black-browed Albatross	6	11.1	Tickell & Pinder (1975)
9.	Grey-headed Albatross	6	11.5	Tickell & Pinder (1975)
12.	Sooty Albatross	7	9.5	Berruti (1979)
13.	Light-mantled Sooty Albatross	5	15	Thomas *et al.* (1983)
14.	Southern Giant Petrel	9	6.5	Conroy (1972)
		12.7	5.1	Hunter (1984b)
15.	Northern Giant Petrel	7	8	Mougin (1975)
16.	Antarctic Fulmar	11	4	Mougin (1975)
17b.	Northern Fulmar	12–19	3.5	Mougin (1967)
19a.	Cape Pigeon	15	3.6	Mougin (1968)
		12	3.7	Pinder (1966)
20.	Snow Petrel	13	3.8	Mougin (1968)
		5	8.4	Brown (1966)
		8	5.7	Isenmann (1970)
21.	Kerguelen Petrel	3	13	Mougin (1975)
		4	12	Jouventin *et al.* (1985)
22b.	Grey-faced Petrel	3	15.6	Imber (1976b)
31b.	Soft-plumaged Petrel	4	12.5	Jouventin *et al.* (1985)
38b.	Dark-rumped Petrel	3–4	16.5	Simons (1985)
42.	Mottled Petrel	4	13.1	Warham *et al.* (1977)
43.	Bonin Petrel	6	8.8	Grant *et al.* (1983)
50.	Blue Petrel	5	8.6	Jouventin *et al.* (1985)
51c.	Salvin's Prion	5–6	5.9	Jouventin *et al.* (1985)
52c.	Antarctic Prion	15	3	Tickell (1962)
54.	Fairy Prion	(10)	5.6	Richdale (1965b)
		(20)	2.4	Harper (1976)
58A.	Shoemaker	5	9.6	Mougin (1970b)
		6–7	7.5	Jouventin *et al.* (1982)
69.	Sooty Shearwater	(10)	5	Richdale (1963)
		(5)	9.4	Warham *et al.* (1982)
70.	Short-tailed Shearwater	4	(12.5)	Marshall & Serventy (1956b)
72a.	Manx Shearwater	6–12	6	Harris (1966)
79b.	Wilson's Storm Petrel	(16)	2.5	Beck & Brown (1972)
82e.	White-faced Storm Petrel	(10)	4.5	Richdale (1965a)
83a.	Black-bellied Storm Petrel	(13)	3	Beck & Brown (1971)
86a.	British Storm Petrel	12.5	2.6	Davis (1957)
99b.	Fork-tailed Storm Petrel	15–19	2.3–2.6	Simons (1981b)
102.	Georgian Diving Petrel	21	2.4	Jouventin *et al.* (1985)
103b.	Common Diving Petrel	(55)	1	Richdale (1965a)
		35	1.3	Jouventin *et al.* (1985)

[a] *Figures in parentheses are approximate.*

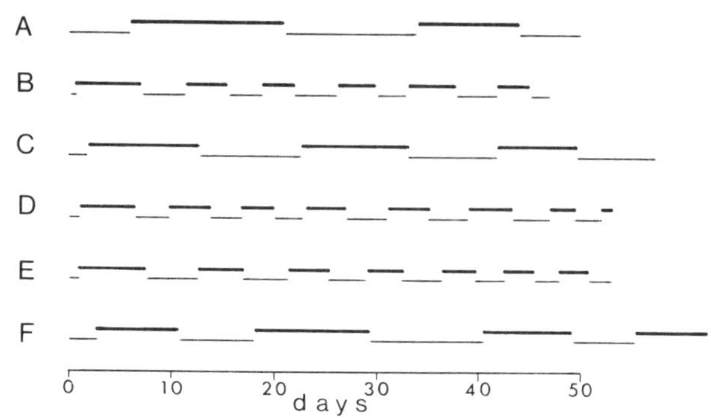

Figure 14.2 *Incubation stints by Procellariidae.* (**A**) L.brevirostris; (**B**) F.glacialoides; (**C**) P.aequinoctialis; (**D**) D.capense; (**E**) P.nivea; (**F**) M.halli. *Heavy bars = males; light bars = females. From Mougin (1975).*

birds having a greater capacity for fasting than small ones. The Forked-tailed Storm Petrels seldom fasted for more than 4 days and Simons (1981b) found that the mean weight loss of six adults was about 20% over a 3-day shift and 30% over a 5-day one, the latter being rare.

While there is some link between body size and shift lengths, with storm petrels and prions sitting for relatively short spells, the longest are not restricted to albatrosses. Gadfly petrels like *P.macroptera* and *P.inexpectata* may also undertake protracted fasts, despite their average size (Table 14.1). Particularly long stints have been recorded for *D.irrorata* and *P.palpebrata* of 20 and 24 days respectively (Harris, 1973; Thomas *et al.*, 1983) and 20, 23 and 25 days for *P.macroptera, P.phaeopygia* and *M.halli* (Imber, 1976b; Simons, 1986; Warham, 1962), but all are much shorter than the maximal second spells of *D.immutabilis* and *D.nigripes* of 38 and 39 days (Rice & Kenyon, 1962). The prize goes to the male Laysan Albatross that fasted for 58 days during its first spell on the egg (Fisher, 1971, p.26).

b. Fasting. During the female's first post-egg stay at sea she feeds up to replenish her energy store in readiness for her first long incubation fast.

Both sexes have to deposit adequate fat reserves if they are to last on the egg for the required time and continue to renew those reserves during each spell at sea. A bird that fails to reach the minimum weight will have to leave the egg prematurely and nesting may fail. The relative fat deposits necessary will be lower in large species and these are also those whose exodus periods tend to be short (Table 12.1).

c. Food supply. Long incubation stints allow petrels to travel correspondingly far in search of food. There is circumstantial evidence that some do that. For example, Rice & Kenyon (1962) had an off-duty Black-footed Albatross captured 22 days later 3700 km away. Warham *et al.* (1977) pointed out that the 13-day stint of Mottled Petrels would enable them to feed among the pack-ice off Antarctica where the birds are common during the breeding season and from the composition of their food Imber (1976b) deduced that breeding *P.macroptera* might feed as far as 650 km from their nests along the Subtropical Convergence. Of course, this winter-breeding species might take foods present at only low densities and part of the long time involved might be spent closer to the nest but requiring longer foraging to build up reserves.

Petrels that change over frequently on the egg must feed at restricted distances, the most obvious instance being the coastal feeding *P.urinatrix*. This visits its nest daily (Table 14.1). When food supply varies between seasons or colonies or bad weather interferes with feeding differences in shift lengths between populations or seasons would be expected. In his study of *O.furcata* Simons (1981b) found that in 1977 the mean shift length was 2.6 ± 1.0 days (n = 103), in 1978 2.3 ± 0.9 days (n = 151); $P < 0.05$, with more temporary egg desertion in 1977. Similarly, among Cory's Shearwaters Mougin *et al.* (1987b) noted significant differences at the same colony: in 1982 shifts averaged 7.4 days long, in 1980, 9.7 days. These authors ascribed the variations to changes in the food supply.

Other potential causes of shift-length variability among birds at a particular place and time are individual variations in feeding competence (presumably linked with age), and when one partner takes an excessively long time to replenish reserves the other may be obliged to remain sitting for longer than normal. Fisher (1971, p.26) found that 593 experienced female *D.immutabilis* stayed on their egg after laying for a mean of 1.7 days whereas 73 inexperienced females did so for 3 days, perhaps because the partners of the latter were not on hand to take over.

IV Intermittent incubation and chilling resistance

A *The degree of intermittent incubation*

For most experienced petrels the shift system of incubation runs smoothly and partners return in good time to relieve one another. However, in some pairs and some years, for various reasons, reliefs are delayed and the bird on duty forced to break its fast, doing so in time to reach the feeding grounds before death intervenes. As the eggs are resistant to chilling (see below), when the partner does return it may resume incubation and in due time the egg often hatches. Lockley (1932) was one of the first to record successful breeding despite breaks in incubation.

This intermittent incubation probably occurs in all species to some extent. It is not restricted to one sex, nor to species that lay small eggs [*contra* Whittow (1980)], and neither are all surface-nesters nor all diving petrels continuous incubators as stated by Ricklefs (1984a). Among southern surface-nesting petrels intermittency is often unnoticed because eggs even briefly uncovered are lost to predators. But where there are no avian scavengers neglected eggs may remain visible and viable for many days, and northern albatrosses have long been known to hatch such eggs. Hadden (1941) saw those of *D.immutabilis* hatch after 4–5 days' exposure and eggs of *D.irrorata* abandoned for 6 days also did so (Harris, 1973). The incidence of such events may be lower among albatrosses than among smaller petrels as the former, being larger, can fast longer, but good data are lacking. Interrupted incubation by crevice and burrow nesters is possible because their eggs are hidden from predators.

B *Chilling resistance*

The ability of petrel eggs to withstand chilling, so allowing successful breeding despite temporary desertion, was investigated by Matthews (1954) with those of the Manx Shearwater. He found that their survival was no different when these were chilled for up to 3 days, but with longer periods of neglect survival declined until all of 11 eggs chilled in a burrow for 8 days died.

In his experiments older embryos survived better than young ones but Matthews pointed out that the behaviour of the adults might have been a factor in this — they might have been more prone to keep incubating older eggs because of their greater investment in the breeding episode — rather than that old embryos were the more resistant.

Chilling resistance may be much greater than Matthews found. More recent instances include an egg of *O.castro* at the Galapagos that remained unincubated but alive for 23 days and a chick inside a chipping egg of *O.tethys* that called for 16 days (Harris, 1969b). This author also reviewed the capacity of the eggs of other seabirds to withstand chilling: none survived as well as those of storm petrels.

Pefaur (1974) investigated egg neglect at three nests of *O.oceanicus*, a species in which intermittency had been reported by Roberts (1940). Pefaur's work revealed that, despite repeated absences, the eggs could still hatch. At one successful nest the egg was neglected on eight occasions for, on average, 22.5 h (range 6–42 h), amounting to 30% of the total period monitored. That a bird was missing was quickly apparent in the sharp fall of the burrow temperature to that of the ambient air (Fig. 12.5). This was only +3°C, very much lower than in Matthews's experiments. Beck and Brown (1972) noted that neglect by *O.oceanicus* occurred particularly after incubation spans longer than 60 h whereas their *F.tropica* incubated without breaks (Beck & Brown, 1971); yet pairs of that species studied elsewhere had many breaks (Jouventin *et al.*, 1985), a difference suggesting the influence of local conditions.

Another possible example of the degree of interrupted incubation varying between populations is given by *L.brevirostris*. Mougin (1969) saw none with this bird but Jouventin *et al.* (1985) found it common, with one instance of an egg hatching after being deserted continuously for 11 days.

The most extreme examples of egg neglect come from studies of *O.furcata*. In two consecutive seasons at an Alaskan colony Simons (1981b) found cold eggs in nests on 9.6 days out of a total egg stage of 50.8 days and on 5.2 days out of 46.1 days in the following year. Boersma & Wheelwright (1979) provide the most detailed information on egg neglect in this species, from the same colony as that of Simon's study. Most of their eggs that hatched had been unattended several times for 1 or 2 days (mean 1.7 days; $n = 229$) at a burrow temperature of about 10°C, the average egg being uncovered for a total of 11 days. There was a tendency for the amount of neglect to increase as incubation proceeded and Boersma and Wheelwright noted that adult weights declined as the season progressed, suggesting their being increasingly taxed by the breeding effort. The birds also tended to neglect eggs 1–2 days before the hatch, perhaps to ensure that they had food for the hatchling (Wheelwright & Boersma, 1979).

Although handling and inspections may have increased the neglect (daily handling caused the desertion of 43 nests) and reduced the hatching success of others, disturbance would have been expected to show as complete nesting failure, the birds handled leaving for good the following night. This did not happen: many missing birds re-appeared to resume incubation, often succeeding in raising their chick.

C *Causes and advantages of interrupted incubation*

The principal proximate cause is the delayed return of the partner, but a variety of events can conceivably induce that delay, such as persistent contrary winds, changes in food availability or location that may require foragers to travel further and thus risk more impediments to their return, a temporary blockage of the burrow by snow, etc. And, as Boersma and Wheelwright (1979) pointed out, for species that can only risk landing after dark, the time during which this cover is available may be short in high latitudes: should they be held up at sea, a further delay to await nightfall may ensue.

These authors predicted that egg neglect will be greater in species that are distant foragers, that within a species the degree of egg neglect should increase with latitude, but that at very high latitudes where temperatures are extreme and kill eggs, egg neglect should be low. There are too few data to test these hypotheses and other factors may operate. Wilson's Petrels nesting at or only just above 0°C still exhibit intermittency while any Snow Petrel in inland Antarctica would probably lose an uncovered egg to the skuas even before it was frozen to death! Such egg desertion does occur but is understandably rarely seen (Tickell, 1962, p.22). In addition, even among coastal feeders with daily change-overs, some neglect occurs, e.g. in the diving petrel *P.urinatrix* (Thoresen, 1969; Mougin, 1985).

Mougin (1985) found that on two islands in the Kerguelen Group *P.urinatrix* spent 14% and 8% of the total incubation period off the egg and *P.georgicus* at the same places 16% and 2%, the actual mean lengths of desertions being 1.0–1.4 days. Mougin proposed that with *georgicus* the differences between the two islands were due to differences in food availability, because birds from both islands stayed at sea the same time — for 2.3 ± 0.9 (n = 27) and 2.3 ± 0.7 (n = 43) days, yet at the island with a 16% neglect rate, food gained in 2.3 days allowed the birds to sit for only 1.8 ± 0.8 (n = 33) days, whereas at the other they did so for 2.4 ± 0.7 (n = 107) days.

Nor is temporary desertion absent among species where neglect may lead to embryo death from overheating. Some surface nesting *P.arminjoniana* deserted eggs for up to 6 days (Gardner *et al.*, 1985). In this study the mean length of incubation stints preceding neglect was 12.2 days (n = 5), whereas that for shifts not followed by desertion was 9.3 days. To what extent nestings fail through egg overheating in this and other tropical species has not been recorded, but Shallenberger (1973) found Wedge-tailed Shearwaters in shallow burrows often leaving their eggs on hot days within minutes of returning to them, so cannot have been driven off due to hunger. Egg and burrow temperatures were not recorded and the upper lethal temperature for any petrel egg is, to my knowledge, unknown.

The obvious advantage to a pelagic species heavily dependent on synchronisation of the breeding rhythms of the parents, is that an uncovered egg does not necessarily mean breeding failure, although hatching success of neglected eggs and the survival of resulting chicks probably decrease (Boersma & Wheelwright, 1979). Nonetheless, the insurance allows underweight adults to continue feeding without certain nest failure by their returning underweight and imperilling their own lives and later breeding opportunities. Thus resistance to egg chilling may be a major adaptation of tubenoses to pelagic foraging, permitting them to nest far from their main food source and to use islands highly suited to their nesting needs but which, in some years, might otherwise be too distant from the food supply. And, by being able far away from their nests they lessen competition from other seabirds that must feed closer in. Smaller petrels having lower flight speeds and less energy-storage capacity cannot range as far as the bigger ones and they relieve their partners more often. That even diving petrels, despite their nightly nest reliefs, still experience some egg desertion (Payne & Prince, 1979) may be due to temporary food shortages, competition with other inshore feeders, a reduced capacity to store energy (e.g. no stomach oil — Warham, 1977b), or other reasons.

V Egg desertion

Abandonment of eggs by birds undisturbed by man is a widespread phenomenon among petrels, but good information on really undisturbed populations, particularly of burrowing species, difficult to collect. Eggs may be permanently abandoned at any stage of the incubation cycle, perhaps more at the start when little investment has been incurred, decreasing towards the hatch when change-overs are more frequent and the chick vocal within its shell. The most common cause of desertion is the failure of the mate to reappear and the on-duty bird's resultant decline in weight. The bird that departs may come back in due course to the cold egg, to continue sitting spasmodically before finally leaving.

Some birds continue in relays to cover eggs that fail to hatch long after the normal time. Boersma and Wheelwright (1979) gave the mean interval between the laying and the desertion of *O.furcata* eggs that did not hatch as 60.4 days — 60% more than the normal minimum period.

Temporary desertion is often the prelude to abandonment. A bird sitting long after its natural span may show signs of increasing nervousness and its desertion can be predicted. Fisher (1971, p.38) listed a range of reasons why Laysan Albatrosses deserted their egg, failure of the partner to return being the most important of these, apart from catastrophic losses in storms. Most desertions occurred during the early days of incubation spans (his Table 8). He also found that the sexes were equally likely to desert but that particular individuals were especially prone to do so, abandoning their egg year after year. Such birds would be expected to have high divorce rates, but data are lacking. Fisher (1967) concluded that the minimal survival weights to allow these albatrosses sufficient energy to reach their foraging areas safely were between 1600 and 1900 g (c.f. Fig. 14.3). Weights of the gadfly petrel *P.macroptera* at egg desertion were the lowest recorded while ashore but none was known to have died (Imber, 1976b). However, in their study of this species, Johnstone & Davis (1990) found that desertions were not a response to critically depleted energy supplies.

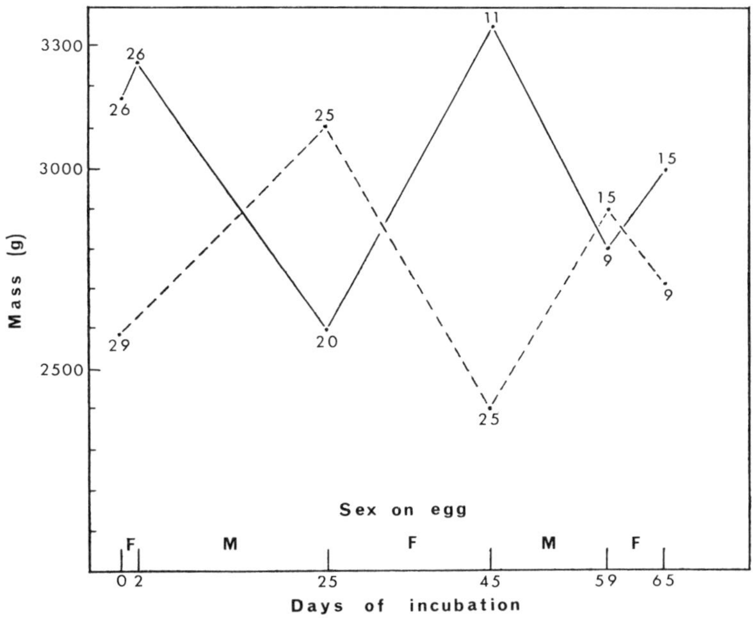

Figure 14.3 *Weight changes in incubating Laysan Albatrosses. Heavy lines = males, dashed lines = females, sample sizes as indicated. The sample for each sex contained the same birds each time, although not all the birds could be found for weighing on each occasion. From Fisher (1967).*

VI Incubation periods

A *'True' and 'apparent' incubation periods*

The clutch size being one, the incubation period of a petrel egg is readily defined as the length of time elapsing between the laying of the egg and the chick's complete freedom from the shell, incubation being continuous (Heinroth, 1922). The last provision is important because eggs subjected to interrupted incubation result in highly variable 'apparent' periods. Tickell (1962) was the first clearly to distinguish the difference, noting that among Antarctic Prions the true period was 44.8 $\pm$ 0.56 days (44–46 days) whereas apparent periods ranged from 42–50 days. This is typical of the degree of variability in the time to hatch continuously covered eggs, the coefficients of variation for those of southern albatrosses and many burrowing species ranging from about 1.25% to 2.5%. Jouventin *et al.* (1985) specified 'real' incubation periods for some petrels: otherwise there are few sets of data for the true periods of burrowing forms. Most published figures are of apparent periods and show wide variances around their means, these in turn being greater than the true means for continuous incubation. Examples reflecting intermittency are the figure of 44–60 days (n = 54) for Audubon's Shearwater given by Harris (1969d) and 48–56 for the Wedge-tailed Shearwater (Shallenberger, 1973). An extreme case is *O.furcata* with an apparent period of 49.8 ± 6.7 days (n = 33; coeff. of varn. = 13.5%) but a true one of about 38.6 days (Boersma & Wheelwright, 1979), the apparent period being directly proportional to the number of days neglected (Fig. 14.4).

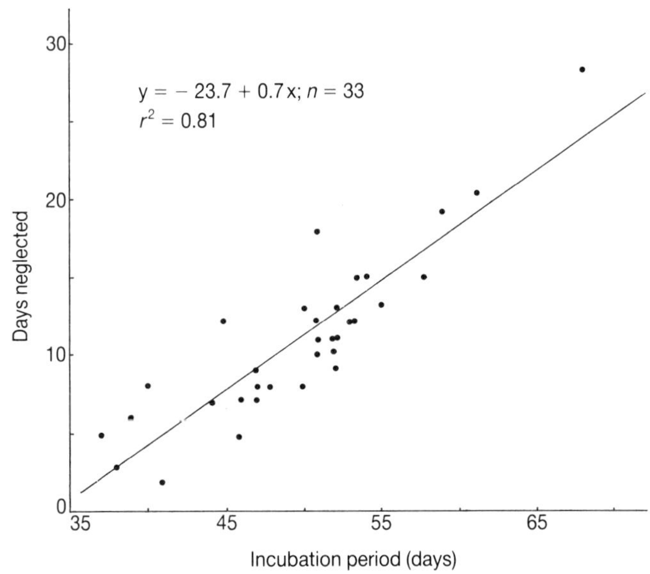

Figure 14.4 *Relationship between egg neglect and apparent incubation period among Fork-tailed Storm Petrels. From Boersma & Wheelwright (1979).*

In general, the true period is likely to be among the lowest of the range of values and, for cool climate species where the development of the uncovered egg soon ceases, an estimate of the true figure is given by deducting the time uncovered from the total time to hatching. In warm climates suspension of incubation may not necessarily result in significant cessation of development but studies have not been made under those conditions.

B *Variability of incubation periods*

A basic factor affecting the lengths of incubation periods is body size, large petrels taking longer to hatch their eggs, everything else being equal, than small ones. This is in line with the patterns of incubation generally as established by Furbringer

(1888), Heinroth (1922), Huxley (1927), Worth (1940) and others. In consequence, incubation periods of petrels also correlate with other variables linked to body size, e.g. egg size, LB^2 index, standard wing length, yolk and birth weights (Warham, 1968). For their size petrels take an unusually long time to hatch their eggs. The domestic hen and the Wedge-tailed Shearwater lay similar sized eggs, the former hatches them in 21, the latter 52 days. The differences are even greater when small petrels are compared with other similar-sized birds.

Figures 14.5 and 14.6 show log:log plots of incubation periods and female body masses, egg masses, LB^2 and yolk masses, the data being drawn from Table 14.2. The figures used have been selected using fresh eggs, true incubation periods, female body weights or weights of mixed-sex samples where sexual dimorphism is low, and drawn from the same populations or subspecies. Sample sizes may be large but some for incubation periods are small and could include inflated values due to undetected intermittency.

The general pattern is for incubation periods to be highly correlated in an allometric manner with these four variables, the smaller species taking relatively longer to hatch their eggs than the larger ones (Warham, 1964).

Those that diverge most from the general pattern are the fulmarine petrels, six of

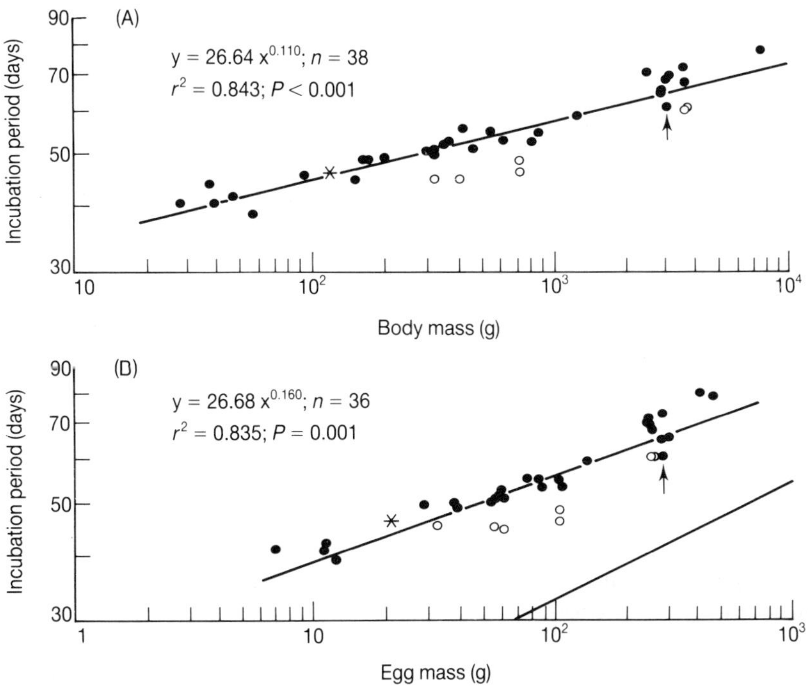

Figure 14.5 (**A**) *Incubation periods and body masses of 38 petrels. Open circles = fulmars; star =* Pelecanoides georgicus; *plot for* D.irrorata *arrowed.* (**B**) *Incubation periods and egg masses for 36 petrels. Lower line is part of generalised curve for other birds from Grant* et al. *(1982b). Symbols as in Fig. 14.5A.*

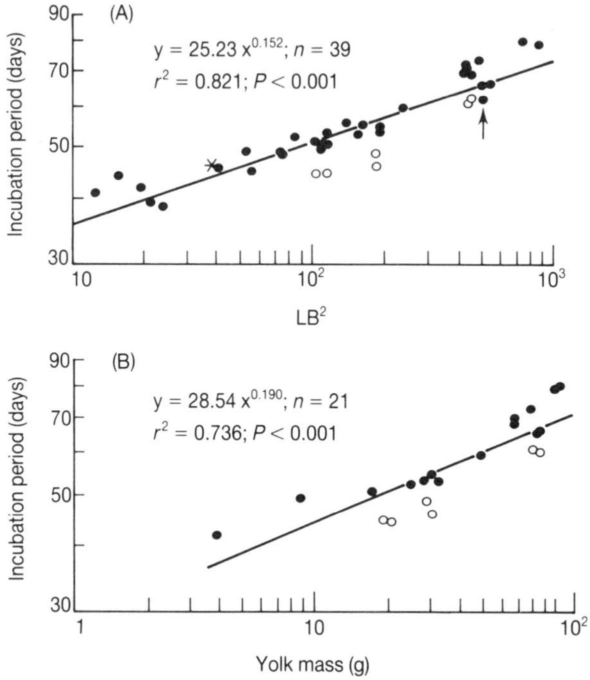

Figure 14.6 (**A**) *Incubation periods and* LB^2 *for 39 petrels. Symbols as in Fig. 14.5A.* (**B**) *Incubation periods and yolk masses for 21 petrels. Symbols as in Fig. 14.5A.*

the seven species being plotted in these figures. All hatch their eggs more quickly than do similar sized petrels laying similar-sized eggs. I lack data for *Thalassoica* but Orton (1968) indicated that 13 weeks elapse from laying to fledging, so that its development is probably rapid also. Kamanev (1968) gave one determination - 46 days. The value of 55 days given by Mehlum *et al.* (1986) was estimated from egg dimensions and is probably in excess of the true period by 8 or 9 days. I concluded that the relatively rapid rates of egg development were adaptations by fulmars to breeding in high latitudes where the season is short, the days long, and food abundant (Warham, 1968) (Chapter 15, Section VIC below).

Another value that stands out in these graphs is the position of the *D.irrorata*, which, for an albatross, hatches its egg in record time — 60.5 days (Harris, 1973). While there may well be advantages to this bird in spending a shorter time on the egg, its environmental conditions are in sharp contrast to those of the fulmars, it being a tropical, not a polar species; nor does it seem likely that high ambient temperatures speed development because other albatrosses nesting on hot coral atolls incubate for 5 or 6 days longer.

There are few accurate figures for the incubation period of diving petrels — Table 15.3. The figure of 46.4 days (from 10 determinations) for *P.georgicus* given by Payne & Prince (1979) plots close to the logarithmic regression lines despite laying small eggs for a petrel of its size (Chapter 13, Section IVC) above. Jouventin *et al.*

(1985) also estimated 46 days as the real incubation period of that bird and 50 days for the heavier *P.urinatrix*. The available data for prions also fit well and Croxall's belief that diving petrels and prions have very long incubation periods for their sizes may have been due to his use of apparent incubation periods (Croxall, 1984). The Kerguelen Petrel *L.brevirostris* plots with the rest of the petrels and not with the fulmars despite the fulmarine links suggested by Imber (1985a).

C *Incubation periods and body masses*

Using the data for the 38 species plotted in Fig.14.5A:

$$y = 26.64x^{0.110} \qquad \text{(i)}$$

where y = incubation period in days and x = body mass in g; $r^2 = 0.843$.
However, if the six fulmars are excluded the equation becomes:

$$y = 26.33x^{0.116} \qquad \text{(ii)}$$

where $r^2 = 0.935$.

Equation (ii) may be compared with $y = 27.07x^{0.114}$ for an earlier plot for 15 non-fulmarine petrels (Warham, 1968) and equation (i) with $y = 29.78x^{0.096}$ (Jouventin & Mougin, 1981) and $y = 32.4x^{0.12}$ (Croxall, 1984). Both these last two equations were, however, derived from less carefully selected data, used masses of mixed sexes and included some 'apparent' incubation periods. For predicting the incubation period of a non-fulmarine petrel from its body mass, equation (ii) would be most suitable.

As shown in Fig. 14.5A, most of the albatrosses have longer incubation periods than predicted by equation (ii) and, at the other end of the size scale, the plots for two of the storm petrels are also rather divergent. Some, such as the figure of 44 days for *O.homochroa*, may include some egg neglect, while the figure used for *O.oceanicus* is of 40 days whereas a more recent true value of 35.7 days by Jouventin *et al.* (1985) plots on the regression line. Nevertheless in this and other similar allometric relationships, better data may show that a suite of curves best represents the facts rather than a single one.

D *Incubation periods and egg variables*

1 *Egg mass*

Using all the data of Fig. 14.5B:

$$y = 26.68x^{0.160} \qquad \text{(iii)}$$

where y = incubation period in days, x = mean egg mass in g; $n = 36$; $r^2 = 0.835$.
If the six fulmars are excluded:

$$y = 26.85x^{0,164} \qquad \text{(iv)}$$

where $n = 30$; $r^2 = 0.923$. The fulmars are still shown to have short incubation

Table 14.2 Weights, incubation and nestling periods of petrels

	Species	Body weight (g)	Egg weight (g)	LB^2	Yolk weight (g)	Incubation period (days)	Nestling period (days)
1a.	Wandering Albatross	7686(22F)	471(10)	874(54)	100.5	78.4(163)	278(35)
2b.	Royal Albatross		417(41)	748(53)	107	79.3(35)	240(64)
3.	Galapagos Albatross	3040(13F)	284(25)	505(28)		60.5(5)	
5.	Black-footed Albatross	2990(104F)	305(79)	542(83)	88.9	65.6(75)	
6.	Laysan Albatross	2990(104F)	285(88)	506(105)	84.9	65.0(432)	160 (many)
7a.	Black-browed Albatross	3694(94F)	257(117)	453(117)	68.4	68.1(72)	116(?)
9.	Grey-headed Albatross	3624(95F)	276(63)	490(63)	79.6	72.4(103)	141(?)
10b.	Yellow-nosed Albatross	2060(33)	199.8(47)	352(47)		71.3(3)	115?
11a.	Buller's Albatross	3012(31)	250(100)	426(75)	68.5	69.1(59)	167(44)
12.	Sooty Albatross	2500(176)	243.5(31)	437(67)		70.7(40)	164
13.	Light-mantled Albatross	3150(13)	243(22)	429(22)		69.9(22)	141?
14.	Southern Giant Petrel	3798(78F)	260.6(15)	455(49)	81.6	60.6(90)	123(15M) 117(18F)
15.	Northern Giant Petrel	3724(43F)	256.4(8)	446(46)	86.4	60.1(8)	114(30M) 110(26F)
16.	Antarctic Fulmar	734(11F)	103(17)	118(17)	30.5	46(11)	52.6(17)
17b.	Northern Fulmar	729(88F)	101(34)	189(100)	28.7	48.6(67)	53.2(47)
19a.	Cape Pigeon	407(22F)	62(46)	115(46)	21.4	44.5(13)	48(4)
20.	Snow Petrel	330(10F)	57.4(40)	104(40)	19.4	44.8(17)	46.7(30)
21.	Kerguelen Petrel	331(29)	55.5(8)	109(22)		49.7(3)	61(4)
22b.	Great-winged Petrel	555(60F)	86(60)	163(26)	30.1	54.6(36)	118(38)
31a.	Soft-plumaged Petrel	302(72)	57.1(22)	109(36)	17.1	50.2(6)	91(2)
34.	Cahow					52.4(15)	92(32)
38b.	Hawaiian Petrel	434(38)	76.9	140(30)		55.3(11)	111(26)

42.	Mottled Petrel	329(42)	61.6(10)	117(62)		50.2(21)	
43.	Bonin Petrel	176(76F)	39.2(12)	72.5(12)		48.7(19)	82(5)
50.	Blue Petrel	202(215)	38.3(20)	73.7(86)		49(7)	53.0(41)
51c.	Salvin's Prion	161.8(325)	34.7(9)	64.4(24)		50.3(6)	52–63(11)
52c.	Antarctic Prion	153(12F)	32.8(23)	56.4(42)		44.8(10)	50.2(22)
53.	Thin-billed Prion	129(4)	30.6(10)	57.5(63)		46.5(2)	49(11)
54.	Fairy Prion	132(100)		47.9(100)	9.4		50(139)
56.	Bulwer's Petrel	94.9(42)		41.3(146)		45.5(2)	62(57–67)
58a.	Shoemaker	1280(26F)	134.3(32)	239(56)		58.9(14)	98(9)
60.	Parkinson's Petrel	747(5F)	98.8(9)	177(9)		56.5(1)	107(6)
62b.	Cory's Shearwater	877(26F)	104.3(46)	190(46)		54.5(88)	96.8(78)
64.	Wedge-tailed Shearwater	369(100)	60.0(27)	114(36)	25.1	52.4(17)	109(33)
69.	Sooty Shearwater	819(299)	105.9(32)	192(77)	33.3	52.7(9)	96.7(53)
70.	Short-tailed Shearwater	634(54F)	86(15)	157(100)	28.0	53(24)	94(18)
71.	Christmas Shearwater	360(18F)		85(51)		52(5)	96(6)
72a.	Manx Shearwater	469(23F)	58.4(10)	103(83)		51(43)	70(53)
77c.	Audubon's Shearwater	163(13F)	29.2(30)	52.4(125)	8.9	49(54)	75(41)
78c.	Little Shearwater	195(15)		67(17)		54(4)	72(3)
79b.	Wilson's Storm Petrel	40.0(33F)	11(8)	21.2(15)		40.2(8)	60(12)
82c.	White-faced Storm Petrel	47.2(100)		24.3(100)	4.9		57.3(40)
83a.	Black-bellied Storm Petrel	56.5(15F)	15(9)	27.0(8)		39.1(1)	
86a.	British Storm Petrel	28(50)	7(15)	12.6(100)		40.6(36)	70(53)
89.	Madeiran Storm Petrel	41.8(389)	8.5(28)	17.4(72)			78(25)
91a.	Leach's Storm Petrel	48.5(17)	11.2(20)	19.1(87)	3.9	41.8(19)	61.2(20)
97.	Ashy Storm Petrel	38.1(83)		15.4(46)		44(?)	84(?)
99b.	Fork-tailed Storm Petrel	58.6(33)	12.6(93)	23.9(183)		38.6(33)	58.4(20)
102.	Georgian Diving Petrel	121(71)	21.1(19)	38.7(32)		46.3(10)	50.5(8)
103c.	Common Diving Petrel	141(52)	22.1(19)	39.9(31)			54.6(10)

Female weights indicated by (F); samples of mixed sex ().

periods relative to their egg masses despite their heavy shells. Equation (iii) may be compared with that given for 38 tubenoses by Grant *et al.* (1982b) of $y = 31x^{0.13}$; $r^2 = 0.67$, another equation based on data that included apparent incubation periods. For predictive purposes equation (iv) would be most appropriate.

As in Fig. 14.5A many of the albatrosses plot above the line, taking longer to hatch their eggs than the regression equation predicts and here intermittency seems unlikely for reasons explained earlier.

2 *Incubation periods and* LB² *index*

This method of presentation (Fig. 14.6A) permits the inclusion of data where incubation periods are known but fresh egg weights are not:

$$y = 25.32x^{0.152} \quad \text{(v)}$$

where y = incubation period in days and $x = LB^2$, egg dimensions being measured in cm; $n = 39$; $r^2 = 0.821$. If the fulmars are excluded the equation becomes:

$$y = 25.25x^{0.157} \quad \text{(vi)}$$

with $n = 33$; $r^2 = 0.909$.

3 *Incubation periods and yolk masses*

Few values for yolk masses of fresh eggs are available (Table 13.4) but, again there appears to be a linear relationship between the logarithms of the two variables as shown in Fig. 14.6B:

$$y = 28.54x^{0.190} \quad \text{(vii)}$$

where y = mean incubation period in days and x = mean yolk mass in g and with $n = 21$ (fulmars included); $r^2 = 0.736$. Comparison of this figure with that of Fig. 13.12 indicates that the short periods of the fulmars are not a consequence of their having relatively small yolks which would obligate hatching to be earlier.

4 *Incubation periods and birth weights*

Body weights being proportional to egg weights and these in turn to hatchling weights (Chapter 15, Section IVB1), an allometric interdependence exists between incubation periods and birth weights. Figure 14.7 shows this relationship on logarithmic axes for 26 species for which there are adequate data (see Table 15.4).

As before, the fulmars need shorter incubation periods to produce hatchlings of similar sizes to those of other similar-sized tubenoses. The regression equations are:

$$y = 0.00000014x^{4.98} \quad \text{(viii)}$$

with $n = 26$, $r^2 = 0.758$, $p < 0.0001$; where y = birth weight in g and x = incubation period in days.

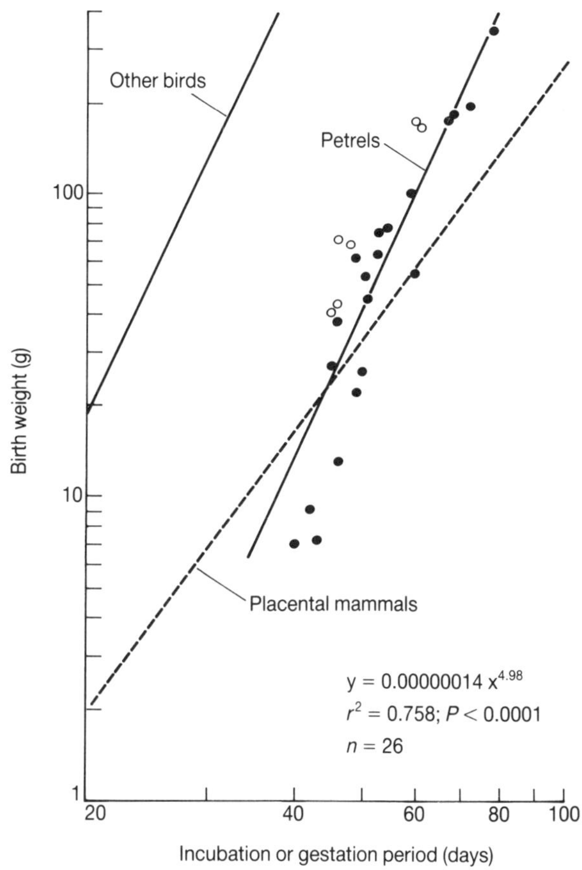

Figure 14.7 *Birth weights and incubation periods of petrels and birds in general, and birth weights and gestation periods of placental mammals. Points for fulmars are shown with open circles. From Warham (1968) after Needham (1931).*

If the fulmars are omitted

$$y = 0.000000009x^{5.61} \quad \text{(ix)}$$

where $n = 20$, $r^2 = 0.900$, $P < 0.0001$.

Also plotted are sections from the curves of birth weight *v.* incubation periods for birds in general and *v.* gestation periods for placental mammals taken from Needham (1931). Although some of the data he used were faulty, these curves do underline the special position of the Procellariiformes *vis-à-vis* other birds. In their development to birth they more resemble the mammals than other birds and it may be significant that the low body temperatures of petrels more resemble those of placentals than those of birds in general. A similar correspondence holds between the development times to birth/unit weight of adult for petrels and placental mammals (Warham, 1968).

E *Other variables affecting incubation periods*

Where reliable figures are available, the range of variation in the time required to hatch a continually incubated petrel egg is small. Variation might be even less if observations were more intense, e.g. by automatic monitoring, but in practice most past determinations come from daily or twice-daily checks to get laying and hatching dates, the results usually accurate only to ± 0.5–1.0 days. Heating of the eggs may also not be entirely continuous — some birds may leave them to take the air at the tunnel entrance or to drive away intruders more than others, surface nesters ventilate their egg more often, and so on.

These various relationships may also hold intra-specifically, providing another source of variability. For example, Imber (1976b) reported that heavy females of *P.macroptera* produced heavy eggs ($r = 0.47$; $P<0.001$; $n = 60$) and these would be expected to take longer to hatch. Quite apart from egg size *per se*, experience of the adults may play a part. Analysing incubation period variations of Laysan Albatrosses, Fisher (1971, p.24) found that early laying and experienced females took slightly longer to hatch their eggs than late laying, mainly inexperienced ones. A possible reason was that the late layers laid smaller eggs. Another possibility is that some form of social facilitation was involved — the time between pipping and hatching might have been shorter in late hatching eggs, and this period forms a considerable proportion of the total incubation time (see Section IXA below).

In his work on *M.giganteus* Conroy noted that the incubation periods were shorter in pairs with the least number of change-overs, as shown in Fig. 14.8. This was true also of *D.capense* at Signy island but not of that bird at Pointe Geologie

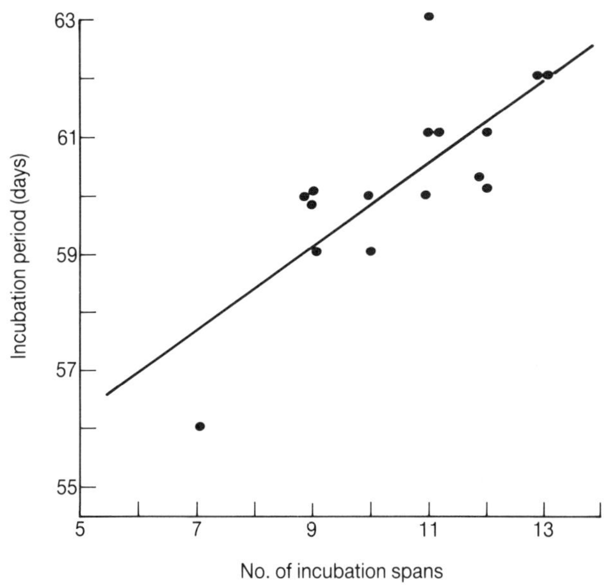

Figure 14.8 *Relationship between incubation period and the number of incubation spells in* Macronectes giganteus. *From Conroy (1972).*

(Conroy, 1972, p.25). Apart from the increased cooling resulting from more change-overs, it is hard to see why incubation should have been longer unless the pairs that switched duties more frequently were naturally restless and uncovered their eggs more often, allowing incubation temperature to fall.

On the data available there is little indication that incubation is faster among species of warm climates, rather than the contrary when the fulmars are considered: differences, if detectable, seem related to other factors such as differences between subspecies in mean body size.

A few seemingly atypically short figures for particular birds, like that of 43 days for a Northern Fulmar (Dunnet, pers. comm.) and 42 days for an Antarctic Prion (Tickell, 1962, p.24) may perhaps have been due to the retention of the formed egg in the oviduct.

VII Incubation temperatures

A *Introduction*

Petrel eggs are incubated on a range of substrates, some cold and damp, others warm and dry; some in unlined nests, others in well constructed ones of leaves or feathers. It might be expected that birds incubating under cold, damp conditions would take longer to hatch their eggs or have higher body and egg temperatures to offset heat losses to the substrate, and this would be more important at a late stage in development, for in its early days the embryo lies at the top of the egg close to its heat source; later metabolic heat production by the embryo becomes important.

The degree to which the egg is insulated from the ambient conditions has not been measured and doubtless varies with the species and particular situations. In large forms like giant petrels and albatrosses the egg seems to be totally invested by the brood pouch and adjacent feathers, an effect that is less obvious among small petrels. Nonetheless, the influences of ambient wind and temperature conditions are probably small, even very small. The only determination of temperatures near an egg is that of Saiz and Hajek (1968) who checked thermistor readings in a nest of *M.giganteus* every 6 h, finding that the temperature beside the egg was not affected by the wind or temperature outside.

The brood patch conforms to the shape and size of the egg and in *D.immutabilis* 24% of the egg surface was in contact (Grant *et al.*, 1982b), in *M.giganteus* about 35% (Tirrell & Murrish, 1979a). Presumably birds with larger contact areas would hatch them more quickly than those where contact was less extensive.

B *Egg and brood patch temperatures*

The temperature of the brood patch is similar to that of the deep body temperature and higher than that of the egg. Ricklefs & Rahn (1979) gave the following mean figures for *O.leucorhoa* — temperature of body 38.3°C, of brood patch 36.6°C, and of egg 32.5°C — i.e. a gradient of 4.1°C between the patch and the egg. Rahn & Huntington (1988) also gave figures for that species — body temperature 38.3°C;

brood patch 35.8°C and egg. 33.4°C. For *D.nigripes* and *D.immutabilis* Grant *et al.* (1982b) graphed continuous traces that show shallower gradients between brood patch and egg as was also shown for *P.pacificus* whose brood patch temperature averaged 37.6 ± 0.5°C (n = 30), 59 eggs 35.0 ± 1.5°C (Whittow *et al.*, 1982). Tirrell & Murrish (1979a) recorded lower temperatures than these for eggs of *M.giganteus*, i.e. 29.7–31.6°C.

It is difficult to measure the temperatures of eggs beneath a petrel. Early attempts often involved dummy eggs with implanted thermistors and many experimenters have not allowed for the heat generated by the embryo itself as it matures, an effect that is considerable (Eycleshymer, 1907). Grant *et al.* (1982b) found that the temperatures at the centres of Laysan Albatross eggs increased 9% between early and late incubation. On 2 December, 15 December, 10 January and 21 January mean values were 34.2 ± 0.4°C, 34.7 ± 0.2°C, 36.5 ± 0.2°C, and 37.3 ± 0.3°C, the last two figures for embryos aged 37–42 and 43–50 days respectively.

Several researchers using plaster eggs have found them accepted by the birds whereas Tickell (1962, p.22) failed to get Antarctic Prions to do that. Implantation of measuring devices should ideally allow normal development to a viable chick and without further handling of the adults. This seems to be difficult to achieve, the wires prevent or hinder any egg turning and the embryos often die. None of those (ages unknown) of Leach's Storm Petrels studied by Williams & Ricklefs (1984) survived, so that their figures are likely to be low. They gave mean values of 35.9°C, 35.7°C and 35.5°C for eggs of *O.leucorhoa*, *O.oceanicus* and *M.giganteus*. Boersma (1982) did have one egg of *O.furcata* hatch after a probe insertion, but it took 60 days instead of the usual 38.

C *Body temperatures*

Body temperatures of petrels are low. Those of 31 species averaged 38.78°C ± 0.17°C (Warham, 1971) and more recent figures are similar. Gross (1935) suggested that the eggs might be getting less heat than those of groups that can hatch them sooner. Warham also suggested that body temperatures could help explain the long incubation periods despite continuous incubation, but noted that a few figures for petrel egg temperatures were available and that some were so low as to suggest cooling before measurement. Drent (1975), misquoting Warham, reviewed known egg and brood patch temperatures, added his own figures for *O.leucorhoa*, and concluded that petrel eggs are not incubated at temperatures greatly different from those of most other birds. However, he was comparing the temperatures of eggs being continuously covered with those of skuas, gulls and passerines subject to fluctuating temperature regimes, and it was left to Ricklefs (1984a) to compare the times taken by petrels and those of other seabirds under continual incubation. These were mainly Pelecaniformes, their egg temperatures similar to those of Procellariiformes, but with rather shorter incubation periods.

This later work suggests that incubated petrel eggs usually do not experience temperatures low enough to slow down development significantly if continuously covered. Some of the data, e.g. the value of 32.5°C for eggs of *O.leucorhoa*, may have been due to faulty techniques, as pointed out by Ricklefs (1984a), who recorded a mean of 35.9°C using improved methods (William & Ricklefs, 1984).

However, the matter is not resolved. For example, that low body temperatures may be a factor affecting incubation periods is supported by work on *Pterodroma hypoleuca* (Grant *et al.*, 1982a). They ascribed the low vapour pressure within the egg to its low central temperature (33.8°C), well below the average temperature for eggs in general of 36.2°C (Drent, 1975), and the long incubation period to low rates of water loss, low water vapour conductance of the shell and low egg temperature. The eggs used, however, were less than 2 weeks old, so that metabolic heat production would have been low, and if the values for Laysan Albatross eggs are any guide, would have risen later. Wink *et al.* (1987) found a considerable gradient between body and egg temperatures in *C.d.diomedea*. Three males averaged 36.7°C, three females 37.6°C and five eggs 29.6 ± 0.6°C.

A low central egg temperature of 30.2 ± 2.6°C (n = 15) was reported by Boersma (1982) for *O.furcata*. One of these eggs hatched after experiencing a mean temperature of 25.5°C (23.8–26.2°C) although Veeck & Kenagy (1980), in laboratory tests, thought that development ceased at 26°C. Boersma pointed out that the damp, cold substrate on which the egg was incubated could act as a heat sink, and that the naturally moist burrow was made more so by the bird's respiration. She considered that the low temperature was a basic reason for the long true incubation period, quite apart from effects of intermittency. Significantly, the chicks can enter a torpid state when chilled or hungry (see Chapter 15, Section VC4).

Ricklefs (1984a) disputed Boersma's ideas on the effects of temperature on incubation and pointed out that a half-grown embryo of *M.giganteus* died after chilling to 3°C for 48 h. Whether this is typical of the resistance of those chicks was not determined, but in any event fulmar eggs take a relatively short time to hatch and resistance to chilling may have been of less value in their evolution since, even if they evolved without skuas as egg predators, they did evolve with other giant petrels and these eat uncovered eggs as eagerly as skuas do!

VIII Weight changes during incubation

A *Eggs*

Loss of weight by petrel eggs during incubation is mainly through loss of water. Most lose 0.3–0.5% of their initial masses daily, amounting to 14–19% over the incubation periods as a whole (table in Grant *et al.*, 1982b) and during most of this time the rate of loss is rather constant. It increases sharply once the egg is starred and direct contact between the embryo and the outside air established. Otherwise the rate of water loss and of oxygen uptake is slow compared with those of similar-sized eggs of other birds due to reduced conductance through the shell and reduced functional pore area, both features of their prolonged incubation, see e.g. Grant *et al.* (1982a, b) and Whittow *et al.* (1982).

B *Fasting birds*

An incubating tubenose starts its stint with good fat reserves and during incubation the body masses of the parents tend to vary inversely as the sitting birds neither eat nor drink. With the Laysan Albatross Fisher (1967) found that, over all, females gained weight, despite the fluctuations, whereas males had lost weight by the end of the incubation (Fig. 14.3), partly because the males averaged 35 days on the egg, the females only 30 days. Weight loss during long spans exceeded 20% of the initial weights according to Rice & Kenyon (1962).

Mass declines during incubation have been followed in a number of species; by Croxall (1982b), Grant (1984), by Despin & Mougin (1988) and others. Croxall pointed out that weight loss during incubation fasts correlated with body weight, being relatively small in large species. Daily losses range from *c.* 6.0% of the initial weights for small storm petrels, to *c.* 2.0% in medium-sized shearwaters and only *c.* 0.8% in the great albatrosses. The highest figure determined was for *O.leucorhoa* of 7.5% per day (Ricklefs *et al.*, 1986). For the large shearwater *Calonectris d.borealis* the weight loss was 19 g day^{-1} for the males (the bigger sex) and 17 g day^{-1} for the females. At sea the males gained weight at 24 g day^{-1} the females 18 g day^{-1}, or about 2.7% of their initial weights daily (Despin & Mougin, 1988). Male and female *D.melanophrys* and *D.chrysostoma* lost about 1–2% of their initial weights daily while on the egg (Prince *et al.*, 1981).

Weight loss during fasting tends to be more rapid at the start of the fast and in heavier individuals, presumably because these are initially digesting stomach contents rather than recycling fat reserves; later rates of loss tend to stabilise, partly because defaecation ceases. Despin & Mougin (1988) attempted to show that the rate of loss for petrels of warmer climes is less than that of cold ones. However, they included a very high figure of 5.9% per day for *P.turtur* in their cold water group and one for *P.inexpectata* in their warm water one, both in the wrong groups. Furthermore, the figure for *P.turtur* was high because the birds were not in a post-absorbtive state (Harper, pers. comm.). More and better data are needed to support their hypothesis. However, these authors showed the logarithms of the daily losses of mass are correlated with those of the body masses and Fig. 14.9 and Table 14.3 bring together the available data. The regression line is:

$$y = 0.216x^{0.629}$$

with $r^2 = 0.946$; $n = 19$; where y = mean daily loss of body mass (g) and x = mean body mass (g).

Bearing in mind the difficulties in non-automatic weighing of incubating birds, the varying amounts of unabsorbed food likely to be present, the risks of desertion and other effects on behaviour, the above correlation is high. The decline during a fast by post-absorptive birds occurs through gaseous exchange and faecal, urinary and evaporative water losses.

IX Hatching

A *The pre-hatch period*

The time elapsing between the first sign of shell fracture and the liberation of the chick from the shell is usually rather long, 3–6 days. Shorter times have been recorded, e.g. for *D.e.sanfordi* Richdale (1952) gave seven times varying from 1 day

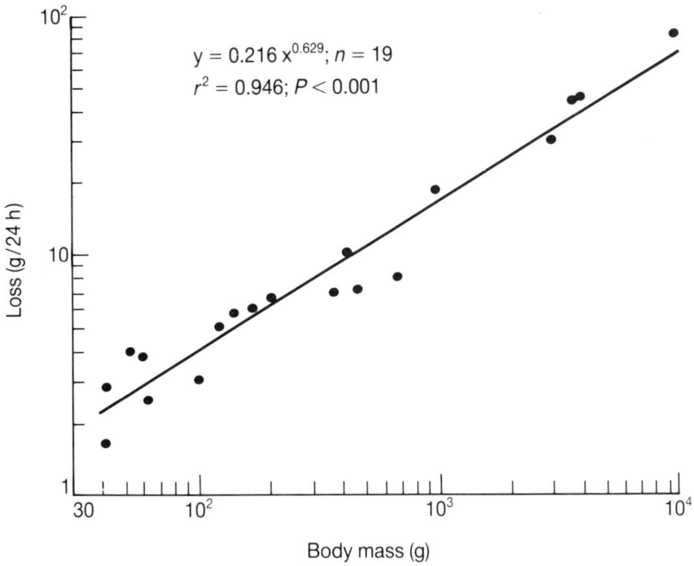

Figure 14.9 *Loss of weight of 19 species of fasting petrels during incubation.*

Table 14.3 Weight losses during incubation fasts

	Species	(g)	Daily loss (g)	% Daily loss	Source
1b.	Wandering Albatross	9800	82.5	0.8	Croxall & Ricketts (1983)
6.	Laysan Albatross	2996	29.8	1.0	Rice & Kenyon (1962)
7a.	Black-browed Albatross	3810	45.5	1.2	Prince *et al.* (1981)
9.	Grey-headed Albatross	3690	44.0	1.2	Prince *et al.* (1981)
22b.	Grey-faced Petrel	668	8.0	1.2	Imber (1976b)
38b.	Hawaiian Petrel	459	7.1	1.5	Simons (1985)
42.	Mottled Petrel	370	6.9	1.9	Warham *et al.* (1977)
43.	Bonin Petrel	202	6.5	3.2	Grant & Whittow (1983)
48b.	Pycroft's Petrel	148	5.6	3.8	Dunnet (1985)
57.	Bulwer's Petrel	102	3.0	2.9	Jouventin & Mougin (1981)
62b.	Cory's Shearwater	986	18.3	1.9	Despin & Mougin (1988)
72a.	Manx Shearwater	430	10.0	2.3	Harris (1966)
77c.	Audubon's Shearwater	168	6.0	3.6	Harris (1969d)
79b.	Wilson's Storm Petrel	42	2.8	6.7	Beck & Brown (1972)
83a.	Black-bellied Storm Petrel	62	2.5	4.0	Beck & Brown (1971)
89.	Madeiran Storm Petrel	42	1.6	3.8	Harris (1969b)
91a.	Leach's Storm Petrel	53	4.0	7.5	Ricklefs *et al.* (1986)
99b.	Fork-tailed Storm Petrel	59	3.8	6.4	Simons (1981b)
102.	Georgian Diving Petrel	125	5.0	4.0	Roby & Ricklefs (1983)

8.5 h to 4 days 10 h. This pre-hatching phase has been closely followed in *P.pacificus* by Pettit & Whittow (1983). They described the sequence of events as (1) the star-fracturing or 'external pipping' of the shell, usually at the blunt end. This starts 47 days after the egg is laid. Stage (2) follows one day later when the embryo's bill penetrates the air cell at the blunt pole — 'internal pipping'. Two days later stage (3) is reached when a distinct 'pip-hole' appears in the shell while (4) follows 2 days later again — the egg hatches. It is during these pre-hatching stages that the rate of weight decline accelerates due to an abrupt increase in water loss. The sequence is different in albatrosses and non-procellariiform birds: internal pipping occurring followed by star fracture of the shell (Whittow, pers. comm.).

Pettit & Whittow (1983) presented comparable data for three other petrels (Table 14.4) and these show that the time between external pipping and hatching represents a greater proportion of the incubation period in the smaller species, that the water loss during pipping is relatively greater in small eggs, and that the main variation in water loss consists of a difference in that lost during the pre-pipping stages and that lost from external pipping to the hatch.

B *The hatch*

At the time of hatching either parent may be present, although Imber (1976b) found female *P.macroptera* usually on duty then. Many chicks free themselves from the shell by day or by night, only Richdale (1952, p.32) and Brown (1966) considered that this occurred only by day — in *D.e.sanfordi* and *P.nivea* respectively. Albatross chicks peep lustily for several days before external starring and this could provide a cue to the parents to shorten their off-duty periods, perhaps to ensure they have food for the hatchling. This may not be necessary, however, as a Royal Albatross that had fasted for 10 days fed its newly hatched chick (Richdale, 1965a, p.73) and Richdale's work on small petrels suggests that these may have the same ability, but perhaps disgorging only stomach oil.

When incubating albatrosses occasionally rise to stretch themselves or to ventilate their egg, they almost invariably bend down to touch or look at it, and usually voice a few calls — 'talking to the egg'. Fisher (1971, p.61) found that these actions became more frequent as hatching approached, perhaps forming the first step in recognition of the parent's voice by the chick and vice versa. Harper (1976) heard Fairy Prions calling in response to the cheeps of unattached chicks and suggested that such communication might be essential in triggering the switch from incubation to chick care. He also found that eggs without active occupants around hatching time were soon abandoned although, as we have seen, petrels will sit on dead eggs long after they should have hatched. Harper also noticed that adult Fairy Prions tended to collect nesting material around the egg at about the time when the chick began to call.

Despite pre-hatching communication, when the chick starts to break out the adult often appears to ignore it although becoming more restless and sitting in a more erect attitude. Such behaviour is easily observed among albatrosses, but only with difficulty among burrowing forms. At a TV-monitored test of *P.phaeopygia*

Table 14.4 Water loss from pipped eggs of four species of petrel

	Fresh egg mass (g)	External pipping-hatching interval as % of total incubation	Water loss during the interval between external pipping-hatching		Total water loss during incubation as % of fresh egg mass	Reference
			As % of total water loss	As % of fresh egg mass		
Bonin Petrel (*Pterodroma hypoleuca*)	39.20	12.1	34.9	6.5	18.6	Grant *et al*. (1982a)
Wedge-tailed Shearwater (*Puffinus pacificus*)	60.04	9.2	29.0	5.0	17.2	Whittow *et al*. (1982)
Laysan Albatross (*Diomedea immutabilis*)	284.8	4.9	8.8	1.4	16.0	Grant *et al*. (1982b)
Black-footed Albatross (*Diomedea nigripes*)	304.9	5.6	11.1	1.8	16.1	Grant *et al*. (1982b)

From Pettit & Whittow (1983).

Simons (1985) watched events continuously over 4 days of the hatch. About 2 days before chick emergence the adult became more alert, moving the egg frequently with its beak and sitting high as if standing slightly, with wings held out a little to either side. As the chick broke through, the parent helped indirectly by adjusting the position of the egg and standing so as not to restrict the emerging bird. Albatrosses behave similarly. But with *O.furcata* the same author watched an adult assist the chick by placing its bill inside the partly opened shell and shaking it from side to side (Simons, 1981b). This is the only record I can find of a petrel actively helping at the birth of its chick.

The egg shells are usually ignored, crushed into the nest with the other nest materials, if any, or scratched out to the sides of the nest or chamber by parents, perhaps to be returned later by the chick to form part of the nest lining, e.g. in *P.carneipes* (photo opposite). Reports of petrels deliberately shifting shells from the nest most probably refer to incidental removal during scratching-out activity.

CHAPTER 15
The Chick Stage

Fleshy-footed Shearwater in nest chamber.
Photo: author.

Contents

I Introduction

On the altricial–precocial scale of development (Nice, 1962), petrels fit best into the semi-precocial category because of their slow growth and need of parental feeding, a precocious ability to acquire homeothermy and of being able to shift around on their legs almost as soon as their down is dry (Warham, 1983). Ricklefs *et al.* (1980b), however, regarded the chick of *O.leucorhoa* as developmentally precocial because of the 'water index' (a measure of the maturity of a tissue) of the legs and integument, which for the petrel neonate, was 4.1 compared with values of 3–4 for semiprecocial and 6–8 for altricial birds. Diving petrel chicks are perhaps less precocial than other burrow-bred species, requiring brooding longer than them because of their thin down. Albatrosses and fulmars, but not apparently other surface-nesting tubenoses, are also less precocial in needing brooding for longer than chicks in burrows.

Success in chick rearing, though perhaps less than in incubation, depends on the dove-tailing of the activities of the two parents and I know of only three instances where a single parent has raised a chick: Davis (1957) reported a Manx Shearwater and Ollason & Dunnet (1982) two fulmars doing that. Brown & Adams (1984) also described a female Wandering Albatross that for 256 days brought up a chick single handed which eventually fledged. Abandoned chicks are sometimes fed by passers-by and Lefebvre (1977) recounted a Laysan Albatross chick that evidently fledged after being tended by no less than six strange adults.

And, as for incubation, chick rearing is a slow process. Despite but a single mouth to feed, even the smaller petrels take at least 47 days to fledge their chick, despite their higher provisioning rates than those of large petrels. The chick of *H.pelagicus*, adult weight 28 g, does not fly until about 70 days old (Table 14.2) whereas a 25 g House Sparrow can rear three to four chicks in just 14 days. Of course this bird gets its food within a few metres of its nest, and delivers many meals daily whereas the parent petrels often fly far and individual parents come back no more than once every 24 h, most of them much less often.

The rearing of petrel chicks can be looked on as occurring in two inter-digitating stages. In the first the hatchling is brooded and guarded by one or other parent, getting its small meals mostly from the on-duty bird, in the second stage it remains alone, being fed at intervals by both parents working independently. The guard stage among burrowing species is short, often only one or two days, but is much longer among most species nesting in the open.

The energetic constraints posed by the procellariiform life-style and chick-rearing procedures are not as yet unravelled but Ricklefs (1983) provided an initial model and indicated the kinds of data needed for a more quantitative evaluation.

II The hatchling

Once dry, the neonate is well insulated by a coat of loose, rather ragged down. In most species two downs develop in the same follicle, the initial protoptiles being succeeded by a mesoptile down and the series finishing with the teleoptiles or definitive feather. The rachis of each is morphologically continuous with the next stage, the barbs of one being in the same series as those of the other but separated from them by the calamus (which lacks barbs) and the protoptiles remain attached to the tips of the mesoptiles and sometimes both may remain attached to the teleoptiles before being lost through wear. In the prion *P.desolata* the mesoptiles become visible below the protoptiles on the eighth day (Tickell, 1962, p.30) while

Richdale (1963, p.27) stated that in *P.griseus* they usually show on the seventh day, but in *P.urinatrix* on the sixth. In Antarctic procellariids protoptile replacement occurs when about 15% of the nestling period has passed (Mougin, 1975, p.119). In contrast to *P.griseus*, the mesoptiles appeared in chicks of *P.pacificus* at 22.6 ± 3.7 days (n = 23) according to Pettit *et al.* (1984b).

Percival (1942) compared the juvenile downs of the Royal Albatross and Fairy Prion with those of some other birds and concluded that the double downs of penguins and petrels are not homologous with the protoptiles and mesoptiles of ducks and swans. He called the double down of petrels a diplotile, its parts the pseudo-protoptile and pseudo-mesoptile and considered the succession of downs in petrels a primitive feature.

Not all petrels grow two downs. At least two of the long-legged storm petrels, *O.oceanicus* and *P.marina*, have but one (Roberts, 1940); Richdale, 1965a) and that unusually long. These single downs would be haploptiles in Percival's terminology. The situation in the other genera of the Oceanitinae, *Fregetta* and *Nesofregetta* seems undetermined and the biological significance of double downs unclear.

The powder-puff appearance of the newly dry chick may hide bare, flesh-coloured patches on the face and sides of the neck and there may be a bald patch on the crown, e.g. in *H.pelagicus* (Lockley, 1932) and *P.marina*. These fleshy patches darken later and in many species the down is thin on the head. Hatchlings often have pale legs and these usually darken with age.

Petrel downs are usually grey or white, often paler ventrally. The mesoptiles may be of a different shade to the protoptiles, e.g. in *P.urinatrix* where the primary down is white, the secondary grey, so that while both are present the chick presents a mottled appearance. More usually the overall pattern is quite uniform, one species with a marked pattern is that of *M.halli* (Fig. 3.4). In *D.nigripes* and *D.immutabilis* the chicks have a peppercorn appearance due to the white tips of the down feathers and this appearance persists into the mesoptile stage (Rice & Kenyon, 1962).

Aberrantly plumaged chicks are not rare, particularly in museum collections. A few Mottled Petrel chicks had whitish chins and one a white belly instead of being medium grey all over (Warham *et al.*, 1977). With some other species two or more plumage phases can be distinguished. There are light and dark phases of the downy chicks of *D.bulleri* at the Snares Islands but the most extreme example is furnished by the young of the Galapagos Albatross whose downy young may be pale and almost white, dark brown or of intermediate shades (figures in Loomis, 1918). These variants show well in the protoptile stage, but are not restricted to that stage. Harris (1973) found that the proportions of the three phases varied from colony to colony and from season to season. Chicks of *P.pacificus* in New South Wales had quite variable down: most were grey, but shades between that and a rusty red were recorded, 15 out of 343 being this colour (Swanson & Merritt, 1974). Except for the fulmars *M.giganteus* and *M.halli* and *F.glacialis*, where adult plumage seems to correspond well with that of the downy state, the significance of these various polymorphisms and their relationship to the adult plumage patterns is unknown.

The sequences of feather eruption have been charted for a number of species, usually to help in ageing chicks. Davis (1957) found that he could age 80% of his storm petrel chicks to the nearest 5 days by feather development. The remainder consisted of underfed and retarded birds.

Charts of the moult of down and feather development have also been published for *D.nigripes, D.immutabilis, M.giganteus, D.capense, P.nivea, F.glacialoides, P.desolata, P.turtur, P.phaeopygia, P.aequinoctialis, P.griseus, P.pacificus, O.oceanicus, P.marina, F.tropica, O.castro* and *P.urinatrix* (Rice & Kenyon, 1962; Conroy, 1972; Brown, 1966; Mougin, 1968, 1975; Simons, 1985; Pettit *et al.*, 1984b; Richdale, 1963, 1965b; Beck & Brown, 1972; Beck & Brown, 1971; Allan, 1962).

Unfortunately not all the descriptions are accurate enough for comparisons between species — the appearance of teleoptile sheaths or 'pin' feathers often not being distinguished from the feather vanes proper. In general, the first true feathers appear on the scapulars, then on the remiges followed by those on the rectrices and wing coverts with those on the breast, face, belly and head appearing later. Primaries first burst through their quills when about 30% of the chick's time in the nest has passed, the first true rectrices after 35–40% of that time has gone.

Most chicks have their eyes closed at birth but after 24 h many are bright-eyed, e.g. in *P.urinatrix, D.epomophora, P.griseus, P.pacificus, P.inexpectata, P.neglecta* and fulmars, others have their eyes closed, but in burrowing species this may be a response to bright light and chicks tend to be reluctant to open their eyes during their early days. However, *O.oceanicus* evidently keeps its eyes closed until day 5 or 6, *F.tropica* until day 10 (Beck & Brown, 1972; 1971) whereas some *O.furcata* have their eyes open at day 2 (Boersma *et al.*, 1980, p.274).

The shape of the hatchling's bill may differ considerably from that of the adult but little information is available. At birth albatrosses have a short, stubby, downturned bill in which the ungues occupy nearly half its length (Fig. 15.1). Prion chicks, on the other hand, start with thin, narrow beaks which get deeper

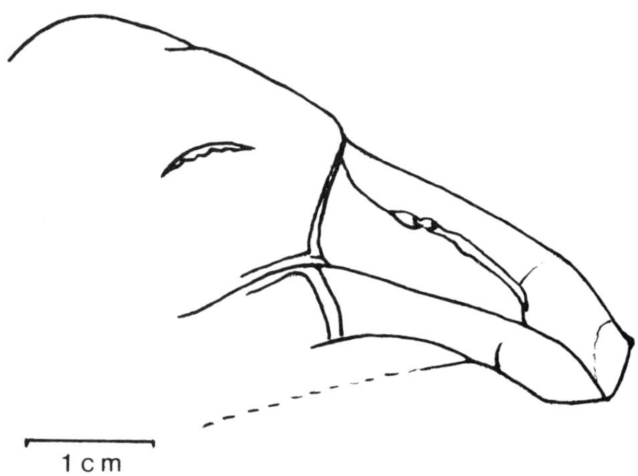

Figure 15.1 *Bill of 65-day-old Buller's Albatross 4 days from hatching.*

and wider as they develop. Fleming (1941a) drew attention to their resemblance to those of adult storm petrels.

The egg tooth is whitish and prominent, the time taken for it to disappear varying considerably. With the Common Diving Petrel the mean for 30 chicks was 12.8 ± 2.8 days (10–19 days) and for nine Fairy Prions 12 days (Richdale, 1965a,b), in *P.desolata* 10 days, in 18 chicks of *P.inexpectata* 8 days (3–16 days), in those of *P.hypoleuca* 9.9 ± 2.0 days and in *P.pacificus* 13.6 ± 1.6 days for 22 chicks (Tickell, 1962; Warham *et al.*, 1977; Pettit *et al.*, 1982b, 1984b). Chicks of species that are dark-billed as adults usually have these parts paler initially, darkening during development.

Few descriptions of the first feedings of petrel chicks are available and most suggest that the early meals mainly consist of stomach oil, e.g. with *P.pacificus* and *P.hypoleuca* (Pettit *et al.*, 1984b, 1982b). Rice and Kenyon (1962) believed that chicks of *D.immutabilis* and *D.nigripes* are fed exclusively on oil during their first days and Richdale (1942, p.256) saw a 24-h-old Royal Albatross given a liquid meal by its female parent after she had fasted for 10 days. With some species small meals may be given soon after the down is dry and more than one meal on the first day but Tickell (1968, p.33) found that newborn *D.exulans* were not fed during their first 24 h and given very little during the next 3 days. In contrast, hatchlings of the albatross *P.palpebrata* are fed frequent very small, mostly liquid, meals (Thomas *et al.*, 1983).

Harper (1976) reported that Fairy Prions returning to their newborn chick carried maximum loads of food (*c.*20 g) although the chicks needed but 5–10 g. One parent could supply that while the other sometimes stood by, both leaving still heavy with food. These chicks did not take feeds from both parents on the same night until they were 8–9 days old.

III The guard stage

The guard stage is defined as the period between the freeing of the chick from its shell and the time at which it is first left alone in the nest. This stage can again be divided into a brooding period when the nestling is heated by a parent and which merges into a post-brooding period when the parent is still on guard but the chick spends increasingly less time beneath it, often sitting uncovered alongside. Once the chick is homeothermic the constraints on guarding are likely to be the chick's vulnerability to predators and/or the nutritional state of the parent(s) and offspring, e.g. that one parent is unable to supply the chick's needs or parental reserves have fallen to a critical level (Croxall, pers. comm.). In general, there is a considerable variation in brooding and guarding times from one pair to another at the same colony, some leaving as soon as the chick relinquishes their shelter, when there is no non-brooding stage. The status of such birds is not known — they could be inexperienced pairs or inefficient foragers, for example. The time spent guarding may also vary with external conditions. Thus Harris (1969d) found that when feeding conditions were good *P.lherminieri* chicks were brooded for their first week, if bad they were left immediately and only visited to be fed.

A *Surface nesters*

There is a sharp difference in the lengths of the guard stage between surface-nesting albatrosses and fulmars and petrels that use burrows or crevices. Chicks on the former groups take an appreciable time to gain the ability to maintain their body temperature and are guarded for some time thereafter; chicks of the burrowing species are homeothermic within 1 or 2 days of birth and their parents soon leave.

Many figures are available on the duration of the guard stage but few clearly indicate the length of time that the chick is actually covered by the parent. Conroy (1972) and Hunter (1984b) do that for *Macronectes*. Hunter found that at South Georgia *M.halli* brooded for 19.7 ± 2.5 days and then stayed at the nest for a further 12.5 ± 7.7 days (0–30 days) whereas *M.giganteus* at the same place brooded for 18.4 ± 2.2 days and then left after only 2.4 ± 2.2 days (0–7 days) more. Hunter pointed out that *M.halli* fed on locally available seal carrion and could therefore prolong its nest attendance whereas its sibling species had a more varied diet and took little nearby food. Hunter also found that the chicks were particularly vulnerable to starvation and exposure in the non-brooding period, those of *giganteus* suffering more than *halli*. These latter may also have benefitted from their more sheltered sites here as elsewhere (Table 3.2). Male parents of both species spent more time guarding than the females did, but only with *halli* was the difference significant.

Guard stage lengths for other surface nesters are rather similar to those for giant petrels and include 31 days brooding + 3 days non-brooding for *D.exulans* (Mougin, 1970a) and 21.0 ± 2.1 days (n = 23) guarding by *P.fusca* (Berruti, 1979). Despite their warmer environments, the albatrosses of the North Pacific have guard stages of lengths comparable with those in the south. *D.irrorata* is 'brooded for several weeks' and 94 *D.immutabilis* chicks were first left unattended at 17.2 ± 2.5 days (12–25 days), 24 *D.nigripes* at 19.1 ± 2.7 days (13–25 days) according to Rice & Kenyon (1962). Both these albatrosses resumed guarding intermittently during the following 10 days or so, and this has been noted among other tubenoses in response to bad weather. Among diving petrels resumption of guarding (and possibly brooding) is common and one *P.urinatrix* chick that was guarded for 14 consecutive days had a parent beside it on day 22 (Richdale, 1965a, p.33). More often, however, once a chick is unguarded it is not brooded again.

In his work on Northern and Southern Fulmars Mougin (1967) found that *F.glacialis* brooded its chick for about 10 days (5–15 days) and remained with it for a further 4 days, the guard stage totalling 11–28 days. The southern birds brooded for about 12 days (8–16 days), non-brooding for 12 more days, so that the guard stage lasted 17–38 days. Mougin explained the difference to the more severe conditions at Terre Adelie, the chill factor there being 660 cal m^{-2} h^{-1}, that at the Scottish site 260 cal m^{-2} h^{-1}.

B *Underground nesters*

Guard periods of 2–3 days are typical of burrowing species, brooding time not being related to body size even in cold climates. Some chicks are left on the night of the day on which they are born when they can only have been heated for a few hours, e.g. in *P.inexpectata* (Warham *et al.*, 1977), *P.tenuirostris* (Farner & Serventy, 1971) and *G.nereis* (Despin *et al.*, 1972). Beck & Brown (1971) thought that *F.tropica* was not brooded at all, but despite a cold climate with snow, the chicks fledged. Perhaps more remarkable are those of Wilson's Storm Petrel guarded at Signy Island for 3–5 days (Beck & Brown, 1972) and for much the same time in Antarctica (Lacan, 1971) despite the more rigorous climate there. Chicks of the largest burrowing petrel, the Shoemaker, were guarded for 3.8 ± 1.4 days (2–8 days) — 24 determinations by Hall (1987) and those of *P.macroptera* for 1.7 days (12–72 h; n = 36; Imber, 1976b).

Rather surprisingly, parents of the storm petrel *H.pelagicus* tend to stay quite a long time with their chicks — on average 6.2 days (3–10 days) according to Davis (1957). Those of a larger storm petrel, *O.furcata,* in Alaska were covered for about 5.3 days, and the 12 that were left unattended on days 2 and 3 all died (Boersma *et al.*, 1980). Simons (1981b) found that even when guarding had ended adults of this bird often brooded young chicks for up to 3 h at night when they returned to feed them but stopped doing that when the chicks were 3 weeks old. Simons (1985) attributed the short guard period of about 2 days given to the *P.phaeopygia* chick to the size of the parents' food reserves and their ability to feed the chick.

The chick's early competence in thermoregulation and the ameliorated microclimate of the burrow allow the parents to leave early and replenish their energy stores after the fasts of incubation, permit a greater foraging range and help ensure that the chick will get meals from both. Ricklefs (1983) has discussed aspects of the energetics involved.

The main exceptions to the brief guard stages among burrowing tubenoses are the diving petrels. Richdale (1965a) showed that one or other adult *P.urinatrix* stays with their chick for 7–14 days, most being covered for about 10 days and those in the warm climate of New Zealand's North island are also guarded for 10–15 days (Thoresen, 1969). Richdale ascribed the long guard stage of diving petrels to the very short down of their heads and backs. Being coastal feeders there would seem to be less need for the rapid acquisition of homeothermy among these birds. At South Georgia the subspecies *exsul* was guarded for 11–12 days and, at the same place the average age of unbrooded *P.georgicus* was 6.7 days (Roby & Ricklefs, 1983), i.e. about 5 days less than *exsul*.

It might be expected that the surface nesting petrels of the tropics like *Puffinus nativitatis, Pterodroma ultima, neglecta* and *arminjoniana* would also have extended guarding periods, but such information as is available suggests that, like their burrowing relatives, they leave their chicks alone at a very early age. Gardner *et al.* (1985) stated that *P.arminjoniana* 'incubates' its chick for its first 5 days and surface nesting *P.macroptera,* most of which lay underground, guard their chicks for but a few days — even tiny ones will sleep in the rain beside but not beneath their parents. Those *P.pacificus* that nest on the surface also cease guarding after 2 or 3 days. Surface-nesting chicks in warm climatic zones do not generally remain in the open exposed to the sun, but retreat to cooler places under vegetation, rocks or coral breccia.

C *Parental roles during chick guarding*

The tendency to return to the nest more often that develops late in incubation, carries over into the chick stage. Brooding spells are considerably shorter than those during incubation except among diving petrels. More frequent changeovers may perhaps be necessary to make sure that the chick gets adequate food, although later in the post-guard stage the time between visits may increase considerably and yet the chicks get substantial meals, perhaps gleaned from further afield.

Diving petrels and British Storm Petrels change guard nightly when brooding chicks, but typical guard spells of albatrosses and fulmars last 2 to 3 days. Longer ones are infrequent, although *D.exulans*, *D.immutabilis* and *D.nigripes* may occasionally stay with their chick for 5 or 6 days and spells of 10, 12 and 10 days have been recorded respectively (Tickell, 1968; Rice & Kenyon, 1962),

Rice & Kenyon (1962) found that the guard stage of *D.immutabilis* was made up of 6.0 ± 1.3 spans (3–10), that for *D.nigripes* of 7.6 ± 1.6 spans (5–12). The number for the *Phoebetria* albatrosses seems to be similar but there are more with the great albatrosses. About 11 are usual among *D.exulans* at South Georgia (Tickell, 1968, p.30) and a rather similar pattern is given by Richdale (1952) for *D.epomophora*. For *M.giganteus* Conroy (1972, p.37) gave a mean of 8.7 changes of guard (6–11) at 24 nests, each span averaging 2.3 days.

Guard duties generally appear to be fairly evenly distributed between the sexes but good data are scarce. Male *D.exulans* stayed longer than their partners (Fig. 15.2) but the difference was small compared with the differences between individual birds (Tickell, 1968, p.29). In *D.epomophora* 56% of 263 spans were done by the males (Richdale, 1952, p.74). Among giant petrels Hunter (1984b) found that female *M.halli* undertook 42% and 41% of the brooding and non-brooding spells respectively, their partners significantly more, while for *M.giganteus* the differences were in the same direction but not significant — sample sizes were smaller. As with *D.exulans*, the range of variation between birds was considerable, e.g. some males of both species did not stay once their chick no longer needed brooding; no female did that.

Brooding adults tend to be more restless than incubating ones and from time to time lift themselves off the chick and bend down to preen it, usually uttering low calls as when addressing the egg.

D *Attainment of homeothermy*

That chicks of burrowing petrels can be left untended when but a few days old, whereas those of fulmars and albatrosses need guarding for weeks rather than days, is not a consequence of the former's protection from predators as compared with an unguarded chick above ground: chicks of burrowing species simply become homeothermic much sooner than the others. Such an early ability to regulate their body temperatures is most unusual among nestlings. Even those of the North Pacific albatrosses (breeders in the cooler winter months) need parental heat at first. Their body temperatures while brooded tend to be higher than those of their parents (Howell & Bartholomew, 1961a), a situation also applying to

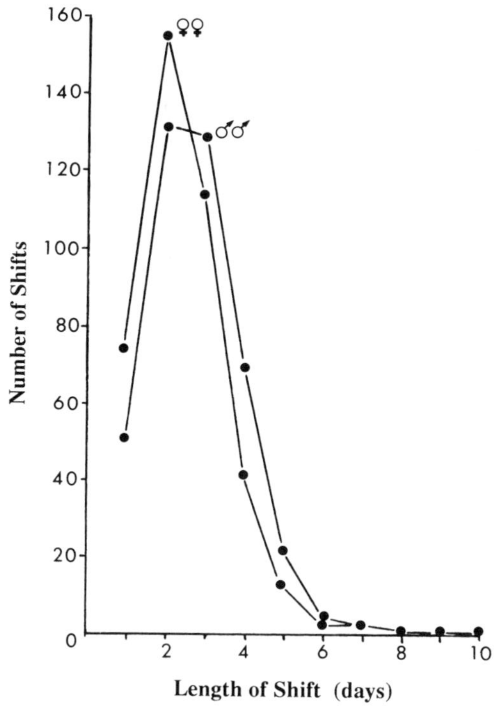

Figure 15.2 *Guard duties of* D.exulans. *From Tickell (1968).*

P.tenuirostris in Bass Strait and to *P.hypoleuca* at Midway Atoll (Farner & Serventy, 1959; Pettit *et al.*, 1982b).

The development of thermoregulatory ability has been followed by exposing known age chicks of some southern procellariids and of *D.exulans*, *P.palpebrata* and *D.nigripes* to known ambient temperatures and monitoring the resulting changes (Mougin, 1967, 1968, 1970a,b,c, 1975; Rice & Kenyon, 1962). Others have measured chick temperatures *in situ*, e.g. with *P.hypoleuca* (Howell & Bartholomew, 1961b; Pettit *et al.*, 1982b), *P.tenuirostris* (Farner & Serventy, 1959), *P.puffinus* (Bech *et al.*, 1982), *P.pacificus* and *P.nativitatis* (Howell & Bartholomew, 1961b), *O.oceanicus* (Roberts 1940), *O.leucorhoa* (Ricklefs *et al.*, 1980a), *O.furcata* (Boersma, 1986) and *Pelecanoides* (Ricklefs & Roby 1983).

Among southern albatrosses the body temperatures of brooded chicks are rather similar to those of their parents. Newly hatched Wandering Albatrosses, for example, while varying widely (31.0–39.3°C), averaged 37.6°C (n = 244), the same as those of resting adults (Mougin, 1970a). But removed from cover the chicks' temperatures dropped rapidly. They were unable to maintain steady temperatures until the end of the guard stage, the mean for 1040 measurements then being 38.1°C (35.3–39.8°C), well within the range for adult birds. The situation in *P.palpebrata* (Mougin, 1970c) and *D.immutabilis* and *D.nigripes* appears to be rather similar. Rice & Kenyon (1962) found that young *nigripes* could not maintain normal temperatures (37–38°C) in the shade without shivering until 18–20 days old. Howell & Bartholomew (1961a) found that brooded, newly hatched and dry chicks of *immutabilis* and *nigripes* had temperatures averaging 38.8 and 38.9°C, higher than those of their parents brooding under overcast skies (37.5 and 38.1°C), and they emphasised that the heat stress of chicks at nests in the open is considerable, post-guarded ones panting vigorously and using the webs of their raised feet as heat radiators. An 89 g chick of *P.nativitatis* (a surface layer) placed in direct sunlight, panted vigorously while its temperature rose to stabilise at 42–43°C.

Among the fulmarine petrels *D.capense, F.glacialis, F.glacialoides* and *M.giganteus* adult temperatures of about 38.7°C were attained by the time the chicks were left on their own, despite quite wide differences in the ambient conditions (Prevost, 1964; Mougin, 1967, 1975).

Figures for the emancipation of chicks of burrowing species from parental warmth are mostly rather similar. Farner & Serventy (1959) found that 2-day-old *P.tenuirostris* in burrows that were 15°C cooler than they were, could hold their body temperatures almost at adult level, with a tendency to increase above that up to day 4 but thereafter to maintain a lower figure. Manx Shearwater chicks alone in burrows at 5°C were homeothermic at around 34°C (Bech *et al.*, 1982). Lone Shoemakers 1–3 days old had temperatures (37.6°C, $n = 145$) like the adults (37.4°C, $n = 94$) when measured by Mougin (1970b). Eleven 1-day-old *P.hypoleuca* had average temperatures of 37.2 ± 0.5°C, seven 3-day-olds 38.3 ± 0.5°C and 26 that were 15 to 60 days old averaged 38.2 ± 0.4°C — the same as those of the parent birds (Pettit *et al.*, 1982b).

Nestlings of small burrowing petrels appear to thermoregulate as efficiently at an early age as do those of the larger underground nesters. On the Antarctic breeding grounds 33 lone *O.oceanicus* had a mean body temperature of 37.8°C, only slightly below that of three adults at the same site (Mougin, 1968) and of 38.8°C for 10 measured by Roberts (1940). Both *O.furcata* and *O.leucorhoa* have some thermoregulatory capability from hatching but are not homeothermic until about 5 days old and their temperatures are subject to a lot of variation, apparently because of variations in feeding (Boersma, 1986; Ricklefs *et al.*, 1980a). Boersma found that lively, well-fed chicks of *O.furcata* averaged 37.4 ± 2.0°C ($n = 838$), much the same as adults by day of 37.3 ± 2.3°C. Likewise a 5-day-old chick of *P.hypoleuca* that had missed meals for 2 days had a body temperature of 24°C at an ambient temperature of 20.5°C, but the following day, after a good meal, its temperature was back to 38.2°C (Pettit *et al.*, 1982b).

The slower development of thermoregulation among diving petrels is shown by comparisons with *P.desolata* (Ricklefs & Roby, 1983) and by Thoresen's figures for *P.urinatrix*. Ten chicks at the end of the guard stage and 10–15 days old had reached the minimum body temperature of adult birds, but not until they were 22–26 days old did the mean value reach 38°C, the average figure for adults at rest (Thoresen, 1969). The birds studied by Ricklefs and Roby, however, reached adult temperatures sooner.

A facet of the petrel chick's thermoregulatory apparatus that has thus far escaped examination is the insulating role of subdermal fat deposits as suggested by Lack (1968) and others. Even starved chicks seem to retain such a layer.

IV Nestling behaviour

Very young chicks of surface nesting petrels are completely covered in the parent's brood pouches, being treated like an egg except when the old bird rises as if to give the nestling a chance to breathe. Almost invariably, as the chick's head is uncovered, it gives a tremendous yawn, perhaps to ventilate the far reaches of its lungs.

Whether they nest on the surface or below it, parent birds meticulously preen the heads and backs of their young, using rapid nibbling movements of their mandibles. As the chicks get larger they often reciprocate, preening the heads, necks and flanks of their attendants. Richdale (1965a, p.24) reported that parent diving petrels, orientating to their young by touch in the dark, tend to run their bills over the foreheads and necks of the chicks and wear off the down prematurely, giving the nestling a closely cropped head and a topknot.

During its first day the petrel chick is inert but some becomes increasingly active and those in burrows can shift about when but a few days old. Chicks brooded in surface nests are less precocious, but gradually spend less and less time beneath their parents, become increasingly adventurous, and may make excursions from their nests while the on-duty adult remains there. Despite this, much of the nestling's time is spent sleeping and this continues during the post-guard stage (Warham, 1956, 1958; Simons & Whittow, 1984). The latter found that young *P.phaeopygia* spent 65% of their time resting quietly, 25% of their time sleeping, their oxygen intake then being greatly reduced and metabolic rates 20–30% below those when alert, thus saving energy. Awake and alone petrel chicks self-preen, scratch their heads with a foot brought up from under the wing and yawn frequently. Nestlings in burrows tend to become alert at dusk, perhaps in response to the nightime chorus overhead, to the prospect of a meal, or to an internal clock.

A chick's activity appears to depend partly on the time since its last meal. After a feed it may be quiescent for long periods but becomes increasingly active during the run-up to its next meal but, if this is delayed, the chick tends to become inactive again and in some instances torpidity ensues (see Section C3 below). Weather also influences activity. Southern albatrosses turn their backs to wind, hail, rain and snow and allow the latter to pile up around them, perhaps increasing their insulation against wind chill. Unseasonably warm days in the subantarctic appear to be more distressing than cold ones to healthy, well-padded, young albatrosses and these pant heavily, there generally being no shade to which they can retreat, whereas those of the North Pacific can often find shade under nearby *Scaevola* bushes or *Casuarina* trees.

A *Nest building*

From an early age some chicks drag in twiglets, leaves, roots, feathers and other loose items which they tuck in around them to sit in neatly line cups often more substantial than those during incubation. This activity usually takes place in complete darkness and is widespread among shearwaters — e.g. in *P.puffinus, assimilis, carneipes* and *tenuirostris* (Warham, 1952, 1955, 1958, 1960). When the chicks start to venture out after dark they may also drag back loose debris, e.g. coloured threads placed near the entrance eventually become part of the nests.

A more advanced form of this activity is seen among post-guard chicks of the great and North Pacific albatrosses. These tend to build their own 'play' nests not far from those in which they were born (Richdale, 1952, p.77). Such nests can be quite substantial if local materials are adequate. However, on the reappearance of a parent the chicks hurry back to their proper nests. Activity of this kind appears to have no immediate value but Black-footed Albatross chicks, which are typically reared in hollow excavations in loose sand, tend to kick this away from them in windy weather using similar motions to those used by parents before egg laying. These actions help to prevent the young birds being engulfed in sand but are not always successful.

B *Defensive activity*

Petrel nestlings discourage attack and intrusions in a variety of ways. Albatross chicks draw themselves up to their full height and rotate to face the threat, clopping their bills or snapping them quickly. If attacked they will reach forward and slurp stomach oil and food in the general direction of the disturbance. Fulmarine chicks make coughing and lunging noises at the source of alarm and are usually ready to spit volleys of stomach oil. In contrast to the splatterings of albatross chicks, fulmar discharges are usually well aimed. Many chicks of fulmarine petrels have large stores of this lipid residue from their food, but at about 40 kJ g^{-1} (Warham *et al.*, 1976) and metabolisable (Roby *et al.*, 1986), oil discharge might be expected to a last line of defence.

Northern Fulmar chicks have long been famous for the readiness to spit oil and Duffey (1951) found that they tend to do so at any bird that approaches, their own parents included, until, after about 3 weeks these seem to be recognised. The spitting response to sudden movement nearby seems to be innate and tiny fulmar chicks or starved ones may make spitting motions without ejecting oil.

There is virtually no information on the defensive repertoire of petrels apart from hissing, intimidatory lunges and spitting actions but those that are aggressive when adult, e.g. *P.aequinoctialis,* bite and scratch vigorously as chicks. None of these tactics seems of much avail for burrowing species attacked by cats or mustelids. Their effectiveness against natural enemies like skuas has not been evaluated.

V Chick feeding

A *Feeding procedure*

Whether nesting on the surface or below ground the procedure for passing food to the chick is similar throughout the order (Richdale, 1939; Warham, 1952, 1955, 1956, 1958, 1960; Simons, 1981) but action in underground nesting chambers are hard to observe without proper facilities and most feeding ceases when a light is turned on. Richdale's description of feeding by the Royal Albatross is typical of all.

The hungry chick pecks the sides of the parent's bill at the gape, calling persistently. The adult lowers its head but lifts its tail and folded wings somewhat and moves these quickly up and down as the neck distends and, with violent muscular contractions and sluicing sounds, the food is pumped from the proventriculus. As the adult opens its beak the chick inserts its own crosswise, its gape pressed hard into that of the parent bird as shown in Fig. 15.3. The chick's lower mandible forms a trough so that, as the parent squirts the food through its restricted throat, it is diverted sideways into the young bird's mouth, the latter's lower mandible and throat vibrating with a scissoring action of the mandibles. The chick withdraws to swallow the food while the parent gulps back undelivered material. Richdale believed that by constricting the opening of its throat the parent could control the size of the solid items released with the oil mash, larger items being given as its beak widens and the chick follows suit to take them. Food is seldom dropped during transfers.

Figure 15.3 *Fleshy-footed Shearwater feeding chick. From Warham (1958).*

This crossed-bill feeding method is used by very young chicks, e.g. by a 24-h-old *P.belcheri* and 3-day-old *P.macroptera* (Strange, 1980; Warham, 1956), but some newly hatched birds may simply make thrusting movements as they pick out items from the parent's mouth. Young diving petrels do this (Richdale, 1965a, p.32) as do

P.desolata and *P.turtur* (Tickell, 1962, p.27; Harper, 1976). Sagar (1979) found cross-billed feeding was used by Cape Pigeon chicks until they were about 40 days old, thereafter they usually faced the parent and took food directly from its gape.

Fulmars make low clucking sounds as they shuffle towards their chick. Shearwaters and gadfly petrels may call as they enter the entrance tunnel at which the chick awakens and responds with its high-pitched begging cries. Many parents meticulously preen the head and back of the nestling before delivering a meal and this tends to stimulate the chick to solicit. In deep burrows the feeding process takes place in complete darkness, the chicks first experiencing daylight on the day they fledge. Tactile, acoustic and presumably olfactory stimuli must be important for successful food transfer under these conditions. Preening may help the parent to orientate to its chick. The latter tends to jab rather wildly around the head of the adult bird which often draws the nictitating membranes over its eyes (Warham, 1952, p.1251; 1960, plate 8). These membranes also shield them during preening of the chick's loosening down. If necessary the eyes are closed completely.

A succession of feeds is usually delivered. Parent *P.tenuirostris* gave from 5 to 28 during a single parental visit (Warham, 1960). Harper (1976) saw Fairy Prions give meals lasting 2–9 min and involving 18–20 deliveries. Feeding episodes of Snow Petrels took about 15 min with regurgitations at about 10 min^{-1}, only small amounts of food passing initially (Brown, 1966).

B *Parental roles in chick feeding*

Each parent feeds the chick, both in the guard and in the post-guard stage. Their visits seem to be independent of each other and the alternating regime of incubation tends to carry over to the nestling stage. Except perhaps with diving petrels, visits to feed the young are more frequent than change-overs on the egg. Ricklefs (1983) calculated that intervals between feeds were about one third of the length of those between incubation spells.

Among the giant albatrosses (easily sexed on plumage characters) the tendency is for the males to be slightly more attentive. Tickell (1968, p.35) watched a group of marked *D.exulans* and found that the males fed 63 times (53%) and the females 55 times (47%). During Richdale's 16-day post-guard watch at a nest of *D.epomophora* the male spent 4 h 18 min with the chick, the female 2 h 13 min (Richdale, 1952, p.81). Whether the larger males actually provide more food than the females is not recorded.

Data on the roles of the sexes in feeding chicks in nests below ground are limited, partly because of the difficulty of identifying the sexes at night. Beck & Brown (1972) found that the male *O.oceanicus* was the more regular attendant of its chick. At 10 nests there were 166 visits by the males, 51 by the females and five when both were found together. Despite this, the females were thought to bring the major share of the food, their partners being more concerned with keeping the burrows clear of snow. During 2 separate weeks of continual observations by Simons (1981b) it was found that the parents of one *O.furcata* chick shared equally in feeding it.

Information is also sparse concerning the times of day at which the chicks are fed. Both Richdale (1952) and Tickell (1968) found that the giant albatrosses were fed almost exclusively by day, usually before midday. Daytime feeding seems to be common among mollymawks (pers. obs.) but few have watched these birds after dark, and during their brief guard stage the young of burrowing petrels get meals by day or night. Later they are fed only by night except for a few tropical species like *O.tethys* (Harris, 1969b).

C *Feeding frequencies*

Newborn petrels get small and quite frequent meals, the duty parent being able to supply oil and food for some time after leaving the sea. The chick is fed whenever a change of guard occurs. Exceptions, like the instance described by Richdale (1952, p.23), seem rare, though perhaps the result of inadequate observation. He suggested that visits without feeding the chick were part of a 'warming up period' of transition behaviour between that appropriate to incubation or guarding and that for foraging. And in surface-nesting tropical species visits to nests without supplying the chick are perhaps not unusual (c.f. Ricklefs, 1984b).

In the post-guard stage a parent seldom delays long before feeding and surface nesters usually leave soon after having done that. Such parents are thus at the nest for minutes rather than for hours, although, as in most behaviour, there is much individual variation. Burrowing petrels may stay longer. Although these too usually feed their chick soon after reaching the nest; some may go straight back to sea, others rest and sleep either in the nest chamber itself or outside on the ground nearby (Warham, 1955, 1958). Such birds presumably depart with the main pre-dawn exodus.

Because of these brief and rather infrequent visits it is evidently only by chance that once guarding has ended the parents meet at the nest and their chicks get double meals. This is particularly true of large species and with shearwaters and gadfly petrels: among storm petrels, prions and diving petrels, that feed their chicks frequently, such meetings are more common (Fig. 15.4).

1 *Problems in measuring meal sizes and feeding frequencies*

There is a considerable body of information concerning feeding frequencies, much of it based on regular weighings of chicks rather than on direct observation or automated recording. Such work is very time-consuming if statistically adequate data are needed. Even then, because of the rapid loss of weight experienced by the chick, morning and evening weighings may not reveal small meals and the sizes of large ones are underestimated. The importance of this was shown in the study of *P.turtur* by Harper (1976). The food intake of this very tame bird had been painstakingly calculated 40 years before by Richdale (1965b). He used night and morning weighings to obtain average meal sizes of 15.0 $\pm$ 0.6 g (n = 316). The same methods with Harper's birds gave meal sizes of 15.3 $\pm$ 1.4 g (n = 57). However, this worker was able to weigh both parents and chicks before and after feeding and thus get meal sizes directly and before digestion had begun. He also back-checked by comparing the weight gains of chicks against the weight losses of

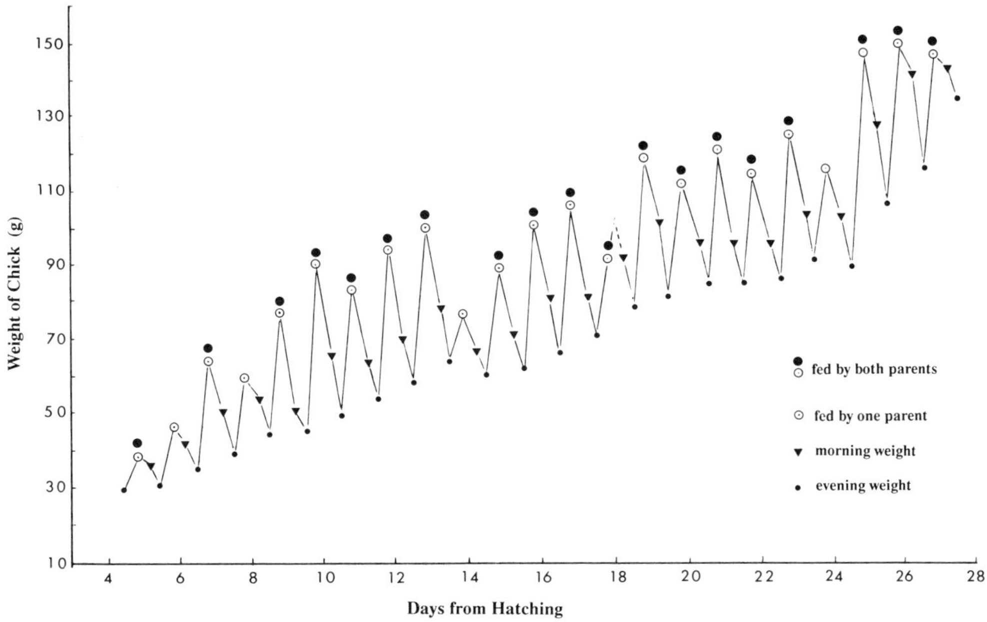

Figure 15.4 *Daily growth and feeding of a Fairy Prion chick. From Harper (1976), courtesy of* New Zealand Journal of Zoology.

their parents. In a typical example the mean loss of a particular parent over 15 weighings was 26.6 ± 1.2 g, the chick's gain 25.1 ± 1.6 g, the 1–2 g difference not being due to spillage. The difference between the 57 true and 57 apparent meal sizes calculated from morning and evening weights was large, amounting to about 120%, viz 33.7 ± 1.8 g *v.* 15.3 ± 1.4 g.

Loss of weight overnight due to excretion and respiration is thus very substantial, in the case of *P.turtur* totalling 55% of the weight of the food given. As Harper pointed out, night and morning weight changes have been widely used to estimate not only meal sizes, but also the frequency of parental visits, the lengths of chick fasts and of desertion periods, and to assess the limitations imposed by provisioning rates on the adults' foraging ranges.

Such problems tend to be less acute with larger birds and surface nesters which can be watched or recorded continuously and feeding frequencies established directly, e.g. with weighings taken before and after a meal, preferably automatically, e.g. as done by Ricketts & Prince (1984). Richdale (1963, p.37), in his twice-daily weighings of Sooty Shearwaters made no allowance for 'the small drop in weight due to respiration and excretion'. He gave no figure for the difference but it could well have been significant.

Regular weighings and measurements of chicks can provide a general picture of growth and development, but for detailed studies of metabolic costs and energetics true figures for meal sizes are essential. Attempts to overcome the problems by estimating meal sizes from indirect weighings by calculating mass loss with time, as done by Simons & Whittow (1984) and Ricklefs *et al.* (1985) involve more assumptions.

For small burrowing petrels the problem is particularly acute where direct observation is difficult and metabolic and excretory losses are high. And many, perhaps most, adult petrels will not tolerate repeated handlings such as Harper's birds accepted. Furthermore, if the chicks are weighed frequently at night, the deduced meal sizes will be depressed when adults find the investigator at their nest: many then fly off and the chick misses a meal. The study of weight changes of diving petrels by Payne & Prince (1979) is the only one I know of where this possible source of error is noted, although Lack (1968, p.275) did caution of the possibility. See also Zach (1988). For accurate assessment of meal sizes, instrumentation to record automatically (a) the time and direction of each passage of the birds through the access tunnel, (b) their weights coming and going, (c) the chicks' weight before and after a visit and (d) a video- or other recording of behaviour in the nest chamber, seems to be needed. At a less advanced level the risk of overlooking visits because small feeds are undetected can be lessened by fencing burrow entrances: displaced palisades may or may not indicate parental visits, intact ones mean none has occurred.

Most weighings of chicks have been done using a hand-held balance with the animal in a bag. If done from the chick's early days these often become accustomed to the procedure and discharge little oil, food or excreta. Among a random sample of chicks there will be some that never spit, some that stop doing so, and some that always discharge oil or food. These latter will thus experience retarded development and their data be useless for most purposes. And many small petrels also discharge when handled, e.g. *O.furcata*, behaving quite differently to *P.turtur*.

Such problems are customarily ignored. Hunter (1948b) did monitor a control group of giant petrels, finding that those chicks were slightly heavier than regularly handled birds. He stated that the differences were not significant but gave no figures.

2 *Feeding frequencies*

Because of the above problems, many of the figures for frequencies of visits to chicks will be lower than the true value, fasting periods are overestimated and meal sizes underestimated. Data on large, surface-nesting, diurnally active birds are most likely to be correct, although observation has seldom been continuous. With birds nesting underground perhaps that of Simons (1985) on TV monitored nests is most reliable, apart from Harper's work. Simons found that with *P.phaeopygia* the rates peaked during the period of maximum growth at 30–60 days, after which visiting declined until, in the last quarter of the nestling period, the chicks were visited and fed only 3.2 times on average (Table 15.1). Chicks of its congener *P.macroptera* were visited every 3.9 nights and as burrow entrances were fenced by the observer, probably few visits were missed (Imber, 1976b).

Rather similar figures are given for some surface nesting petrels. Southern Giant Petrel chicks 0–5 days old were fed on 98.7% of those days, on 41% when 31–40 days old and on 27% of days at 91–100 days (Conroy, 1972). Southern Fulmars during early chick raising gave feeds on 54–55% of days, in the post-guard stage on 28–37% and at the end of rearing only on 8–9% of the days, figures for the Northern Fulmar being much the same (Mougin, 1975). The general pattern is for visits and

Table 15.1 Estimated feeding rates of Dark-rumped Petrel chicks

Chick age (days)	n^b	Est. food delivered to chick (g)	% Total food	Mean number chick feeding visits	Est. mean food load size (g)	Load as % adult wt
0–30	9	791.7 (53.0)[a]	32	12.5 (1.8)	63.3	14.6
31–60	10	901.5 (173.8)	36	14.3 (4.03)	62.9	14.5
61–90	9	682.6 (79.7)	27	12.3 (3.28)	55.4	12.7
91–120	9	114.7 (67.5)	5	3.2 (1.9)	35.6	8.2
0–Fledging (total)	7	2,501.7 (94.6)	100	44.5 (7.6)	56.2	12.9

[a] Values in parentheses are 1 S.D. [b] Chicks monitored by event recorder.
From Simons (1985).

feeds to be most frequent during the first half of the chick's growth, gradually declining towards fledging (Fig. 15.5).

Tickell & Pinder (1975), who weighed some mollymawk chicks before and after parental visits, and many more once daily, found a regular decline in visits with time both of *D.melanophrys* and *D.chrysostoma*, e.g. for the former from 14.2 visits to chicks 1–20 days old down to 5.8 visits during their last 20 days in the nest. These

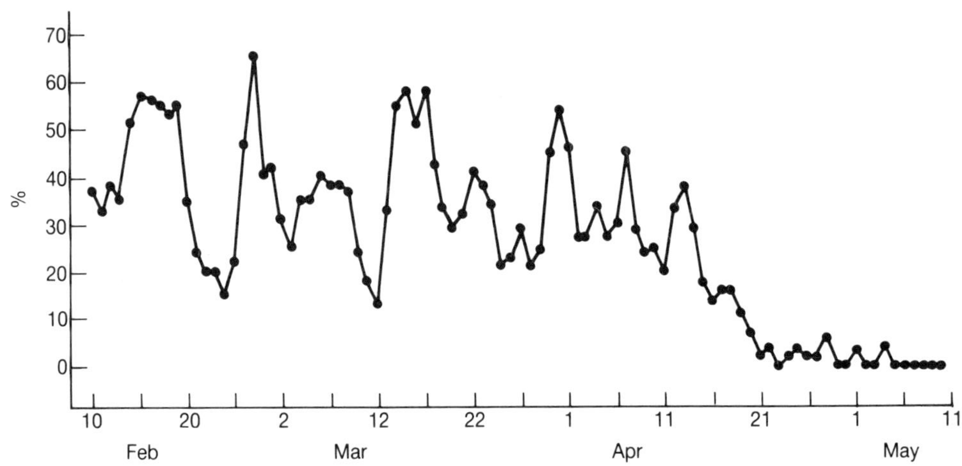

Figure 15.5 *Nightly visits of adult Sooty Shearwaters to 57 chicks towards the end of the breeding season. From Richdale (1963), courtesy of The Zoological Society of London.*

authors recognised that their methods underestimated the weights of meals and failed to separate two small meals from a single large one. While the frequency of feeds to nestling *D.melanophrys* fell, the mean daily meals fed increased to a maximum between 61 and 80 days.

Small petrels give more parental attention than large ones. At burrows of *O.furcata* fitted with event recorders the number of feeding visits from hatching to day 30 was 0.7 ± 0.6 day^{-1} ($n = 100$) in 1977 and 1.1 ± 0.7 day^{-1} ($n = 92$) in 1978, the food delivered each time being 8.0 ± 4.0 g ($n = 75$) in 1977 and 8.0 ± 3.8 g ($n = 85$) in 1978 (Simons, 1981b). Some large petrel chicks also get much attention. Cory's Shearwaters on Crete were visited every 0.8 days and on half these occasions were fed by both parents (Round & Swann, 1977), so that in those circumscribed seas they must have foraged close to the breeding grounds.

There may be some exceptions to the general rule that visits and meals decrease towards fledging, Harper (1976) reporting that the quantity of food given to Fairy Prion chicks did not decline as they aged and that the weight loss towards fledging was not due to reduced provisioning.

3 *Fasting*

Like their parents, the chicks of tubenosed birds can withstand long fasts. Even those of small petrels such as *H.pelagicus* (28 g) and *O.leucorhoa* (49 g) fast for 6–7 days 'without observable effect' (Lockley, 1932; Ricklefs *et al.*, 1985). Chicks of large albatrosses often have to go without food for 5–12 days (Tickell, 1968, p.41) and one of *D.exulans* did so for 51 days, lost 44% of its initial weight, but still fledged. Fasts of mollymawk chicks, although occasionally lasting 12 days, were shorter. For chicks of *D.melanophrys* 0–20 days old 85% of fasts were of only 1 day, and even among older chicks these still comprised 50% of the fasts. Fasts of *D.chrysostoma* chicks were longer: in 1–20 day olds 66% of fasts were 2–4 days long compared with only 15% for *melanophrys* chicks of the same age, $P < 0.001$ (Tickell & Pinder, 1975). They suggested that *D.chrysostoma* either travels further for food or spends more time selecting its prey than the other bird.

Most fasting arises from the tardy return of a parent, perhaps due to changed climatic or feeding conditions at sea or adverse winds, possibly far from the nest. Climatic deterioration on the breeding grounds may also be a factor, such as snow blockage. This is a common cause of mortality among Snow Petrels in some areas. The adults may manage to dig through soft snow to reach their chicks, but if this has melted and frozen again digging may be impossible; the chicks not already suffocated then starve to death. Richdale (1952) found that fog around the breeding places delayed the return of Royal Albatrosses, but what effect fog has on the provisioning of chicks in burrows is uncertain. Most researchers have concluded that bright moonlight has little deterrent effect on parental visits but Oka *et al.* (1987) hypothesised that there is a semilunar rhythm to parental visits of *P.tenuirostris*, perhaps linked to a semilunar pattern in surface swarming of their euphausid prey.

Fasting chicks lose weight. With those of *O.oceanicus* that were snowed in, Wasilewski (1986) found losses proportional to the weight at the start of starvation. In chicks starved 4–5 days at burrow temperatures between $-6°C$ and $-10°C$ the decrease in weight (y) in g day^{-1} in relation to the initial weight (x) in g was given by:

$$y = 0.0975x - 2.29$$

with $r = 0.95$ and $P < 0{,}001$.

Otherwise there are few curves showing weight losses of starved chicks over time (except during final desertion), but what are available suggest that the daily rate of loss is high initially as stomach contents are digested, then falls for a lengthy period to a rather constant figure while lipid is consumed. In some cases the rate may then increase as a switch to burning protein occurs. Tickell (1968, p.35) plotted the losses of two *D.exulans* chicks and other data are those of Payne & Prince (1977), Ricklefs *et al.* (1985) and Croxall *et al* (1988).

4 *Torpor*

Petrel chicks that are starved become cold, often shiver, and may become apparently lifeless but, as various observers have found, often recover when warmed. Such revivals are most often noted in small species — in *H.pelagicus, O.castro, O.oceanicus, P.desolata,* for example — but also in some larger ones like *P.griseus* (Warham *et al.*, 1982). Davis (1957) commented on a storm petrel chick moribund after 6 days without food but which was remarkably lively the morning after a meal, that its survival had been helped by a general slowing down of its metabolism. Perrins *et al.* (1973) found that chicks of *P.puffinus* did not go torpid when starved but their temperatures fell. Chicks of *P.phaeopygia* also became partially torpid and lethargic, shivering but often recovering quickly when brooded (Simons & Whittow, 1984).

True torpidity due to extreme hypothermia produces chicks that appear dead, with no visible breathing. Their resistance to chilling reflects that of the egg (Chapter 14, Section IVB).

The ability to reduce metabolic rates is probably widespread among petrels but may be little used in mild climates. The condition has been closely examined in only one species, *O.furcata,* breeding in cold, damp burrows in Alaska (Wheelwright & Boersma, 1979; Boersma, 1986).

Boersma found that unbrooded chicks of this species had body temperatures (T_b) of 37.4 ± 2.0°C (36–39°C; $n = 838$), depending on the ambient temperature (T_a) in the burrow and this varied considerably with its type and situation (but not with its depth). Cold, starved chicks became sluggish and some torpid, with T_b dropping by 10°C and even down to 10.6°C. Young chicks survived lower T_b's than old ones. Ten 1- to 3-day-olds survived drops to 10.6–14.2°C, in twelve 8- to 27-day-olds T_b fell to 18.8–26.3°C. Small chicks required parental heat to recover, but Boersma believed that larger ones were capable of self-arousal and sensing the increased warmth consequent on the appearance of a parent in the nest chamber, even though that bird would presumably not try to brood. Once active the chicks could be fed and T_b maintained.

Torpor was seen as a last resort, occurring infrequently and often ending in death. As the chick grows its insulation increases, it thermoregulates more easily, surface area to volume ratios and maintenance costs decline, so that it can withstand fasting better, but, if persistently starved, older chicks also lapse into hypothermia.

The subsequent history of chicks that have experienced torpidity is unknown but at least one of Davis's storm petrels survived: it fledged at a rather low weight after a record nestling period of 73 days and was captured off France by a fishing vessel (Davis, 1957, p.378).

D *Meal sizes*

A meal consists of a series of feeds given at one time and because of the problems outlined above there are very few sets of complete data. The figures are more reliable for large species and for surface nesters, as already noted.

Even with daily weighings that underestimate meal weights, it has long been known that petrel chicks may accept gargantuan meals, sometimes doubling their weight overnight. Some 19- to 28-day-old *P.griseus* did that (Richdale, 1963, p.41) and a 29-day-old *P.inexpectata* weighing 360 g was 99% heavier at 715 g the next morning (Warham *et al.*, 1977).

Such mammoth meals doubtless often come from feedings by both parents on the same day and Richdale considered that those of more than 300 g fed to *P.griseus* were of this nature. Double meals of *P.turtur* were approximately twice the weight of single ones (Harper, 1976) and double meals of *P.macroptera* were calculated to occur once in 14 meals, single ones being as much as 160 g (24% of adult weight), double ones 160–225 g (Imber, 1976b). A 110 g meal given to a Dark-rumped Petrel chick was 26% of the adult's weight and thought to be close to the carrying capacity of the bird (Simons & Whittow, 1984).

These authors, although monitoring burrow activity automatically, did not weigh their *P.phaeopygia* chicks at short intervals but corrected for weight losses from previous determinations of these, allowing 15 g day^{-1} for 0- to 15-day-olds, 20 g day^{-1} for 16- to 60-day-olds, 25 g day^{-1} for 61- to 91-day-olds and 13 g day^{-1} for older chicks. Meal sizes fell after 60 days and averaged 13% of the mean adult weight (Table 15.1).

Croxall *et al.* (1988) used similar methods for examining meals given by *O.oceanicus*, arriving at average meals of 7.0 g (20% of adult body weight) and they concluded that meals of 15–25% of body weight are typical for storm petrels. Young *O.furcata* weighed before and after meals got 3–19 g at a meal, the parents being able to deliver loads equivalent to 29% of their body weights (Simons, 1981b). Mean sizes of *P.georgicus* and *P.urinatrix* were given as 22.3 g and 26.0 g (21% and 20% of adult weights) by Ricklefs *et al.* (1985) and double meals were not involved. True meal sizes for three *P.turtur* chicks fed by single parents averaged 25.1, 15.7 and 18.7 g representing 18%, 12% and 14% of the total adult mean weight on landing of 135.8 g (Harper, 1976). The average meal of *P.aequinoctialis* chicks was calculated at 190 g, i.e. 46% of a 10-day-old chick's weight, 10% of that of a 92-day-old and about 15% of mean adult weight (Berruti *et al.*, 1985).

Meal sizes estimated from 4-hourly weighings of *P.nativitatis* and *P.alba* chicks that had reached adult weight averaged 48.2 and 49.4 g respectively (14.2% and 18.2% of adults' weights), although daily intake was higher at 54.8 g and 37.1 g

(Ricklefs, 1984b). Those of 40–60 g *O.leucorhoa* weighed at 3 h intervals averaged 8.1 g, 8.7 g and 10.1 g in samples of light, medium and heavy chicks, i.e. 18%, 19% and 22% of adult weight (Ricklefs *et al.*, 1985).

Feeds again representing about 18.5% and 16% of adult weight were given to chicks of the mollymawks *D.melanophrys* and *D.chrysostoma*. These were weighed automatically every 10 min and got 680 g and 600 g at a time, or 37% and 39% of the chicks' pre-meal weights (Ricketts & Prince, 1984).

Giant albatrosses seem to deliver relatively smaller meals. Tickell (1968, p.35) weighed 53 *D.exulans* chicks before and after being fed. Meal weights ranged from 113 g to 2112 g, this latter evidently given by a single bird and amounting to about 24% of the parent's weight. Tickell also gave mean sizes during Wandering Albatrosses' last 100 days as nestlings. They varied from 660–880 g (7.6%–10.2% of mean adult weight) and declined to 400–520 g in the 20 days before departure.

A Royal Albatross chick had 15 meals (from 110 feeds) averaging 610 g during 16 days. The mean amount from a single regurgitation was 85 g. Meal sizes ranged from 225 to 1930 g, but the last was a double meal given in 11 feeds during which the chick gained 35% of its initial weight and then refused further food (Richdale, 1952, p.79).

Contrary to the findings with many other species, Harris (1966) and Harper (1976) found that the size of meals given to *P.puffinus* or *P.turtur* varied little with the season or with the size of the chick. Likewise meal size in *O.leucorhoa* was basically constant, as was meal frequency, from day 10 (Ricklefs, pers. comm.).

Estimates of the total food intake include that of Simons & Whittow (1984) who calculated that the average young *P.phaeopygia* received 2.5 kg — Table 15.1. Young Wandering Albatrosses got 51–82 kg from the end of brooding to departure (Tickell, 1968, p.38). Tickell & Pinder (1975) calculated that *D.melanophrys* chicks got about 20 kg during their time in the nest and Berruti *et al.* (1985) that the Shoemaker chick ingested about 8.6 kg during its 96 days ashore.

VI Nestling periods

A *Introduction*

For tubenoses the nestling or fledgling period can be defined as the time elapsing from the chick's birth to its reaching the ocean. This is, in effect, its age at its first flight, as most fledglings fly straight out to sea. Average periods vary from 45 to 278 days according to species (Table 15.4).

Precise figures for nestling periods are not as easily obtained as those for incubation periods, published values seldom being accurate to more than ± 0.5 days. Furthermore, the fledgling date assumed may not be that on which the chick reaches the water, as some may alight on land and delay final departure. This is probably quite common among surface nesters but is easily overlooked, because fledglings tend to come down closer to the coast and well away from their colonies. Errors are even more likely with chicks reared under ground; if failing to get away before dawn at their first attempt they may take refuge in burrows at lower altitudes until the onset of darkness allows a new attempt. In practice, the nestling period is usually taken as the age at which the full-grown chick disappears.

Reported figures will thus tend to underestimate true ages at departure, and in some tabulations calculated rather than precisely determined figures are included. This has not been done in Table 15.4.

B *Variability*

Nestling periods are much more variable than true incubation periods, partly because the embryo develops in very uniform surroundings and contains its own energy supply whereas the chicks have to cope with fluctuating microclimates and food supplies. Warham (1968) found that for 12 species with adequate data the mean coefficient of variation of nestling periods was 4.7; for a small sample of true incubation periods this statistic was 1.7. Mougin (1975) pointed out that nestling periods were less variable among Antarctic tubenoses than among subantarctic ones — a difference mostly due to Antarctic fulmarines which are constrained by the shortness of the summer during which they can breed.

Some of the variation stems from environmental conditions, e.g. El Nino effects. Violent gales tend to delay departures. Steady breezes may stimulate take-offs; for example, chicks of *D.epomophora* needed a wind speed of at least Beaufort Scale 3 for their first flight (Richdale, 1952, p.87). Calms inhibit departures and if these persist birds may simply walk to the sea and swim away. Bright moonlit nights may also inhibit nestlings from leaving their burrows although apparently not with *P.vittata* or *P.turtur* (Richdale, 1965b, pp.111, 134). At high latitudes delays may arise through snow blockage of burrows or nest recesses.

Variation in nestling periods within a species has often been ascribed to variation in provisioning by parents, with light-weight starving chicks staying longer in the nest. Badly starved ones may die there or wander away and die elsewhere, but some that fledge under weight survive to return to their natal colony. Mougin (1975, p.122) explained differences in nestling periods of *M.giganteus* over three seasons at Terre Adelie of 119 (112–128), 111 (104–118) and 114 (108–120) days as due to varying feeding rates between the years. Nestling periods for *P.phaeopygia* in Hawaii over 3 successive years were 112.8 ± 2.9 days (n = 10); 106.9 ± 3.7 days (n = 7) and 110.1 ± 4.5 days (n = 9) again suspected to be governed by differing provisioning rates over these seasons (Simons, 1985).

At least with some species, the length of the chick's time in the nest does not depend on whether it has been well fed or not, e.g. Richdale's 'heavy' *P.griseus* chicks were in their burrows for 96.8 ± 2.90 days (n = 35), his 'light' ones for 96.7 ± 4.5 days (n = 18) (Richdale, 1963, p.53). Fledging periods are also independent of chick weights in *P.puffinus* (Brooke, 1986b).

Among the greatest range of nestling periods is that of the continuous breeding *P.lherminieri* at the Galapagos Islands. The mean for 41 chicks was 75 days, ranging from 62 to 100 days, the longer periods at least being ascribed to temporary food shortages and reduced feeding (Harris, 1969d). The same author also showed that the two populations of *O.castro* breeding there took different times to rear their chicks. Nineteen pairs feeding chicks in the hot season took 70 ± 4.0 days, 25 in the cold season, for no apparent reason, taking 78 ± 9.5 days (Harris, 1969b, p.124). A very wide range of nestling periods is reported for another sedentary storm petrel, *O.homochroa*, in California. Its chicks are in their burrows from 66 to 119 days (Ainley *et al.*, 1990).

With other petrels mean nestling periods during different seasons have been similar. Fairy Prion chicks fledged at 50 ± 2.5 days (43–56 days) during two successive seasons but the mean departure dates in an early year of 15.6 ± 4.5 February were significantly different from that in a normal one of 19.8 ± 4.1 February (Richdale, 1965b, p.110).

That part of the variation in times that chicks are nestbound is related to the varying 'quality' of the parents as shown by the amount of food they deliver, is suggested by an analysis of the nestling periods of Royal Albatrosses made by Robertson & Wright (1973). They confirmed Richdale's suggestion that established pairs fledge their chicks at similar ages at each breeding episode, so that the individual variation in nestling period was much less than that of the population. Over six seasons the chicks of one pair were 250 (243–257) days ashore, for those of another one the mean time was 224 (216–232) days, the population mean being 240 days, ranging from 216–257 days. Brooke (1986b) has also suggested that pair quality is reflected in the tendency of established breeding *P.puffinus* to produce lighter or heavier chicks with little variability in their nestling periods, and he provides evidence that the within-pair constancy is hereditary.

Another possible source of variability relates to the position of the nest. Chicks of *P.tenuirostris* from the peripheries of colonies were lighter and less developed than those in the centres, suggesting that the latter belonged to established pairs and likely to fly sooner (Naarding, 1980).

A particularly clear example of nestling periods varying between populations is provided by *O.oceanicus* (Table 15.2). Here, despite an increase in body size with latitude, the most southerly population of larger birds rears its chicks in only 48 days — substantially faster than do populations at lower latitudes. This appears to be an adaptation to polar conditions as with fulmars (see Section C1 below).

Belopol'skii (1961) cited information suggesting that *F.glacialis* at Novaya Zembla fledge their chicks more rapidly than those on Kharlov Island. Evidently the petrels lay when the coasts are icebound and fly far to rich feeding beyond the ice edge, rather in the style of inland nesting Snow Petrels. No recent or first-hand information on this is known to me.

C *Nestling periods and body size*

Larger species of petrels take longer to rear their chicks than do smaller species, bigger ones having to grow more. The relationship between body size (as measured by mass) and nestling periods is not linear but allometric, the rate of change in nestling period with adult body mass being constant (Fig. 15.6). Nestling periods come from Table 15.4, the masses from samples of mixed sex, with means for 43 being plotted:

$$y = 23.66x^{0.208 \pm 0.18} \qquad \text{(i)}$$

with $r^2 = 0.571$, $P < 0.0001$; where y = mean nestling period in days and x = mean body mass in g. If the six fulmarine species are omitted then:

$$y = 22.72x^{0.229 \pm 0.14} \qquad \text{(ii)}$$

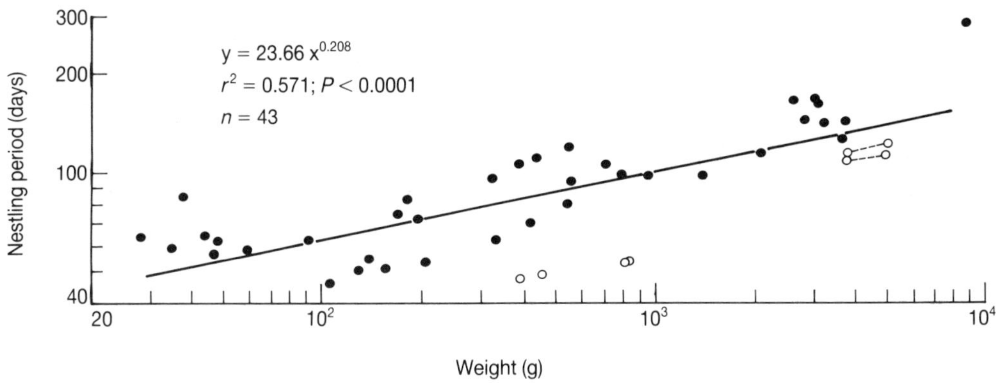

Figure 15.6 *Nestling period* v. *adult body weight — mean values for 43 species. Open circles are points for fulmars.*

with $r^2 = 0.732$, $P < 0.0001$, $n = 37$.

Despite the variability of nestling times for all species and some rather small sample sizes, the correlation in logarithmic terms is high. And because nestling periods are allometrically related to body masses they are also related to other variables linked to body mass such as standard wing lengths or egg dimensions. Using data for 24 species Warham (1968) found that for egg capacities and nestling periods r^2 was 0.909, $P < 0.001$ and, for 19 non-fulmars:

$$y = 0.496x + 52.61 \quad \text{(iii)}$$

where y = mean nestling period in days, x = mean egg capacity in ml.

In Fig. 15.6 the plots for the six fulmars are indicated separately: these birds raise their chicks much more quickly than do other petrels of similar size — as they also hatched their eggs in record time (Fig. 14.5). Hence the time taken from laying to fledging is, for a tubenose, very short. *F.glacialis* and *F.glacialoides* take 102 and 99 days, *D.capense* 93 days (cf. equal sized *P.puffinus* taking 121 days), the large form of *P.nivea* 92 days (cf. similar-sized *L.brevirostris* taking 112 days) and the total times for *M.giganteus* and *M.halli* of 181 and 172 days being shorter than those of similar-sized mollymawks. Data for *T.antarctica* (not plotted) of 45 days incubation and 42 days in the nest (Croxall, 1984) suggest very fast development even for a fulmar and need confirmation.

This general speeding up of development in fulmars has been supposed to represent an adaptation to breeding in short polar summars (Warham, 1968; Lack, 1968; Croxall, 1984). The birds have access to a temporary bloom of food with continuous daylight giving them more time to feed themselves and their chicks.

The other polar petrel is *O.oceanicus*. Adelie Land birds taking only 91 days have the shortest development time of any petrel. This species does not plot below the regression line in Fig. 15.6 because the data used there were for the Signy Island population (Table 15.2). Had Lacan's figure for Terre Adelie been used, *O.oceanicus* would be the only storm petrel to plot below that line. However, by being

Table 15.2 Nestling periods of Wilson's Storm Petrel at different locations

Location	*n*	Adult wt. (g)	*n*	Nestling period (days)	References
S. Georgia (54°00′S)	34	34.9±2.7	?	78 (64–97)	Croxall *et al.* (1988) Copestake & Croxall (1985)
Signy I. (60°40′S)	226	38.3±2.3	12	60 (54–69)	Beck & Brown (1972)
King George I. (62°10′S)	276	39.0	14	59 (55–65)	Wasilewski (1986)
Adelie Land (66°40′S)	50	43.2	13	47.7 (46–51)	Lacan, 1971

substantially larger and yet raising its chick more quickly, the Antarctic population appears to exhibit a parallel adaptation to that of the fulmars.

Other petrels with relatively short nestling periods that plot below the regression line of Fig. 15.6 are the prions and the Blue and Kerguelen Petrels.

The fulmars do not have the shortest nestling period. This is achieved by the diving petrel *P.georgicus* which, at South Georgia, are nest-bound for only 45.5 ± 1.5 days (n = 18) according to Payne & Prince (1979) (Table 15.3). With an incubation period of 46.4 ± 1.6 days (n = 16), this gives a total of 93 days — very fast development for a 103 g petrel. The same bird at The Crozets (46°S) appears to be smaller but to take longer, with a 'true' incubation period of about 46 days making 96 days in all (Jouventin *et al.*, 1985) (Table 15.3). It is difficult to know whether these birds have speeded up development or whether the rates are characteristic of the family: little is known of the basic breeding statistics of the two South American species. Payne & Prince noted that the *georgicus* chick had

Table 15.3 Incubation and nestling periods of diving petrels

P.urinatrix		*P.georgicus*		
Incubation period (days)	Nestling period (days)	Incubation period (days)	Nestling period (days)	References
c. 50.5[a]	53.6±3.9[a] (n=10)	*c.* 46	50.3±2.7 (n=8)	Jouventin *et al.* (1985) (Crozet Is.)
—	53.9±2.6[a] n=16)	46.4±1.6 (n=10)	45.5±1.5 (n=18)	Payne & Prince (1979) (S. Georgia)
54[b] (n=1)	50.5[b] (n=16)	—	—	Thoresen (1969) (Northern New Zealand)
c. 56[c]	53.7±2.5[c] (n=54)	—	—	Richdale (1965a) (Southern New Zealand)

[a] Subspecies *exsul*. [b] Subspecies *urinatrix*. [c] Subspecies *chathamensis*.

thicker down than the *urinatrix* ones breeding nearby, and that *georgicus* faced greater thermal challenges, effectively restricting it to a shorter breeding season.

Apart from *D.exulans*, the species plotting furthest above the regression line of Fig. 15.6 are all tropical or warm water forms — *D.immutabilis* and *D.nigripes*, the gadfly petrels *P.phaeopygia* and *P.macroptera*, the tropical shearwaters *P.nativitatis* and *P.pacificus*, and the Ashy Petrel *O.homochroa*. Galapagos populations of *O.castro* also take rather a long time to rear their chicks whereas those at Ascension Island, plotted in the figure, conform to the trend line. As far as they go, these data lend support to the view that the chicks of petrels breeding in the tropics tend to stay in their nests longer than do those of colder seas.

The giant albatross *D.exulans* takes longer over chick development than the regression of equation (i) predicts. Figures for *D.epomophora* are few and not plotted but suggest a similar situation for that species. This may reflect a decrease in provisioning during the long southern winters and/or metabolic demands on chicks that often get plastered with snow and drenched with rain.

As is true of incubation (Chapter 14, Section VIE) the link with body size seems to extend intra-specifically. In both *Macronectes* Hunter (1984b) showed that the male chicks took longer to fledge on average than the females — 123 *v.* 117 days for *giganteus* and 114 *v.* 110 days for *halli*. Both sexes are shown in Fig. 15.6. Presumably, with adequate data, a similar sexual difference could be detected in other markedly sexually dimorphic species such as other fulmars and *Calonectris*.

D *Nestling periods and incubation periods*

Petrels with short nestling periods tend to have short incubation periods as shown by Lack (1967). For 38 species for which adequate data are available:

$$y = 4.29x \pm 2.19 - 137 \qquad \text{(iv)}$$

with $r^2 = 0.797$; $P < 0.0001$, where y = mean nestling period and x = incubation period, both in days (Fig. 15.7).

The correlation between these variables may be a causal one. A small egg would be expected to hatch sooner if only because less material has to be converted to tissue, while the size of the hatchling will vary with the size of its egg, and the time to complete development to adult size would be expected to relate to the size at hatching. Hence with small petrels that produce large eggs relative to their body sizes, a greater proportion of the time from laying to fledging would be expected to be devoted to incubation than with a large petrel laying an absolutely larger but proportionately smaller egg that produces a relatively smaller chick. In that case the proportion of the development time devoted to incubation should be smaller.

This is generally true. The larger species of Fig. 15.7 lie to the right of and above the smaller ones. Thus the nestling period of *D.exulans* is 3.6 times that of its incubation period, for the medium-sized *P.tenuirostris* only about 1.8 times as long, while the chick of *O.furcata* is only in its nest for 1.5 times the time it is in the egg. Nearly all species take longer to raise their chick than to hatch their egg.

In Fig. 15.7 the birds that plot closest to the origin are not only the smallest ones — the storm petrels — but also the much larger fulmars *P.nivea*, *F.glacialis*,

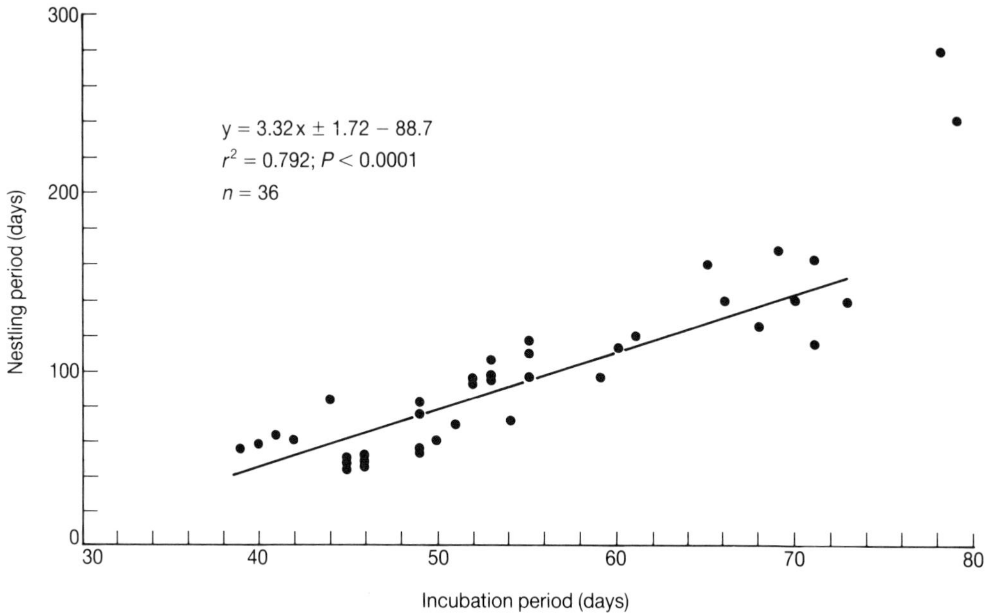

Figure 15.7 *Nestling period* v. *incubation period — mean values for 38 species.*

F.glacialoides and *D.capense* and the prion *P.desolata*. Fulmars take very little longer to raise their chicks than to hatch their eggs. As with Fig. 15.6 the giant albatrosses plot far above the regression line — their nestling periods are inordinately long. The equation for Fig. 15.7 omitting these two species is:

$$y = 3.32x \pm 1.72 - 88.7 \qquad \text{(v)}$$

with $n = 36$, $r^2 = 0.792$, $P < 0.001$.

The position of *Pelecanoides* is peculiar: it plots with the medium-sized fulmars in Fig. 15.7. It has long been suspected that the incubation and nestling times of *urinatrix* were almost the same (Warham, 1968) but the data were poor. For that species they still are inadequate: figures for two species are summarised in Table 15.3. This shows that for *georgicus* at South Georgia incubation lasts marginally longer than chick rearing — the only example of such a reversal recorded among petrels. At The Crozets, however, the 'true' incubation period is similar to than at South Georgia but the nestling period longer. Diving petrels lay small eggs (Chapter 13, Section IVB) and from the egg dimensions given by Richdale, Thoresen, Payne & Prince, and Jouventin, calculated incubation times, using $y = 25.25x^{0.157}$ [where y = incubation period (days) and $x = LB^2$], are all of 44 or 45 days — more in line with those from field observations of *P.georgicus*, a slightly smaller bird. Presumably the short incubation period of that bird is mostly accounted for by the relatively small egg, but, all things being equal, raising the smaller hatchling should take longer despite inshore feeding, but this does not seem to happen, at least with *georgicus* at South Georgia, although Croxall (1984) thought so. Better data for *P.urinatrix* are needed.

E *Parental weight changes during chick rearing*

The body weights of parent tubenoses feeding chicks vary during that time but there are no figures of sexed birds weighed at short intervals to allow trends to be closely related to activity.

The best information is for the albatross *D.immutabilis*. Fisher (1967) weighed parent birds four times between hatching and fledging. Males were consistently the heavier with the weights of both sexes falling during the guard stage and, when the chicks reached their maxima, rising again continuing to do so during the last 3 months of chick care: while the chicks lost weight the parents gained it. Fisher thought that as the birds came to their nest less often at this stage they could be tapping more distant food sources than before and expending less energy in flying between the nesting and foraging areas. When they finally left the males averaged 2880 ± 66 g (n = 19), the females 2660 ± 55 g (n = 21) compared with their weights on arrival 8.5 months before of 3310 ± 42 g (n = 50) and 2990 ± 70 g (n = 20).

Figures for *M.giganteus* are rather similar, as far as they go, but, in contrast to *immutabilis*, both sexes put on weight during incubation and they were quite heavy at the hatch. Thereafter their weights fell, at least during the early part of chick care (Conroy, 1972, p.16). Weight declines during chick rearing seem to occur among diving petrels (Thoresen, 1969). With *P.puffinus* the breeders tended to gain weight during the chick's first month of life (Brooke, 1978d) and, unlike *D.immutabilis*, did so well before these reached their peak weights. Brooke suggested that weight increases of Manx Shearwaters late in the season reflected premigratory fat deposition whereas the stabilisation of body weights that occurred with *P.tenuirostris* as chick rearing was completed, was not considered to be a consequence of the laying down of fat (Lill & Baldwin, 1983). They concluded that either the migrants fed extensively en route or fed intensively in home waters during a short post-desertion period.

Both parents and chicks of Wilson's Storm Petrels at South Georgia increase their body weights before fledging (Beck & Brown, 1972), so boosting energy stores for migration. Copestake & Croxall (1985) suggested that they did this by delaying laying and slowing down chick development (Table 15.2), while capitalising on the lengthening nights for nocturnal feeding.

VII Comparing development rates

A *Weight-specific rates*

Nestling periods themselves are measures of the rates of development of chicks (and were so used by Lack, 1968), but as these are size-related, for inter-specific comparisons they need to be scaled accordingly.

One method, using the whole of the chick period, is to compare nestling times per unit weight of adult, and figures for 43 species are shown on logarithmic axes in Fig. 15.8. Also presented for comparison is the curve for incubation periods per unit weight of adult. The data come from Table 15.4. The regression equations are:

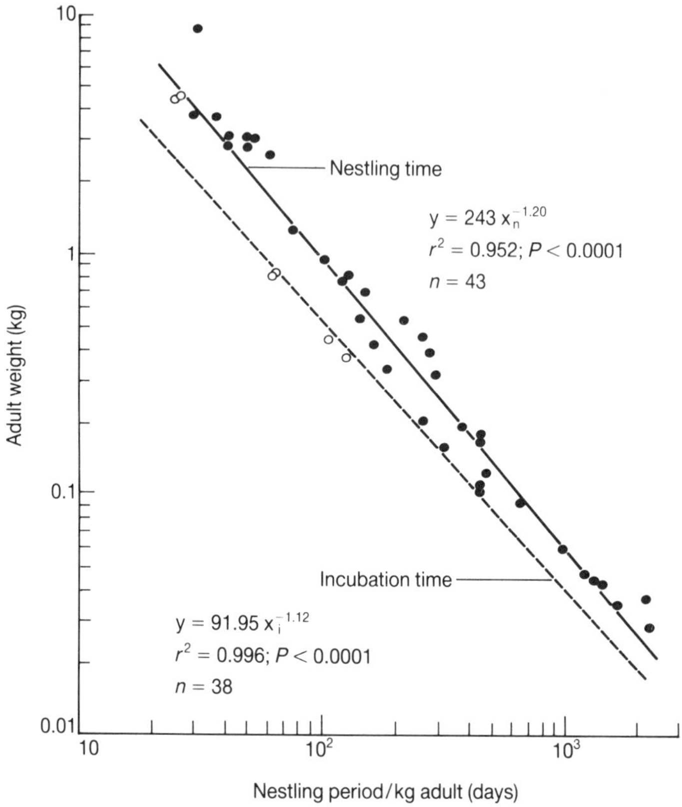

Figure 15.8 *Nestling times/unit weight of adult. Points for fulmars open circles. Dashed line shows incubation time/unit weight of adult. After Warham (1968).*

$$y = 243.3x_n^{-1.20 \pm 0.27} \quad \text{(vi)}$$

with $n = 43$, $r^2 = 0.952$, $P < 0.0001$; where y = mean body weight in kg, x_n = mean nestling period kg^{-1} adult in days.

If six fulmarine species are omitted:

$$y = 349.8x_n^{-1.26 \pm 0.23} \quad \text{(vii)}$$

with $n = 37$, $r^2 = 0.970$, $P < 0.0001$.

The equation linking incubation time per unit weight of adult is:

$$y = 91.95x_i^{-1.12 \pm 0.07} \quad \text{(viii)}$$

with $n = 38$, $r^2 = 0.996$, $P < 0.0001$; where y = mean adult weight in kg, x_i = mean incubation period kg^{-1} adult in days.

The slopes of equations (vi) and (vii) are not significantly different but their intercepts are, i.e. the rates of production per unit weight of adult are less during chick raising than during incubation. The points for weight-specific nestling times are also more dispersed than those of the weight-specific incubation times (r^2 = 0.952 *v.* 0.996).

As measured by the times to produce unit weight of adult, the large species are more efficient than the small ones. An albatross like *D.melanophrys* takes about 30 days to produce 1 g of albatross; theoretically *H.pelagicus* would take 2250 days to make 1 kg of storm petrel.

Species points that deviate most from the trend lines are that for *D.exulans* and those for the smaller fulmars. The former is less efficient than equation (vi) would predict, taking 32 days instead of a predicted 16 to produce 1 kg of adult — the long over-wintering and possibly reduced provisioning may be factors here. Poor data for *D.epomophora* suggest a similar situation for that species, whereas the points for both lie almost on the regression line for incubation time/unit weight of adult. They produce unit weight during incubation more rapidly than any other petrel. The smaller fulmars are more efficient than other similar-sized tubenoses both during incubation and during chick rearing, as already discussed.

B *Growth curves*

Curves showing chick growth for many species of Procellariiformes have been published, based on repeated measurements from hatching to flying. Typically these show changes in weight and in length of wing, tail, bill, tarsus and middle toe with claw. Examples are given in Figs 15.9 and 15.10. Published curves include those of Richdale, 1952 for *D.epomophora*; of Berruti (1979) for *Phoebetria*; of Hunter (1984b) for *Macronectes*; of Pinder (1966) for *D.capense*; of Simons (1985) for *P.phaeopygia*; of Hall (1987) for *P.aequinoctialis*; of Pettit *et al.* (1984b) for *P.pacificus*; Oka (1989) for *P.tenuirostris*; Prince & Copestake (1990) for *P.turtur*; Beck & Brown (1972) for *O.oceanicus* and Boersma *et al.* (1980) for *O.furcata*. Many other workers give only the curves for weight *v.* age.

During much of the chick's stay in the nest it lays down fat subdermally and around the viscera and in those species that have been used as human food the process of 'trying out' chicks for their fat has been widespread. Generally, at the end of the harvest, the birds yield little fat or stomach oil. Few determinations of lipid levels have been made but Ricklefs *et al.* (1980a) found a steady increase of about 0.4 g day^{-1} of lipid in Leach's Petrel chicks up to about day 60, with no decline at the end of the study to the much lower adult value. These authors also analysed the growth of skeletal elements, weight, wing, tail, various other body components, and water content of these birds.

Typically growth curves for petrel chicks, smoothed to reduce the abrupt rises and falls as in Fig. 15.4, are of the S-shaped sigmoid form as for other vertebrates including man (Needham, 1931). Increases are small at first, build up to a maximum at a point of inflection, and then decrease to an asymptotic value. There is much inter- and intra-specific variation in the amplitude and time scales of these curves due to major factors like body size, and for comparative purposes some form of fitted equation is needed. The logistic, Gompertz and von Bertalanffy equations have been used for comparing growth as demonstrated by weight changes (Ricklefs, 1968) but the first two have been found most appropriate for petrels.

Table 15.4 Development statistics of petrels

No.	Species[a]	Body[b] wt (g)	Inc. per. (days)	Inc. time (kg^{-1}) adult (days)	Nestl. per. (days)	Nestl. time kg^{-1} adult (days)	Birth wt (g)	Chick max (g)	Chick max/ adult wt (%)	Fledg. wt (g)	Fledg. wt/ adult wt (%)	% N.P. to max wt	$t_{10}-t_{90}$ (days)	K[c]	References
1b.	*D.exulans*	8677	78	9.0	278	32	352	11700	135	8933	103	81	164	0.026	Tickell (1968)
															Croxall (1984)
2b.	*D. epomophora*		79		240		340								Richdale (1952)
5.	*D.nigripes*	3212	66	20.5	140	44									Rice & Kenyon (1962)
6.	*D.immutabilis*	3092	65	21.0	160	52									Rice & Kenyon (1962)
7a.	*D.melanophrys*	3660	68	18.6	125	34	181	4700	123	3505	92	69	62; 57	0.048	Ricketts & Prince (1968)
															Weimerskirch *et al.* (1986)
9.	*D.chrysostoma*	3731	73	19.6	141	38	195	4655	125	3355	90	62	60	0.042	Ricketts & Prince (1968)
10b.	*D.chlororhynchos*	2060	71	34.5	115	56		3410	166	2479	120	63	54		Weimerskirch *et al.* (1986)
11a.	*D.bulleri*	3012	69	22.9	167	55									Warham (unpubl.)
12.	*P.fusca*	2600	71	27.3	164	63			115				69		Weimerskirch *et al.* (1986)
13	*P.palpebrata*	2800	70	25.0	141	50	176	3360	120			57	62; 67	0.043	Thomas *et al.* (1983)
															Weimerskirch *et al.* (1986)
14.	*M.giganteus*	4467	61	13.7	120	27	167	6000	134			71	70	0.047	Hunter (1984b)
															Croxall (1984)
15.	*M.halli*	4390	60	13.7	112	26	174	5700	130			85	59	0.040	Hunter (1984b)
															Croxall (1984)
16.	*F.glacialoides*	802	46	57.4	53	66	71	1200	150	880	110	63	30	0.121*	Mougin (1967)
															Croxall (1984)
17b.	*F.glacialis*	813	46	59.7	53	65	68	1120	138	860	106	70			Mougin (1967)
															Ollason & Dunnet (1978)
19a.	*D.capense*	452	45	100	48	106	43	690	154			77	27	0.194*	Mougin (1975)
															Croxall (1964)
20.	*P.nivea*	388	45	116	47	127	40					71	23	0.183*	Mougin (1975)
															Isenmann (1970)
21.	*L.brevirostris*	331	50	151	62	187	42	410	124						Jouventin *et al.* (1985)
22b.	*P.macroptera*	542	55	102	118	218				542	100				Imber (1976b)
23.	*P.lessonii*	780			102	131		1000	128						Mougin (1975)
34.	*P.cahow*		52		92										Wingate (pers. comm.)
38a.	*P.phaeopygia*	434	55	128	111	258	60	529	146			65			Simons & Whittow (1984)
42.	*P.inexpectata*	329	51	154			53								Warham *et al.* (1977)

43.	*P. hypoleuca*	182	49	271	82	453		283	155	206	113	61			Pettit *et al.* (1982b)
50.	*H. caerulea*	202	49	243	53	262		210	105	176	88	66	33	0.136*	Fugler *et al.* (1987)
															Croxall (1984)
51c.	*P. vittata*	160	50	315			26	186	117	154	97	75	31		Derenne & Mougin (1976)
52c.	*P. desolata*	159	45	283	51	318	27	220	138				28	0.151*	Tickell (1962)
															Croxall (1984)
54.	*P. turtur*	132,110			50	378	17	150	136	101	92	60	26	0.145*	Harper (1976)
															Richdale (1965b)
56.	*B. bulwerii*	92			62	674									Amerson & Shelton (1976)
58a.	*P. aequinoctialis*	1394	59	42	98	70	93	1687	121	1215	87	84	62	0.043	Jouventin *et al.* (1985)
															Hall (1987) Croxall (1984)
60.	*P. parkinsoni*	704			107	152				757	108				Imber (1987)
62b.	*C. diomedea*	946	55	58	97	103	77			855	90				Zino *et al.* (1987)
63.	*C. leucomelas*	545			80	147									Yoshida (1981)
64.	*P. pacificus*	390	53	136	109	279		544	139	423	108	75			Pettit *et al.* (1984b)
69.	*P. griseus*	787	53	67	97	123	75	901	115	622	79	73			Richdale (1963)
															Warham *et al.* (1982)
70.	*P. tenuirostris*	559	53	95	94	168	63	808	144	616	110				Serventy *et al.* (1971)
															Lill & Baldwin (1983)
71.	*P. nativitatis*	324	52	161	96	296									Amerson & Shelton (1976)
72a.	*P. puffinus*	420	51	121	70	167	45	561	134	420	100	71			Harris (1966)
															Perrins *et al.* (1973)
77c.	*P. lherminieri*	168	49	292	75	446	22	186	111			67			Harris (1969d)
78f.	*P. assimilis*	195	54	277	72	369									Glauert (1946)
79a.	*O. oceanicus*	35	(40)	1143	78	2230	6.7	49	140	35	100	62	(40)		Croxall *et al.* (1988)
79b.	*O. oceanicus*	38	40	1052	60	1579	7	61	157	55	145	95	43	0.077	Beck & Brown (1972)
82b.	*P. marina*	47			57	1213	9	65	138	54	114	60			Richdale (1965a)
86a.	*H. pelagicus*	28	41	1450	63	2250		42	150	34	121	79			Davis(1957)
89.	*O. castro*	44			64	1471	7.5	75	170	49	111	66			Allan (1962)
91a.	*O. leucorhoa*	48	43	896	61	1271	7.3	69	144	58	121	82			Watanuki (1985)
															Ricklefs *et al.* (1985)
97.	*O. homochroa*	38	(42)		84	2211	9.2	48		42	110			0.105*	Ainley *et al.* (1990)
99.	*O. furcata*	59	39	654	58	990		85	145	73	124	86			Simons (1981b)
102.	*P. georgicus*	103	46	467	46	442	13	144	140	104	101	72	24	0.165*	Payne & Prince (1979)
															Croxall (1985)
103c.	*P. urinatrix*	141,123	51	358	54	383	16	157	128	131	107	63	30	0.129*	Jouventin *et al.* (1985)
															Payne & Prince (1979

[a] For subspecies see Checklist, Appendix 1; [b] Sexes Combined; [c] All K_G except * which are K_L.

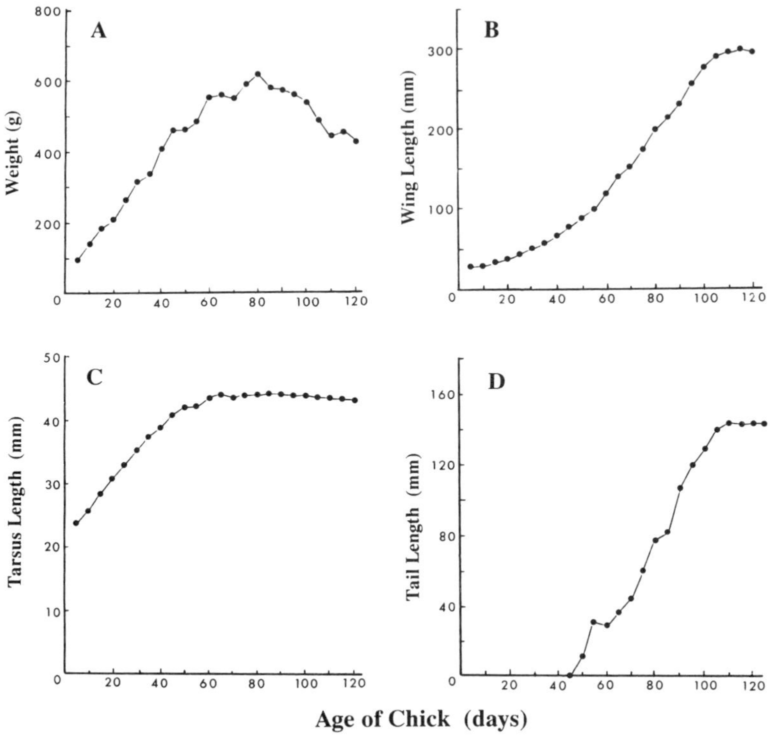

Figure 15.9 *Growth curves for* P.phaeophygia *chicks:* **A.** *weight,* **B.** *wing length,* **C.** *tarsus length,* **D.** *tail length. From Simons (1985).*

1 *Curves for weight*

Chick weights are inherently more variable than are dimensions of body parts because most are not of birds in a post-absorptive state and stomach contents are not only variable but may comprise a very large proportion of the recorded weight. Another source of error is the chick's habit of swallowing bits of vegetation and small pebbles, the 'ballast stones' of uncertain function.

Petrel chicks weigh about 0.74 $\pm$ 0.05 times the weight of the fresh egg (means for 28 spp.), or $0.41LB^2$, but as most hatchling weights have been taken, not precisely at birth, but sometime on day 1, small meals will often be included and true birth weights (Table 15.4) slightly below these values.

At hatching chicks weigh 4–6% of the weights of adult albatrosses and giant petrels, 8–10% of those of other fulmars, 10–16% of those of gadfly petrels, 16–17% of prions, 9–13% of shearwaters and 16–20% of the weights of adult storm petrels. These values reflect the relationships between adult weight and egg weights (Chapter 13, Section IVB). The neonate contains a lot of water, 80% of its weight according to Ricklefs (1979), who pointed out that this was similar to figures for other semi-precocial or precocial chicks but less than those of altrical birds.

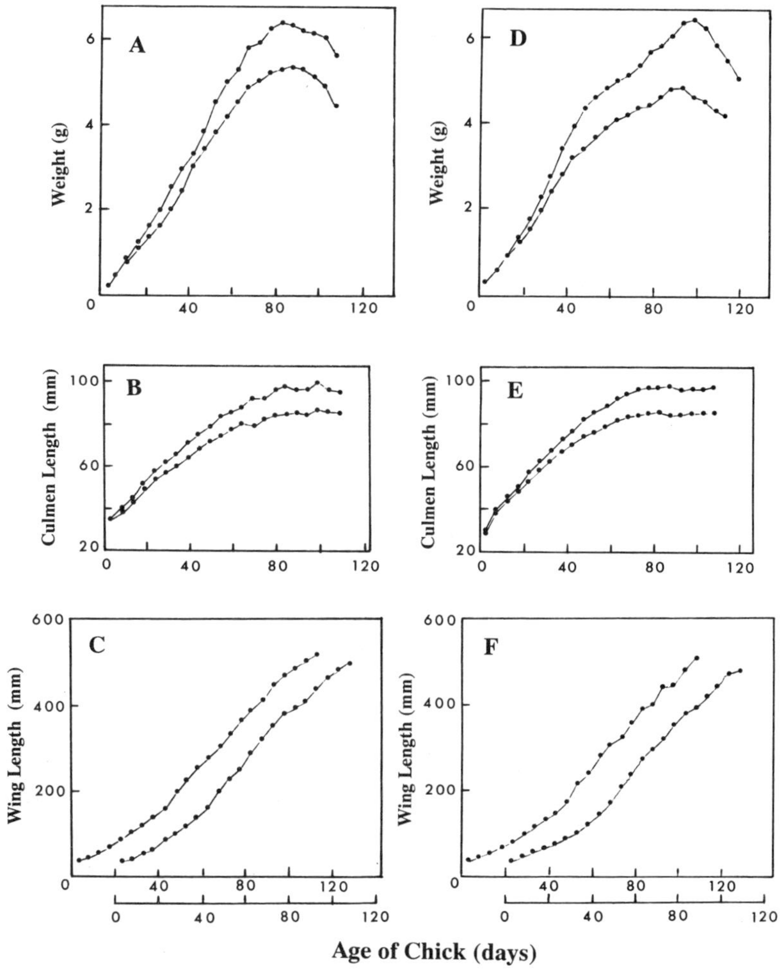

Figure 15.10 *Growth of giant petrel chicks.* **A–C** M.giganteus *(n = 11 males, 17 females);* **D–F** M.halli *(n = 16 males, 15 females). Upper curves are for males. From Hunter (1984b), courtesy of The Zoological Society of London.*

The logistic equation has often been used for fitting curves to weight data. It assumes an inflection point when 50% of the asymptomatic weight has been reached and can be used to calculate K, a constant for comparing the growths of species or of different animals (see Ricklefs, 1968). Another statistic used for comparisons is an inverse of the growth rate, the time taken to grow from 10%–90% of the asymptotic value, i.e. covering the 'linear' portion of the curve. Neither of these methods is ideal (Croxall, 1984; Zach, 1988) but have the benefit of simplicity for first approximations. Some values for these statistics are given in Table 15.4. The 'K' values in that table are not all strictly comparable, some being based on Gompertz equation (K_G) — where an inflection point at 37% of asymptotic weight is assumed, and others (L_L) on logistic analyses.

The figures for K vary from 0.026 for the giant albatross *D.exulans*, which grows more slowly than any other species listed, to 0.194 for *D.capense*. Indeed all the values for fulmars, except for *Macronectes*, are high, as expected of their fast growth already demonstrated. High values are also shown for the prions and diving petrels. Values for t_{10-90} reflect these K values and also highlight the much more rapid growth of Wilson's Storm Petrel in Antarctica compared with the Signy Islands population (Table 15.2) (Croxall, 1984, p.574).

To avoid problems in comparing different types of growth constants Croxall & Gaston (1988) have taken the mean growth rates in g day^{-1} from the t_{10-90} statistic when the rate is maximal and plotted these against mean adult body weights for 27 species, finding that:

$$y = 0.118x^{0.76} \qquad \text{(ix)}$$

with $r^2 = 0.94$, $P < 0.01$; where y = growth rate in g day^{-1} and x = adult body weight in g. Because of the tight fit of the points to the curve they conclude that variations in rates of growth as measured by weight are unlikely to account by themselves for variations in nestling periods.

Another approach is that of Pennycuick *et al.* (1984). Using figures for nine species of southern petrels, they described methods of calculating foraging ranges from appropriately scaled data on the weights and energy contents of meals and developed a dimensionless growth constant for comparisons between species.

Petrel chicks usually rise to between 130–150% of the mean weights of adults (Table 15.4) and much of the variation in those data is probably due to small sample sizes, seasonal differences in food provisioning, and similar factors. The high value of 166% was reported for the mollymawk *D.chlororhynchos* at the Crozet Islands, where *D.melanophrys* chicks also peaked at 146% of adult weight (Weimerskirch *et al.*, 1986). Storm petrel chicks also seem to reach consistently high weights — 140–170% of those of the adult birds — even up to 192% in *O.oceanicus* (Beck & Brown, 1972). Peak weights may be important, because with some species, e.g. *P.puffinus* (Perrins *et al.*, 1973), those that reach the highest maxima also fledge at the highest weights and survive better.

Values for the weights of chicks at the point of making their first flights are also summarised in Table 15.4 and inter-specific comparisons can again be made of the ratios of these weights to those of the adults. Probably most of the figures will err on the high side, because very few have intercepted chicks on their way to sea as Harper (1976) did with *P.turtur*; most will be lighter on their maiden flights than at their last weighings. The values tend to be around 100% of the adult weight with a few fledging as low as 80% of that. Again, storm petrels seem to fly when quite heavy — all the values being above those for the adult birds — and this may be made possible by their low wing loadings (Warham, 1977a).

Conroy (1972) measured chicks of *M.giganteus* and having sexed a sample by dissection, plotted separate curves for males and females. Hunter (1984b), apparently using Conroy's criteria, produced a more detailed set of curves (Fig. 15.10) and for both giant petrels. These show that weights and bill lengths began diverging significantly when the chicks were only about 20 days old, i.e. having traversed about 17% of their time to fledging. The wings of the two sexes diverged far less, as expected, the wings of adult males being only about 105% longer than those of females.

2 *Weight recession*

Petrel chick weights are unusual among birds in that the maxima attained exceed adult weights and then decline to fledging (Fig. 15.9). Standard equations therefore, only cover growth up to the attainment of the maxima: there is no true asymptote. The final stages of this weight recession can be compared using the ratios of chick maxima and weights at fledging to adult weights (Table 15.4).

The decline in weight before a chick flies usually coincides with a fall off in parental visits and smaller meals (Table 15.1), even desertion in some cases (Section VIII below). The point at which the weight decline begins varies considerably from chick to chick but is usually reached after 60–80% of the nestling period has elapsed (Table 15.4). *Phoebetria* albatross chicks however, attained their maxima when only 50–60% of their long stays in the nest had passed (Berruti, 1979; Thomas *et al.*, 1983).

During this final stage body weights decline whether or not the chicks are deserted by their parents, and growth is mainly of integument and bone. At sea the young bird must find its own food, perhaps in stormy conditions, and needs well developed flight feathers. According to Ricklefs (1979) the skin and feathers of adult *O.leucorhoa* make up nearly half of its dry weight. In the face of declining provisioning, petrel chicks are thought to metabolise much remaining stomach oil (old chicks seldom spit, presumably because they have nothing to disgorge — (Warham, 1976b)) and then depot fat and perhaps protein reserves, as with penguins (Cherel *et al.*, 1989). Lill & Baldwin (1983) estimated that departing *P.tenuirostris* chicks had a mean fat content of 9–14% giving them a potential flight duration of 20–30 h and a range of 620–940 km, so that feeding in local waters or during migration would be essential. Ricklefs *et al.* (1980a) found that nestling *O.leucorhoa* accumulated lipid throughout most of their time in the nest. Chicks at 55 and 59 days old contained 30 g and 35 g respectively; these workers estimated that 15 g of lipid would supply the chicks' energy needs for up to a week. However, it was found that there was no feather or muscle growth among those chicks during their weight recession: they lost weight in order to fly (Ricklefs, pers. comm.).

The ability of the chicks to climb above adult weight and to store fat helps them survive if the parent's food deliveries fall off and allows adult trans-equatorial migrants to depart early. However, chicks of non-migratory petrels also reach weights above those of their parents and store fat. Presumably this acts as an energy source to tide them over their first critical days and early attempts at foraging. Extra energy is also required during the chicks' final days on land to power their greatly increased activity then, while reduced body weights will facilitate take-off at departure.

3 *Curves for body dimensions*

Data sets for the growth of body parts are fewer than those for weight and show that the legs and feet grow the fastest, reaching asymptotic values by the time that 55–70% of the nestling period has elapsed. The long-legged storm petrels apparently take longer but the data are sparse. Richdale (1965a) found that of all the body parts measured, only the tarsus and middle toe with claw had reached adult size in *P.marina*. Indeed, he noted that fledgling *P.griseus* had slightly longer claws than the adults, presumably because they were unworn, a feature seen with other young tubenoses. Early development of the hind limbs allows petrel chicks some mobility from an early age.

The beak is the next to reach asymptotic length, doing so after 60–90% of nestling period has passed.

Development of wings and tail tends to lag behind. These often barely reach adult lengths before the chick's first flight. Both light and heavy fledglings have similarly developed wings, only badly starved ones showing retardation (Harris, 1966a) so that there is a premium on being able to fly even if under weight, and some such survive.

A detailed analysis of the organ development of Leach's Storm Petrel was provided by Ricklefs *et al.* (1980b). They found that the weight of the hatchling relative to that of the adult was high (16%) — as would be expected of the relatively very large egg — so that the body components had to grow correspondingly less to gain adult size. The viscera had to grow to 3–4 times their initial size, the integument 8 times, the wings 14 times, and the pectoral muscles 59 times. These latter developed very quickly, reaching half adult size in 30 days, and these workers believed that, together with the legs, they play a major role in heat production, that their rate of growth at any time controls the pace of the chick's growth and that the whole system of organ development is a consequence of the precocial mode of development and not a slow, energy-saving adaptation due to food and/or nutrient limitation on the lines of Lack's hypothesis.

In this species the femora were full size at 20 days, the tibia and tarsus reached that at 40 and 50 days by when the humerus and radius had ceased growing. In the legs and wings bone mineralisation was complete before fledging whereas this was then but two-thirds complete in the skull (cf. Sugimori *et al.*, 1985) for *P.tenuirostris*).

VIII The 'desertion' period

This refers to the abandonment of their chick by the parent birds towards the end of its stay in the nest. The chick completes its development alone, losing weight in the process, eventually flying away unaided. With one exception, *Macronectes*, only chicks of a few highly migratory petrels have true desertion periods, and a small proportion even of these is not abandoned.

A *The starvation theory*

The widely held belief that Wandering and Royal Albatrosses were deserted by their parents and starved during their last months in the nest, surviving through the southern winter on their body fat, was disproved by Richdale (1939) for Royal Albatrosses and by Carrick *et al.* (1960) for *D.exulans*. Filhol (1885) saw Royal Albatross chicks being fed when only a few weeks from flying but his observations were subsequently ignored. Richdale (1954, 1965b) provides a history of the myth of albatross starvation.

B *Species with no desertion*

In general, where parents have been proved to visit their nests after the chicks have left, nearly all have been non-migratory species. They include *P.urinatrix*, *P.marina*, *P.vittata*, *P.turtur*, *O.castro*, *O.furcata*, *O.leucorhoa*, *D.capense*, *F.glacialoides*, *P.macroptera*, *P.pacificus*, *P.lherminieri*, both *Phoebetria* albatrosses and *D.melanophrys*, *D.chrysostoma* and *D.immutabilis*. At least at the sites where the studies were made, all the chicks deserted their parents not *vice versa*.

No albatross appears to desert its young; their chicks usually leave between parental visits. One contrary record concerns *D.chlororhynchos* at Nightingale Island where Rowan (1951) saw no adults ashore for at least a fortnight before the last chicks were flying. Although few adults, perhaps none, might be expected to be seen during their short visits to the nesting grounds then, this apparently complete absence is strange, and it may be that Rowan's observations were too brief to catch the parents on land.

Burger (1980) stated that in more than half of the albatrosses for which data are available there is some parental feeding after the chicks can fly and before they leave their colony. A mean figure of 10 days post-flying feeding is given. However, except possibly after short, exploratory and perhaps involuntary flights during pre-departure exercising, there appears to be no clear instance of petrel chicks, once having flow properly, returning to their nests and being fed, i.e. there is no direct evidence for any post-fledging care.

Several of the species that feed their chicks to the end breed in the tropics. The Wedge-tailed Shearwater *P.pacificus* is an example. Both northern and southern populations have no desertion period, although the Australian ones do migrate some distance northwards after nesting. Amerson & Shelton (1976) said that the *Bulweria bulwerii* chick was abandoned for several days before fledging but they did not state that they looked for the parents after the chicks had gone.

Non-desertion is not restricted to birds that feed their chicks often like the diving petrels and storm petrels: one with the most extended of attentive periods, *P.macroptera*, also visits its nest after its chick has gone (Imber, 1976b). Whether other tropical petrels that migrate to the North Pacific like *P.solandri* desert their chicks, is not known, though Oliver (1955) stated that *P.neglecta* did that.

C *True desertion*

1 *Verification problems*

To establish the existence of true desertion requires showing that parents do not come back to the nest after the chicks have left and this has seldom been done. Usually by chick departure time non-breeders have gone, the colonies are quiet, superficially appearing deserted, for the parents now come in infrequently to deliver small meals and do not linger. Without automatic monitoring it is especially difficult to demonstrate desertion among burrowing petrels. Simply fencing the entrances is of little value as obstructions are displaced by the chicks during their nocturnal explorations (see Section B1 below). By fencing entrances after the chicks have left it may be possible to check whether the parents have gone as well: a displaced lattice means something, most likely a parent, has passed through. However, chicks from higher ground could do that, so can skinks, rabbits or other animals and with some petrels like *L.brevirostris* there is a resurgence of nocturnal activity and temporary occupation of burrows by adult, probably non-breeding birds — the so-called 'protogamic' behaviour (Despin *et al.*, 1972) that could confuse the result.

Those petrels for which a desertion period has been established include *P.griseus, P.tenuirostris, P.puffinus, C.diomedea* and *M.giganteus*, all but the last being notable post-nuptial migrants.

The extent of parental desertion varies greatly from pair to pair. Burger (1980, p.390) stated that Richdale (1963) reported no desertion in *P.griseus* whereas (p.56) he wrote that in a sample of 58 'heavy' chicks feeding ceased for a mean of 11.9 ± 5.6 days (0–27 days) before they flew: he fenced off burrow entrances to detect post-fledging activity. Likewise for *P.tenuirostris* desertion lasted for a mean of 14 days, ranging from 1–23 days (Serventy *et al.*, 1971) and for 68 *P.puffinus* the figure given by Harris (1966) was 8.5 days (2–15 days). Zino *et al.* (1987) found that during 908 'chick-days' at the end of the breeding season of *C.diomedea*, there were but 29 cases where weight increases occurred suggesting chick feeding, and only four of these were on the night before departure.

In other words, a few chicks even of the trans-equatorial migrants are not deserted even though the majority is. Some of this variability might relate to parental experience and feeding efficiency. A bird under stress at the end of the season might switch to replenishing its own reserves in readiness to migrate, leaving its chick to its own devices; a more experienced parent still in good condition might not need to do that.

Lill & Baldwin (1983) found that at the time of chick desertion adult *P.tenuirostris* have little fat and weigh only 5% above their rather low weight at first landfall. They suggested that desertion may be necessary to let the parents feed up locally before migrating.

The situation with *O.oceanicus*, a famous long-distance migrant, is rather ambiguous. It might have been expected that a small bird breeding in a short summer would leave its chick to complete its development alone while the parents begin their northwards movement, but both Roberts (1940) and Beck & Brown (1972) found adults (presumably the correct parents) in burrows after their young had left. More information is needed on this aspect of their biology.

Burger (1980, p.393) proposed that procellariids do not have a true desertion period but that the chicks leave during inattentive periods which tend to be long at the end of the breeding time, due, according to this author, to a food shortage requiring parents to forage over long distances. Richdale (1965b, p.151) showed that for his 'heavy' Sooty Shearwater chicks the lengths of desertion periods were indeed the same as the previous inattentive ones and so no hardship for the chicks. Lighter ones had to fast much longer, however, and in any event, his post-fledging checks showed that the parents had gone. The chicks, heavy or light, were unfed and really were deserted.

It may be noted that I can find no direct evidence for the belief that birds which leave their chicks actually begin their migratory movements forthwith. The flocks of shearwaters streaming north from New Zealand and the Bass Strait no doubt include breeders, failed breeders and non-breeders but these are inseparable at sea. A study of parent birds by satellite telemetry may be needed to clarify their movements at this time.

3 *Giant petrels*

Conroy (1972, p.36) documented desertion by *M.giganteus* at the South Orkney Islands. Birds fledging before 10 May were deserted for 6.85 ± 4.1 days, those after that data for an average of 14.4 ± 5.3 days. He believed that parents stopped feeding their chicks in early May regardless of their age, sex or weight, so this was not a case of the chicks leaving during inattentive periods. If correct, this would be the only fulmar with a true desertion period and the more peculiar as the breeders do not migrate but attend their nests frequently thoughout the winter (Warham, 1962; Conroy, 1972). Perhaps at this rather southerly station these large birds have difficulty in fitting their breeding cycles into the short southern summer and leave early to feed up for the moult before the onset of winter. Such behaviour would be expected to be more obvious in Antarctica but has not been reported from there.

IX Pre-fledging behaviour

A *Food refusal*

Many studies show that in their last days the nestlings receive less food than before, even though in some instances the parent's visits do not decline markedly. At the same time the chick's weight is falling — an event that has often been used as evidence for chick desertion. It was Richdale again (1939) who first reported petrel chicks refusing food. He also saw the demands of diving petrel chicks decline towards the end of their stay.

This tendency to refuse food has been noted elsewhere, e.g. some *P.phaeopygia* chicks were deserted for as long as 3 weeks, while others were fed small meals just before they flew. Most began to refuse food 1–2 weeks before that event (Simons, 1985). In the congeneric *P.macroptera* advanced chicks were visited but little food transferred (Imber, 1976b).

Such behaviour suggests that the chicks themselves determine when they are ready to go and Simons (1985) pointed out that if parents continued to feed until their young refused more, a range of attendance patterns could be expected at the end of the nestling period, with chicks that had matured fast and stored adequate fat tending to reject food and be deserted, while less well advanced ones might beg and be fed up to departure. Harris's twinning experiments with *P.lherminieri* in which, by switching chicks, he extended nestling periods from a mean of 75 days to 103–120 days (Harris, 1969d) supports the view that the time of flying is determined by the chick and signalled by its begging pattern. This would help explain the wide range of desertion times seen even among long-distance migrants.

Experiments on giving chicks of *O.leucorhoa* supplementary food by Ricklefs *et al.* (1987) support the view that petrel chicks regulate their own food intake. Triglyceride, protein and water, were injected directly into the proventriculi of chicks and their growth compared with that of control birds. None of the batches given food supplements or water grew significantly faster than the controls or made more weight, so that the treated chicks somehow regulated their intake of natural food or increased their metabolic rates to burn up surplus energy. There was no evidence for that, their body temperatures being normal. Presumably the treated chicks begged less but the experimenters did not check on that: they thought that the adults responded primarily to their chicks' demands by adjusting the frequency of their feeds rather than meal sizes.

B *Final explorations*

As a petrel chick gets older the more active it becomes and the more time it spends awake. Except for some mollymawk chicks ensconced on tall pillar nests that are difficult to regain when vacated, and those 'imprisoned' in rock crevices or narrow ledges, chicks whether on the surface or below it, become more venturesome. During a 16-day watch by a Royal Albatross nest the chick, then about 100 days old, left its nest 31 times, travelled about 65 m and stayed away for 8 h 53 min. During 176 h of daylight it slept for 58 h (Richdale, 1952, p.82).

1 *Surface nesters*

With its weight falling in the last days before leaving the nest permanently, the chick becomes still more restless. Many young albatrosses wander extensively but probably not venturing beyond visual range of their nest. They beat their wings and lift onto their toes while standing on rising ground. On windy days they exercise more often, facing the breeze, perhaps getting raised from their feet and making short flights. They return to their nests, however, from time to time and do so in haste, with calling, should a parent arrive while they are away. Giant petrels behave similarly.

Almost fully feathered mollymawks may shift from their nests where these are on level ground to exercise their wings: those on pillar nests do so while standing there. In large colonies many chicks will rise in unison to face wind gusts and to beat their wings. Sooty albatrosses and fulmars, however, are mostly restricted from shifting from their nesting ledges and must exercise without any extensive exploratory behaviour. For these a fall usually means death by starvation, chicks being ignored if not on their nests. Wing-flapping may begin quite early; with *D.capense* when they are but 3 weeks old, but increasing towards fledging time (Pinder, 1966). Williamson *et al.* (1954) described the fledging of a group of *F.glacialis* and drew attention to attempts by adults to occupy the nest sites and dislodge the chicks, inducing these to spit oil in defence and to use a cackling head-waving action like those of mature birds. However, the intruders were not marked and were not necessarily the parents.

2 *Burrow nesters*

Pre-fledging activities of chicks from burrows are more difficult to see, but the young birds tend to sit at their tunnel entrances. During this late stage the trend for young shearwaters to carry bits of vegetation back to their nest chambers increases. They often attack vegetation nearby, pulling at the leaves and scattering them to either side of their bodies in incipient nest-building. Fulmars on grassy slopes may do the same (Williamson *et al.*, 1954). With gadfly petrels down rubbed off around the entrances suggests that open air activity starts some time before fledging, but Perrins *et al.* (1973) found that, after surfacing, most young *P.puffinus* left within 3 to 4 days. During this exploratory phase marked chicks shifted some metres from their burrows, returning before dawn. Sooty Shearwater chicks were found on the surface for 5.9 $\pm$ 3.8 nights (0–14 nights) during their last days ashore (Richdale, 1965a, p.35). On bright moonlit nights many chicks stayed below.

The most detailed descriptions of activity at this stage are those of Simons on *P.phaeopygia* using marked birds and night-vision equipment. Wing flapping began in the burrows about 2 months before fledging, emergence 2 to 3 weeks before departure. The chicks explored a 10 m radius of their nests, climbing rocks that could serve as launching platforms and exercised their wings (Simons, 1985). Other gadfly petrels and shearwaters behave like this, but all return to their burrows to be fed and are safely back before daylight. During this exploratory period the young birds find the best places locally for final take-off and also, presumably, become imprinted with the features of their natal environs. As the translocation experiments of Serventy (1967) shows, such imprinting helps explain the philopatric tendencies of tubenoses.

Perrins *et al.* (1973) found that the weights of young Manx Shearwaters taken from the surface varied during the course of the night. The first to appear after dark were slightly lighter than those appearing later; they suggested that the lighter birds were slightly older and about to leave for the sea, the heavier ones younger and not yet ready to fledge.

Wandering nestlings may be noisy. Rowan (1952) reported that chicks of *P.gravis* meander around 'croaking incessantly'. Those of *P.pacificus* are also quite vocal both in their burrows and when outside, while chicks of *P.urinatrix* ready to fly, trail about uttering repeated and monotonous calls (Richdale, 1965a, p.35).

X Departure

The spread of fledging dates tends to follow those of laying and hatching but is greater than either. Chicks of sedentary species like *P.alba*, whose laying extends over months, also depart over a wide range of dates. Antarctic species, except for *O.oceanicus*, tend to fledge the bulk of their chicks over a few days. The main departure of trans-equatorial migrant shearwaters is also truncated, but the data are poor. Richdale (1963, p.100) found that his 'heavy' *griseus* disappeared between

19 April and 10 May, mean date 1 May, but he did not determine when the last ones left.

A *First flight*

Petrels mostly fledge when there is wind enough to give them extra lift for take-off and they leave from some nearby elevation, however slight, that helps them to clear the ground in their first few m in the air. David Wingate met this need when a young Cahow *Pterodroma cahow* climbed up his legs to launch itself from the top of his head, this being the only raised place available (Wingate, 1972). Petrels in forests often gain height by climbing trees in the style of their parents (e.g. Harper, 1976) and as these leave detectable scent trails on frequently used trunks, the nestlings may be helped to choose the right trees by their odour. Richdale (1965a) found a diving petrel chick 4 m from its burrow persistently climbing a plant to get 45 cm above the ground: he ascribed this to a need to find a high point from which to get launched.

Departure as a whole is relatively sedate even in big colonies, with little of the frenetic activity characteristic of the height of the season and the pre-dawn outflow of adults from preferred launching points. Spread out over the departure slopes and perched on tussock tumps, Sooty Shearwater chicks beat their wings as they test the breeze, their take-offs orderly and rather cryptic. On quiet nights wind gusts are signalled by a sudden flurry of flickering wings as batches of fledglings leap to the air and disappear into the night (Warham *et al.*, 1982). Fairy Prion chicks begin wing actions and excessive preening about 52 h before leaving, often after their evening meal, so that they have about 7 h to reach the sea before sunrise. Most do so before midnight (Harper, 1976).

A feature of the fledging of Laysan Albatrosses noted by Fisher & Fisher (1969) was that nestlings reared near each other tended to disappear on the same day, as if they had left as a group. Likewise birds found together on the beaches (see below) were often from associated groups on the colonies. Simultaneous departure could have been triggered by some form of social stimulation or be a consequence of a subcolony synchrony in laying.

B *Strandings*

Most chicks get to sea on their first flight. Some fail and may shelter in a burrow lower down until the next night when they try again. This may result in burrows high up being mostly empty while those below are heavily occupied. Downes *et al.* (1959) noted fully feathered *M.giganteus* fledglings by the shoreline far from the colonies and Warham (1962) a marked chick about 500 m from its nest. On the low atolls of the North Pacific the albatrosses lack elevated take-off points and may end up crash-landing on the beaches. If there is little wind such birds may become stranded and die, or, if they swim away, get eaten by sharks.

Departure of *D.immutabilis* from Midway Atoll was followed by Fisher & Fisher (1969) after banding 9073 chicks on the various colonies. Of these 24% were resighted on the beaches. The distribution of these 'beached' birds was not random: most tended to go to the southern and eastern sides of the atoll where the shallows enclosed by the fringing reef are narrow and deep water close, rather than to northern beaches adjacent to the lagoon. Only 10% of the stranded birds went there. The majority shifted ESE against the prevailing wind and towards the sound of the surf on the nearest reef (Fig. 15.11).

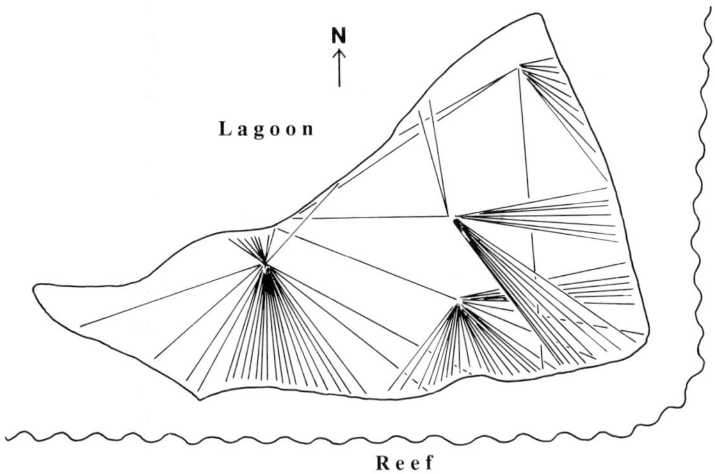

Figure 15.11 *Direction of movement of marked Laysan Albatross fledglings from their colonies to the beaches at Eastern Island, Midway Atoll. Most moved towards deep water. From Fisher & Fisher (1969).*

These Hawaiian albatross fledglings have no enemies on land, southern tubenoses have predators eager to kill stranded birds, so that the availability of high ground and suitable launching points may be an important necessity for successful breeding here, much less so in the North Pacific.

There is little information on how long birds that fail to reach the sea at their first try take to get there. The marked Laysan Albatrosses studied by the Fishers were stranded for 1 or 2 days before disappearing and presumably getting away successfully. Those that stayed on the beaches for 4 or 5 days died there.

During persistent calms chicks may try to walk to the sea. Lockley (1930) described young *P.puffinus* after dark proceeding in what he nicely termed 'a flying walk', pattering along and sustaining themselves with quickly beaten wings. He noted that they seemed to know the nearest direction of the sea. They did not avoid obstacles but tried to clamber over them and pushed through dense vegetation — typical behaviour of young petrels taking the shortest route on their crucial first journey to the water. Their process was sporadic as they paused at intervals for rest. Those that reached the cliffs blundered straight over, then, catching the updraft, flapped and glided down to the waves. Rowan (1952) described fledging Greater Shearwaters waddling down through the tussocks on Nightingale Island in the late afternoon and evening, some taking off and alighting before reaching the coast. Some birds that entered the sea without taking wing had difficulty getting through the surf.

When flying out on foggy nights young disorientated birds may be attracted to bright lights and become stranded on land or on ships, often to die or be eaten by cats. Well known examples include those of *P.huttoni* in New Zealand and *P.auricularis* and *P.phaeopygia* in Hawaii (Harrow, 1976; Telfer *et al.*, 1987). Contrary winds at fledging time may also result in high mortality with young birds being driven far inland.

C *Behaviour at sea*

Successful first flights generally carry fledglings straight out to sea. In light airs some may come down offshore and Northern Fulmars may congregate in small groups and have been recorded swimming hard towards the open sea, others eventually getting airborne and passing out of sight. According to Fisher (1952) most *F.glacialis* fledglings spend some days afloat near their cliffs in small groups at which time the Faeroese collect them for food. Colour-banded Cape Pigeon chicks flew straight out to sea and none was sighted inshore among other Cape Pigeons although looked for (Pinder, 1966). Harper (1976) saw both *P.turtur* and *P.macroptera* chicks drinking copiously on first reaching the water and suggested that thirst might be a proximate factor in determining the fledging date. Similar behaviour seems not to have been seen among well-studied species like *F.glacialis*.

Once at sea the fledgling petrels must fend for themselves. There is no evidence that their parents take any notice of them and the sighting of supposedly young Wandering Albatrosses billing with adult birds at sea is no indication of kinship.

XI The fledgling

On flying a substantial proportion of fledglings still retains remnants of down, usually on the neck and thighs: such birds turn up on the decks of ships at sea or are 'wrecked' inland by storms. Half of 40 *P.marina* chicks fledged still carrying wisps of down, those that left first having more than later ones, but this was not the case with *P.turtur* (Richdale, 1965a, p.75; 1965b, p.111).

Otherwise young petrels look very like their parents except in a few species which have a distinctive juvenal plumage like *D.exulans*, *D.albatrus*, *D.irrorata*, *Macronectes* and, to some extent, the mollymawks. In the hand fledglings can often be identified by their plumage as this is often the only time in their lives where all their feathers are fresh and unbraded. Later, because body and wing moults are asynchronous, their plumage contains a mixture of old and new feathers, and in dark-plumaged birds the old ones have a brownish cast. Abraded primaries and rectrices also help in identifying older birds. In many newly fledged birds there are pale edges to some of the feathers, especially on the wing coverts, mantle and scapulars and these may create a scalloped effect which disappears with wear.

Other clues to age are in the bill. Fledglings have rather soft beaks and in diving petrels and storm petrels the nasal tubes tend to be less prominent than in adult birds. Young mollymawks fledge with dark blackish or brownish beaks which take months to acquire the yellow bands and red tips of mature birds and the head plumage is often darker than in the adults (Tickell, 1969; Warham *et al.*, 1974).

XII Breeding success

Some petrel nestings fail as soon as the egg is laid and failures continue to fledging. In general, losses during the egg stage are higher than those during chick rearing. Eggs are more often lost early in incubation than later and more small chicks die than large ones. Mortality arises through many causes — flooding, gales, egg breakage due to trampling by the adults, infertility, egg and chick desertion, predation and disturbance. Success may be affected by the breeding experience of the adults and the length of their pair bond and these in turn may determine the date of laying, early layers being the more successful, e.g. in *F.glacialis* and *P.puffinus* (Ollason & Dunnet, 1986; Brooke, 1978a).

Nest sanitation may be important. Streaks of faeces radiate from surface and ledge nests during the late chick stage and large nestlings of the giant albatrosses can shoot excreta to 1.5 m from the nest rim. Chicks in burrows usually defaecate away from the entrance opening and their subterranean chambers do not become foul even in damp substrates: possibly nest inquilines help. Wheelwright & Boersma (1979) found carrion-feeding fungus beetles, Leiodidae, in 39% of *O.furcata* burrows and while these may have cleaned the nest, they also tunnelled into the bodies of torpid chicks and killed them. Active chicks were also stressed when so attacked. Wheelright & Boersma concluded that the insulative benefits of well-lined nests were outweighed by the disadvantages of beetle attack and by the cooling effects on the chicks of the damp linings. Womersely (1955) identified 14 species of mites in nests of *P.tenuirostris,* five probably ectoparasitic, the rest living on detritus and excreta. Detailed accounts of the faunas of surface and underground nests, which also include fleas, staphylinid and other beetles and dipteran larvae, seem not to have been made.

There are practical difficulties in collecting reliable data on hatching success (the proportion of eggs laid that hatch); fledging success (the proportion of chicks that fledge) or overall breeding success (the proportion of eggs laid to chicks that fly). To get these statistics the eggs must be counted, preferably close to the lay, before some have been lost without trace. This is specially difficult with burrowing species and those that have long laying periods, for then the adults are most susceptible to disturbance. If, as often happens, losses through observer disturbance are ignored, the sample will be biased in favour of older birds, these tending to be more persistent that recent recruits. Handling of eggs or chicks for weighing or other purposes also imposes some unnatural mortality. In consequence, birds used for mortality and success data need to be separate from those used for other information. To act as a check on the study sample a control group is needed where, once egg depositions have been established, no further close inspections are made until fledging starts.

Because of such problems there are few sets of data where the fates of good samples of undisturbed eggs have been followed through to fledging. Many studies give hatching or fledging success, although the latter is often only taken to the point at which banding is done, mortalities to final departure being missed. The longest studies are those on *F.glacialis* and *P.tenuirostris* by Ollason & Dunnet (1986, 1988); Serventy & Curry (1984). Both extended over 32–34 seasons but both have suffered losses due to observer effects. Other figures are given for southern procellariids by Mougin (1975), Conroy (1972), Hunter (1984b) and Hall (1987); for albatrosses by Tickell (1968), Fisher (1971), Berutti (1979), Woodward (1972), Harris (1973) and Croxall & Prince (1979); for shearwaters by Brooke (1978a), and Byrd *et al.* (1983); and for storm petrels by Harris (1969b), Morse & Buchheister (1979) and Ainley *et al.* (1990).

Breeding success in populations distant from human disturbance nonetheless varies from year to year at a particular colony and from colony to colony in the same year because of differing environmental and other factors.

Examples of variations at South Georgia over four seasons were given by Croxall & Prince (1979) for two mollymawks. For *D.chrysostoma* overall success averaged 54%, range 47%–67%, for *D.melanophrys* 41%, range 16%–67%. At the same place success rates for *D.exulans* climbed from 53.7% in 1980 to 72.4% in 1990 (Croxall *et al.*, 1990). For one group of *D.irrorata* about 82% of nests were successful in one year but only 19% in the next (Harris, 1973).

Data exemplifying colony-to-colony variation in the same season include those for *M.giganteus* by Conroy (1972, p.44). In four colonies with a total of 174 eggs, 41% yielded fledglings, the range between colonies varying from 26% to 58%. Least successful was a colony surrounded by four skua territories, these birds being the major cause of egg and chick loss. Further mortality arose from the exposed habitat of that colony, chicks being killed or disabled by heavy snow. Mougin (1975, p.143) reported similar effects of bad weather on this and other fulmars in Antarctica, yet one group of *M.giganteus* there experienced 100% success.

Where unnatural predators are present mortality may be so high that breeding success is nil, and if such predation persists extinction eventually follows. This appeared to be the destiny of a major population of *P.parkinsoni* until the last-minute eradication of feral cats. In 1979 38.7% of the Maui Island breeding *P.phaeopygia* studied by Simons (1985) reared their chick; in 1980, after reducing the local mongooses and cats, 71.4% of the eggs produced flying young.

The data do not suggest that any group or family has higher overall success rates than others, but there are inadequate figures for the small burrowing petrels that are often believed to suffer high losses. These figures suggest that only 40–50% of eggs yield flying young. Small groups in favourable situations may do particularly well. For instance, 93% of 15 eggs of *P.pacificus* laid in nesting boxes were successful, but only 80% of 93 eggs laid in natural burrows (Byrd *et al.*, 1983). Hunter (1984b) gave very high values for giant petrels studied over 3 years: 82.5 $\pm$ 11.5% (n = 108) for *M.halli*; 71.5 $\pm$ 10.3% (n = 171) for *giganteus*. Figures given by Conroy (above) and Mougin (53% for *giganteus* in Adelie Land, 29% for both species at The Crozets) are much lower. However, Hunter did not count eggs until half-way through incubation by which time significant numbers could have been lost.

Observer effects depressing success are probably slight in birds like albatrosses and some gadfly petrels and prions, high among giant petrels and shearwaters such as *P.carneipes*, *P.griseus* and *P.tenuirostris* and many storm petrels. Hall (1987), who recognised the problem of disturbance, found that only 12% of the Shoemakers in his main study group of 41 birds reared chicks from their eggs, whereas 54% of the 26 pairs at one seldom visited did that.

Some information relating success to pair bonding have been given. For 906 newly established pairs of *F.glacialis* overall success averaged 50%, for 997 established pairs 63% succeeded (Ollason & Dunnet, 1986). Brooke (1978a) showed a similar effect among *P.puffinus* (success rates varied between 62% and 75% over four seasons), with established pairs fledging chicks from 78% of 191 eggs laid, new pairs only 63% of 77 eggs. The difference was attributed to the lower hatching success of naive pairs, both categories being equally successful in raising chicks. In this species early layers did best as Harris (1969b) found for the storm petrel *O.castro*, whose overall success was about 30%.

Lifetime reproductive success has recently been estimated for *P.tenuirostris* based on 418 birds followed over 25 years: on average they each produced 4.8 eggs and fledged 2.7 young (Wooller *et al.*, 1988). These authors found that 27% of their birds produced no chick, 19% only one and a mere 8% of those that completed their reproductive years produced 53% of all the young birds that returned to breed on their natal island. With increased breeding experience chick production also went up: birds that made only one attempt in their lifetime produced 0.35 chicks, those that tried for 20 years or more produced 11.4 fledglings. A parallel study of this population relates success to age and breeding experience of the parents (Wooller *et al.*, 1989). Overall success increased from 40–45% initially to a 75–80% maximum and then declined to 55–60% in ageing birds. Those that lived longest produced more young, especially early in their reproductive years. Wooller *et al.* believed that the breeding performance was influenced by age, the cumulative breeding experience, their remaining breeding potential and the differential survival of good and poor breeders.

XIII Discussion

Lack (1967, 1968) regarded the slow development of tubinares as a consequence of the food supply often being distant and patchy, Antarctic specues raising chicks quickly because food is plentiful while it lasts, tropical ones taking longer as the food supply there is poorer. In other words, the chicks grow slowly to adjust to the energy needs of breeding. In Lack's view the main function of the chick's fat is to tide it over temporary periods of shortage, a view still widely held. He also hypothesised that chick feeding declines towards the end of the season because the food supply is falling, resulting in the departure from the colonies, first of the failed non-breeders, and later of the breeders which, if migratory, may desert their chicks. Growth rates were seen as varying in response to a wide range of conditions, clutch size being fixed.

Ricklefs disputed Lack's food-limiting explanation partly on the results of his study of Leach's Storm Petrel whose chicks he found were fed regularly and not starved. His analysis of Richdale's data on feeding frequencies of *P.griseus* revealed that nights of feeding or not feeding were highly significantly associated, suggesting that while the birds may fly to distant feeding grounds during prolonged absences, food given to the chicks is found from close at hand (Ricklefs *et al.*, 1985). Furthermore, meal sizes for Leach's Petrels did not vary with the length of time a parent had been away. The chicks grew at about the same rate as similar sized galliforms, were similarly capable of some mature functions at birth, while the accumulation of tissue during the last half of the nestling period accounted for only some 5% of the total energy required by the chick, so that growth rates could evidently be increased considerably without major effects on the energy budget. At least in *O.leucorhoa*, according to Ricklefs, fat stores far exceed those needed by the chicks during their 1–3 day fasts. Instead of being determined by food availability, Ricklefs and his co-workers visualised chick feeding as being regulated by the food demands of the chick, long spans between meals resulting when parents have to feed far from their nests but, by bringing bigger meals less often, save energy in its transportation.

Lack (1968) considered that the one-egg clutch of Procellariiformes evolved because they could not find food quickly enough to make two eggs and because their pelagic feeding on patchy food incurs heavy travel costs so that they could only raise a single chick anyway. Alternatively the single-egg clutch has been viewed as an adaptation to reduce reproductive output (Wynne-Edwards, 1962). In attempts to test these ideas a series of experiments has been conducted by following the fates of artificially created sets of twins.

Six species have been used. In only one of these tests, and that the most extensive, was the productivity of the twinned pairs better than controls with single chicks (Perrins *et al.*, 1973). In some, e.g. with *D.immutabilis* and *P.tenuirostris* (Rice & Kenyon, 1962; Norman & Gottsch, 1969) the introduced siblings may have been too large and discriminated against by the foster parents. In the albatross and *O.castro* (Harris, 1969b) one chick became dominant and got most of the food, and this could have occurred in other trials with burrowing birds, unbeknown to the experimenters. However, in five artificially twinned nests of *O.furcata* all additions were accepted and although no twins fledged, two of the originals and two of the introduced ones did so (Boersma *et al.*, 1980).

In an early experiment parents of Manx Shearwater twins fledged only 0.3 young/pair as opposed to single chick control pairs that fledged 0.95 young (Harris, 1966). The twinned parents brought more food more often but each chick was fed less often than were those in single-chick nests. Harris thought that the survival of the four from twinned nests that flew was doubtful, although one of these underweight chicks did eventually return to the colony (Perrins *et al.*, 1973).

In a later experiment these latter researchers created 42 sets of twins using newly hatched *P.puffinus* chicks. Thirty-three of these pairs raised both chicks (1.79 chicks/pair) whereas 45 controls with only their own chick raised but 0.93 chicks/pair. Here, perhaps because the twins were created when they were very young, both were accepted before the adults had come to recognise their own chicks by voice, behaviour or smell. Nor evidently, did inter-sibling competition operate significantly to the chicks' detriment. Although twins had lower weights on making their first flights (Fig. 15.12) their 23% post-fledgling survival was not

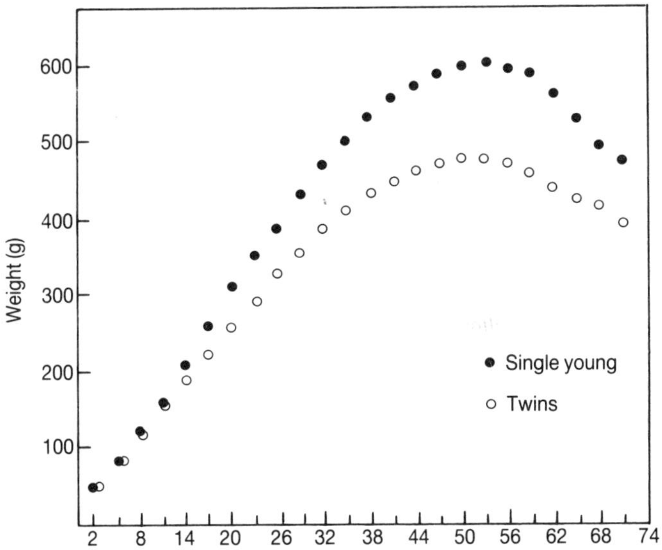

Figure 15.12 *Growth of twinned Manx Shearwater chicks. From Perrins* et al. *(1973).*

significantly lower than the 29% of the single controls.

This experiment appears to support Ricklef's belief that food availability does not limit chick growth (at least in good seasons) although Perrins *et al.* (1973) suggested that in their experiment the limiting factor was the female's inability to make two eggs — otherwise two egg clutches would have evolved long ago (and double brood patches also).

References

AINLEY, D.G. (1977). Feeding methods in seabirds: a comparison of polar and tropical nesting communities in the eastern Pacific Ocean. In *Adaptations Within Antarctic Ecosystems* (G. Llano, ed.) pp. 669–685. Gulf Publishing, Houston, Texas.

AINLEY, D.G. (1980). Geographic variation in Leach's Storm-petrel. *Auk* **97**, 837–853.

AINLEY, D.G. & BOEKELHEIDE, R.J. (1990). Leach's Storm-petrel and Ashy Storm-petrel. In *Seabirds of the Farallon Islands* (D.G. Ainley & R.J. Boekelheide, eds). Stanford University Press, Stanford.

AINLEY, D.G. & MANOLIS, B. (1979). Occurrence and distribution of the Mottled Petrel. *West Birds* **10**, 113–123.

AINLEY, D.G., MORRELL, S. & LEWIS, J.T. (1975). Patterns in the life histories of storm petrels on the Farallon Islands. *Living Bird* **13**, 295–312.

AINLEY, D.G., LEWIS, T.J. & MORRELL, S. (1976). Molt in Leach's and Ashy Storm-petrels. *Wilson Bull* **88**, 76–95.

AINLEY, D.G., O'CONNOR, E.F. & BOCKELHEIDE, R.J. (1984). The marine ecology of birds in the Ross Sea, Antarctica. *Orn Mongr* **32**, American Ornthologists Union, Washington.

AINLEY, D.G., HENDERSON, R.P. & STRONG, C.S. (1990). Leach's Storm-petrel and Ashy Storm-petrel. In *Seabirds of the Farallon Islands* (D.G. Ainley & R.J. Boekelheide, eds) pp. 128–162. Stanford University Press, Stanford.

ALBIN, E. (1740). *Natural History of Birds* (Suppl) Vol. 3. Author, London.

ALEXANDER, W.B., FALLA, R.A., JOUANIN, C., MURPHY, R.C., SALOMONSEN, F., VOOUS, K.H., WATSON, G.E., BOURNE, W.R.P., FLEMING, C.A., KURODA, N.H., ROWAN, M.K., SERVENTY, D.L., TICKELL, W.L.N., WARHAM, J., & WINTERBOTTOM, J.M. (1965). The families and genera of the petrels and their names. *Ibis* **107**, 401–405.

ALLAN, R.G. (1962). The Madeiran Storm Petrel *Oceanodroma castro*. *Ibis* **103b**, 274–295.

AMERICAN ORNITHOLOGISTS' UNION (1983). *Checklist of North American Birds* 6th edn. American Ornithologists Union, Washington.

AMERSON, A.B. & SHELTON, P.C. (1976). The natural history of Johnston Atoll Central Pacific Ocean. *Atoll Res Bull* **192**, 1–479.

AMIET, L. (1958). The distribution of *Diomedea* in eastern Australian waters. *Notornis* **7**, 219–230.

ANDERSEN, K. (1895). *Diomedea melanophrys* in the Faeroe Islands. *Proc R Phys Soc* **13**, 91–114.

ANDERSON, A. (1982). The establishment and growth of a new Fulmar colony on sand dunes. *Bird Study* **29**, 189–194.

ANTHONY, A.W. (1900). Nesting habits of the Pacific coast species of the genus *Ruffinus*. (*sic*) *Auk* **17**, 247–252.

AR, A. & YOM-TOV, Y. (1978). The evolution of parental care in birds. *Evolution* **32**, 655–669.

AR, A., RAHN, H. & PAGANELLI, C.V. (1979). The Avian Egg: Mass and Strength. *Condor* **81**, 331–337.

ASHMOLE, N.P. (1971). Sea bird ecology and the marine environment. In *Avian Biology* (D.S. Farner, J.R. King & K.C. Parkes, eds) Vol. 1, pp. 224–286. Academic Press, New York and London.

ASHMOLE, N.P. & ASHMOLE, M.J. (1967). Comparative feeding ecology of sea birds of a tropical oceanic island. *Peabody Mus Nat Hist Bull* **24**, 1–131.

ASTHEIMER, L.B. & GRAU, C.R. (1990). A comparison of yolk growth rates in seabird eggs. *Ibis*, **132**, 380–394.

ASTHEIMER, L.B., PRINCE, P.A. & GRAU, C.R. (1985). Egg formation and the pre-laying period of Black-browed and Grey-headed Albatrosses *Diomedea melanophris* and *D. chrysostoma* at Bird Island, South Georgia. *Ibis* **127**, 523–529.

AUDUBON, J.J. (1838) *Ornithological Biography*, etc. Vol. IV. Black, Edinburgh.

AUSTIN, O.L. (1952). Notes on some petrels of the North Pacific. *Bull Mus Camp Zool Harvard* **107**, 391–407.

AVERY, G. (1982). Results of beach patrols conducted in southern Africa in 1981. *Cormorant* **10**, 87–96.

BAILEY, R. (1966). The sea-birds of the southeast coast of Arabia. *Ibis* **108**, 224–264.

BAILEY, R.S. (1968). The pelagic distribution of sea-birds in the western Indian Ocean. *Ibis* **110**, 493–519.

BAILEY, R.S., POCKLINGTON, R. & WILLIS, P.R. (1968). Storm-petrels *Oceanodroma* spp. in the Indian Ocean. *Ibis* **110**, 27–34.

BAKER, A.J. & COLEMAN, J.D. (1977). The breeding cycle of the Westland Black Petrel (*Procellaria westlandica*). *Notornis* **24**, 211–231.

BAKER, R.R. (1978). *The Evolutionary Ecology of Animal Migration*. Hodder and Stoughton, London.

BALTZ, D.M. & MOREJOHN, G.V. (1977). Food habits and niche overlap of seabirds wintering on Monterey Bay, California. *Auk* **94**, 526–543.

BANG, B.G. (1966). The olfactory apparatus of tubenosed birds (*Procellariiformes*). *Acta Anat* **65**, 391–415.

BANNERMAN, D.A. (1914). The distribution and nidification of the Tubinares in the North Atlantic islands. *Ibis* (10th Ser) **2**, 438–494.

BANNERMAN, D.A. (1968). *History of the Birds of the Cape Verde Islands*. Oliver and Boyd, Edinburgh.

BARRAT, A. (1974). Note sur le Petrel Gris *Procellaria cinerea*. *Com Natn Fr Rech Antarct* **33**, 19–23.

BARRAT, A., BARRE, H & MOUGIN, J-L. (1976). Donnees ecologiques sur les Grands Albatross *Diomedea exulans* de l'ile de la Possessian (Archipel Crozet). *Oiseau Revue Fr Orn* **46**, 143–155.

BARRE, H. 1976. *Pterodroma lessonii* (Garnot) a l'ile de la Possessian (Ile Crozet). *Com Natn Fr Rech Antarct* **40**, 61–75.

BARROWCLOUGH, G.F., CORBIN, K.W. & ZINK, R.M. (1981). Genetic differentiation in the Procellariiformes. *Comp Biochem Physiol* **69B**, 629–623.

BARTLE, J.A. (1968). Observations on the breeding habits of Pycroft's Petrel. *Notornis* **15**, 70–99.

BARTLE, J.A. (1985). Westland Black Petrel. In *The Complete Book of New Zealand Birds* p. 91. Reader's Digest, Sydney.

BARTLE, J.A. (1987). Westland Black Petrel research notes, 10–29/4/87. *OSNZ News* *44*, 5.

BARTLE, J.A. (1990). Sexual segregation of foraging zones in procellariiform birds: implications of accidental capture on commercial fishery longlines of Grey Petrels (*Procellaria cinerea*). *Notornis* **37**, 146–150.

BARTON, D. (1979). Albatrosses in the western Tasman Sea. *Emu* **79**, 31–35.

BECH, C., BRENT, R., PEDERSEN, P.F., RASMUSSEN, J.G. & JOHANSEN, K. (1982). Temperature regulation in chicks of the Manx Shearwater *Puffinus puffinus*. *Ornis Scand* **13**, 206–210.

BECH, C., MEHLUM, F. & HAFTORN, S. (1988). Development of chicks during extreme cold conditions: the Antarctic Petrel (*Thalassoica antarctica*). *Acta XIX Congr Int Orn* **2**, 1447–1456.

BECK, J.R. (1969). Food, moult and age of first breeding in the Cape Pigeon, *Daption capensis* Linnaeus. *Br Antarct Surv Bull* **21**, 33–44.
BECK, J.R. (1970). Breeding seasons and moult in some smaller Antarctic petrels. In *Antarctic Ecology* (M.W. Holdgate, ed.) Vol. 1, pp. 542–550. Academic Press, London.
BECK, J.R. & BROWN, D.W. (1971). The breeding bilogy of the Black-bellied Storm Petrel *Fregetta tropica*. *Ibis* **113**, 73–90.
BECK, J.R. & BROWN, D.W. (1972).The biology of Wilson's Storm Petrel, *Oceanites oceanicus* (Kuhl), at Signy Island, South Orkney Islands. *Br Antarct Surv Scient Rep* **69**, 1–54.
BEEBE, W. (1926). *The Arcturus Adventure*. Putnam, New York.
BEEBE, W. (1935). Rediscovery of the Bermuda Cahow. *Bull N Y Zool Soc* **38**, 187–190.
BELOPOL'SKI, L.O. (1961). *Ecology of Sea Colony Birds of the* Barents Sea. Israel Program of Scientific Translation, Jerusalem.
BENNETT, G. (1860). *Gatherings of a Naturalist in Australasia etc*. Van Voorst, London.
BENT, A.C. (1922). Life histories of North American petrels and pelicans and their allies. *US natn Mus Bull* **121**, 1–335.
BERRUTI, A. (1979). The breeding biologies of the Sooty albatrosses *Phoebetria fusca* and *P. palpebrata*. *Emu* **79**, 161–175.
BERRUTI, A. (1981a). The status of the Royal Penguin and Fairy Prion at Marion Island, with notes on feral cat predation on nestlings of large birds. *Cormorant* **9**, 123–128.
BERRUTI, A. (1981b). Displays of the sooty albatrosses *Phoebetria fusca* and *P. palpebrata*. *Ostrich* **52**, 98–103.
BERRUTI, A. & HUNTER, S. (1986). Some aspects of the breeding biology of Salvin's Prion *Pachyptila vittata salvini* at Marion Island. *Cormorant* **13**, 98–106.
BERRUTI, A., ADAMS, N.J. & BROWN, C.R. (1985). Chick energy balance in the Whitechinned Petrel, *Procellaria aequinoctialis*. In *Antarctic Nutrient Cycles and Food Webs* (W.R. Siegfried, P.R. Condy & R.M. Laws, eds) pp. 460–465. Springer, Berlin.
BIERMANN, W.H. & VOOUS, K.H. (1950). *Birds Observed and Collected during the Whaling Expeditions of the 'Willem Barendsz' in the Antarctic, 1946–1947 and 1947–1948*. Brill, Leiden.
BILLINGS, S.M. (1968). Homing in Leach's Petrel. *Auk* **85**, 36–43.
BOARD, R.G., TULLETT, S.G. & PERROTT, H.R. (1977). An arbitrary classification of the pore systems in avian eggshells. *J Zool Lond* **182**, 251–265.
BOERSMA, P.D. (1982) Why some birds take so long to hatch. *Am Nat* **120**, 733–750.
BOERSMA, P.D. (1986). Body temperature, torpor, and growth in chicks of Fork-tailed Storm Petrels (*Oceanodroma furcata*). *Physiol Zool* **59**, 10–19.
BOERSMA, P.D. & DAVIES, E.M. (1987). Sexing monomorphic birds by vent measurements. *Auk* **104**, 779–783.
BOERSMA, P.D. & WHEELWRIGHT, N.T. (1979). Egg neglect in the Procellariiformes: reproductive adaptations in the Folk-tailed Storm-Petrel. *Condor* **81**, 157–165.
BOERSMA, P.D. WHEELWRIGHT, N.T., NERINI, M.K. & WHEELWRIGHT, E.S. (1980). The breeding biology of the Fork-tailed Storm-Petrel (*Oceanodroma furcata*). *Auk* **97**, 268–282.
BONAPARTE, C.-L. (1857). *Conspectus Generum Avium* Vol. 2. Brill, Leiden.
BOURNE, W.R.P. (1953). On the races of the Frigate Petrel *Pelagodroma marina* (Latham) with a new race from the Cape Verde Islands. *Bull Br Orn Club* **73**, 79–82.
BOURNE, W.R.P. (1955). The birds of the Cape Verde Islands. *Ibis* **97**, 508–556.
BOURNE, W.R.P. (1957). Additional notes on the birds of the Cape Verde Islands, with particular reference to *Bulweria mollis* and *Fregata magnificeus*. *Ibis* **99**, 182–190.
BOURNE, W.R.P. (1960). Petrels of the Indian Ocean. *Sea Swallow* **13**, 26–39.
BOURNE, W.R.P. (1962). In *Handbook of North American Birds* (R.S. Palmer, ed.) Vol. 1, pp. 204–205. Yale University Press, New Haven and London.
BOURNE, W.R.P. (1964). The relationship between the Magenta Petrel and the Chatham

Island Taiko. *Notornis* **11**, 139–144.
BOURNE, W.R.P. (1966). The plumage of the fulmars of St. Kilda in July. *Bird Study* **13**, 209-213.
BOURNE, W.R.P. (1967a). Long-distance vagrancy in the petrels. *Ibis* **109**, 141–167.
BOURNE, W.R.P. (1967b). Subfossil petrel bones from the Chatham Islands. *Ibis* **109**, 1–7.
BOURNE, W.R.P. (1968). Notes on the diving petrels. *Bull Br Orn Club* **88**, 77–88.
BOURNE, W.R.P. (1975). The lachrymal bone in the genus *Bulweria*. *Ibis* **117**, 535.
BOURNE, W.R.P. (1977). Half a pair of Black-browed Albatrosses. *Br Birds* **70**, 301-303.
BOURNE, W.R.P. (1983a). The Soft-plumaged Petrel, the Gon-gon and the Freira, *Pterodroma mollis, P.feae* and *P.madeira*. *Bull Br Orn Club* **103**, 52–58.
BOURNE, W.R.P. (1983b). The appearance and classification of the *Cookilaria* petrels. *Sea Swallow* **32**, 65–71.
BOURNE, W.R.P. (1985). Petrel. In *A Dictionary of Birds* (B. Campbell & E. Lack, eds) pp. 451-456. Poyser, Calton.
BOURNE, W.R.P. (1986). Recent work on the origin and suppression of bird species in the Cape Verde Islands, especially the shearwaters, the herons, the kites and the sparrows. *Bull Br Orn Club* **106**, 163–170.
BOURNE, W.R.P. (1987a). The classification and nomenclature of the petrels. *Ibis* **129**, 404.
BOURNE, W.R.P. (1987b). The affinities, breeding behaviour and distribution of Jouanin's Petrel *Bulweria fallax*. *Bull Br Orn Club* **107**, 4–6.
BOURNE, W.R.P. & DAVID, A.C.F. (1983). Henderson Island, Central South Pacific, and its birds. *Notornis* **30**, 233–252.
BOURNE, W.R.P. & ELLIOTT, H.F.I. (1965). The correct scientific name for the Kerguelen Petrel. *Ibis* **107**, 548–550.
BOURNE, W.R.P. & WARHAM, J. (1966). Geographical variation in the giant petrels of the genus *Macronectes*. *Ardea* **54**, 45–67.
BOURNE, W.R.P., MACKRILL, E.J., PATERSON, A.M. & YESOU, P. (1988). The Yelkonan Shearwater *Puffinus (puffinus?) yelkouan*. *Br Birds* **81**, 306–319.
BOWDLER, S. (1974). An account of an archaeological reconnaissance of Hunters Island, N.W. Tasmania 1973/4. *Rec Queen Victoria Mus Launceston* **54**, 1–22.
BRADFIELD, J.R.G. (1959), Radiographic studies on the formation of the hen's egg shell. *J Exp Biol* **28**, 125–140.
BRADLEY, J.C., WOOLLER, R.D. SKIRA, I.J. & SERVENTY, D.L. (1989). Age-dependent survival of breeding Short-tailed Shearwaters *Puffinus tenuirostris*. *J Anim Ecol* **58**, 175–188.
BRASIL, L. (1917). Notes sur la faune ornithologique de l'Oceanie. *Bull Mus Natn Hist Nat Paris* **23**, 429–441.
BRETAGNOLLE, V. (1988a). Cycles de presence et rhythmes d'activite chez cinq especes de petrels antarctiques. *Oiseau Revue Fr Orn* **58**, 44–58.
BRETAGNOLLE, V. (1988b). Social behaviour of the Southern Giant Petrel. *Ostrich* **59**, 116–125.
BRETAGNOLLE, V. (1990) Behavioural affinities of the Blue Petrel *Halobaena caerulea*. *Ibis* **132**, 102–123.
BRETAGNOLLE, V. (in press). Calls of Wilson's Storm Petrel: functions, individual and sexual recognitions, and geographic variation. *Behaviour*.
BRETAGNOLLE, V., ZOTIER, R. & JOUVENTIN, P. (1990). Comparative population biology of four prions (Genus *Pachyptila*) from the Indian Ocean and consequences for their taxonomic status. *Auk* **107**, 305–316.
BREWSTER, W. (1881). Critical notes on a petrel new to North America. *Bull Nuttall Orn Club* **6**, 91–97.
BRIGGS, K.T. & CHU, E.W. (1986). Sooty Shearwaters off California: distribution, abundance and habitat use. *Condor* **88**, 355–364.

BROOK, D. & BECK, J.R. (1972). Antarctic Petrels, Snow Petrels and South Polar Skuas breeding in the Theron Mountains. *Br Antarct Surv Bull* **27**, 131–137.

BROOKE, M. (1990). *The Manx Shearwater*. Poyser, London.

BROOKE, M. de L. (1978a). Some factors affecting the laying date, incubation and breeding success of the Manx Shearwater, *Puffinus puffinus*. *J Anim Ecol* **47**, 477–495.

BROOKE, M. de L. (1978b). The dispersal of female Manx Shearwaters. *Ibis* **120**, 546–551.

BROOKE, M. de L. (1978c). A test for visual location of the burrow by Manx Shearwaters *Puffinus puffinus*. *Ibis* **120**, 347–349.

BROOKE, M. de L. (1978d) Weights and measurements of the Manx Shearwater, *Puffinus puffinus*. *J Zool Lond* **186**, 359–374.

BROOKE, M. de L. (1978e). Sexual differences in the voice and individual vocal recognition in the Manx Shearwater (*Puffinus puffinus*). *Anim Behav* **26**, 622–629.

BROOKE, M. de L. (1986a). The vocal systems of two nocturnal petrels, the White-chinned *Procellaria aequinoctialis* and the Grey *P.cinerea*. *Ibis* **128**, 502–512.

BROOKE, M. de L. (1986b). Manx Shearwater chicks: seasonal, parental and genetic influences on the chick's age and weight at fledging. *Condor* **88**, 324–327.

BROOKE, M. de L. (1987). Population estimates and breeding biology of the petrels *Pterodroma externa* and *P. longirostris* on Isla Alejandro Selkirk, Juan Fernandez Archipelago. *Condor* **89**, 581–586.

BROOKE, R.K. (1981). Modes of moult of flight feathers in albatrosses. *Cormorant* **9**, 13–18.

BROOKE, R.K. & FURNESS, B.L. (1982). Reversed modes of moult of flight feathers in the Blackbrowed Albatross *Diomedea melanophris*. Cormorant **10**, 27–30.

BROTHERS, N.P. (1979). Tasman Island, Tasmania, *Corella* **3**, 55–57.

BROTHERS, N.P. (1984). Breeding, distribution and status of burrow nesting petrels at Macquarie Is. *Aust Wildl Res* **11**, 113–131.

BROWN, C.R. & ADAMS, N.J. (1984). Female Wandering Albatross *Diomedea exulans* raising a chick on its own at Marion Island. *Cormorant* **12**, 103–104.

BROWN, D.A. (1966). Breeding Biology of the Snow Petrel *Pagodroma nivea* (Forster). *Aust Natn Antarct Res Exped Sci* (Rep B) (1) *Zool* **89**, 1–63.

BROWN, L.E. & FISHER, H.T. (1966). Electrophoretic study of blood proteins of some procellariform birds. *Auk* **83**, 111–116.

BROWN, R.G.B. (1970). Fulmar distribution: a Canadian perspective. *Ibis* **112**, 44–51.

BROWN, R.G.B. (1980). The field identification of Black and Markham's Storm-Petrels. *Am Birds* **34**, 368.

BROWN, R.G.B. (1981). Seabirds in northern Peruvian waters, November–December 1977. *Bol Inst Mar Peru* (Vol Extra) 34–37.

BROWN, R.G.B. *Revised Atlas of Eastern Canadian Seabirds*. *Can Wildl Serv*, Dartmouth.

BROWN, R.G.B. (1988). The wing-moult of fulmars and shearwaters (Procellariidae) in Canadian Arctic waters. *Can Field-Nat* **102**, 203–208.

BROWN, R.G.B., BOURNE, W.R.P. & WAHL, T.R. (1978). Diving by Shearwaters. *Condor* **80**, 123–125.

BROWN, R.G.B., BARKER, S.P., GASKIN, D.E. & SANDEMAN, M.R. (1981). The foods of Great and Sooty Shearwaters *Puffinus gravis* and *P.griseus* in eastern Canadian waters. *Ibis* **123**, 19–30.

BROWN, R.S., NORMAN, F.I. & EADES, D.W. (1986). Notes on Blue and Kerguelen Petrels found beach-washed in Victoria, 1984. *Emu* **86**, 228–238.

BUCK, P. (1949). *The Coming of the Maori*. Whitcombe and Tombs, Wellington.

BULLER, W.L. (1894). On the birds observed during a voyage from New Zealand to England. *Trans Proc NZ Inst* **26**, 182–199.

BURGER, A.E. (1978). Interspecific breeding attempts by *Macronectes giganteus* and *M.halli*. *Emu* **78**, 234–235.

BURGER, J. (1980). The transition to independence and postfledging parental care in seabirds. In *Behavior of Marine Animals* (J. Burger, B.L. Olla & H.E. Winn, eds) Vol. 4, pp. 367–447. Plenum, New York.

BUTLER, N. (1882). *The History of the Bermundaes or Summer Islands* (L.H. Lefroy, ed.). Franklin, New York.

BYRD, G.V., MORIARTY, D.I. & BRADY, B.G. (1983). Breeding biology of Wedge-tailed Shearwaters at Kilauea Point, Hawaii. *Condor* **85**, 292–296.

BYRD, G.V., SINCOCK, J.L., TELFER, T.C., MORIARTY, D.I. and BRADY, B.G. (1984). A cross-fostering experiment with Newell's race of the Manx Shearwater. *J Wildl Management* **48**, 163–168.

B.R.S.M. (1941). New facts about the Manx puffins. *J Manx Mus* **5**, 50.

CAMPBELL, A.G. (1933). The White-faced Storm-Petrel. *Emu* **33**, 86–92.

CARRICK, R. & DUNNET, G.M. (1954). Breeding of the fulmar *Fulmarus glacialis*. *Ibis* **96**, 356–370.

CARRICK, R.C., KEITH, K. & GWYNN, A.M. (1960). Fact and fiction on the breeding of the Wandering Albatross. *Nature Lond* **188**, 112–114.

CARTER, M. (1984). Preliminary notice of an invasion and wreck of Blue Petrels, *Halobaena caerulea*, and Kerguelen Petrels, *Pterodroma brevirostris*, during August 1984 in southern Australia. *Australas Seabird Grp Newsl* **21**, 17–19.

CHALONER, J. (1656). *A Short Treatise of the Isle of Man*. Streater, London.

CHENEVAL, J. (1984). Les oiseaux aquatiques (Gaviiformes a Anseriformes) du gisement. Aquitanien de Saint-Gerand-le-Puy (Allier, France): revision systematique. *Palaeovertebrata* **14**, 33–115.

CHEREL, Y., STAHL, J.-C. & LE MAHO, Y. (1987). Ecology and physiology of fasting in King Penguin chicks. *Auk* **104**, 254–262.

CHU, E.W. (1984). Sooty Shearwater off California: diet and energy gain. In *Marine Birds: Their Feeding Ecology and Commerical Fisheries Relationships* (D.N. Nettleship, G.A. Sanger & P.F. Springer, eds) pp. 64–71. Ministry of Supply and Services, Ottawa.

CLANCEY, P.A., BROOKE, R.K. & SINCLAIR, J.C. (1981). Variation in the current nominate subspecies of *Pterodroma mollis* (Gould) (Aves: Procellariidae). *Durban Mus Novit* **12** 203–213.

CLAPP, R.B. (1971). A specimen of Jouanin's Petrel from Lisianski Island, North-western Hawaiian Islands. *Condor* **73**, 490.

CLAPP, R.B. & WIRTZ, W.O. (1975). The natural history of Lisianski Island, Northwestern Hawaiian Islands. *Atoll Res Bull* **186**, 1–196.

CLARKE, M. & PRINCE, P.A. (1980). Chemical composition and calorific value of food fed to mollymauk chicks *Diomedea melanophris* and *D.chrysostoma* at Bird Island, South Georgia. *Ibis* **122**, 488–494.

CLARKE, M. & PRINCE, P.A. (1981). Cephalopod remains in regurgitations of Black-browed and Grey-headed Albatrosses at South Georgia. *Br Antarct Surv Bull* **54**, 1–7.

CLARKE, M.P., CROXALL, J.P. & PRINCE, P.A. (1981). Cephalopod remains in regurgitations of the Wandering Albatross *Diomedea exulans* L. at South Georgia. *Br Antarct Surv Bull* **54**, 9–21.

CLAUGHER, D. (1976). A trematode associated with the death of the White-faced Storm Petrel (*Pelagodroma marina*) on the Chatham Islands. *J Nat Hist* **10**, 633–641.

COCK, G.L.A. & SCHNEIDER, D.C. (1982). Duration of ship following by Wandering Albatrosses. *Cormorant* **10**, 105–107.

COLE, C.F. (1913). A visit to Babel Island, the nesting place of *Puffinus tenuirostris brevicaudus* Gould. *Ibis* (10th Ser) **1**, 509–512.

COLLINS, J.W. (1884). Notes on the habits and methods of capture of various species of sea birds that occur on the fishing banks off the eastern coast of North America, and which are used as bait for catching codfish by New England fishermen; *Rep US Commissioner Fish and Fisheries for 1882*, 311–335.

CONDON, H.T. (1939). The cranial osteology of certain Tubinares. *Trans R Soc S Aust* **63**, 311–328.

CONE, C.D. (1964). A mathematical analysis of the dynamic soaring flight of the albatross with ecological interpretations. *Virginia Inst Mar Sci Spec Scient Rep* **50**, 1–104.

CONROY, J.W.H. (1972). Ecological aspects of the biology of the Giant Petrel, *Macronectes giganteus* (Gmelin), in the maritime Antarctic. *Br Antarct Surv Sci Rep* **75**, 1–74.

COOK, J. (1777). *A Voyage Towards the South Pole and Round the World. Performed in His Majesty's Ships the 'Resolution' and 'Adventure'*, etc. Strahan and Cadell, London.

COOPER, W.J., MISKELLY, C.M., MORRISON, K. & PEACOCK, R.J. (1986). Birds of the Solander Islands. *Notornis* **33**, 77–89.

COPESTAKE, P.C. & CROXALL, J.P. (1985). Aspects of the breeding biology of Wilson's Storm Petrel *Oceanites oceanicus* at Bird Island, South Georgia. *Br Antarct Surv Bull* **66**, 7–17.

COPESTAKE, P.G., CROXALL, J.P. & PRINCE, P.A. (1988). Use of cloacal sexing techniques in mark-recapture estimates of breeding population size in Wilson's Storm petrel *Oceanites oceanicus* at South Georgia. *Polar Biol* **8**, 271–279.

COUES, E. (1864–66). A critical review of the family Procellariidae. *Proc Acad Nat Sci Philad* **1864**, 72–91, 116–144; **1866**, 25–33, 134–197.

COULSON, J.C. (1963). Egg size and shape in the Kittiwake (*Rissa tridactyla*) and their use in estimating age composition of populations. *Proc Zool Soc Lond* **140**, 211–227.

COULSON, J.C. & HOROBIN, J.M. (1972). The annual re-occupation of breeding sites by the fulmar. *Ibis* **114**, 30–42.

COWAN, A.N. (1979). Ornithological studies at Casey, Antarctica, 1977–1978. *Aust Bird Watcher* **8**, 69–90.

COWAN, A.N. (1981). Size variation in the Snow Petrel (*Pagodroma nivea*). *Notornis* **28**, 169–188.

COX, J.B. (1980). Some remarks on the breeding distribution and taxonomy of the prions (Procellariidae: *Pachyptila*). *Rec S Aust Mus* **18**, 91–121.

CRACRAFT, J. (1981). Toward a phylogenetic classification of the recent birds of the world (Class Aves). *Auk* **98**, 681–714.

CRAMP, S. & SIMMONS, K.E.L. (eds). (1977). *The Birds of the Western Palearctic* Vol. 1. Oxford University Press, Oxford.

CRAMP, S., BOURNE, W.R.P. & SAUNDERS, D. (1974). *The Seabirds of Britain and Ireland*. Collins, London.

CROCKETT, D.E. (1975). Kermadec Islands Expedition Reports: The Wedge-tailed Shearwater (*Puffionus pacificus pacificus*) in the northern Kermadecs. *Notornis* **22**, 1–9.

CROCKETT, D.E. (1984). Chatham Island Taiko Expedition 1983–84. *OSNZ News* **30**, 2.

CROSSIN, R.S. (1974). The storm petrels (Hydrobatidae). In *Pelagic Studies of Seabirds in the Central and Eastern Pacific Oceans* (W.B. King, ed.). *Smithson Contr Zool* **158**, pp. 1–277.

CROXALL, J.P. (1979). Distribution and population changes in the Wandering Albatross *Diomedea exulans* at South Georgia. *Ardea* **67**, 15–21.

CROXALL, J.P. (1982a). Sexual dimorphism in Snow Petrels *Pagodroma nivea*. *Notornis* **29**, 171–180.

CROXALL, J.P. (1982b). Energy costs of incubation and moult in petrels and penguins. *J Anim Ecol* **51**, 177–194.

CROXALL, J.P. (1984). Seabirds. In *Antarctic Ecology* (R.M. Laws, ed) Vol. 2, pp. 533–616.

Academic Press, London.
CROXALL, J.P. & GASTON, A.J. (1988). *Patterns of Reproduction in High-latitude Northern- and Southern-Hemisphere Seabirds. Acta XIX Congr Int Orn* **1**, 1176–1194.
CROXALL, J.P. & HUNTER, I. (1982). The distribution and abundance of burrowing seabirds (Procellariiformes) at Bird Island, South Georgia: II. South Georgia Diving Petrel *Pelecanoides georgicus. Br Antarct Surv Bull* **56**, 69–74.
CROXALL, J.P. & PRINCE, P.A. (1979). Antarctic seabird and seal monitoring studies. *Polar Rec* **19**, 573–595.
CROXALL, J.P. & PRINCE P.A. (1980). Food, feeding ecology and ecology and ecological segregation of seabirds at South Georgia. *Biol J Linn Soc* **14**, 103–131.
CROXALL, J.P. & RICKETTS, C. (1983). Energy costs of incubation in the Wandering Albatross *Diomedea exulans. Ibis* **125**, 33–39.
CROXALL, J.P., RICKETTS, C. & PRINCE, P.A. (1984a). Impact of Seabirds on marine research, especially Krill, of South Georgia waters. In *Seabird Energetics* (G.C. Whittow & Rahn, eds) pp. 285–317. Plenum, New York.
CROXALL, J.P., PRINCE, P.A., HUNTER, I., MCINNES, S.J. & COPESTAKE, P.G. (1984b). The seabirds of the Antarctic Peninsula, islands of the Scotia Sea, and Antarctic Continents between 80° and 20°W: their status and conservation. In *Status and Conservation of the World's Seabirds* (J.P. Croxall, P.G.H. Evans & R.W. Schreiber, eds) pp. 637–666. Tech. Publ. 2, ICBP, Cambridge.
CROXALL, J.P., HILL, H.J., LIDSTONE-SCOTT, R., O'CONNELL, M.J. & PRINCE, P.A. (1988). Food and feeding ecology of Wilson's Storm Petrel *Oceanites oceanicus* at South Georgia. *J Zool London* **216**, 83–102.
CROXALL, J.P., ROTHERY, P. PICKERING, S.P.C. & PRINCE, P.A. (in press). Reproductive performance, recruitment and survival of Wandering Albatrosses *Diomedea exulans* at Bird Island, South Georgia *J Anim Ecol* **59**.
CRUZ, F. & CRUZ, J.B. (1990). Breeding, morphology, and growth of the endangered Dark-rumped Petrel. *Auk* **107**, 317–326.
CUNNINGHAM, R.O. (1871). *Notes on the Natural History of the Strait of Magellan and West Coast of Patagonia*, etc. Edmonston and Douglas, Edinburgh.
DARWIN, C. (1872). *The Origin of Species by means of Natural Selection, or the Preservation of Favoured Races in the Struggle for Life*. Murray, London.
DAVIES, R.H. (1846). Some account of the habits and natural history of the Sooty Petrel (Mutton Bird). *Tasm J Nat Sci* **2**, 13–16.
DAVIS, P. (1957). The breeding of the Storm Petrel. *Br Birds* **50**, 85–101, 371–384.
DAWSON, W.L. (1923). *The Birds of California* Vol. 4. South Moulton Company, San Diego.
DE JUANA, A.E., VARELA, J. & WITT, H.H. (1984). The conservation of seabirds at the Chafarinas Islands. In *Status and Conservation of the World's Seabirds* (J.P. Croxall, P.G.H. Evans & R.W. Schreiber, eds) pp. 363–370. Tech. Publ. 2, ICBP, Cambridge.
DE NAUROIS, R. (1969). Notes breves sur l'avifaune de l'Archipel du Cap-Vert. Faunistique, endemisme, ecologie. *Bull Inst Fondamentale Afr Noire* **31** (Ser 1) 143–218.
DE NAUROIS, R. (1978). Procellariidae reproducteurs en Nouvelle-Caledonie. *CR Acad Sci Paris* (Ser D) **287**, 269–271.
DE NAUROIS, R. & ERARD, C. (1979). L'identite subspecifique des populations neo-caledoniennes de *Pterodroma rostrata* Peale 1848. *Oiseau Revue Fr Orn* **49**, 235–239.
DE NAUROIS, R. & PREVOST, J. (1972). Disposition helicoidal de l'intestin chez certaines espèces de pètrels du genre *Pterodroma*. *Alauda* **40**, 93–97.
DERENNE, P. & MOUGIN, J.-L. (1976). Les Procellariiformes a nidification hypogee de l'ile aux Cochons (Archipel Crozet, 46° 06°S, 50° 14'E). *Com Natn Fr Rech Antarct* **40**, 149–175.

DESPIN, B. (1977). Biologie du Damier du Cap, *Daption capense* a l'Ile de la Possessian (Archipel Crozet). *Oiseau Revue Fr Orn* **47**, 149–157.

DESPIN, B. & MOUGIN, J.-L. (1988). Evaluation de la depense energetique et de la consommation alimentaire du Puffin Cendre *Calonectris diomedea borealis* d'apres l'etude de la decroissance ponderale an cours du jeune. *Oiseau Revue Fr Orn* **58**, 28–43.

DESPIN, B., MOUGIN, J.-L. & SEGONZAC, M. (1972). Oiseaux et Mammiferes de l'Ile de L'Est. *Com Natn Fr Rech Antarct* **31**, 1-106.

DEVILLERS, P. & TERSCHUREN, J.A. (1980). Les petrel geants (*Macronectes*) des Iles Falkland et du Sud de l'Amerique du Sud. *Gerfaut* **70**, 447–454.

DIONYSIUS (200 A.D.?). Ornithiaca (Ixeutica seu de aucupio libri tres in epitomen metro solutam reducti). Paraphrase in *Codex Vindobonensis Med. Gr. 1*, Folios 464–485. (Also facsimile edn. in *Dioscurides Herbal*, same codex, Akad. Druck, Graz 1970).

DIXON, C. (1885). Ornithology of St. Kilda. *Ibis* (5th Ser) **3**, 69–97.

DIXON, C.C. (1933). Some observations on the albatrosses and other birds of the southern oceans. *Trans R Canad Inst* **19**, 117–139.

DOWNES, M.C., EALEY, E.H.M., GWYNN, A.M. & YOUNG, P.S. (1959). The birds of Heard Island. *Aust Natn Antarct Res Exp Rep* **B1** *Zool*. 1–135.

DOWNIE, J.C., Hinshaw, V. & Laver, W.G. (1977). The ecology of influenza. Isolation of type 'A' influenza viruses from Australian pelagic birds. *Aust J Exp Biol Med Sci* **55**, 635–643.

DRENT, R. (1975). Incubation. In *Avian Biology* (D.S. Farner & J.R. King, eds) Vol. 5, pp. 333–420. Academic Press, New York and London.

DU BOST, F.D. and SEGONZAC, M. (1976). Note complementaire sur le cycle reproducteur du Grand Albatross (*Diomedea exulans*) de l'lle de la Possession, Archipel Crozet. *Comm Natn Fr Rech Antarct* **40**, 53–60.

DUFFEY, E. (1951). Field studies on the Fulmar *Fulmarus glacialis*. *Ibis* **93**, 237–245.

DUFFY, D.C., HAYS, C. & PLENGE, M.A. (1984). The conservation of Peruvian seabirds. In *Status and Conservation of the World's Seabirds* (J.P. Croxall, P.G.H. Evans & R.W. Schreiber, eds) pp. 245–259. Techn. Publ. 2, ICBP, Cambridge.

DUNNET, G.M. (1985). Pycroft's Petrel in the breeding season at Hen and Chickens Islands. *Notornis* **32**, 5–21.

DUNNET, G.M. & ANDERSON, A. (1961). A method of sexing living Fulmars in the hand. *Bird Study* **8**, 119–126.

DUNNET, G.M. & OLLASON, J.C. (1978). The estimation of survival rate in the Fulmar, *Fulmarus glacialis*. *J Anim Ecol* **47**, 507–520.

DUNNET, G.M. & OLLASON, J.C. (1982). The feeding dispersal of Fulmars *Fulmarus glacialis* in the breeding season. *Ibis* **124**, 359–361.

DUNNET, G.M., ANDERSON, A. & CORMACK, R.M. (1963). A study of the survival of adult Fulmars with observations on the pre-laying exodus. *Br Birds* **56**, 2–18.

DUNNET, G.M., OLLASON, J.C. & ANDERSON, A. (1979). A 28-year study of breeding Fulmars *Fulmarus glacialis* in Orkney. *Ibis* **121**, 293–300.

DU TETRE, J.-B. (1667). *Histoire Generale des Antilles, habitees par les Francois*. Tolly, Paris.

DYER, P.K. & HILL, G.J.E. (in press). A solution to the problem of determining occupancy status of Wedge-tailed Shearwater burrows. *Emu*.

EAGLE, M. (1980). Black-winged Petrels on Portland Island. *Notornis* **27**, 171–175.

EALEY, E.H.M. (1954). Analysis of stomach contents of some Heard Island birds. *Emu* **54**, 204–210.

EDWARDS, G. (1747). *A Natural History of Birds* Part II. Author, London.

ELLIOTT, C.C.H. (1970a). Additional notes on the sea-birds of Gough Island. *Ibis* **112**, 112–114.

ELLIOTT, C.C.H. (1970b). Ecological considerations and the possible significance of weight variations in the chicks of the Great Shearwater on Gough Island. *Ostrich* Suppl. 8, 385–396.

ELLIOTT, H.F.I. (1957). A contribution to the ornithology of the Tristan da Cunha Group. *Ibis* **99**, 545–586.

ELWES, R. (1859). Note on the breeding and mode of capture of the Short-tailed Petrel, or Mutton-bird (*Puffinus obscurus*), in the islands of Bass's Straits. *Ibis* (1st Ser) **1**, 397–399.

ELY, C.A. & CLAPP, R.B. (1973). The natural history of Laysan Island, Northwestern Hawaiian Islands. *Atoll Res Bull* **171**, 1–361.

ENTICOTT, J.W. & O'CONNELL, M. (1985). The distribution of the spectacled form of the White-chinned Petrel (*Procellaria aequinoctialis conspicillata*) in the South Atlantic Ocean. *Br Antarct Surv Bull* **66**, 83–86.

EVANS, G.R. (1973). Hutton's Shearwaters initiating local soil erosion in the Seaward Kaikoura Range. *NZ J Sci* **16**, 637–642.

EVANS, P.G.H. (1984). The seabirds of Greenland: their status and conservation. In *Status and Conservation of the World's* Seabirds (J.P. Croxall, P.G.H. Evans & R.W. Schreiber, eds) pp. 49–84. Tech. Publ. 2, ICBP, Cambridge.

EYCLESHYMER, A.C. (1907). Some observations and experiments on the natural and artificial incubation of the common Fowl. *Biol Bull* **12**, 360–375.

EYTON, T.C. (1867). *Osteologia Avium; or a Sketch of the Osteology of Birds*. Hobson, Wellington (Salop).

FALLA, R.A. (1934). The distribution and breeding habits of petrels in northern New Zealand. *Rec Auckland Inst Mus* **1**, 245–259.

FALLA, R.A. (1937). Birds. *BANZ Antarct Res Exped 1929–1931 Rep* (Ser B) **2**, 1–288.

FALLA, R.A. (1940). The genus *Pachyptila* Illiger. *Emu* **40**, 218–236.

FALLA, R.A. (1942). Review of the smaller Pacific forms of *Pterodroma* and *Cookilaria*. *Emu* **42**, 111–118.

FALLA, R.A. (1946). An undescribed form of the Black Petrel. *Rec Canterbury Mus* **5**, 111–113.

FARNER, D.S. & SERVENTY, D.L. (1959). Body temperature and the ontogeny of thermoregulation in the Slender-billed Shearwater. *Condor* **61**, 426–433.

FEFER, S.T., HARRISON, C.S., NAUGHTON, M.B. & SHALLENBERGER, R.J. (1984). Synopsis of results of recent seabird research conducted in the Northwestern Hawaiian Islands. In *Proceedings of 2nd Symposing on Resource Investigations in the Northwestern Hawaiian Islands* (R.W. Grigg & K.Y. Tanoue, eds) Vol. 1, pp. 9–76. University Hawaii Sea Grant College, Honolulu.

FILHOL, H. (1885). Oiseaux. In *Mission de l'Ile Campbell* Vol. 1. pp. 35–64. Acad France, Paris.

FINCH, D.W., RUSSELL, W.C. & THOMPSON, E.V. (1978). Pelagic birds in the Gulf of Maine. *Am Birds* **32**, 140–155, 281–294.

FISCHER, K. (1985). Ein albatrosartiger Vogel (*Diomedeoides minimus* nov. gen., nov. sp., Diomedeoididae nov. fam., Procellariiformes) aus dem Mitteloligozän bei Leipzig (GDR). *Mitt Zool Mus Berl* **61** *Suppl Ann Orn* **9**, 113–118.

FISHER, H.I. (1967). Body weights in Laysan Albatrosses *Diomedea immutabilis*. *Ibis* **109**, 373–382.

FISHER, H.I. (1968). The 'two-egg clutch' in the Laysan Albatross. *Auk* **85**, 134–136.

FISHER, H.I. (1969). Eggs and egg-laying in the Laysan Albatross, *Diomedea immutabilis*. *Condor* **71**, 102–119.

FISHER, H.I. (1971). The Laysan Albatross: its incubation, hatching and associated behaviors. *Living Bird* **10**, 19–78.

FISHER, H.I. (1972). Sympatry of Laysan and Black-footed Albatrosses. *Auk* **89**, 381–402.

FISHER, H.I. (1975a). Longevity of the Laysan Albatross, *Diomedea immutabilis*. Bird-banding **46**, 1–6.

FISHER, H.I. (1975b). Mortality and survival in the Laysan Albatross, *Diomedea immutabilis*. *Pacif Sci* **29**, 279–300.

FISHER, H.I. (1975c). The relationship between deferred breeding and mortality in the Laysan Albatross. *Auk* **92**, 433–441.

FISHER, H.I. (1976). Some dynamics of a breeding colony of Laysan Albatrosses. *Wilson Bull* **88**, 121–142.

FISHER, H.I. and FISHER, M.L. (1969). The visits of Laysan Albatrosses to the breeding colony. *Micronesica* **5**, 173–201.

FISHER, H.I. and FISHER, J.R. (1972). The oceanic distribution of the Laysan Albatross *Diomedea immutabilis*. *Wilson Bull* **84**, 7–27.

FISHER, J. (1952). *The Fulmar*. Collins, London.

FISHER, J. (1966). The Fulmar population of Britain and Ireland, 1959. *Bird Study* **13**, 5–76.

FLEMING, C.A. (1941a). The phylogeny of the prions. *Emu* **41**, 134–155.

FLEMING, C.A. (1941b). Notes on neozelanic forms of the subgenus *Cookilaria*. *Emu* **41**, 69–80.

FLEMING, C.A. (1950). Some South Pacific sea-bird logs. *Emu* **49**, 169–188.

FLEMING, C.A. (1982). *George Edward Lodge. The Unpublished New Zealand Bird paintings*. Nova Pacifica, Wellington.

FLINDERS, M. (1801). *Observations on the Coasts of Van Diemen's Land, on Bass's Strait and its Islands, and on part of the Coasts of New South Wales*. Nicol, London.

FORBES, W.A. (1882). Report on the anatomy of the petrels (*Tubinares*), collected during the voyage of H.M.S. Challenger. *Zool Challenger Exped* **4 (11)**, 1–64.

FORSTER, J.A. (1777). *A Voyage round the World in His Brittannic* Majesty's Sloop 'Resolution', *Commanded by Capt. James Cook, during the years 1772, 3, 4 and 5*. White, Robson, Elmsly, Robinson, London.

FORSTER, J.R. (1785). Memoire sur les Albatros. *Mem Math Phys Paris (Acad Sci)* **10**, 563–572.

FOWLER, J.A. & SWINFEN, R. (1984). Scottish storm petrels in Iceland. *Scott Birds* **13**, 52.

FRINGS, H. & FRINGS, M. (1961). Some biometric studies on the albatrosses of Midway Atoll. *Condor* **63**, 304–312.

FUGLER, S.R., HUNTER, S., NEWTON, I.P. & STEELE, W.K. (1987). Breeding biology of Blue Petrels *Halobaena caerulea* at the Prince Edward Islands. *Emu* **87**, 103–110.

FULLAGAR, P. (1976). Cabbage Tree Island, New South Wales. *Aust Bird Bander* **14**, 94–97.

FULLAGAR, P.J., McKEAN, J.L. & VAN TETS, G.F. (1974). In *Environmental Survey of Lord Howe Island* (H.F. Recher & S.S. Clark eds) pp. 55–63. Government Printer, Sydney.

FURBRINGER, M. (1888). Untersuchungen zur Morphologie und Systematik der Vogel, zugleich ein Beitrag zur Anatomie der Stutz — und Bewegungsorgane. *Bijd Dierk K Zool Genoot Nat Art Mag Amsterdam* **15**, 1–1751.

FURNESS, R.W. & TODD, C.M. (1984). Diets and feeding of Fulmars *Fulmarus glacialis* during the breeding season: a comparison between St Kilda and Shetland. *Ibis* **126**, 379–387.

GABRIEL, J. (1912). Further notes on the Mutton-birds of Bass Strait. *Victoria Nat* **28**, 206–212.

GAIN, L. (1914). Oiseaux antarctiques. *Deux Exp Antarct Fr 1908–1910 Sci Nat* **5 (4)**, 1–200.

GALLAGHER, M.D. (1960). Bird notes from Christmas Island, Pacific Ocean. *Ibis* **102**, 489–502.

GARDNER, A.S., DUCK, C.D. & GREIG, S. (1985). Breeding of the Trindade Petrel *Pterodroma arminjoniana* on Round Island, Mauritius. *Ibis* **127**, 517–522.

GARDNER, M. (1987). Pink coloration on Little Gulls. *Br Birds* **80**, 628.

GIBSON, J.D. (1963). Third report of the New South Wales Albatross Study Group. *Emu* **63**, 215–223.

GIBSON, J.D. (1967). The wandering albatross (*Diomedea exulans*): results of banding and observations in New South Wales coastal waters and the Tasman Sea. *Notornis* **14**, 47–57.

GIBSON, J.D. & SEFTON, A.R. (1955). Notes on some albatrosses of coastal New South Wales. *Emu* **55**, 44–48.

GILL, F.B. (1967). Observations on the pelagic distribution of seabirds in the Western Indian Ocean. *Proc US Natn Mus* **123**, 1–33.

GILL, F.B., JOUANIN, C. & STORER, R.W. ((1970). Notes on the seabirds of Round Island, Mauritius. *Auk* **87**, 514–521.

GILLHAM, M.E. (1961). Alteration of the breeding habitat by sea-birds and seals in Western Australia. *J Ecol* **49**, 289–300.

GILLHAM, M.E. (1963). Breeding habitats of the White-faced Storm Petrel (*Pelagodroma marina*) in eastern Bass Strait. *Pap Proc R Soc Tasmania* **97**, 33–41.

GIOL, A. (1957). Cattura di *Bulweria fallax*, Jouanin. *Riv Ital Orn* **27**, 118–121.

GLAUERT, L. (1946). The Little Shearwater's year. *Emu* **46**, 187–192.

GODFREY, G.K. (1954). Tracing field roles (*Microtus agrestis*) with a Geiger-Muller counter. *Ecology* **35**, 5–10.

GODMAN, F. du CANE. (1907–10). *A Monograph of the Petrels*. Witherby, London.

GOULD, J. (1841). Birds. In *Zoology of the Voyage of the Beagle during the years 1832–1836* (C. Darwin, ed.) Vol. 2. Smith Elder, London.

GOULD, P.J. (1983). Seabirds between Alaska and Hawaii. *Condor* **85**, 286–291.

GRANT, G.S. (1984). Energy cost of incubation to the parent seabird. In *Seabird Energetics* (G.C. Whittow & H. Rahn, eds) pp. 59–71. Plenum Press, New York.

GRANT, G.S. & WHITTOW, G.C. (1983). Metabolic cost of incubation in the Laysan Albatross and Bonin Petrel. *Comp Biochem Physiol* **74A**, 72–82.

GRANT, G.S. & WHITTOW, G.C. (1984). Metabolic rate of Laysan Albatross and Bonin Petrel chicks on Midway Atoll. *Pacif Sci* **38**, 170–176.

GRANT, G.S., PETTIT, T.N., RAHN, H., WHITTOW, G.C. AND PAGANELLI, C.V. (1982a). Regulation of water loss from Bonin Petrel (*Pterodroma hypoleuca*) eggs. *Auk* **99**, 236–242.

GRANT, G.S., PETTIT, T.N., RAHN, H., WHITTOW, G.C. & PAGANELLI, C.V. (1982b). Water loss from Laysan and Black-footed Albatross eggs. *Physiol Zool* **55**, 405–414.

GRANT, G.S., PAGANELLI, C.V., PETTIT, T.N., WHITTOW, G.C. & RAHN, H. (1982c). Determination of fresh egg mass during natural incubation. *Condor* **84**, 121–122.

GRANT, G.S., WARHAM, J., PETTIT, T.N. & WHITTOW, G.C. (1983). Reproductive behavior and vocalizations of the Bonin Petrel. *Wilson Bull* **95**, 522–539.

GRAU, C.R. (1976). Ring structure of avian egg yolk. *Poult Sci* **55**, *1418–1422.*

GRAU, C.R. (1984). Egg Formation. In *Seabird Energetics* (G.C. Whittow & H. Rahn, eds) pp. 35–57. Plenum, New York.

GRAY, G.R. (1840). *List of the Genera of Birds*. British Museum of Natural History, London.

GRAY, J. (1968). *Animal Locomotion*. Weidenfeld and Nicholson, London.

GREW, N. (1681). *Museum regalis societatis. Or a catalogue and description of the natural and artificial rarities belonging to the Royal Society and preserved at Gresham College*. Author, London.

GRIFFIN, D.R. (1940). Homing experiments with Leach's Petrels. *Auk* **57**, 61–74.

GRIFFITHS, A.M. (1983). Factors affecting the distribution of the Snow Petrel (*Pagodroma nivea*) and the Antarctic Petrel (*Thalassoica antarctica*). *Ardea* **71**, 145–150.

GRIFFITHS, A.M., SIEGFRIED, W.R. & ABRAMS, R.W. (1982). Ecological structure of a pelagic seabird community in the Southern Ocean. *Polar Biol* **1**, 39–46.

GROSS, W.A.O. (1935). The life history cycle of Leach's Petrel (*Oceanodroma leucorhoa leucohroa*) on the outer sea islands of the Bay of Fundy. *Auk* **52**, 382-399.

GRUBB, T.C. (1973). Colony location by Leach's Petrel. *Auk* **90**, 78-82.

GRUBB, T.C. (1974). Olfactory navigation to the nesting burrow in Leach's Petrel (*Oceanodroma leucorrhoa*). *Anim Behav* **22**, 192-202.

GRUBB, T.C. (1979). Olfactory guidance of Leach's Storm Petrel to the breeding island. *Wilson Bull* **91**, 141-143.

GUILLOTIN, M. & JOUVENTIN, P. (1980). Le Petrel des Neiges a Pointe Geologie. *Gerfaut* **70**, 51-72.

GUZMAN, J.R. & MYRES, M.T. (1983). The occurrence of Shearwaters (*Puffinus* spp.) off the west coast of Canada. *Can J Zool.* **61**, 2064-2077.

HADDEN, F.C. (1941). Midway Islands. *Hawaiian Planters' Rec* **45**, 179-221.

HAGEN, Y. (1952). Birds of Tristan da Cunha. *Results Norw Scient Exped Tristan da Cunha 1937-1938*. No. 20. Kommisjon Hos Jacob Dybwad. Oslo.

HAGEN, Y. (1982). Migration and longevity of Yellow-nosed Albatrosses *Diomedea chlororhynchos* banded on Tristan da Cunha in 1938. *Ornis Scand* **13**, 247-248.

HALL, A.J. (1987). The breeding biology of the White-chinned Petrel *Procellaria aequinoctialis* at South Georgia. *J Zool Lond* **212**, 605-617.

HALL-CRAGGS, J. & SELLAR, P.J. (1976). Distinguishing characteristics in the burrow-calling of Storm and Leach's Petrels. *Br Birds* **69**, 293-297.

HAMER, K. & READ, H. (1987). Patterns of return to land in a colony of Cory's Shearwater *Calonectris diomedea* on Selvagem Grande. *Seabird* **10**, 3-11.

HAMLET, M.P. & FISHER, H.I. (1967). Air sacs of respiratory origin in some procellariiform birds. *Condor* **69**, 586-595.

HANEY, J.C. (1985). Band-rumped Storm-petrel occurrences in relation to upwelling off the coast of the south-eastern United States. *Wilson Bull* **97**, 543-547.

HANEY, J.C. (1987). Aspects of the pelagic ecology and behavior of the Black-capped Petrel (*Pterodroma hasitata*). *Wilson Bull* **99**, 153-168.

HANEY, J.C. & MCGILLIVARY, P.A. (1985). Aggregations of Cory's Shearwaters (*Colonectris diomedea*) at Gulf Stream fronts. *Wilson Bull* **97**, 191-200.

HARPER, P.C. (1972). The field identification and distribution of the Thin-billed Prion (*Pachyptila belcheri*) and the Antarctic Prion (*Pachyptila desolata*). *Notornis* **19**, 140-175.

HARPER, P.C. (1973). The field identification and supplementary notes on the Soft-plumaged Petrel (*Pterodroma mollis* Gould, 1844). *Notornis* **20**, 193-201.

HARPER, P.C. (1976). Breeding biology of the Fairy Prion (*Pachyptila turtur*) at the Poor Knights Islands, New Zealand. *NZ. J Zool* **3**, 351-371.

HARPER, P.C. (1978). The plasma proteins of some albatrosses and petrels as an index of relationships in the Procellariiforms. *NZ. J Zool* **5**, 509-549.

HARPER, P.C. (1980). The field identification and distribution of the prions (genus *Pachyptila*), with particular reference to the identification of storm-cast material. *Notornis* **27**, 235-286.

HARPER, P.C. (1983). Biology of the Buller's Shearwater (*Puffinus bulleri*) at the Poor Knights Islands, New Zealand. *Notornis* **30**, 299-318.

HARPER, P.C. (1987). Feeding behaviour and other notes on 20 species of Procellariiformes at sea. *Notornis* **34**, 169-192.

HARPER, P.C. & FOWLER, J.A. (1987). Plastic pellets in New Zealand storm-killed prions (*Pachyptila spp.*), 1958-1977. *Notornis* **34**, 65-70.

HARPER, P.C., WATSON, G.E. & ANGLE, J.P. (1972). New records of the Kerguelen Petrel (*Pterodroma brevirostris*) in the South Atlantic and Pacific Oceans. *Notornis* **19**, 56-60.

HARRIS, M.P. (1965). Puffinosis among Manx Shearwaters on Skokholm. *Br Birds* **58**,

426–434.
HARRIS, M.P. (1966). Breeding biology of the Manx Shearwater *Puffinus puffinus*. *Ibis* **108**, 17–33.
HARRIS, M.P. (1969a). Age at breeding and other observations on the Waved Albatross *Diomedea irrorata*. *Ibis* **111**, 97–98.
HARRIS, M.P. (1969b). The biology of storm petrels in the Galapagos Islands. *Proc Calif Acad Sci* (4th Ser) **37**, 95–166.
HARRIS, M.P. (1969c). Breeding Seasons of sea-birds in the Galapagos Islands. *J Zool Lond* **159**, 145–165.
HARRIS, M.P. (1969d). Food as a factor controlling the breeding of *Puffinus lherminieri*. *Ibis* **111**, 139–156.
HARRIS, M.P. (1970). The biology of an endangered species the Dark-rumped Petrel (*Pterodroma phaeopygia*) in the Galapagos Islands. *Condor* **72**, 76–84.
HARRIS, M.P. (1972). Inter-island movements of Manx Shearwaters. *Bird Study* **19**, 167–171.
HARRIS, M.P. (1973). The biology of the Waved Albatross *Diomedea irrorata* of Hood Island, Galapagos. *Ibis* **115**, 483–510.
HARRIS, M.P. & HANSEN, L. (1974). Sea-bird transects between Europe and Rio Plate, South America in autumn 1973. *Dansk Orn Foren Tidsskr* **68**, 117–137.
HARRIS, S.W. (1974). Status, chronology, and ecology of nesting storm petrels in north western California. *Condor* **76**, 249–261.
HARRISON, C.J.O. (1977a). Non-passerine birds of the Ipswichian Interglacial from the Gower caves. *Trans Br Cave Res Assoc* **4**, 441–442.
HARRISON, C.J.O. (1977b). The limb osteology of the diving petrels and Little Auk as evidence of the retention of characters in morphologically convergent species. *Ardea* **65**, 43–52.
HARRISON, C.J.O. (1986) The affinities of the extinct Aldabra Gadfly Petrel *Pterodroma kurodai*. *Ibis* **128**, 297.
HARRISON, C.J.O. & WALKER C.A. (1978). Pleistocene bird remains from Aldabra Atoll Indian Ocean. *J Nat Hist* **12**, 7–14.
HARRISON, C.S., HIDA, T.S. & SEKI, M.P. (1983). Hawaiian seabird feeding ecology. *Wildl Monogr* **85**, 1–71.
HARRISON, C.S., NAUGHTON, M.B. & FEFER, S.I. (1984). In *Status and Conservation of the World's Seabirds* (J.P. Croxall, P.G.H. Evans & R.W. Schreiber, eds) pp. 513–526. Tech. Publ. 2, ICBP, Cambridge.
HARROW, G. (1976). Some observations of Hutton's Shearwater. *Notornis* **23**, 269–288.
HASEGAWA, H. (1978). Recent observations on the Short-tailed Albatross, *Diomedea albatrus*, on Torishima. *J Yamashina Inst Orn* **10**, 58–69.
HASEGAWA, H. (1982). The breeding status of the short-tailed Albatross *Diomedea albatrus* on Torishima, 1979/80–1980/81. *J Yamashina Inst Orn* **14**, 16–24.
HASEGAWA, H. (1984a). [*Short-tailed Albatross — the White-winged Wanderer over the Sea.*] Heibon Sha Co., Tokyo.
HASEGAWA, H. (1984b). Status and conservation of seabirds in Japan, with special attention to the Short-tailed Albatross. In *Status and Conservation of the World's Seabirds* (J.P. Croxall, P.G.H. Evans & R.W. Schreiber, eds) pp. 487–500. Tech. Publ 2, ICBP, Cambridge.
HATCH, S.A. (1983). Mechanism and ecological significance of sperm storage in the Northern Fulmar with reference to its occurrence in other birds. *Auk* **100**, 593–600.
HATCH, S.A. (1987). Copulation and male guarding in the Northern Fulmar. *Auk* **104**, 450–461.
HAYASHI, T. (1983). [Straying of Streaked Shearwaters *Calonectris leucomelas* with special reference to seasonal winds.] *Tori* **32**, 21–29.

HAYWARD, G.D. (1987). Betalights: an aid in the nocturnal study of owl foraging habitat and behavior. *J Raptor Res* **21**, 98–102.

HECTOR, J. (1895). In the anatomy of flight of certain birds. *Trans NZ Inst* **27**, 285–287, 651–653.

HECTOR, J.A.L. (1988). Reproductive endocrinology of albatrosses. *Acta XIX Congr Int Orn* **2**, 1702–1709.

HECTOR, J.A.L. & GOLDSMITH, A.R. (1985). The role of prolactin during incubation: comparative studies of three *Diomedea* albatrosses. *Gen Comp Endocrinol* **60**, 236–243.

HECTOR, J.A.L., FOLLETT, B.K. & PRINCE, P.A. (1986). Reproductive endocrinology of the Black-browed Albatross *Diomedea melanophris* and the Grey-headed Albatross *D.chrysostoma*. *J Zool Lond* **208**, 237–253.

HEINROTH, O. (1922). Die Beziehungen zwischen Vogelgewicht, Eigewicht, Gelegegewicht und Brutdauer. *J Orn* **70**, 172–285.

HELLYER, N.R., KINSKY, F.C. & IMBER, M.J. (1973). A Soft-plumaged Petrel from the Bay of Plenty. *Notornis* **20**, 71–72.

HEMMINGS, A.D. & BAILEY, E.C. (1985). Pursuit diving by Northern Giant Petrels at the Chatham Islands. *Notornis* **32**, 330–331.

HEPBURN, I.R. & RANDALL, R.E. (1975). Nest site distribution of the fulmar (*Fulmarus glacialis*) within the Monach Isles National Nature Reserve, Outer Hebrides, U.K. *J Biogeogr* **2**, 223–228.

HERMES, N., EVANS, O. & EVANS, B. (1986). Norfolk Island birds: a review 1985. *Notornis* **33**, 141–149.

HILLER, A., WAND, U., KAMPF, H. & STACKEBRANDT, W. (1988). Occupation of the Antarctic Continent by petrels during the past 35,000 years: inferences from a C^{14} study of stomach oil deposits. *Polar Biol* **9**, 69–77.

HINDWOOD, K.A. & SERVENTY, D.L. (1941). The Gould Petrel of Cabbage Tree Island. *Emu* **41**, 1–20.

HOLDGATE, M.W. (1963). Observations of birds and seals at Anvers Island, Palmer Archipelago, in 1955–57. *Br Antarct Surv Bull* **2**, 45–51.

HOLYOAK, D.T. & THIBAULT, J.C. (1984). Contribution a l'etude des oiseaux de Polynesie orientale. *Mem Mus Nat Hist Paris* (Ser A) *Zool* **127**, 1–209.

HOWARD, H. & BARNES, L.G. (1987). Middle Miocene marine birds from the foothills of the Santa Ana Mountains, Orange County, California. *Contrib Sci* **383**, 1–9.

HOWELL, T.R. & BARTHOLOMEW, G.A. (1961a). Temperature regulation in Laysan and Black-footed Albatrosses. *Condor* **63**, 185–197.

HOWELL, T.R. & BARTHOLOMEW, G.A. (1961b). Temperature regulation in nesting Bonin Petrels, Wedge-tailed Shearwaters and Christmas Island Shearwaters. *Auk* **78**, 343–354.

HUBER, L.N. (1971). Notes on the migration of the Wilson's Storm Petrel *Oceanites oceanicus* near Eniwetok Atoll, Western Pacific Ocean. *Notornis* **18**, 38–42.

HUDSON, R. (1966). Adult survival estimates for two Antarctic fulmars. *Br Antarct Surv Bull* **8**, 63–73.

HUMPHREY, P.S. & LIVEZEY, B.C. (1983). Giant petrels (*Macronectes giganteus*) nesting in Chubut, Argentina. *Gerfaut* **73**, 3–8.

HUNT, G.L., GOULD, P.J., FORSELL, D.J. & PETERSON, H. (1981). Pelagic distribution of marine birds in the eastern Bering Sea. In *The Eastern Bering Sea Shelf: Oceanography and Resources* Vol. 2, pp. 689–718. University Washington Press, Seattle.

HUNTER, J. (1793). *An Historical Journal of the Transactions at Port Jackson and Norfolk Island, etc.* Stockdale, London.

HUNTER, S. (1983a). The food and feeding ecology of the giant petrels *Macronectes halli* and *M.giganteus* at South Georgia. *J Zool Lond* **200**, 521–538.

HUNTER, S. (1983b). Interspecific breeding in giant petrels at South Georgia. *Emu* **82** (Suppl), 312–314.

HUNTER, S. (1984a). Moult of the giant petrels *Macronectes halli* and *M. giganteus* at South Georgia. *Ibis* **126**, 119–132.

HUNTER, S. (1984b). Breeding biology and population dynamics of giant petrels *Macronectes* at South Georgia (Aves: Procellariiformes). *J. Zool Lond* **203**, 441–460.

HUNTER, S. (1984c). Movements of giant petrels *Macronectes* spp. ringed at South Georgia. *Ringing & Migration* **5**, 105–112.

HUNTER, S. (1987). Species and sexual isolating mechanisms in sibling species of giant petrels *Macronectes*. *Polar Biol* **7**, 295–301.

HUTTON, F.W. (1865). Notes on some of the birds inhabiting the Southern Ocean. *Ibis* (New Ser) **1**, 276–298.

HUTTON, F.W. (1895). On *Majaqueus aequinoctialis* from Antipodes Island. *Trans Proc NZ Inst* **27**, 177.

HUXLEY, J.S. (1927). On the relation between egg-weight and body weight in birds. *J Linn Soc Lond* **36**, 457–466.

IMBER, M.J. (1971). Filoplumes of petrels and shearwaters. *NZ J Mar Freshw Res* **5**, 396–403.

IMBER, M.J. (1973). The food of Grey-faced Petrels (*Pterodroma macroptera gouldi* (Hutton)), with special reference to diurnal vertical migration of their prey. *J Anim Ecol* **42**, 645–662.

IMBER, M.J. (1975). Behaviour of petrels in relation to the moon and artificial lights. *Notornis* **22**, 302–306.

IMBER, M.J. (1976a). Comparison of prey of the Black *Procellaria* petrels of New Zealand. *NZ J Mar Freshw Res* **10**, 119–130.

IMBER, M.J. (1976b). Breeding biology of the Grey-faced Petrel *Pterodroma macroptera gouldi*. *Ibis* **118**, 51–64.

IMBER, M.J. (1981). Diets of stormpetrels *Pelagodroma* and *Garrodia* and of prions *Pachyptila* (Procellariiformes). In *Proceedings of the Symposium on Birds of the Sea and Shore* (J. Cooper, ed.) pp. 63–88. African Seabird Group, Cape Town.

IMBER, M.J. (1983). The lesser petrels of Antipodes Islands, with notes from Prince Edward and Gough Island. *Notornis* **30**, 283–298.

IMBER, M.J. (1984a). Migration of White-faced Storm-Petrels *Pelagodroma marina* in the South Pacific and the status of the Kermadec subspecies. *Emu* **84**, 32–35.

IMBER, M.J. (1984b). Trematode anklets on white faced Stormpetrels *Pelagodroma marina* and Fairy Prions *Pachyptila turtur*. *Cormorant* **12**, 71–74.

IMBER, M.J. (1985a). Origins, phylogeny and taxonomy of the gadfly petrels *Pterodroma* spp. *Ibis* **127**, 197–229.

IMBER, M.J. (1985b). Black Petrel. In *Complete Book of New Zealand Birds* p. 90. Reader's Digest, Sydney.

IMBER, M.J. (1987). Breeding ecology and conservation of the Black Petrel (*Procellaria parkinsoni*). *Notornis* **34**, 19–39.

IMBER, M.J. & JENKINS, J.A.F. (1981). The New Caledonian Petrel. *Notornis* **28**, 149–160.

IMBER, M.J. & LOVEGROVE, T.G. (1982). Leach's Storm Petrels (*Oceanodroma l.leucorhoa*) prospecting for nest sites at the Chatham Islands. *Notornis* **29**, 101–108.

IMBER, M.J. & NILSSON, R.J. (1980). South Georgian Diving Petrels (*Pelecanoides qeorqicus*) breeding on Codfish Island. *Notornis* **27**, 325–330.

IMBER, M.J. & RUSS, R. (1975). Some foods of the Wandering Albatross (*Diomedea exulans*). *Notornis* **22**, 27–36.

IREDALE, T. (1911). An additional note on the birds of Lord Howe and Norfolk Islands. *Proc Linn Soc NSW* **35**, 773–781.

IREDALE, T. (1914). The surface breeding petrels of the Kermadec Group. *Ibis* (10th Ser) **2**,

423–436.

ISENMANN, P. (1970). Contribution a la biologie de reproduction du Petrel des Neiges (*Pagodroma nivea* Forster). Le problem de la petite et de la grande forme. *Oiseau Rev Fr Orn* **40** (No. Spec) 99–134.

JACKSON, R. (1958). The Westland Petrel. *Notornis* **7**, 230–233.

JACKSON, S. (1988). Diets of the White-chinned Petrel and Sooty Shearwater in the southern Benguela Region, South Africa. *Condor* **90**, 20–28.

JACOB, J. & HOERSCHELMANN, H. (1982). Chemotaxonomische untersuchungen zur systematic der Rohrennasen (Procellariiformes). *J Orn* **123**, 63–84.

JAMES, P.C. (1984a). The status and conservation of seabirds in the Mediterranean. *In Status and Conservation of the World's Seabirds* (J.P. Croxall, P.G.H. Evans & R.W. Schreiber, eds) pp. 371–375. Tech Publ **2**, ICBP, Cambridge.

JAMES, P.C. (1984b). Sexual dimorphism in the voice of the British Storm Petrel *Hydrobates pelagicus*. *Ibis* **126**, 89–92.

JAMES, P.C. (1985). The vocal behaviour of the Manx Shearwater *Puffinus puffinus*. *Z Tierpsychol* **67**, 269–283.

JAMES, P.C. (1986). How do Manx Shearwaters *Puffinus puffinus* find their burrows? *Ethology* **71**, 287–294.

JAMES, P.C. & ROBERTSON, H.A. (1985). First record of Swinhoe's Storm Petrel *Oceanodroma monorhis* in the Atlantic Ocean. *Ardea* **73**, 105–106.

JAMESON, W. (1958). *The Wandering Albatross*. Hart-Davis, London.

JARRY, G. (1969). Notes sur les oiseaux nicheurs de Tunisie. *Oiseau Revue Fr Orn* **39**, 112–120.

JEHL, J.R. (1972). On the cold trail of an extinct petrel. *Pacif Discovery* **25**, 24–29.

JEHL, J.R. (1973). The distribution of marine birds in Chilean waters in winter. *Auk* **90**, 114–135.

JEHL, J.R. (1974a). The near-shore avifauna of the middle American west coast. *Auk* **91**, 681–699.

JEHL, J.R. (1974b). The distribution and ecology of marine birds over the continental shelf of Argentina in winter. *Trans San Diego Soc Nat Hist* **17**, 217–234.

JEHL, J.R. (1974c). Aerial feeding by a shearwater. *Auk* **91**, 188–189.

JENKINS, J.A.F. (1974). Local distribution and feeding habits of Buller's Shearwater (*Puffinus bulleri*). *Notornis* **21**, 109–120.

JENKINS, J.A.F. (1982). A note on the winter distribution of the White-headed Petrel. *Sea Swallow* **31**, 37–38.

JOHNSTONE, G.W. (1974). Field characters and behaviour at sea of giant petrels in relation to their oceanic distribution. *Emu* **74**, 209–218.

JOHNSTONE, G.W. (1977). Comparative feeding ecology of the giant petrels *Macronectes giganteus* (Gmelin) and *M.halli* (Mathews). In *Adaptations within Antarctic Ecosystems* (G. Llano, ed.) pp. 647–668. Gulf Publishing, Houston.

JOHNSTONE, G.W. (1978). Interbreeding by *Macronectes halli* and *M.giganteus* at Macquarie Island. *Emu* **78**, 235.

JOHNSTONE, G.W. (1982). Zoology. In *Expedition to the Australian Territory of Heard and McDonald Islands 1980* (C. Veenstra & J. Manning, eds) pp. 33–39. Tech. Rep. 31, Department of National Development, Canberra.

JOHNSTONE, R.E. (1979). Three more records of the Kerguelen Diving-Petrel in Western Australia. *West Aust Nat* **14**, 133.

JOHNSTONE, R.M. & DAVIS, L.S. (1990). Incubation routines and foraging-trip regulation in the Grey-faced Petrel *Pterodroma macroptera gouldi*. *Ibis* **132**, 14–20.

JOHNSTONE, R.M. & NIVEN, B.E. (1989). Sexing Grey-faced Petrels by discriminant analysis of measurements. *Notornis* **36**, 261–265.

JONES, D.M., SAMOUR, J.H., KNIGHT, J.A. & FFINCH, J.M. (1984). Sex determination of monomorphic birds by fibreoptic endoscopy. *Vet Rec* **115**, 596–598.

JONES, F.W. (1936a). The wanderings of the albatrosses. *Emu* **36**, 103–105.

JONES, F.W. (1936b). Notes on the breeding of the Short-tailed Shearwater, *Puffinus tenuirostris*, in 1936. *S Aust Orn* **13**, 223–228.

JONES, F.W. (1937). The diving petrel of Bass Strait. *Victorian Nat* **54**, 51–55.

JONES, M.J. (1986). Breeding synchrony of Cory's Shearwater *Calonectris diomedea* on Selvagem Grande. *Ibis* **128**, 423–426.

JOUANIN, C. (1955). Une nouvelle espece de Procellariide. *Oiseau Rev Fr Orn* **25**, 155–161.

JOUANIN, C. (1957). Les procellariides melaniques signales en Mer d'Oman. *Oiseau Revue Fr Orn* **27**, 12–27.

JOUANIN, C. (1964). Le comportement en juillet des Petits Puffins, *Puffinus assimilis baroli* (Bonaparte) de l'Ile Selvagem Grande. *Bolm Mus Munic Funchal* **18**, 142–157.

JOUANIN, C. (1970). Le Petrel Noir de Bourbon, *Pterodroma aterrima* Bonaparte. *Oiseau Rev Fr Orn* **40**, 48–68.

JOUANIN, C. & GILL, F.B. (1967). Recherche du Petrel de Barau, *Pterodroma baraui*. *Oiseau Rev Fr Orn* **37**, 1–19.

JOUANIN, C. & MOUGIN, J.-L. (1979). Order Procellariiformes. In *Checklist of Birds of the World* (E. Mayr & G.W. Cottrell, eds) Vol. I, 2nd ed, pp. 48–121. Mus Comp Zool Cambridge, Mass.

JOUANIN, C. & PAULIAN, P. (1960). Recherches sur des ossements d'oiseaux provenant de l'ile Nouvelle-Amsterdam (Ocean Indien). *Proc 12th Int Orn Congr* 368–372.

JOUANIN, C. & ROUX, F. (1966). La colonie de Puffin Cendres *Calonectris diomedea borealis* (Cory) de Selvagem Grande. *Bolm Mus Munic Funchal* **20**, 14–28.

JOUANIN, C., MOUGIN, J.-L., ROUX, F. & ZINO, A. (1979). Le Petrel de Bulwer *Bulweria bulwerii* dans l'archipel de Madere et aux iles Selvagens. *Oiseau Revue Fr Orn* **49**, 165–184.

JOUVENTIN, P. (1990). Shy Albatrosses *Diomedea cauta salvini* breeding on Penguin Island, in the Crozet Archipelago, Indian Ocean. *Ibis* **132**, 126–127.

JOUVENTIN, P. & MOUGIN, J.-L. (1981). Les strategies adaptatives des oiseaux de mer. *Rev Ecol (Terre et Vie)* **35**, 217–272.

JOUVENTIN, P. & VIOT, C.-R. (1985). Morphological and genetic variability of Snow Petrels *Pagodroma nivea*. *Ibis* **127**, 430–441.

JOUVENTIN, P. & WEIMERSKIRCH, H. (1984). L'Albatross Fuligineux a Dos Sombre *Phoebetria fusca*, example de strategie d'adaptation extreme a la vie pelagique. *Rev Ecol (Terre et Vie)* **39**, 401–429.

JOUVENTIN, P. & WEIMERSKIRCH, H. (1988). Demographic strategies of southern albatrosses. *Acta XIX Congr Int Orn* **1**, 857–868.

JOUVENTIN, P. & WEIMERSKIRCH. H. (1990). Satellite tracking of Wandering Albatrosses. *Nature, Lond* **343**, 746–748.

JOUVENTIN, P., MONICAULT, G. DE & BLOSSEVILLE, J.M. (1981). La danse de l'Albatross, *Phoebetria fusca*. *Behaviour* **78**, 43–80.

JOUVENTIN, P., MOUGIN, J.-L., STAHL, J.-C. & WEIMERSKIRCH, H. (1982). La segregation ecologique entre les oiseaux des Iles Crozet. Donnees preliminaries. *Com Natn Fr Rech Antarct* **51**, 457–467.

JOUVENTIN, P., ROUX, J.-P., STAHL, J.C. & WEIMERSKIRCH, H. (1983). Biologie et frequence de production chez l'Albatros a Bec Jaune (*Diomedea chlororhynchos*). *Gerfaut* **73**, 161–171.

JOUVENTIN, P., STAHL, J.C., WEIMERSKIRCH, H. & MOUGIN, J.-L. (1984). The seabirds of the French subAntarctic islands & Adelie Land, their status and conservation. In *Status and Conservation of the World's Seabirds* (J.P. Croxall, P.G.H.

Evans & R.W. Schreiber, eds) pp. 609–625. Tech Publ. 2, ICBP, Cambridge.

JOUVENTIN, P., MOUGIN, J.-L., STAHL, J.-C. & WEIMERSKIRCH, H. (1985). Comparative biology of the burrowing petrels of the Crozet Islands. *Notornis* **32**, 157–220.

JOUVENTIN, P., MARTINEZ, J. & ROUX, J.P. (1989). Breeding biology and current status of the Amsterdam Island Albatross *Diomedea amsterdamensis*. *Ibis* **131**, 171–182.

KADAR, Z. (1978). *Survivals of Greek Zoological Illuminations in Byzantine Manuscripts*. Akad, Kiado, Budapest.

KAMENEV, V.M. (1968). [Ornothological observations in the Mirny area in 1966/67.] *Inform Byull Sov Antarkt Eksp* **67**, 36–37.

KAY, G.T. (1953). The Fulmar's bill. *Scott Nat* **65**, 125–127.

KEPLER, C.B. (1967). Polynesian rat predation on nesting Laysan Albatrosses and other Pacific seabirds. *Auk* **84**, 426–430.

KEPLER, C.B. & KEPLER, A.K. (1980). The birds of Molokini Island, Maui. *Elepaio* **40**, 147–149.

KEPLER, C.B., KEPLER, A.K. & SIMONS, T.R. (1984). Hawaii's seabird islands, No. 1: Moke´ehia. *Elepaio* **44**, 71–74.

KERRY, K.R. & COLBACK, G.C. (1972). Light-mantled Sooty Albatrosses on Macquarie Island. *Aust Bird Bander* **10**, 61–62.

KERRY, K.R., HORNE, R.S.C. & DORWARD, D.F. (1983). Records of the Short-tailed Shearwater *Puffinus tenuirostris* in Antarctic waters. *Emu* **83**, 35–36.

KIDDER, J.H. & COUES, E. (1876). Contributions to the natural history of Kerguelen Island, made in connection with the United States Transit-of-Venus Expedition, 1874–75. II. Oology etc. *Bull US Natn Mus* **3**, 7–20.

KING, W.B. (1970). The trade wind zone oceanography pilot study Part VII: Observations of sea birds March 1964 to June 1965. *US Fish Wildl Serv Spec Scient Rep Fish Washington* **586**.

KING, W.B. (1974). Wedge-tailed Shearwater (*Puffinus pacificus*). In *Pelagic Studies of Seabirds in the Central and Eastern Pacific Ocean* (W.B. King, ed.) pp. 53–95. Smithson Contr Zool, *Washington* **158**.

KINSKY, F.C. (ed). (1970). *Annotated Checklist of the Birds of New Zealand*. Reed, Wellington.

KINSKY, F.C. & HARPER, P.C. (1968). Shrinkage of bill width in skins of some *Pachyptila* species. *Ibis* **110**, 100–102.

KLEMM, R.D. (1969). Comparative myology of the hind limb of procellariiform birds. *South Illinois Univ Monogr Sci Ser.* **2**, 1–269.

KURODA, N. (1954). *On the Classification and Phylogeny of the Order Tubinares, Particularly the Shearwaters* (Puffinus). Author, Tokyo.

KURODA, N. (1955). On the osteology of the Gadfly-Petrels, *Pterodroma*. *Annot Zool Japonenses* **28**, 171–177.

KURODA, N. (1960). Notes on the breeding seasons in the Tubinares (Aves). *Jap J Zool* **12**, 449–464.

KURODA, N. (1967). Morpho-anatomical analysis of parallel evolution between the Diving Petrel and Ancient Auk, with comparative osteological data on other species. *Misc Rep Yamashina Inst Orn Zool* **5**, 111–137.

KURODA, N. (1983). Some osteological notes on Procellariiformes. *Tori* **32**, 41–61.

KURODA, N. (1986). On the intestinal twistings in Gadfly-Petrels and comparative notes on the digestive tract in Procellariiformes. *Jap J Orn* **35**, 1–14.

LABAT, J.-B. (1722). *Noveau voyage aux Isles de l'Amerique contenant l'histoire naturelle de ces pays, etc*. Vol. 1. Paris.

LACAN, F. (1971). Observations ecologiques sur le Petrel de Wilson (*Oceanites oceanicus*) en Terre Adelie. *Oiseau Rev Fr Orn* (No. Spec) **41**, 65–89.

LACK, D. (1954). *The Natural Regulation of Animal Numbers*. Clarendon Press, Oxford.
LACK, D. (1966). *Population Studies of Birds*. Clarendon Press, Oxford.
LACK, D. (1967). Interrelationships in breeding adaptations as shown by marine birds. *Proc XIV Int Orn Congr* 3–42.
LACK, D. (1968). *Ecological Adaptations for Breeding in Birds*. Methuen, London.
LARSEN, J.W. (1967). *The Dark-rumped Petrel in Haleakala Crater*. National Park Service US Department of the Interior. Hawaii.
LATHAM, J. (1785). *A General Synopsis of Birds* Vol. 3. Author, London.
LEE, D.S. (1979). Second record of the south Trinidad Petrel (*Pterodroma arminjoniana*) for North America. *Am Birds* **33**, 138–139.
LEFEBVRE, E.A. (1977). Laysan Albatross breeding behavior. *Auk* **94**, 270–274.
LE GRAND, G., EMMERSON, K. and MARTIN, A. (1984). The status and conservation of seabirds in the Macaronesian Islands. In *Status and Conservation of the World's Seabirds* (J.P. Croxall, P.G.H. Evans & R.W. Schreiber, eds) pp. 377–391. Tech Pub 2, ICBP, Cambridge.
LENSINK, C.J. (1984). The status and conservation of seabirds in Alaska. In *Status and Conservation of the World's Seabirds* (J.P. Croxall, P.G.H. Evans & R.W. Schreiber, eds) pp. 13–27. Tech. Pub. 2, ICBP, Cambridge.
LILL, A. & BALDWIN, J. (1983). Weight changes and the mode of depot fat accumulation in migratory Short-tailed Shearwaters. *Aust J Zool* **31**, 891–902.
LINNAEUS, C. (1745). Storm-vaders-fogelen Beskrifven af Carl Linnaeus. *Kongl Swensk Vetens Acad handl* **6**, 93–96.
LINNAEUS, C. (1758). *Systema naturae per regna tria naturae, secundum classes, ordines, genera, species cum characteribus, differentiis, synonymis, locis*. Editio decima, reformata Vol. 1. Laurentii Salvii, Holmiae.
LITTLER, F.M. (1910). *A Handbook of the Birds of Tasmania and its Dependencies*. Author, Launceston.
LIVERSIDGE, R. & LE GRAS, G.M. (1981). Observations of seabirds off the eastern Cape, South Africa, 1958–1963. In *Proceedings of the Symposium on Birds of the Sea and Shore* (J. Cooper, ed.). African Seabird Group, Cape Town.
LOCKLEY, R.M. (1930). On the breeding-habits of the Manx Shearwater, with special reference to its incubation- and fledging-periods. *Br Birds* **23**, 202–218.
LOCKLEY, R.M. (1932). On the breeding habits of the Storm-Petrel, with special reference to its incubation- and fledging-periods. *Br Birds* **25**, 206–211.
LOCKLEY, R.M. (1942). *Shearwaters*. Dent, London.
LOCKLEY, R.M. (1952). Notes on the birds of the Berlengas (Portugal), the Desertas and Baixo (Madeira) and the Salvages. *Ibis* **94**, 144–158.
LOCKWOOD, W.B. (1954). Linguistic notes on 'Fulmar'. *Br Birds* **47**, 336–339.
LONNBERG, E. (1921). In *The Natural History of Juan Fernandez Islands* (C. Skottsberg, ed.) pp. 1–17. Almqvist and Wiksells, Uppsala.
LOOMIS, L.M. (1918). A review of the albatrosses, petrels, and diving petrels. *Proc Calif Acad Sci* (4th Ser) **2** (2), 1–187.
LOVENSKIOLD H.L. (1960). The Snow Petrel *Pagodroma nivea* nesting in Dronning Maud Land. *Ibis* **102**, 132–134.
LOVENSKIOLD, H.L. (1963). *Avifauna Svalbardensis*. Norsk Polarinst, Oslo.
LOWE, P.R. (1925). On the classification of the Tubinares or Petrels. *Proc Zool Soc Lond* **1925**, 1433–1443.
LOWE, P.R. & KINNEAR, N.B. (1930). Birds *Br Antarct ("Terra Nova") Exped 1910 Nat Hist Rep Zool* **4** (5), 103–193.
LUDERS, D.J. (1977). Behaviour of Antarctic Petrels and Antarctic Fulmars before laying. *Emu* **77**, 208–214.

MACDONALD, M.A. (1977a). Adult mortality and fidelity to mate and nest-site in a group of marked fulmars. *Bird Study* **24**, 165–168.

MACDONALD, M.A. (1977b). The pre-laying exodus of the Fulmar *Fulmarus glacialis* (L.). *Ornis Scand* **8**, 33–37.

MACDONALD, M.A. (1980). The winter attendance of fulmars at land in NE Scotland. *Ornis Scand* **11**, 23–29.

MAESAKO, Y. (1985). Community structure of *Machilus thunbergii* forests disturbed by birds (*Calonectris leucomelas*: Streaked Shear-water) on Kanmurijima Island, Kyoto Prefecture, Japan. *Japan J Ecol* **35**, 387–400.

MAHER, W.J. (1962). Breeding biology of the Snow Petrel near Cape Hallett, Antarctica. *Condor* **64**, 488–499.

MARSHALL, A.J. & SERVENTY, D.L. (1956a). Moult adaptation in relation to long-distance migration in petrels. *Nature Lond* **177**, 943.

MARSHALL, A.J. & SERVENTY, D.L. (1956b). The breeding cycle of the Short-tailed Shearwater, *Puffinus tenuirostris* (Temminck), in relation to trans-equatorial migration and its environment. *Proc Zool Soc Lond* **127**, 489–510.

MARSHALL, A.J. & SERVENTY, D.L. (1959). Experimental demonstration of an internal rhythm of reproduction in a trans-equatorial migrant (the Short-tailed Shearwater *Puffinus tenuirostris*). *Nature Lond* **184**, 1704–1705.

MARTENS, F. (1675). *Friedrich Martens von Hamburg Spitzbergische oder Groenland-ische Reise beschreibung*. Schultz, Hamburg.

MARTIN, M. (1698). *A Late Voyage to St. Kilda, the Remotest of the Hebrides* etc. Brown and Goodwin, London.

MATHEWS, G.M. (1912–13). *The Birds of Australia* Vol. 2. Witherby, London.

MATHEWS, G.M. (1928). *The Birds of Norfolk and Lord Howe Islands and the Australian South Polar Quadrant*. Witherby, London.

MATHEWS, G.M. (1933). On *Fregetta* Bonaparte and allied genera. *Novit Zool* **39**, 34–54.

MATHEWS, G.M. (1936). The ossification of certain tendons in the patagial fan of Tubinares. *Bull Br Orn Club* **56**, 45–50.

MATHEWS, G.M. (1942). New genus. *Emu* **41**, 305.

MATHEWS, G.M. (1948). Systematic notes on petrels. *Bull Br Orn Club* **68**, 155–170.

MATHEWS, G.M. & IREDALE, T. (1915). On some petrels from the north-east Pacific Ocean. *Ibis* (10th Ser) **3**, 572–609.

MATTHEWS, G.V.T. (1954). Some aspects of incubation in the Manx Shearwater *Procellaria puffinus*, with particular reference to chilling resistance in the embryo. *Ibis* **96**, 432–440.

MATTHEWS, L.H. (1929). The birds of South Georgia. *Discovery Rep* **1**, 563–592.

MATTHEWS, L.H. (1949). The origin of stomach oil in the petrels, with comparative observations on the avian proventriculus. *Ibis* **91**, 373–392.

MATTHEWS, L.H. (1951). *Wandering Albatross*. Macgibbon and Kee, London.

MAYAUD, N. (1931). Contribution a l'etude de la mue des puffins. *Alauda* (2nd Ser) **2**, 230–249.

MAYAUD, N. (1932a). Considerations sur la morphologie et al systematique de quelques puffins. *Alauda* **4**, 41–78.

MAYAUD, N. (1932b). Sur la mue des puffins. *Alauda* **4**, 110–112.

MAYAUD, N. (1934). Considerations sur la morphologie et la systematique de quelques puffins (addendum). *Alauda* **6**, 87–95.

MEETH, P. & MEETH, K. (1977). Blue Petrels *Halobaena caerulea* off Peru. *Ardea* **65**, 90–91.

MEHLUM, F., RAHN, H., BECH, C. & HAFTORN, S. (1986). Interrelationships between egg dimensions, pore numbers, incubation time, and adult body mass in Procellariiformes with special reference to the Antarctic Petrel *Thalassoica antarctica*.

Polar Res **5**, 55–58.
MEHLUM, F., GJESSING, Y., HAFTORN, S. & BECH, C. (1988). Census of breeding Antarctic Petrels *Thalassoica antarctica* and physical features of the breeding colony at Svarthamaren, Dronning Maud Land, with notes on breeding Snow Petrels *Pagodroma nivea* and South Polar Skuas *Catharacta maccormicki*. *Polar Res* **6**, 1–9.
MENDELSOHN, J. (1981). Movements of prions *Pachyptila* spp. and low pressure systems at Marion Island. In *Proceedings of the Symposium on Birds of the Sea and Shore* (J. Cooper, ed.) pp. 223–231. African Seabird Group, Cape Town.
MERCIER, G. (1987). Le Fulmar (*Fulmarus glacialis*) estivant et nicheur en Picardie. *Avocette* **11**, 15–40.
MERTON, D.V. (1970). Kermadec Islands Expedition Reports: a general account of birdlife. *Notornis* **17**, 147–199.
MESETH, E.H. (1975). The dance of the Laysan Albatross *Diomedea immutabilis*. *Behaviour* **54**, 217–257.
MILLEDGE, D. (1977). One year's observations of seabirds in continental shelf waters off Sydney, N.S.W. *Corella* **1**, 1–12.
MILLER, L. (1940). Observations on the Black-footed Albatross. *Condor* **42**, 229–238.
MILLER, L.H. (1951). A Miocene petrel from California. *Condor* **53**, 78–80.
MILLS, E.L. (1968). Observations of the Ringed Storm-petrel off the North-West coast of South America. *Condor* **70**, 87–88.
MILNE-EDWARDS, A. (1867–68). *Recherches anatomiques et paleontologiques pour servir a l'histoire des oiseaux fossiles de la France* Vol. 1. Masson, Paris.
MILNE-EDWARDS, A. (1874). Observations sur les oiseaux fossiles des Faluns de Saucats et de la Mollasse de Leognan. *Bibltque Ec Ht Etud, Paris, Sect Sci Nat,* **11** (3), 1–12.
MONTAGUE, T.L. (1984). The food of Antarctic Petrels (*Thalassoica antarctica*). *Emu* **84**, 244–245.
MONTEVECCHI, W.A., KIRKHAM, I.R., ROBY, D.D. & BRINK, K.L. (1983). Size, organic composition, and energy content of Leach's storm-petrel (*Oceanodroma leucorhoa*) eggs with reference to position in the precocial-altricial spectrum and breeding ecology. *Can J Zool* **61**, 1456–1463.
MONTGOMERY, H.H. (1892). Some account of the mutton birds, or Sooty Petrels (*Nectris brevicaudus*) etc. *Pap Proc R Soc Tasm* **1891**, 1–9.
MONTGOMERY, H.H. (1896). A night in a petrel rookery. *Pap Proc R Soc Tasm* **1896** v–vii.
MOORS, P.J. (1980). Southern Great Skuas on Antipodes Island, New Zealand: observations on food, breeding, and growth of chicks. *Notornis* **27**, 133–146.
MORGAN, W.L. & RITZ, D.A. (1982). Comparison of the feeding apparatus in the mutton-bird, *Puffinus tenuirostris* (Temminck) and the Fairy Prion, *Pachyptila turtur* (Kuhl) in relation to the capture of krill, *Nyctiphanes australis*. *J Exp Mar Biol Ecol* **59**, 61–76.
MORSE, D.H. & BUCHHEISTER, C.W. (1977). Age and survival of breeding Leach's Storm-petrels in Maine. *Bird-Banding* **48**, 341–349.
MORSE, D.H. & BUCHHEISTER, C.W. (1979). Nesting patterns of Leach's Storm-Petrels on Matinicus Rock, Maine. *Bird-Banding* **50**, 145–148.
MOUGIN, J.-L. (1967). Etude ecologique des deux especes de fulmars le Fulmar Atlantique (*F.glacialis*) et le Fulmar Antarctique (*F.glacialoides*). *Oiseau Rev Fr Orn* **37**, 57–103.
MOUGIN, J.-L. (1968). Etude ecologique de quatres especes de petrel antarctiques. *Oiseau Rev Fr Orn* (No. Spec) **38**, 1–51.
MOUGIN, J.-L. (1969). Notes ecologiques sur le Petrel de Kerguelen *Pterodroma brevirostris* de l'Ile de la Possession (Archipel Crozet). *Oiseau Rev Fr Orn* (No. Spec) **39**, 58–81.
MOUGIN, J.-L. (1970a). Observations ecologiques sur les Grand Albatros (*Diomedea exulans*) de l'Ile de la Possession (Archipel Crozet) en 1968. *Oiseau Rev Fr Orn* (No. Spec) **40**, 16–36.

MOUGIN, J.-L. (1970b). Le petrel a Menton Blanc *Procellaria aequinoctialis* de l'Ile de la Possession (Archipel Crozet). *Oiseau Rev Fr Orn* (No. Spec) **40**, 62–96.

MOUGIN, J.-L. (1970c). Les albatross fuligineux *Phoebetria palpebrata* et *P.fusca* de l'Ile de la Possession (Archipel Crozet). *Oiseau Rev Fr Orn* (No. Spec), **40**, 37–61.

MOUGIN, J.-L. (1971). Note complementaire sur le Petrel a Menton Blanc *Procellaria aequinoctialis* de l'Ile de la Possession (Archipel Crozet). *Oiseau Rev Fr Orn* **41**, 82–83.

MOUGIN, J.-L. (1975). Ecologie comparee des Procellariidae Antarctiques et Subantarctiques. *Com Natn Fr Rech Antarct* **36**, 1–195.

MOUGIN, J.-L. (1977). Nidification a l'ile Marion (46° 53'S, 37° 52"E) d'un Grand Albatros (*Diomedea exulans* L.) ne a l'ile de la Possession, Archipel Crozet (46° 25'S, 51° 45'E). *CR Acad Sci Paris* **284**, 2277–2280.

MOUGIN, J.-L. (1985). Petrels, Petrels-tempete et Petrels plongeurs de l'ile de Croy, iles Nuageuses, Archipel des Kerguelen (48° 38' 15"S, 68° 38' 30"E). *Oiseau Rev Fr Orn* **55**, 313–349.

MOUGIN, J.-L. (1989). Donnees preliminaries sur la structure et la dynamique de la population de Petrels de Bulwer *Bulweria bulwerii* de l'ile Selvagem Grande (30° 09'N, 15° 52'W). *CR Acad Sci Paris* (Ser III) **308**, 103–106.

MOUGIN, J.-L. (in press). Evaluation de la depense energetique et de la consommation alimentaire du Petrel de Bulwer *Bulweria bulwerii* d'apres l'etude de la decroissance ponderale au cours du jeune. *Bocagiana.*

MOUGIN, J.-L. & PREVOST, J. (1980). Evolution annuelle des effectifs et des biomasses des oiseaux antarctiques. *Rev Ecol* (*Terre et Vie*) **34**, 101–133.

MOUGIN, J.-L. & STAHL, J.-C. (1982). Essai de denombrement des Puffins Cendres *Calonectris diomedea borealis* de l'ile Selvagem Grande (30° 09'N, 15° 52'W) en 1980. *Bocagiana* **63**, 1–17.

MOUGIN J.-L., ROUX, F., JOUANIN C. & STAHL, J.-C. (1984a). Quelques aspects de la biologie de reproduction du Puffin cendre *Calonectris diomedea borealis* des iles Selvagens (30° 09'N, 15° 52'W). *Oiseau Rev Fr Orn* **54**, 229–246.

MOUGIN J.-L., JOUANIN, C., ROUX, F. & STAHL, J.-C. (1984b). Les annees sabbatiques des Puffins cendres *Calonectris diomedea borealis* reproducteurs de l'ile Selvagem Grande, ocean Atlantique nord-oriental (30° 09'N, 15° 52'W). *CR Acad Sci Paris* (Ser 3) **299**, 147–150.

MOUGIN J.-L., ROUX, F., STAHL, J.C. & JOUANIN, C. (1984c). L'evolution des effectifs des Puffins Cendres *Calonectris diomedea borealis* de l'Ile Selvagem Grande (30° 09'N, 15° 52'W) de 1980 a 1983. *Bocagiana* **75**, 1–8.

MOUGIN J.-L., ZINO, F., BISCOITO, M., DESPIN, B. & ROUX, F. (1986a). Quelques observations concernant la paraide chez le Puffin Cendre *Calonectris diomedea borealis* de l'Ile Selvagem Grande (30° 09'N, 15° 52'W). *Bol Mus Mun Funchal* **38**, 5–15.

MOUGIN J.-L., JOUANIN, C., DESPIN, B. & ROUX, F. (1986b). The age of first breeding of Cory's Shearwater on Selvagem Grande and problems of ring loss. *Ringing & Migration* **7**, 130–134.

MOUGIN J.-L., JOUANIN, C. & ROUX, F. (1987a). Structure et dynamique de la population du Puffins Cendres *Calonectris diomedea borealis* de l'ile Salvagem Grande (30° 09'N, 15° 52'W). *Oiseau Rev Fr Orn* **57**, 201–225.

MOUGIN J.-L., JOUANIN, C. & ROUX, F. (1987b). Les parametres controlant la reussite de l'incubation chez le Puffin Cendre *Calonectris diomedea borealis* de l'Ile Selvagem Grande (30° 09'N, 15° 52'W). *Bocagiana* **112**, 1–11.

MOUGIN, J.-L., JOUANIN, C. & ROUX, F. (1987c). Les annees sabbatiques des Puffins Cendres *Calonectris diomedea borealis* de l'Ile Selvagem Grande (30° 09'N, 15° 52'W). Influence du sexe et de l'age. *Oiseau Rev Fr Orn* **57**, 368–381.

MOUGIN J.-L., DESPIN, B., JOUANIN, C. and ROUX, F. (1987d). La fidelite an partenaire et au nid chez le Puffin Cendre, *Calonectris diomedea borealis,* de l'Ile Selvagen Grande. *Gerfaut* **77**, 353–369.

MOUGIN J.-L., JOUANIN, C. & ROUX, F. (1988a). Les differences d'age et d'experience entre partenaires chez la Puffin Cendre *Calonectris diomedea borealis* de l'Ile Selvagem Grande (30° 09'N, 15° 52'W). *Oiseau Rev Fr Orn* **58**, 113–119.

MOUGIN J.-L., JOUANIN, C. & ROUX, F. (1988b). Les migrations du Puffin Cendre *Calonectris diomedea. Oiseau Rev Fr Orn* **58**, 303–318.

MOUGIN J.-L., JOUANIN, C. & ROUX, F. (1988c). L'influence des voisins dans la nidification du Puffin Cendre *Calonectris diomedea. C R Acad Sci Paris* **307**, 195–198.

MUNDAY, B.L. (1966). 'Limey disease' of mutton birds (*Puffinus tenuirostris*). In *Diseases of Tasmania's Free-living Animals* pp. 12–16. Research Bulletin no. 5, Dept of Agriculture, Tasmania.

MUNRO, G.C. (1940). Birds of Hawaii — Wedgetailed Shearwater. *Elepaio* **1**, 7–8.

MUNRO, G.C. (1941). The Ao, Newell's Puffin. *Elepaio* **2**, 9–11.

MURPHY, R.C. (1918). A study of the Atlantic *Oceanites. Bull Am Mus Nat Hist* **38**, 117–146.

MURPHY, R.C. (1923). Notes on a small collection of birds from the Azores. *Ibis* (11th Ser) **5**, 44–49.

MURPHY, R.C. (1924). Birds collected during the Whitney South Sea Expedition. II. *Am Mus Novit* **124**, 1–13.

MURPHY, R.C. (1928). Birds collected during the Whitney South Sea Expedition. IV. *Am Mus Novit* **322**, 1–5.

MURPHY, R.C. (1929). Birds collected during the Whitney South Sea Expedition, X. *Am Mus Novit* **370**, 1–17.

MURPHY, R.C. (1936). *Oceanic Birds of South America.* Macmillan and American Museum of Natural History, New York.

MURPHY, R.C. (1951). The populations of the Wedge-tailed Shearwater (*Puffinus pacificus*). *Am Mus Novit* **1512**, 1–21.

MURPHY, R.C. (1952). The Manx Shearwater, *Puffinus puffinus,* as a species of world-wide distribution. *Am Mus Novit* **1586**, 1–21.

MURPHY, R.C. & HARPER, F. (1921). A review of the Diving Petrels. *Bull Am Mus Nat Hist* **44**, 495–554.

MURPHY, R.C. & PENNOYER, J.M. (1952). Larger petrels of the genus *Pterodroma. Am Mus Novit* **1580**, 1–43.

MURPHY, R.C. & SNYDER, J.P. (1952). The '*Pealea*' phenomenon and other notes on storm petrels. *Am Mus Novit* **1596**, 1–16.

MURTON, R.K. & WESTWOOD, N.J. (1977). *Avian Breeding Cycles.* Clarendon Press, Oxford.

NAARDING, J.A. (1980). Study of the Short-tailed Shearwater *Puffinus tenuirostris* in Tasmania. *Natn Parks Wildl Serv Tasm Rep* 1–78.

NAARDING, J.A. (1981). Study of the Short-tailed Shearwater *Puffinus tenuirostris* in Tasmania. *Natn Parks Wild Serv Tasm Tech Rep* **81/3** 1–34.

NAKAMURA, K. (1974). [On a mass accidental death of the Streaked Shearwater in Sagami Bay.] *Bull Nanagawa Prefectural Mus* **7**, 71–79.

NAKAMURA, K. (1982). Distribution of the gadfly petrels of the genus *Pterodroma* in the Antarctic and subAntarctic regions of the Australian Sector, Austral summer 1981. *Trans Tokyo Univ Fish* **5**, 203–211.

NAKAMURA, K. & TANAKA, Y. (1977). Distribution and migration of two species of the genus *Pterodroma* in the North Pacific. *Misc Rep Yamashina Inst Orn* **9**, 112–120.

NAKAMURA, K., HORI, H. & OSAKA, Y. (1983a). A mass inland drift of Leach's Storm-petrels by a typhoon in the autumn of 1979. *Bull Kanagawa Prefect Mus (Nat Sci)* **14**, 37–44.

NAKAMURA, K., TANAKA, Y. & HASEGAWA, M. (1983b). Distribution Status of the Wilson's Storm Petrel *Oceanites oceanicus* in Japanese waters. *Bull Biogeogr Soc Japan* **38**, 125–128.

NEEDHAM, J. (1931). *Chemical Embryology* Vol. 1. Cambridge University Press, London and New York.

NEWELL, N.D. (1949). Phyletic size increase, an important trend illustrated by fossil invertebrates. *Evolution* **3**, 103–124.

NEWMAN, D.G. (1978). Tuataras and Petrels. *Wildlife — a Review* **9**, 16–23.

NICE, M.M. (1962). Development of behavior in precocial birds. *Trans Linn Soc NY* **8**, 1–211.

NICHOLS, J.T. & MOWBRAY, L.L. (1916). Two new forms of petrels from the Bermudas. *Auk* **33**, 194–195.

NICHOLSKY, A. (1890). Sur la forme de l'oeuf des oiseaux. *Rev Scient Nat St Petersb* **1890**, 105–113.

NITZSCH, C.L. (1833). *System der Pterylographie*. Burmeister, Halle.

NORMAN, F.I. (1985). Localised declines in colonies of the Short-tailed Shearwater: an explanation. *Pap Proc R Soc Tasm* **119**, 103–107.

NORMAN, F.I. & BROWN, R.S. (1987). Notes on the Common Diving-Petrels *Pelecanoides urinatrix* found beach-washed in Victoria. *Emu* **87**, 179–185.

NORMAN, F.I. & GOTTSCH, M.D. (1969). Artificial twinning in the Short-tailed Shearwater *Puffinus tenuirostris*. *Ibis* **111**, 391–393.

NORRIS, A.Y. (1965). Observations of seabirds in the Tasman Sea and in New Zealand waters in October and November, 1962. *Notornis* **12**, 80–105.

OGI, H. (1984a). Feeding ecology of the Sooty Shearwater in the western subarctic North Pacific Ocean. In *Marine Birds: their Feeding Ecology and Commercial Fisheries Relationships* (D.N. Nettleship, G.A. Sanger & P.F. Springer, eds) pp. 78–84. Ministry of Supply and Services, Canada.

OGI, H. (1984b). Seabird mortality incidental to the Japanese salmon gill-net fishery. In *Status and Conservation of the World's Seabirds* (J.P. Croxall, P.G.H. Evans & R.B. Schreiber, eds) pp. 717–722. Tech. Publ. **2**, ICBP, Cambridge.

OGI, H., KUBODERA, T. & NAKAMURA, K. (1980). The pelagic feeding ecology of the Short-tailed Shearwater (*Puffinus tenuirostris*) in the subarctic Pacific region. *J Yamashina Inst Orn* **12**, 157–182.

OGILVIE-GRANT, W.R. (1896). On the birds observed at the Salvage Islands, near Madeira. *Ibis* (7th Ser) **2**, 41–55.

OKA, N. (1989). Chick growth and development of the Short-tailed Shearwater *Puffinus tenuirostris* in Tasmania. *J Yamashina Inst Orn* **21**, 193–207.

OKA, N. & MARUYAMA, N. (1986). Mass mortality of Short-tailed Shearwaters along the Japanese coast. *Tori* **34**, 97–104.

OKA, N., MARUYAMA, N. & SKIRA, I. (1987). Chick growth and mortality of Short-tailed Shearwaters in comparison with Sooty Shearwaters, as a possible index of fluctuation of Australian Krill abundance. *Proc Natn Inst Polar Res Tokyo Symp Polar Biol* **1**, 166–174.

OKAMOTO, B. (1972). [Calonectris leucomelas *on Kanmurijima Island*.] Komine Shoten, Tokyo.

OLIVER, W.R.B. (1955). *New Zealand Birds* 2nd edn. Reed, Wellington.

OLLASON, J.C. & DUNNET, G.M. (1978). Age, experience and other factors affecting the breeding success of the fulmar, *Fulmarus glacialis*, in Orkney. *J Anim Ecol* **47**, 961–976.

OLLASON, J.C. & DUNNET, G.M. (1980). Nest failures in the fulmar: the effect of observers. *J Field Orn* **51**, 39–54.

OLLASON, J.C. & DUNNET, G.M. (1982). The feeding dispersal of fulmars *Fulmarus*

glacialis in the breeding season. *Ibis* **124**, 359–361.
OLLASON, J.C. & DUNNET, G.M. (1983). Modelling annual changes in numbers of breeding Fulmars, *Falmarus glacialis,* at a colony in Orkney. *J Anim Ecol* **52**, 185–198.
OLLASON, J.C. & DUNNET, G.M. (1986). Relative effects of parental performance and egg quality on breeding success of fulmars *Fulmarus glacialis*. *Ibis* **128**, 290–296.
OLLASON, J.C. & DUNNET, G.M. (1988). Variation in breeding success in fulmars. In *Reproductive Success* (T.H. Clutton-Brock, ed.) pp. 268–287. University of Chicago Press, Chicago.
OLSON, S.L. (1975a). Paleornithology of St. Helena Island, South Atlantic Ocean. *Smithson Contr Paleobiol* **23**, 1–49.
OLSON, S.L. (1975b). Remarks on the generic characters of *Bulweria*. *Ibis* **117**, 111–113.
OLSON, S.L. (1984). Evidence of a large albatross in the Miocene of Argentina (Aves: Diomedeidae). *Proc Biol Soc Wash* **97**, 741–743.
OLSON, S.L. (1985a). Early Pliocene Procellariiformes (Aves) from Langebaanweg, South-western Cape Province, South Africa. *Ann S Afr Mus* **95**, 123–145.
OLSON, S.L. (1985b). An early Pliocene marine avifauna from Duinefontein, Cape Province, South Africa. *Ann S Afr Mus* **95**, 147–164.
OLSON, S.L. (1985c). The Italian specimen of *Bulweria fallax* (Procellariidae). *Bull Br Orn Club* **105**, 29–30.
OLSON, S.L. (1985d). In *Avian Biology* Vol. VIII. (D. Farner, J. King & K.C. Parkes, eds) pp. 79–252. Academic Press, London.
ORTON, M.N. (1968). Notes on Antarctic Petrels. *Emu* **67**, 225–229.
PAGANELLI, C.V., OLSZOWSKA, A. & AR, A. (1974). The avian egg: surface area, volume and density. *Condor* **76**, 319–325.
PALMA, R.L. & PILGRIM, R.L.C.P. (1984). A revision of the genus *Harrisoniella* (Mallophaga: Philopteridae). *NZ J Zool* **11**, 145–166.
PALMER, R.S. (ed.). (1962). *Handbook of North American Birds* Vol. 1. Yale University Press, New Haven and London.
PASCAL, M. (1979). Donnees ecologiques sur l'Albatross a Sourcils Noirs *Diomedea melanophrys* (Temminck) dans l'Archipel des Kerguelen. *Alauda* **47**, 165–172.
PAULIAN, P. (1953). Pinnepedes, Cetaces, Oiseaux des Iles Kerguelen et Amsterdam. *Mem Inst Scient Madagascar*, Ser. A. (**8**), 111–234.
PAYNE, M.R. and PRINCE, P.A. (1979). Identification and breeding biology of the diving petrels *Pelecanoides georgicus* and *P.urinatrix exsul* at South Georgia. *NZ J Zool* **6**, 299–318.
PEFAUR, J.E. (1974). Egg-neglect in the Wilson's Storm Petrel. *Wilson Bull* **86**, 16–22.
PENNYCUICK, C.J. (1960). Gliding flight of the Fulmar Petrel. *J Exp Biol* **37**, 330–338.
PENNYCUICK, C.J. (1982). The flight of petrels and albatrosses (Procellariiformes), observed in South Georgia and its vicinity. *Phil Trans R Soc Lond* **B300**: 75–106.
PENNYCUICK, C.J. & WEBBE, D. (1959). Observations on the Fulmar in Spitzbergen. *Br Birds* **52**, 321–332.
PENNYCUICK, C.J., CROXALL, J.P. & PRINCE, P.A. (1984). Scaling of foraging radius and growth rate in petrels and albatrosses (Procellariiformes). *Ornis Scand* **115**, 145–154.
PERCIVAL, E. (1942). The juvenile plumage of some birds and an interpretation of its nature. *Trans R Soc NZ* **72**, 6–20.
PERRINS, C.M. (1966). Survival of young Manx Shearwaters in relation to their presumed date of hatching. *Ibis* **108**, 132–135.
PERRINS, C.M. & BROOKE, M. de L. (1976), Manx Shearwaters in the Bay of Biscay. *Bird Study* **23**, 295–300.
PERRINS, C.M., HARRIS, M.P. & BRITTON, C.K. (1973). Survival of Manx Shearwaters *Puffinus puffinus*. *Ibis* **115**, 535–348.

PETTIT, T.N. & WHITTOW, G.C. (1983). Water loss from piped Wedge-tailed Shearwater eggs. *Condor* **85**, 107–109.

PETTIT, T.N., GRANT, G.S., WHITTOW, G.C., RAHN, H. & PAGANELLI, C.V. (1982a). Respiratory gas exchange and growth of Bonin Petrel embryos. *Physiol Zool* **55**, 162–170.

PETTIT, T.N., GRANT, G.S. & WHITTOW, G.C. (1982b). Body temperature and growth of Bonin Petrel chicks. *Wilson Bull* **94**, 358–361.

PETTIT, T.N., WHITTOW, G.C. & GRANT, G.S. (1984a). Caloric content and energetic budget of tropical seabird eggs. In *Seabird Energetics* (G.C. Whittow & H. Rahn, eds) pp. 113–137. Plenum, New York.

PETTIT, T.N., BYRD, V.G., WHITTOW, G.C. & SEKI, M.P. (1984b). Growth of the Wedge-tailed Shearwater in the Hawaiian Islands. *Auk* **101**, 103–109.

PHILLIPS, W.W.A. (1955). Wilson Petrel [*Oceanites oceanicus* (Kuhl)] in Indo-Ceylon waters, with special reference to the 1954 southward migration. *J Bombay Nat Hist Soc* **53**, 132–133.

PICKERING, S.P.C. (1989). Attendance patterns and behaviour in relation to experience and pair-bond formation in the Wandering Albatross *Diomedea exulans* at South Georgia. *Ibis* **131**, 183–195.

PILGRIM, R.L.C. & PALMA, R.L. (1982). A list of the chewing lice (Insecta: Mallophaga) from birds in New Zealand. *Natn Mus NZ Misc Ser* **6** pp. 32.

PINDER, R. (1966). The Cape Pigeon, *Daption capensis* Linnaeus, at Signy Island, South Orkney Islands. *Br Antarct Surv Bull* **8**, 19–47.

PITMAN, R.L. (1982). Distribution and foraging habits of Dark-rumped Petrel (*Pterodroma phaeopygia*) in the eastern tropical Pacific. [Abstract]. *Pacif Seabird Grp Bull* **9**, 72.

PITMAN, R.L. (1988). Laysan Albatross breeding in the eastern Pacific — and a comment. *Pacif Seabird Grp Bull* **15**, 52.

PLATH, K. (1969). 'Albie' the Albatross. *Avicult Mag* **75**, 235–237.

PODOLSKY, R.H. & KRESS, S.W. (1989). Factors affecting colony formation in Leach's Storm-Petrel. *Auk* **106**, 332–336.

POWER, D.M. & AINLEY, D.G. (1986). Seabird geographic variation: similarity among populations of Leach's Storm-petrel. *Auk* **103**, 575–585.

POWLESLAND, R.G. (1983). Seabirds found dead on New Zealand beaches in 1981. *Notornis* **30**, 125–135.

POWLESLAND, R.G. (1987). Seabirds found dead on New Zealand beaches in 1985, and a review of *Pterodroma* species recoveries since 1960. *Notornis* **34**, 237–252.

PREVOST, J. (1964). Remarques ecologiques sur quelques Procellarieus antarctiques. *Oiseau Rev Fr Orn* **34** (No Spec), 91–112.

PRINCE, P.A. (1980a). The food and feeding ecology of the Blue Petrel (*Halobaena caerulea*) and Dove Prion (*Pachyptila desolata*). *J Zool Lond* **190**, 59–76.

PRINCE, P.A. (1980b). The food and feeding ecology of the Grey-headed Albatross *Diomedea chrysostoma* and Black-browed Albatross *D.melanophris*. *Ibis* **122**, 476–488.

PRINCE, P.A. (1985). Population and energetic aspects of the relationship between Blackbrowed and Greyheaded Albatrosses and the southern Ocean marine environment. In *Antarctic Nutrient Cycles and Food Webs* (W.R. Siegfried, P.R. Condy & R.M. Laws, eds) pp. 473–477. Springer, Berlin.

PRINCE, P.A. & COPESTAKE, P.G. (1990). Diet and aspects of Fairy Prions breeding at South Georgia. *Notornis* **37**, 59–69.

PRINCE, P.A. & CROXALL, J.P. (1983). Birds of South Georgia: new records and re-evaluation of status. *Br Antarct Surv Bull* **59**, 15–27.

PRINCE, P.A. & FRANCIS, M.D. (1984). Activity budgets of foraging Grey-headed Albatrosses. *Condor* **86**, 297–300.

PRINCE, P.A. & RICKETTS, C. (1981). Relationships between food supply and growth in albatrosses: an interspecies chick fostering experiment. *Ornis Scand* **12**, 207–210.
PRINCE, P.A., RICKETTS, C. & THOMAS, G. (1981). Weight loss in incubating albatrosses and its implications for their energy and food requirements. *Condor* **83**, 238–242.
PRINCE, P.A. & WALTON, D.W.H. (1984). Automated measurement of meal size and feeding frequency in albatrosses. *J Appl Ecol* **21**: 789–794.
PRYOR, M.E. (1968). The avifauna of Haswell Island, Antarctica. *Antarct Res Ser* **12**, 57–82.
PULICH, W. (1982). Documentation and status of Cory's Shearwater in the western Gulf of Mexico. *Wilson Bull* **94**, 381–385.
PYCRAFT, W.P. (1899). Contributions to the osteology of birds. Part III. *Tubinares. Proc Zool Soc Lond* **1899**, 381–411.
PYCRAFT, W.P. (1907). On the systematic position of the petrels. In *A Monograph of the Petrels* (F. Du Cane Godman) pp. xv–xxi. Witherby, London.
RAHN, H. & HUNTINGTON, C.E. (1988). Eggs of Leach's Storm Petrel: O_2 uptake, water loss, and microclimate of the nest. *Comp Biochem Physiol* **91A**, 519–521.
RAHN, H. & WHITTOW, G.C. (1988). Adaptations to a pelagic life: eggs of the albatross, shearwater and petrel. *Comp Biochem Physiol* **91A**, 415–423.
RAHN, H., ACKERMANN, R.A. & PAGANELLI, C.V. (1984). Eggs, yolk and embryonic growth rate. In *Seabird Energetics* (G.C. Whittow & H. Rahn, eds) pp. 89–112. Plenum, New York.
RAHN, H., PAGANELLI, C.V. & SOTHERLAND, P.R. (1985). Initial mass of avian eggs: comparison between measured and calculated values. *J Orn* **126**, 210–212.
RANDALL, R.M. & RANDALL, B.M. (1986). The seasonal occurrence of Leach's Storm Petrel *Oceanodroma leucorhoa* at St. Croix Island, South Africa. *Ostrich* **57**, 157–161.
RECHTEN, C. (1986). Factors determining the laying date of the Waved Albatross *Diomedea irrorata*. *Ibis* **128**, 492–501.
REISCHEK, A. (1886). Observations on *Puffinus gavius* (Forst.), Rain-bird, (Hakoakoa), their habits and habitats. *Trans Proc NZ Inst* **18**, 93–94.
RENSCH, B. (1947). Neuere Probleme der Abstrammungslehre. Enke, Stuttgart.
RHEINHARDT, M.J. (1873). Om vingens anatomiske Bygning Los Stormfugle-Familien, (*Procellaridae* s. *Tubinares*). *Vidensk Medd Naturist Forening* **1873**, 123–138.
RICE, D.W. & KENYON, K.W. (1962). Breeding cycles and behavior of Laysan and Black-footed Albatrosses. *Auk* **79**, 517–567.
RICHARDSON, M.E. (1984). Aspects of the ornithology of the Tristan da Cunha Group and Gough Island, 1972–1974. *Cormorant* **12**, 123–201.
RICHDALE, L.E. (1939). A Royal Albatross nesting on the Otago Peninsula, New Zealand. *Emu* **38**, 467–488.
RICHDALE, L.E. (1942). Supplementary notes on the Royal Albatross. *Emu* **41**, 169–184; 253–263.
RICHDALE, L.E. (1943). The Kuaka, or diving petrel, *Pelecanoides urinatrix* (Gmelin). *Emu* **43**, 24–48.
RICHDALE, L.E. (1944). The Titi Wainui or Fairy Prion, *Pachyptila turtur* (Kuhl). *Trans R Soc NZ* **74**, 32–48; 165–181.
RICHDALE, L.E. (1945). Supplementary notes on the Diving Petrel. *Trans R Soc NZ* **75**, 42–53.
RICHDALE, L.E. (1948). *Maori and Mutton-bird*. Author, Dunedin.
RICHDALE, L.E. (1949). The pre-egg stage in Buller's Mollymawk. *Biol Mongr* **2**. Author, Dunedin.
RICHDALE, L.E. (1950). The pre-egg stage in the albatross family. *Biol Mongr* **3**. Author, Dunedin.
RICHDALE, L.E. (1952). Post-egg period in albatrosses. *Biol Mongr* **4**. Author, Dunedin.

RICHDALE, L.E. (1954). The starvation theory in albatrosses. *Auk* **71**, 239–252.
RICHDALE, L.E. (1957). *A Population Study of Penguins*. Clarendon Press, Oxford.
RICHDALE, L.E. (1963). Biology of the Sooty Shearwater *Puffinus griseus*. *Proc Zool Soc Lond* **141**, 1–117.
RICHDALE, L.E. (1965a). Biology of the birds of Whero Island, New Zealand, with special reference to the diving petrel and the White-faced Storm Petrel. *Trans Zool Soc Lond* **31**, 1–86.
RICHDALE, L.E. (1965b). Breeding behaviour of the Narrow-billed and the Broad-billed Prion on Whero Island, New Zealand. *Trans Zool Soc Lond* **31**, 87–155.
RICHDALE, L.E. & WARHAM, J. (1973). Survival, Pair bond retention and nest-site tenacity in Buller's Mollymawk. *Ibis* **115**, 257–263.
RICKETTS, C. & PRINCE, P.A. (1981). Comparison of growth of albatrosses. *Ornis Scand* **12**, 120–124.
RICKETTS, C. & PRINCE, P.A. (1984). Estimation by use of field weighings of metabolic rate and food conversion efficiency in albatross chicks. *Auk* **101**, 790–795.
RICKLEFS, R.E. (1968). Patterns of growth in birds. *Ibis* **110**, 419–451.
RICKLEFS, R.E. (1979). Adaptation, constraint, and compromise in avian postnatal development. *Biol Rev* **54**, 289–290.
RICKLEFS, R.E. (1983). Some considerations on the reproductive energetics of pelagic seabirds. *Studies in Avian Biology* **8**, 84–94.
RICKLEFS, R.E. (1984a). Prolonged incubation in pelagic seabirds: a comment on Boersma's paper. *Am Nat* **123**, 710–720.
RICKLEFS, R.E. (1984b). Meal sizes and feeding rates of Christmas Shearwaters and Phoenix Petrels on Christmas Island, Central Pacific Ocean. *Ornis Scand* **15**, 16–22.
RICKLEFS, R.E. & RAHN, H. (1979). The incubation temperature of Leach's Storm-petrel. *Auk* **96**, 625–627.
RICKLEFS, R.E. & ROBY, D.D. (1983). Development of homeothermy in the diving petrels *Pelecanoides urinatrix exsul* and *P.georgicus* and the Antarctic Prion *Pachyptila desolata*. *Comp Biochem Physiol* **75A**, 307–311.
RICKLEFS, R.E., ROBY, D.D. & WILLIAMS, J.B. (1986). Daily energy expenditure by adult Leach's Storm-petrels during the nesting cycles. *Physiol Zool* **59**, 649–660.
RICKLEFS, R.E., DAY, C.H., HUNTINGDON, C.E. & WILLIAMS, J.B. (1985). Variability in feeding rate and meal size of Leach's Storm-petrel at Kent Island, New Brunswick. *J Anim Ecol* **54**, 883–898.
RICKLEFS, R.E., WHITE, S.C. & CULLEN, J. (1980a). Energetics of postnatal growth in Leach's Storm-Petrel. *Auk* **97**, 566–575.
RICKLEFS, R.E., WHITE, S. & CULLEN, J. (1980b). Postnatal development of Leach's Storm-petrel. *Auk* **97**, 768–781.
RICKLEFS, R.E., PLACE, A.R. & ANDERSON, D.J. (1987). An experimental investigation of the influence of diet quality on growth rate in Leach's Storm-Petrel. *Am Nat* **130**, 300–305.
RIDDLE, O. (1910). Studies with Sudan III in metabolism and inheritance. *J Exp Zool* **8**, 163–184.
RIDOUX, V. & OFFREDO, C. (1989). The diets of five summer breeding seabirds in Adelie Land, Antarctica. *Polar Biol* **9**, 137–145.
RISTOW, D. & WINK, M. (1980). Sexual dimorphism of Cory's Shearwater. *Il-Merill* **21**, 9–12.
RITCHIE, J. (1930). St. Kilda and the significance of some of its animal inhabitants. *Scott Nat* **1930**, 69–74.
RIVEROS-SALCEDO, J.C. & APARICIO, J.J. (1990). The Peruvian Diving Petrel *Pelecanoides garnotti* in Peru. *Pacif Seabird Grp Bull* **17**, 32–33.
ROBBINS, C.S. & RICE, D.W. (1974). Recoveries of banded Laysan Albatrosses (*Diomedea*

immutablis) and Black-footed Albatrosses (*D.nigripes*). *Smithson Contr Zool* **158**, 232–277.

ROBERTS, B. (1940). The life cycle of Wilson's Petrel *Oceanites oceanicus* (Kuhl). *Br Graham Land Exped Scient Rep* **1** (2), 141–194.

ROBERTSON, C.J.R. (1974). Albatrosses of the Chatham Islands. *Wildlife — a Review*, Wellington, **5**, 20–22.

ROBERTSON, C.J.R. & BELL, B.D. (1984). Seabird status and conservation in the New Zealand region. In *Status and Conservation of the World's Seabirds* (J.P. Croxall, P.G.H. Evans & R.W. Schreiber, eds) pp. 573–586. Tech. Publ. 2, ICBP, Cambridge.

ROBERTSON, C.J. & JENKINS, J. (1981). Birds seen at sea in southern New Zealand waters, February–June 1981. *Australas Seabird Grp Newsl* **16**, 17–29.

ROBERTSON, C.J.R. & KINSKY, F.C. (1972). The dispersal movements of the Royal Albatross (*Diomedea epomophora*) *Notornis* **19**, 289–301.

ROBERTSON, C.J. & VAN TETS, G.F. (1982). The status of birds at the Bounty Islands. *Notornis* **29**, 311–336.

ROBERTSON, C.J.R. & WRIGHT, A. (1973). Successful hand-rearing of an abandoned Royal Albatross chick. *Notornis* **20**, 49–58.

ROBERTSON, H.A. & JAMES, P.C. (1988). Morphology and egg measurements of seabirds breeding on Grand Salvage Island, North Atlantic. *Bull Br Orn Club* **108**, 79–87.

ROBINSON, L.N. (1961). The feeding of Fairy Prions. *Aust Bird Watcher* **1**, 156–157.

ROBY, D.D. & RICKLEFS, R.E. (1983). Some aspects of the breeding biology of the diving petrels *Pelecanoides georgicus* and *P.urinatrix exsul* at Bird Island, South Georgia. *Br Antarct Surv Bull* **59**, 29–34.

ROBY, D.D., PLACE, A.R. & RICKLEFS, R.E. (1986). Assimilation and deposition of wax esters in planktivorous seabirds. *J Exp Zool* **238**, 29–41.

RODHOUSE, P.G., CLARKE, M.R. & MURRAY, A.W.A. (1987). Cephalopod prey of the Wandering Albatross *Diomedea exulans*. *Mar Biol* **96**, 1–10.

ROMANOFF, A.L. & ROMANOFF, A.J. (1949). *The Avian Egg*. Wiley, New York.

ROUND, P.D. & SWANN, R.L. (1977). Aspects of the breeding of Cory's Shearwater *Calonectris diomedea* in Crete. *Ibis* **119**, 350–353.

ROUNSEVELL, D.E. & BROTHERS, N.P. (1984). The status and conservation of seabirds at Macquarie Island. In *Status and Conservation of the World's Seabirds* (J.P. Croxall, P.G.H. Evans & R.W. Schreiber, eds) pp. 587–592. Tech Publ 2, ICBP, Cambridge.

ROUTH, M. (1949). Ornithological observations in the Antarctic seas. *Ibis* **91**, 577–606.

ROUX, J.-P. & MARTINEX, J. (1987). Rare, vagrant and introduced birds at Amsterdam and Saint Paul Islands, southern Indian Ocean. *Cormorant* **14**, 3–19.

ROUX, J.P., JOUVENTIN, P., MOUGIN, J.L., STAHL, J.C. & WEIMERSKIRCH, H. (1983). Un nouvel albatros *Diomedea amsterdamensis* n. sp. decouvert sur l'ile Amsterdam (37° 50'S, 77° 35'E). *Oiseau Rev Fr Orn* **53**, 1–11.

ROUX, J.P., MOUGIN, J.-L. & BARTLE, J.A. (1986). Le Prion de MacGillivray. Donnees taxonomiques. *Oiseau Rev Fr Orn* **56**, 379–383.

ROWAN, M.K. (1951). The Yellow-nosed Albatross, *Diomedea chlororhynchos* Gmelin, at its breeding grounds in the Tristan da Cunha Group. *Ostrich* **22**, 139–159.

ROWAN, M.K. (1952). The Greater Shearwater *Puffinus gravis* at its breeding grounds. *Ibis* **94**, 97–121.

ROWAN, M.K. (1965). Regulation of sea-bird numbers. *Ibis* **107**, 54–59.

ROWAN, A.N., ELLIOTT, H.F.I. & ROWAN, M.K. (1951). The 'spectacled' form of the Shoemaker *Procellaria aequinoctialis* in the Tristan da Cunha Group. *Ibis* **93**, 169–179.

RUMBOLL, M.A.E. & JEHL. J.R. (1977). Observations on pelagic birds in the South Atlantic Ocean in the austral spring. *Trans San Diego Soc Nat Hist* **19**, 1–16.

SAGAR, P.M. (1979). Breeding of the Cape Pigeon (*Daption capense*) at the Snares Islands. *Notornis* **26**, 23–36.

SAGAR, P.M. (1986). The sexual dimorphism of Snares Cape Pigeons (*Daption capense australe*). *Notornis* **33**, 259–263.

SAIZ, F. & HAJEK, E.R. (1988). Observaciones de Temperatura en Nidos de Petrel Gigante *Macronectes giganteus* (Gmelin). *Inst Antart Chileno* **14**, pp. 15.

SALOMONSEN, F. (1965). The geographical variation of the fulmar (*Fulmarus glacialis*) and the zones of marine environment in the North Atlantic. *Auk* **82**, 327–355.

SANGER, G.A. (1974a). Black-footed Albatross (*Diomedea nigripes*). *Smithson Contr Zool* **158**, 96–128.

SANGER, G.A. (1974b). Laysan Albatross (*Diomedea immutabilis*). *Smithson Contr Zool* **158**, 129–153.

SARA, M. (1983). Osservazioni Sulla consistenza numerica e sull'alimentazione della Berta Maggiore (*Calonectris diomedea*), del Canale di Sicilia. *Riv Ital orn* **53**, 183–193.

SCHLATTER, R.P. (1984). The status and conservation of seabirds in Chile. In *Status and Conservation of the World's Seabirds* (J.P. Croxall, P.G.H. Evans & R.W. SCHREIBER, eds) pp. 261–269. Tech Publ 2, ICBP, Cambridge.

SCHLATTER, R.P. & MARIN, M.A. (1983). Breeding of Elliot's Storm Petrel, *Oceanites gracilis*, in Chile. *Gerfaut* **73**, 197–199.

SCHMIDT, W.J. (1967a). Das 'Globularmuster' im Eischalenkalk von *Diomedea*. *Z Zellforsch mikrosk Anat* **77**, 518–533.

SCHMIDT, W.J. (1967b). Das 'Muster' im Eischalenkalk der Vogel dargestellt durch Schliffatzung. *Z Zellforsch mikrosk Anat* **79**, 446–458.

SCHONWETTER, M. (1967). *Handbuch der Oologie* (W. Meise, ed.) Vol. 1. Academie Verlag, Berlin.

SCHRAMM, M. (1983a). The breeding biologies of the petrels *Pterodroma macroptera*, *P.brevirostris* and *P.mollis* at Marion Island. *Emu* **83**, 75–81.

SCHRAMM, M. (1983b). Automatic recording of nest visits by burrow-nesting birds. *J Field Orn* **54**, 192–194.

SCHRAMM, M. (1986). The diet of chicks of Greatwinged, Kerguelen and Softplumaged Petrels at the Prince Edward Islands. *Ostrich* **57**, 9–15.

SCHREIBER, R.W. & ASHMOLE, N.P. (1970). Sea-bird breeding seasons on Christmas Island, Pacific Ocean. *Ibis* **112**, 363–394.

SCHREIBER, R.W. & SCHREIBER, E.A. (1987). Tropical seabirds and the El Nino: 1985 and 1986 perspectives. In *Symposium on Breeding Birds in Captivity* pp. 352–359. Jean Delacour/International Foundation for Conservation.

SEGONZAC, M. (1972). Donnees recentes sur la faune des Iles Saint-Paul et Nouvelle Amsterdam. *Oiseau Rev Fr Orn* **42** (No Spec), 1–68.

SERVENTY, D.L. (1952). Movements of the Wilson Storm-petrel in Australian seas. *Emu* **52**, 105–116.

SERVENTY, D.L. (1956). A method of sexing petrels in field observations. *Emu* **56**, 213–221.

SERVENTY, D.L. (1963). Egg-laying time-table of the Slender-billed Shearwater, *Puffinus tenuirostris*. *Proc XIII Int Orn Congr* **1**, 338–343.

SERVENTY, D.L. (1967). Aspects of the population ecology of the Short-tailed Shearwater *Puffinus tenuirostris*. *Proc XIV Int Orn Congr* 165–190.

SERVENTY, D.L. (1974). The biology behind the mutton-bird industry. *Pap Proc R Soc Tasmania* **107**, 1–9.

SERVENTY, D.L. & CURRY, P.J. (1984). Observations on colony size, breeding success, recruitment and inter-colony dispersal in a Tasmanian colony of Short-tailed Shearwaters *Puffinus tenuirostris* over a 30-year period. *Emu* **84**, 71–79.

SERVENTY, D.L. & WHITTELL, H.M. (1962). *Birds of Western Australia*. Paterson Brokensha, Perth.

SERVENTY, D.L., SERVENTY, V.N. & WARHAM, J. (1971). *The Handbook of Australian Sea-*

birds. Reed, Sydney.

SERVENTY, D.L., GUNN, B.M., SKIRA, I.J., BRADLEY, J.S. & WOOLLER, R.D. (1989). Fledgling translocation and philopatry in a seabird. *Oecologia* **81**, 428–429.

SHALLENBERGER, R.J. (1973). Breeding biology, homing behavior and communication patterns of the Wedge-tailed Shearwater *Puffinus pacificus chlorohynchos*. PhD thesis, University of California.

SHARPE, R.B. (1879). Petrological, botanical, and zoological collections made in Kerguelen's Land and Rodriguez during the Transit of Venus Expeditions, 1874–75. Zoology: Birds. *Phil Trans R Soc Lond* **168**, 101–162.

SHAUGHNESSY, P.D. (1970). Serum proteins of two sibling species of giant petrel (*Macronectes* spp.). *Comp Biochem Physiol* **33**, 721–723.

SHAUGHNESSY, P.D. & FAIRALL, N. (1976). Notes on seabirds at Gough Island. *S Afr J Antarct Res* **6**, 23–25.

SHAW, G. (1826). *General Zoology* etc., Vol. 13 (1), 1–278. Nodder, London.

SHUFELDT, R.W. (1888). Observations upon the osteology of the Order Tubinares and Steganopodes. *Proc US Natn Mus* **2**, 235–315.

SHUFELDT, R.W. (1907). On the osteology of the Tubinares. *Am Nat* **41**, 109–124.

SHUFELDT, R.W. (1916). The bird-caves of the Bermudas and their former inhabitants. *Ibis* (10th Ser) **4**, 623–635.

SHUFELDT, R.W. (1922). A comparative study of some subfossil remains of birds from Bermuda, including the 'Cahow'. *Ann Carnegie Mus Nat His.* **13**, 333–418.

SHUNTOV, V.P. (1974). *Sea Birds and the Biological Structure of the Ocean*. Translation. US Department of Commerce, Springfield.

SIBLEY, C.G., AHLQUIST, J.E. & MONROE, B.L. (1988). A classification of the living birds of the world based on DNA-DNA hybridization studies. *Auk* **105**, 409–423.

SIBSON, R.B. (1949). Visit to Little Barrier. *NZ Bird Notes* **3**, 151–155.

SILYN-ROBERTS, H. & SHARP, R.M. (1985). Preferred orientation of calcite in the ratite and tinamou eggshells. *J Zool Lond* **205A**, 39–52.

SIMONS, T.R. (1981a). A simple event recorder for monitoring cavity-dwelling animals. *Murrelet* **62**, 27–30.

SIMONS, T.R. (1981b). Behavior and attendance patterns of the Fork-tailed Storm-Petrel. *Auk* **98**, 145–158.

SIMONS, T.R. (1984). A population model of the endangered Hawaiian Dark-rumped Petrel. *J Wildl Mgmt* **48**, 1065–1076.

SIMONS, T.R. (1985). Biology and behavior of the endangered Hawaiian Dark-rumped Petrel. *Condor* **87**, 229–245.

SIMONS, T.R. & WHITTOW, G.C. (1984). Energetics of breeding Dark-rumped Petrels. In *Seabird Energetics* (G.C. Whittow & H. Rahn, eds) pp. 159–181. Plenum, New York.

SIMPSON, D.M. (1973). Seabird colonies on the rocky islands of Sento Shosho, 28° 48'N, 123° 36'E, and Sekibo Sho Islet, 25° 55'N, 124° 34'E. *Sea Swallow* **22**, 29.

SKINNER, H.D. & BAUCKE, W. (1928). The Morioris. *Mem Bernice P Bishop Mus* **9**, 343–384.

SKIRA, I.J. (1979). Underwater feeding by Short-tailed Shearwaters. *Emu* **79**, 43.

SKIRA, I.J., WAPSTRA, J.E., TOWNEY, G.N. & NAARDING, J.A. (1985). Conservation of the Short-tailed Shearwater *Puffinus tenuirostris* in Tasmania, Australia. *Biol Conserv* **37**, 225–236.

SLATER, C.A. (1990). First arrival dates at two fulmar *Fulmarus glacialis* colonies in Norfolk. *Bird Study* **37**, 1–4.

SMITH, R.I.L. & PRINCE, P.A. (1985). The natural history of Beauchene Island. *Biol J Linn Soc* **24**, 233–283.

SNOW, D.W. (1965). The breeding of Audubon's Shearwater (*Puffinus lherminieri*) in the Galapagos. *Auk* **82**, 591–597.

SNOW, D.W. & SNOW, B.K. (1966). The breeding season of the Madeiran Storm Petrel *Oceanodroma castro* in the Galapagos. *Ibis* **108**, 283–284.
SOMME, L. (1977). Observations on the Petrel (*Pagodroma nivea*) in Vestfjella, Dronning Maud Land. *Norsk Polarinst Arbok* **1976**, 285–292.
SORENSEN, J.H. (1950). The Light-mantled Sooty Albatross at Campbell Island. *Cape Exp Ser Bull* **8**, 30pp. DSIR, Wellington.
SOTHERLAND, P.R. & RAHN, H. (1987). On the composition of bird eggs. *Condor* **89**, 48–65.
SPARLING, D.W. (1977). Sounds of Laysan and Black-footed Albatrosses. *Auk* **94**, 256–269.
SPEICH, S. & MANUWAL, D.A. (1974). Gular pouch development and population structure of Cassin's Auklet. *Auk* **91**, 291-306.
STAHL, J.C. (1987). Distribution des oiseaux marins dans le sudouest de l'Ocean Indien: donnees preliminaires de la campagne Apsara II — Antiprod III. In *Les rapports des campagnes a la mer MD38/APSARA II — ANTIPROD III a bord du 'Marion Dufresne' 16 Janvier–22 Fevrier 1984* (M. Fontugne & M. Fiala, eds). *Terre Austr Antarct Fr Publ Rech* **84**.
STAHL, J.C., JOUVENTIN, P., MOUGIN, J.-L., ROUX, J.P. & WEIMERSKIRCH, H. (1985). The foraging zones of seabirds in the Crozet Islands sector of the Southern Ocean. In *Antarctic Nutrient Cycles and Food Webs* (W.R. Siegfried, P.R. Condy & R.M. Laws, eds) pp. 478–485. Springer, Berlin.
STANLEY, S.M. (1973). An explanation for Cope's Rule. *Evolution* **27**, 1–26.
STEADMAN, D.W. & OLSON, S.L. (1985). Bird remains from an archaeological site on Henderson Island, South Pacific: man-caused extinctions on an "uninhabited" island. *Proc Natl Acad Sci USA* **82**, 6191-6195.
STEELE, W.K. & KLAGES, N.T. (1986). Diet of the Blue Petrel at Sub-Antarctic Marion Island. *S Afr J Zool* **21**, 253–256.
STEJNEGER, L. (1893). Notes on a third installment of Japanese birds in the Science College Museum, Tokyo, Japan, with descriptions of new species. *Proc US Natn Mus* **16**, 615–638.
STOREY, A.E. (1984). Function of Manx Shearwater calls in mate attraction. *Behaviour* **89**, 73–89.
STOREY, A.E. & GRIMMER, B.L. (1986). Effect of illumination on the natural activities of Manx Shearwaters: colony avoidance or inconspicuous behaviour. *Bird Behaviour* **6**, 85–89.
STOREY, A.E. & LIEN, J. (1985). Development of the first North American colony of Manx Shearwaters. *Auk* **102**, 395–401.
STRANGE, I.J. (1980). The Thin-billed Prion, *Pachyptila belcheri* at New Island, Falkland Islands. *Gerfaut* **70**, 411–445.
STRESEMANN, E. & STRESEMANN, V. (1970). Uber Mauser und Zug von *Puffinus gravis*. *J Orn* **111**, 378–393.
SUDBURY, A.W., TOMPKINS, R.J. & GIBSON, J.D. (1985). Recurring pairs of Wandering Albatrosses at a New South Wales winter feeding ground? *Emu* **85**, 195–197.
SUGIMORI, F., OKA, N. & ISHIBASHI, Y. (1985). The degree of skull ossification as a mean of aging Short-tailed Shearwaters. *J Yamashina Inst Orn* **17**, 159–165.
SWALES, M.P. (1965). The sea-birds of Gough Island. *Ibis* **107**, 17–42; 215–229.
SWANSON, N.M. & MERRITT, F.D. (1974). The breeding cycle of the Wedge-tailed Shearwater on Mutton Bird Island, N.S.W. *Aust Bird Bander* **12**, 3–9.
SZIJJ, L.J. (1967). Notes on the winter distribution of birds in the western Antarctic and adjacent Pacific waters. *Auk* **84**, 366–378.
TAN, N. (1977). [*Streaked Shearwaters and Kanmurijima.*] Tenseisha, Tokyo.
TANAKA, Y. & INABA, F. (1981). [The distribution and migration of the White-necked

petrel, *Pterodroma externa externa*, in the west area of the North Pacific Ocean and the Japanese waters.] *J Yamashina Inst Orn* **13**, 61–68.

TANAKA, Y. & KANEKO, Y. (1983). [Distribution and migration of the Bonin Petrel *Pterodroma hypoleuca* in the Northwest Pacific in relation to sea surface temperatures.] *Tori* **32**, 119–127.

TANAKA, Y., KANEKO, Y. & SATO, S. (1985). [Distribution and migration of smaller petrels of the genus *Pterodroma* in the Northwest Pacific.] *J Yamashina Inst Orn* **17**, 23–31.

TAOKA, M., SATO, T., KAMADA, T. & OKUMURA, H. (1989). Sexual dimorphism of chatter-calls and vocal sex recognition in Leach's Storm-Petrels (*Oceanodroma leucorhoa*). *Auk* **106**, 498–501.

TASKER, C.R. & MILLS, J.A. (1981). A functional analysis of courtship feeding in the Red-billed Gull. *Behaviour* **77**, 222–241.

TELFER, T.C., SINCOCK, J.L., BYRD, G.V. & REED, J.R. (1987). Attraction of Hawaiian seabirds to lights: conservation efforts and effects of moon phase. *Wildl Soc Bull* **15**, 406–413.

TELLERIA, J.L. (1980). Autumn migration of Cory's Shearwater through the Straits of Gibraltar. *Bird Study* **27**, 21–26.

TENNYSON, A.J.D. & TAYLOR, G.A. (1990). Behaviour of *Pterodroma* petrels in response to ''war-whoops''. *Notornis* **37**, 121–128.

THIBAULT, J.-C. (1977). Nouvelles observations sur la disposition helicoidale de l'intestin chez certaines especes du genre *Pterodroma*. *Alauda* **45**, 237.

THIBAULT, J.C. (1985). La reproduction du Puffin Cendre *Calonectris diomedea* en Corse. In *Oiseaux marins nicheurs du Midi et de la Corse* pp. 49–55. Ann Centre Rech Orn Provence, Aix-en-Provence.

THIBAULT, J.-C. & HOLYOAK, D.T. (1978). Vocal and olfactory displays in the genera *Bulweria* and *Pterodroma*. *Ardea* **66**, 53–56.

THOMAS, G. (1982). The food and feeding ecology of the Light-mantled Sooty Albatross at South Georgia. *Emu* **82**, 92–100.

THOMAS, G., CROXALL, J.P. & PRINCE, P.A. (1983). Breeding biology of the Light-mantled Sooty Albatross (*Phoebetria palpebrata*) at South Georgia. *J Zool Lond* **199**, 123–135.

THOMAS, T. (1986). L'effectif des oiseaux nicheurs de l'archipel de Pointe Geologie (Terre Adelie) et son evolution au cours des trentes dernieres annees. *Oiseau Rev Fr Orn* **56**, 349–368.

THOMPSON, D'A.W. (1917). *On Growth and Form*. University Press, Cambridge.

THOMPSON, D.Q. (1951). Notes on distribution of North Pacific albatrosses. *Auk* **68**, 227–235.

THOMSON, A.L. (1965). The transequatorial migration of the Manx Shearwater (Puffin des Anglais). *Oiseau Rev Fr Orn* **35** (No spec), 130–140.

THORESEN, A.C. (1967). Ecological observations on Stanley and Green Islands, Mercury Group. *Notornis* **14**, 182–200.

THORESEN, A.C. (1969). Observations on the breeding behaviour of the diving petrel *Pelecanoides u.urinatrix* (Gmelin). *Notornis* **16**, 241–260.

THRELFALL, W. (1974). Foot injuries in Leach's Storm Petrels. *Wilson Bull* **86**, 65–67.

TICKELL, W.L.N. (1962). The Dove Prion, *Pachyptila desolata* Gmelin. *Falkland Is Depend Surv Scient Rep* **33**, 1–55.

TICKELL, W.L.N. (1967). Movements of Black-browed and Grey-headed Albatross in the south Atlantic. *Emu* **66**, 357–367.

TICKELL, W.L.N. (1968). The biology of the great albatrosses, *Diomedea exulans* and *Diomedea epomophora*. *Antarct Res Ser* **12**, 1–55.

TICKELL, W.L.N. (1969). Plumage changes in young albatrosses. *Ibis* **111**, 102–105.
TICKELL, W.L.N. (1970). Biennial breeding in albatrosses. In *Antarctic Ecology* (M.W. Holdgate, ed.) Vol. 1, pp. 549–557. Academic Press, London and New York.
TICKELL, W.L.N. (1980). The pink ear stain of Wandering Albatrosses *Diomedea exulans*. *Australas Seabird Grp Newsl* **13**, 12–17.
TICKELL, W.L.N. (1984). Behaviour of Blackbrowed and Greyheaded Albatross at Bird Island, South Georgia. *Ostrich* **56**, 64–85.
TICKELL, W.L.N. & GIBSON, J.D. (1968). Movements of Wandering Albatrosses *Diomedea exulans*. *Emu* **68**, 7–20.
TICKELL, W.L.N. & PINDER, R. (1966). Two-egg clutches in Albatrosses. *Ibis* **108**, 126–129.
TICKELL, W.L.N. & PINDER, R. (1967). Breeding frequencies in the albatrosses *Diomedea melanophris* and *D.chrysostoma*. *Nature Lond* **213**, 315–316.
TICKELL, W.L.N. & PINDER, R. (1975). Breeding biology of the Black-browed Albatross *Diomedea melanophris* and Grey-headed Albatross *D.chrysostoma* at Bird Island, South Georgia. *Ibis* **117**, 433–451.
TIMMERMANN, G. (1960). Gruppen-Revisionen bei Mallophagen. II. Genus *Halipeurus* Thompson, 1936. 1. Teil: Die *Halipeurus* — Arten der 'Gadfly Petrels' (Genera *Pterodroma* und *Bulweria*). *Z Parasitenkunde* **20**, 317–334.
TIMMERMANN, G. (1965). Die Federlingsfauna der Sturmvogel und die Phylogenese des procellariiformen Vogelstammes. *Abh Verh Naturw Ver Hamburg, NF* **8**, (Suppl), 1–249.
TIRRELL, P.C. & MURRISH, D.E. (1979a). Vascular anatomy of the brood patch of the giant petrel. *Antarct J US* **14**, 171–172.
TIRRELL, P.C. & MURRISH, D.E. (1979b). Cutaneous blood flow in the giant petrel. *Antarct J US* **14**, 171–172.
TOMPKINS, R.J. (1983a). Purple bill flushes and pink ear marks on Wandering Albatrosses on Macquarie Island. *Australas Seabird Grp Newsl* **18**, 11–15.
TOMPKINS, R.J. (1983b). Fertilisation of Wandering Albatross eggs on Macquarie Island. *Notornis* **30**, 244–246.
TOMPKINS, R.J. (1985). Reproduction and mortality of Wandering on Macquarie Island. *Emu* **85**, 40–42.
TSCHUDI, J.J. VON (1856). Beitrage von geographischen Verbreitung der Meesvogel. *J Orn* **4**, 134–162, 177–191.
TYLER, C. (1969). The snapping strength of the egg shells of various orders of birds. *J Zool Lond* **159**, 65–77.
ULRICH, F. (1904). Zur Kenntnis der Luftsacke bei *Diomedea exulans* und *Diomedea fuliginosa*. *Wiss Ergeb Deutsch Tiefsee-Exped 'Valdivia'* **7**, 319–342.
VALETTE, L.H. (1906). Viaje a las Ilas Orcades Australes. *Anal Minist Agric Republica Argentina* **3**, 3–68.
VAN BENEDEN, P.-J. (1871). Les oiseaux de l'Argile Rupelienne. *Bull Acad R Sci Belgique* (2nd Ser) **32**, 256–261.
VAN FRANEKER, J.A. & MONTAGUE, T. (1987). Recoveries of petrels banded near Casey Station, Wilkes Land, Antarctica, 1984 to 1985. *Corella* **11**, 37–43.
VAN FRANEKER, J.A. & WATTEL, J. (1982). Geographical variation of the Fulmar *Fulmarus glacialis* in the North Atlantic. *Ardea* **70**, 31–44.
VAN OORT, E.D. (1912). On *Aestrelata aterrima* (Bonaparte). *Notes Leyden Mus* **34**, 70.
VAN OORT, G.J. & KRUIJT, J.P. (1953). On the pelagic distribution of some Procellariiformes in the Atlantic and Southern Oceans. *Ibis* **95**, 615–637.
VAN RENSBURG, P.J.J. (1985). The feeding ecology of a decreasing Feral House Cat, *Felis catus*, population at Marion Island. In *Antarctic Nutrient Cycles and Food Webs* (W.R. Siegfried, P.R. Condy & R.M. Laws, eds) pp. 620–624. Springer, Berlin.

VAN RYZIN, M.T. & FISHER, H.I. (1976). The age of Laysan Albatrosses, *Diomedea immutabilis*, at first breeding. *Condor* **78**, 1–9.
VAUGHAN, R. (1980). Notes on Cory's Shearwater (*Calonectris diomedea*) and some other birds on Linosa, Pelagic Isles. *Riv Ital Orn* **50**, 143–154.
VEITCH, C.R. (1985). Methods of eradicating cats from offshore islands in New Zealand. In *Conservation of Island Birds* (P.J. Moors, ed.) pp. 125–141. ICBP, Cambridge.
VERHEYEN, R. (1958). Note sur la classification des Procellariiformes (Tubinares). *Bull Inst R Sci Nat Belgique* **34**, 1–22.
VINSON, J.-M. (1976). Notes sur les Procellariens de l'Ile Ronde. *Oiseau Rev Fr Orn* **46**, 1–24.
VLECK, C.M. & KENAGY, G.J. (1980). Embryonic metabolism of the Fork-tailed Storm Petrel: physiological patterns during prolonged and interrupted incubation. *Physiol Zool* **53**, 32–42.
VOISIN, J.-F. (1968). Les petrels geants (*Macronectes halli* et *Macronectes giganteus*) de l'ile de la Possession. *Oiseau Rev Fr Orn* **38**, 95–122.
VOISIN, J.-F. (1969). L'Albatros Hurleur *Diomedea exulans* a l'Ile de la Possession. *Oiseau Rev Fr Orn* (No Spec) **39**, 82–106.
VOISIN, J.-F. (1976). Observations sur les petrels geants de l'Ile aux Cochons (Archipel Crozet). *Alauda* **44**, 411–429.
VOISIN, J.-F. (1978). Observations sur le comportement des petrels geants de l'Archipel Crozet. *Alauda* **46**, 209–234.
VOISIN, J.-F. (1979). Observations ornithologiques aux Iles Tristan da Cunha et Gough. *Alauda* **47**, 73–82.
VOISIN, J.-F. (1981). A pursuit plunging Wandering Albatross *Diomedea exulans*. *Cormorant* **9**, 136.
VOISIN, J.-F. (1982). Observations on the Falkland Islands giant petrels *Macronectes giganteus solanderi*. *Gerfaut* **72**, 367–380.
VOISIN, J.-F. (1988). Breeding biology of the Northern Giant Petrel *Macronectes halli* and the Southern Giant Petrel *M.giganteus* at Ile de la Possession, Iles Crozet, 1966–1980. *Cormorant* **16**, 65–97.
VOISIN, J.-F. & BESTER, M.N. (1981). The specific status of giant petrels *Macronectes* at Gough Island. In *Proceedings of Symposium on Birds of the Sea and Shore* (J. Cooper, ed.) pp. 215–222. African Seabird Group, Cape Town.
VOOUS, K.H. (1949). The morphological, anatomical, and distributional relationship of the Arctic and Antarctic Fulmars (Aves, Procellariidae). *Ardea* **37**, 113–122.
VOOUS, K.H. (1970). Moulting Great Shearwater (*Puffinus gravis*) in Bay of Biscay. *Ardea* **58**, 265–266.
WALKER, C.A., WRAGG, G.M. & HARRISON, C.J.O. (1990). A new shearwater from the Pleistocene of the Canary Islands and its bearing on the evolution of certain Puffinus shearwaters. *Historical Biol* **3**, 203–224.
WALKER, F.J. (1981). Notes on the birds of Dhofar, Oman. *Sandgrouse* **2**, 56–85.
WARHAM, J. (1952). A family of shearwaters. *Country Life* **111**, 1250–1251.
WARHAM, J. (1955). Observations on the Little Shearwater at the nest. *West Aust Nat* **5**, 31–39.
WARHAM, J. (1956). The breeding of the Great-winged Petrel, *Pterodroma macroptera*. *Ibis* **98**, 171–185.
WARHAM, J. (1957). Additional notes on the Great Winged Petrel *Pterodroma macroptera*. *Ibis* **99**, 511–512.
WARHAM, J. (1958). The nesting of the shearwater *Puffinus carneipes*. *Auk* **75**, 1–14.
WARHAM, J. (1959). The Trinidad Petrel, *Pterodroma arminjoniana*, a new bird for Australia. *Emu* **59**, 153–158.
WARHAM J. (1960). Some aspects of breeding behaviour in the Short-tailed Shearwater.

Emu **60**, 75–87.
WARHAM, J. (1962). The biology of the Giant Petrel *Macronectes giganteus*. *Auk* **79**, 139–160.
WARHAM, J. (1964). Breeding behaviour in Procellariiformes. In *Biologie Antarctique: Premier Symp. S.C.A.R.* (R. Carrick, M. Holdgate & J. Prevost, eds) pp. 389–394. Hermann, Paris.
WARHAM, J. (1966). *The Technique of Wildlife Cinematography*. Focal Press, London and New York.
WARHAM, J. (1967). The White-headed Petrel *Pterodroma lessoni* at Macquarie Island. *Emu* **67**, 1–22.
WARHAM, J. (1968). Biometric studies of birds in the order Procellariiformes. MSc Thesis University Durham, unpubl.
WARHAM, J. (1971). Body temperatures of petrels. *Condor* **73**, 214–219.
WARHAM, J. (1975). Fulmar behaviour. *Scott Birds* **8**, 319–321.
WARHAM, J. (1976). Aerial displays by large petrels. *Notornis* **23**, 255–257.
WARHAM, J. (1977a). Wing loadings, wing shapes, and flight capabilities of Procellariiformes. *NZ J Zool* **4**, 73–83.
WARHAM, J. (1977b). The incidence, functions and ecological significance of petrel stomach oils. *Proc NZ Ecol Soc* **24**, 84–93.
WARHAM, J. (1979). The voice of the Soft-plumaged Petrel (*Pterodroma mollis*). *Notornis* **26**, 357–360.
WARHAM, J. (1982). A distant recovery of a Buller's Mollymawk. *Notornis* **29**, 213–214.
WARHAM, J. (1983). The composition of petrel eggs. *Condor* **85**, 194–199.
WARHAM, J. (1988a). Responses of *Pterodroma* petrels to man-made sounds. *Emu* **88**, 109–111.
WARHAM, J. (1988b). Vocalisations of *Procellaria* petrels, *Notornis* **35**, 169–183.
WARHAM, J. & BELL, B.D. (1979). The birds of Antipodes Island, New Zealand. *Notornis* **26**, 121–169.
WARHAM, J. & BENNINGTON, S.L. (1983). A census of Buller's Albatross *Diomedea bulleri* at the Snares Islands, New Zealand. *Emu* **83**, 112–114.
WARHAM, J. & FITZSIMONS, C.H. (1987). The vocalisations of Buller's Mollymawk, *Diomedea bulleri*, (Aves: Diomedeidae), with some comparative data on other albatrosses. *NZ J Zool* **14**, 65–79.
WARHAM, J. & RICHDALE, L.E. (1973). Survival, pair bond retention and nest-site tenacity in Buller's Mollymawk. *Ibis* **115**, 257–263.
WARHAM, J. & WILSON, G.J. (1982). The size of the Sooty Shearwater population at the Snares Islands, New Zealand. *Notornis* **29**, 23–30.
WARHAM, J., BOURNE, W.R.P. & ELLIOTT, H.F.I. (1974). Albatross identification in the North Atlantic. *Am Birds* **28**, 585–603.
WARHAM, J., WATTS, R. & DAINTY, R.J. (1976). The composition, energy content and function of the stomach oils of petrels (Order, Procellariiformes). *J Exp Mar Biol Ecol* **23**, 1–13.
WARHAM, J., KEELEY, B.R. & WILSON, G.J. (1977). Breeding of the Mottled Petrel. *Auk* **94**, 1–17.
WARHAM, J., WILSON, G.J. & KEELEY, B.R. (1982). The annual cycle of the Sooty Shearwater *Puffinus griseus* at the Snares Islands, New Zealand. *Notornis* **29**, 269–292.
WASILEWSKI, A. (1986). Ecological aspects of the breeding cycle in the Wilson's Storm Petrel, *Oceanites oceanicus* (Kuhl), at King George Island (South Shetland Islands, Antarctica). *Polish Polar Res* **7**, 173–216.
WATANUKI, Y. (1985). Food of breeding Leach's Storm petrels (*Oceanodroma leucorhoa*). *Auk* **102**, 884–886.

WATANUKI, Y. (1986). Moonlight avoidance behavior in Leach's Storm-Petrels as a defense against Slaty-backed Gulls. *Auk* **103**, 14–22.

WATLING, D. & LEWANAVANUA, R.F. (1985). A note to record the continuing survival of the Fiji (MacGillivray's) Petrel *Pseudobulweria macgillivrayi*. *Ibis* **127**, 230–233.

WATSON, G.E. (1968). Synchronous wing and tail molt in diving petrels. *Condor* **70**, 182–183.

WATSON, G.E. & DIVORKY, G.J. (1971). Identification of *Diomedea leptorhyncha* Coues 1866, an albatross with remarkably small salt glands. *Condor* **73**, 487–489.

WATSON, G.E. *et al.* (1971). *Birds of the Antarctic and Subantarctic* (V.C. Bushnell, ed.). *Antarct. Map Folio* Ser. 14. American Geographical Society New York.

WATSON, G.E., LEE, D.S. & BACKUS, E.S. (1986). Status and subspecific identity of White-faced Storm-petrels in the Western North Atlantic Ocean. *Am Birds* **40**, 401–407.

WEIMERSKIRCH, H. (1982). La strategie de reproduction de l'Albatross Fuligineux a dos Sombre. *Comm Natn Fr Rech Antarct* **51** 437–447.

WEIMERSKIRCH, H. & JOUVENTIN, P. (1987). Population dynamics of the Wandering Albatross, *Diomedea exulans*, of the Crozet Islands: causes and consequences of the population decline. *Oikos* **49**, 315–322.

WEIMERSKIRCH, H., JOUVENTIN, P., MOUGIN, J.L., STAHL, J.C. & VAN BEVEREN, M. (1985). Banding recoveries and the dispersal of seabirds breeding in French austral and Antarctic Territories. *Emu* **85**, 22–33.

WEIMERSKIRCH, H., JOUVENTIN, P. & STAHL, J.C. (1986). Comparative ecology of six albatross species breeding on the Crozet Islands. *Ibis* **128**, 195–213.

WEIMERSKIRCH, H., CLOBERT, J. & JOUVENTIN, P. (1987). Survival in five southern albatrosses and its relationship to their life history. *J Anim Ecol* **56**, 1043–1055.

WEIMERSKIRCH, H., ZOTIER, R. & JOUVENTIN, P. (1989). The avifauna of the Kerguelen Islands. *Emu* **89**, 15–29.

WEST, J. & IMBER, M.J. (1985). Some foods of Hutton's Shearwater (*Puffinus huttoni*). *Notornis* **32**, 333–336.

WEST, J.A. & IMBER, M.J. (1986). Some foods of Buller's Mollymawk *Diomedea bulleri*. *NZ J Zool* **13**, 169–174.

WESTERSKOV, K. (1960). Field identification and sex determination of the Royal Albatross. *Notornis* **9**, 1–6.

WHEELWRIGHT, N.T. & BOERSMA, P.D. (1979). Egg chilling and the thermal environment of the Fork-tailed Storm Petrel (*Oceanodroma furcata*) nest. *Physiol Zool* **52**, 231–239.

WHITLEY, G. (1934). The doom of the Bird of Providence, *Pterodroma melanopus* (Gmelin). *Aust Zool* **8**, 42–49.

WHITTOW, G.C. (1980). Physiological and ecological correlates of prolonged incubation in sea birds. *Am Zool* **20**, 427–436.

WHITTOW, G.C. (1983). Physiological ecology of incubation in tropical seabirds. *Studies Avian Biol* **8**, 42–72.

WHITTOW, G.C., ACKERMAN, R.A., PAGANELLI, C.V. & PETTIT, T.N. (1982). Pre-pipping water loss from the eggs of the Wedge-tailed Shearwater. *Comp Biochem Physiol* **72A**, 29–34.

WHITTOW, G.C., SIMONS, T.R. & PETTIT, T.N. (1984). Water loss from the eggs of a tropical sea bird (*Pterodroma phaeopygia*) at high altitude. *Comp Biochem Physiol* **78A**, 537–540.

WHITTOW, G.C., PETTIT, T.N., ACKERMAN, R.A. & PAGANELLI, C.V. (1987). Temperature regulation in a burrow-nesting tropical seabird, the Wedge-tailed Shearwater (*Puffinus pacificus*). *J Comp Physiol* **B157**, 607–614.

WIENS, J.A. & SCOTT, J.M. (1975). Model estimation of energy flow in Oregon coastal seabird populations. *Condor* **77**, 439–452.

WILBUR, H.M. (1969). The breeding biology of Leach's Petrel *Oceanodroma leucorhoa*. *Auk* **86**, 433–442.

WILKINS, G.H. (1923). Report on the birds collected during the voyage of the 'Quest' (Shackleton-Rowett Expedition) to the Southern Atlantic. *Ibis* (11th Ser) **5**, 474–511.

WILKINSON, H.E. (1969). Description of an upper Miocene albatross from Beaumaris, Victoria, Australia, and a review of fossil Diomedeidae. *Mem Natn Mus Victoria* **29**, 41–51.

WILLIAMS, A.J. (1978). Mineral and energy contributions of petrels (Procellariiformes) killed by cats, to the Marion Island terrestrial ecosystem. *S Afr J Antarct Res* **8**, 49–53.

WILLIAMS, A.J. (1984). The status and conservation of seabirds on some islands in the African sector of the Southern Ocean. In *Status and Conservation of the World's Seabirds* (J.P. Croxall, P.G.H. Evans & R.W. Schreiber, eds) pp. 627–635. Tech. Publ. 2, ICBP, Cambridge.

WILLIAMS, A.J., SIEGFRIED, W.R. & COOPER, J. (1982). Egg composition and hatchling precocity in seabirds. *Ibis* **124**, 456–470.

WILLIAMS, J.B. & RICKLEFS, R.E. (1984). Egg temperature and embryo metabolism in some high-latitude procellariiform birds. *Physiol Zool* **57**, 118–127.

WILLIAMSON, K. (1940). The puffins of the Calf Isle. *J Manx Mus* **4**, 178–180, 203–205.

WILLIAMSON, K. (1945). Some new and scarce breeding species on the Faroe Islands. *Ibis* **87**, 550–558.

WILLIAMSON, K., MALCOLM, N., MACDOUGALL, C., NORMAN, D. & YATES, G. (1954). The fledging of a group of young Fulmars. *Scott Nat* **66**, 1–12.

WILLUGHBY, F. (1678). *The Ornithology of Francis Willughby of Middleton in the County of Warwick Esq; etc.* (J. Ray, transl.). Royal Society, London.

WILSON, E. (1907). Aves. *Br Natn Antarct Exped 1901–1904* **2**, *Zool* (pt 2), 1–121.

WILSON, J.A. (1975). Sweeping flight and soaring by albatrosses. *Nature Lond* **257**, 307–308.

WINGATE, D.B. (1964). Discovery of breeding Black-capped Petrels on Hispaniola. *Auk* **81**, 147–159.

WINGATE, D.B. (1972). First successful hand rearing of an abandoned Bermuda Petrel chick. *Ibis* **114**, 97–101.

WINGATE, D.B. (1977). Excluding competitors from Bermuda Petrel nesting burrows. In *Proceedings of Symposium on Management Techniques for preserving Endangered Birds* (S.A. Temple, ed.) pp. 93–102. University of Wisconsin Press and Croom Helm, Madison.

WINK, M., WINK, C. & RISTOW, D. (1982). Brutbiologie mediterraner Gelbschnabel-sturmtaucher (*Calonectris diomedea diomedea*). Seevogel (Spec No) 127–135.

WINK, M. SCHARLAU, W. & RISTOW, D. (1987). Niedrige Ei-und Korpertemperatur ("Hypothermie") bein brutenden Gelbschnabelstrumtauchern (*Calonectris diomedea*). *J Orn* **128**, 334–338.

WOMERSLEY, H. (1955). The Acarina fauna of mutton birds' nests on a Bass Strait island. *Aust J Zool* **3**, 412–438.

WOOD, C.J. (1973). The flight of albatrosses (A computer simulation). *Ibis* **115**, 244–256.

WOODWARD, P.W. (1972). The natural history of Kure Atoll, Northwestern Hawaiian Islands. *Atoll Res Bull* **164**, 318 pp.

WOOLLER, R.D., BRADLEY, J.S., SERVENTY, D.L. & SKIRA, I.J. (1988). Factors contributing to reproductive success in Short-tailed Shearwaters (*Puffinus tenuirostris*). *Acta XIX Congr Int Orn* **1**, 848–856.

WOOLLER, R.D., BRADLEY, J.S., SKIRA, I.J. & SERVENTY, D.L. (1990). The reproductive success of Short-tailed Shearwater *Puffinus tenuirostris* in relation to their age and breeding experience. *J Anim Ecol* **59**, 161–170.

WORMELL, P. (1976). The Manx Shearwaters of Rhum. *Scott Birds* **9**, 103–118.

WORTH, C.B. (1940). Egg volumes and incubation periods. *Auk* **57**, 44–60.

WYNNE-EDWARDS, V.C. (1939). Intermittent breeding of the Fulmar (*Fulmarus glacialis* (L.)), with some general observations on non-breeding in sea-birds. *Proc Zool Soc Lond* **A109**, 127–132.

WYNNE-EDWARDS, V.C. (1953). Leach's Petrels stranded in Scotland in October-November 1952. *Scott Nat* **65**, 167–189.

WYNNE-EDWARDS, V.C. (1962). *Animal Disperson in Relation to Social Behaviour*. Oliver and Boyd, Edinburgh.

YESOU, P. (1982). A propos de la presence remarquable du Puffin Cendre *Calonectris diomedea* pres des cotes du Golfe de Gascogne et de la mer Celtique en 1980. *Oiseau Rev Fr Orn* **152**, 197–217.

YESOU, P. (1985). Nouvelles donnees sur la mue de *Puffinus p. mauretanicus*. *Oiseau Rev Fr Orn* **55**, 177–182.

YESOU, P. (1986). Balearic Shearwaters summering in western France. In *Mediterranean Marine Avifauna* (Medmaravis & X. Monbailliu eds) pp. 515–517. Springer, Berlin.

YOSHIDA, N. (1962). [Breeding ecology of the Streaked Shearwater at Kammurijima I., Maizuru in Kyoto.] *Tori* **17**, 83–108.

YOSHIDA, N. (1973). [A decennial record of fallen Streaked Shearwaters in the Kinki District.] *Tori* **22**, 60–66.

YOSHIDA, N. (1981). [*Seabirds that Climb Trees*.] Ubunsha, Tokyo.

YOUNG, A.S. (1986). Seabird observations from meteorological reports. *Sea Swallow* **35**, 35–37.

YOUNG, E.C. (1978). Behavioural ecology of *lonnbergi* skuas in relation to environment on the Chatham Islands, New Zealand. *NZ J Zool* **5**, 401–416.

YOUNG, E.C., JENKINS, P.E., DOUGLAS, M.E. & LOVEGROVE, T.G. (1988). Nocturnal foraging by Chatham Island skuas. *NZ J Ecol* **11**, 113–117.

YUDIN, K.A. (1957). [Certain adaptive pecularities of the wing in the birds of the order Tubinares.] *Zool Zh* **36**, 1859–1873.

ZACH, R. (1988). Growth-curve analysis: a critical evaluation. *Auk* **105**, 208–210.

ZINK, R.M. (1981). Observations of seabirds during a cruise from Ross Island to Anvers Island, Antarctica. *Wilson Bull* **93**, 1–20.

ZINK, R.M. & ELDRIDGE, J.L. (1980). Why does Wilson's Petrel have yellow on the webs of its feet? *Br Birds* **73**, 385–387.

ZINO, P.A. (1971). The breeding of Cory's Shearwater *Calonectris diomedea* on the Salvage Islands. *Ibis* **113**, 212–217.

ZINO, P.A. (1985). A short history of the Shearwater hunt on the Great Salvage and recent developments on this island. *Bocagiana* **84**, 1–9.

ZINO, P.A., ZINO, F., MAUL, T. & BISCOITO, J.M. (1987). The laying, incubation and fledging periods of Cory's Shearwater *Calonectris diomedea borealis* on Selvagem Grande in 1984. *Ibis* **129**, 393–398.

ZONFRILLO, B. (1986). Diet of Bulwer's Petrel *Bulweria bulwerii* in the Madeiran Archipelago. *Ibis* **128**, 570–572.

ZONFRILLO, B. (1988). Notes and comments on the taxonomy of Jouanin's Petrel *Bulweria fallax* and Bulwer's Petrel *Bulweria bulwerii*. *Bull Br Orn Club* **108**, 71–75.

APPENDIX 1

A Checklist of the Procellariiformes

Order **PROCELLARIIFORMES** Furbringer, 1888

Family DIOMEDEIDAE Gray, 1840

Genus *Diomedea* Linnaeus, 1758. **ALBATROSSES.**

1. ***D.exulans*** Linnaeus, 1758. **WANDERING ALBATROSS.**
   a. *D.e. exulans* Tristan da Cunha Group; Gough I.
   b. *D.e. chionoptera* S. Georgia; Marion; Crozet; Kerguelen and Macquarie Is.
   c. *Unnamed subspecies A* Antipodes and Campbell Is.
   d. *Unnamed subspecies B* Auckland Is.
   e. *D.e. amsterdamensis* Amsterdam I.
2. ***D.epomophora*** Lesson, 1825. **ROYAL ALBATROSS.**
   a. *D.e. epomophora* Campbell and Auckland Is.
   b. *D.e. sanfordi* Taiaroa Head, South I. and Chatham Is. (New Zealand).
3. ***D.irrorata*** Salvin, 1883. **GALAPAGOS** or **WAVED ALBATROSS**
   Hood I. (Galapagos) and La Plata I. (Ecuador).
4. ***D.albatrus*** Pallas, 1769. **SHORT-TAILED ALBATROSS.**
   Torishima I. (Izu Group).
5. ***D.nigripes*** Audubon, 1849. **BLACK-FOOTED ALBATROSS.**
   Torishima I. (Izu Group), Mukojima and Torishima Is. (Bonin Group); Kita-Kojima I. (Senkaku Group); islands of the Hawaiian Leeward Chain.
6. ***D.immutabilis*** Rothschild, 1893. **LAYSAN ALBATROSS.**
   Islands of the Hawaiian Leeward Chain; Kauai I. (Hawaii); Mukojima I. (Bonin Group); Guadalupe I. (Mexico).
7. ***D.melanophrys*** Temminck, 1828. **BLACK-BROWED MOLLYMAWK OR ALBATROSS.**
   a. *D..m. melanophrys* Islands off Tierra del Fuego and Diego de Almagro I. southern Chile; The Falklands, S. Georgia; The Crozets; Kerguelen; Heard and McDonald Is.; Marion and Prince Edward Is.; Macquarie Is.; The Snares (once) and Antipodes Is.
   b. *D.m. impavida* Campbell I.
8. ***D.cauta*** Gould, 1841. **SHY OR WHITE-CAPPED MOLLYMAWK or ALBATROSS.**
   a. *D.c. cauta* Albatross I. and The Mewstone (Tasmania); Auckland Is.
   b. *D.c. salvini* The Snares, Bounty Island and Penguin I. (The Crozets).
   c. *D.c. eremita* Pyramid Rock (Chatham Is.).
9. ***D. chrysostoma*** Forster, 1785. **GREY-HEADED MOLLYMAWK or ALBATROSS.**
   Diego Ramirez Is. (Cape Horn); S. Georgia; Marion and Prince Edward Is.; The Crozets; Kerguelen; Macquarie and Campbell Is.
10. ***D. chlororhynchos*** Gmelin, 1789. **YELLOW-NOSED MOLLYMAWK or ALBATROSS.**
    a. *D.c. chlororhynchos* Tristan da Cunha Group; Gough I.

b. *D.c. bassi* Prince Edward I.; St. Paul and Amsterdam Is.; The Crozets and Kerguelen.

11. ***D.bulleri*** Rothschild, 1893. **BULLER'S MOLLYMAWK or ALBATROSS.**
    a. *D.b. bulleri* Solander and Snares Is.
    b. *D.b. platei* Chatham and Three Kings Is.

Genus *Phoebetria* Reichenbach, 1853. **SOOTY ALBATROSSES.**

12. ***P.fusca*** (Hilsenburg, 1822). **SOOTY ALBATROSS.**
    Tristan da Cunha Group; Gough I.; Marion and Prince Edward Is.; The Crozets; Kerguelen; Amsterdam and St. Paul Is.
13. ***P.palpebrata*** (Forster, 1785). **LIGHT-MANTLED SOOTY ALBATROSS.**
    S. Georgia; Marion and Prince Edward Is.; The Crozets; Kerguelen; Heard I.; Macquarie I.; Auckland Is.; Campbell and Antipodes Is.

Family PROCELLARIIDAE Boie, 1826. **FULMARS, GADFLY PETRELS, PRIONS AND SHEARWATERS.**

Genus *Macronectes* Richmond, 1905. **GIANT PETRELS.**

14. ***M.giganteus*** (Gmelin, 1789). **SOUTHERN GIANT PETREL.**
    Coastal Antarctica; Diego Ramirez and Isla Noir (Chile); Chubut (Argentina); Falkland Is.; S. Georgia; S. Orkney Is.; S. Sandwich Is.; Bouvet I.; Marion and Prince Edward Is.; The Crozets; Kerguelen and Heard Is.; Gough I.
15. ***M. halli*** Mathews, 1912. **NORTHERN GIANT PETREL.**
    Marion and Prince Edward Is.; S. Georgia; The Crozets; Kerguelen; Macquarie I.; Stewart I.; Auckland Is.; Campbell I.; Chatham and Antipodes Is.

Genus *Fulmarus* Stephens, 1826. **FULMARS.**

16. ***F.glacialoides*** (Smith, 1840). **SOUTHERN FULMAR.**
    Antarctic Continent and adjacent islands; S. Sandwich Is.; S. Shetland Is.; Elephant and Bouvet Is.
17. ***F.glacialis*** (Linneaus, 1761). **NORTHERN FULMAR.**
    a. *F.g. glacialis* High Arctic N. Atlantic in N.E. Greenland; Spitzbergen; Devon, Baffin and Bear Is.; Franz Joseph Land; Novaya Zemlya.
    b. *F.g. auduboni* Low Arctic N. Atlantic in N.W. Greenland; Newfoundland; Jan Mayen; Iceland; The Faroes; British Is.; N-W France, Germany (Heligoland).
    c. *F.g. rodgersi* N. Pacific at Commander, Kurile and Aleutian Is.; islands south of the Alaska Peninsula, in Sea of Okhotsk and off Kamchatka.

Genus *Thalassoica* Reichenbach, 1853.

18. ***T.antarctica*** (Gmelin, 1789). **ANTARCTIC PETREL.**
    Antarctic coasts but also up to 350 km inland.

Genus *Daption* Stephens, 1826.

19. ***D.capense*** (Linnaeus, 1758). **CAPE PETREL or CAPE PIGEON.**

a. *D.c. capense* Coastal Antarctica and offshore islands; S. Georgia; S. Sandwich Is.; S. Orkney Is.; Bouvet I.; S. Shetland Is.; The Crozets; Heard Is.; Kerguelen.
b. *D.c. australe* The Snares; Campbell I.; Chatham Is.; The Bounty Is.

Genus *Pagodroma* Bonaparte, 1856.

20. ***P.nivea*** (Forster, 1777), **SNOW PETREL.**
Antarctic Continent and mountains (up to 300 km inland) and offshore islands (Ross, Franklin, Possession, Scott, Balleny); S. Georgia; S. Sandwich Is.; S. Orkney Is.; S. Shetland Is.; Bouvet I.

Genus *Lugensa*[1] Mathews, 1942.

21. ***L.brevirostris*** (Lesson, 1831). **KERGUELEN PETREL.**
Tristan da Cunha Group; Gough I.; Marion I.; Crozet & Kerguelen Is.

Genus *Pterodroma* Bonaparte, 1856. **GADFLY PETRELS.**

22. ***P.macroptera*** (Smith, 1840). **GREAT-WINGED PETREL.**
a. *P.m. macroptera* Tristan da Cunha; Gough I.; Marion and Prince Edward Is.; The Crozets and Kerguelen; islands off the south coast of Western Australia.
b. *P.m. gouldi* North I. of New Zealand, including some headlands and mountains, but mainly on offshore islands.
23. ***P.lessonii*** (Garnot, 1826). **WHITE-HEADED PETREL.**
Auckland, Antipodes and Macquarie Is.; The Crozets; Kerguelen.
24. ***P.incerta*** (Schlegel, 1863). **SCHLEGEL'S PETREL.**
Tristan da Cunha Group; Gough I.
25. ***P.solandri*** (Gould, 1844). **PROVIDENCE PETREL.**
Lord Howe and Philip I. (Norfolk I.).
26. ***P.magentae*** (Giglioli & Salvadori, 1869). **TAIKO or MAGENTA PETREL.**
Chatham I. (New Zealand).
27. ***P.macgillivrayi*[2]** (Gray, 1860). **MACGILLIVRAY'S PETREL.**
Gau I. (Fiji).
28. ***P.rostrata*[2]** (Peale, 1848). **TAHITI PETREL.**
a. *P.r. rostrata* Society and Marquesas Is.
b. *P.r. trouessarti* New Caledonia.
c. *P.r. becki* Bismarck Archipelago (?) and/or Solomon Is. (?).
29. ***P.aterrima*[2]** (Bonaparte, 1857). **MASCARENE or REUNION PETREL.**
Reunion.
30. ***P.ultima*** Murphy, 1949. **MURPHY'S PETREL.**
Austral, Tuomotu and Pitcairn Groups.
31. ***P.mollis*** (Gould, 1844). **SOFT-PLUMAGED PETREL.**
a. *P.m. mollis* Tristan da Cunha Group; Gough and Antipodes Is.
b. *P.m. fusca* Marion and Prince Edward Is.; The Crozets and Kerguelen.
32. ***P.feae*** (Salvadori, 1899). **FEA'S PETREL.**
Cape Verde and Desertas Is.
33. ***P.madeira*** Mathews, 1934. **MADEIRA PETREL.**
Madeira.

34. ***P.cahow*** (Nichols & Mowbray, 1916). **CAHOW or BERMUDA PETREL.**
Castle I. (Bermuda).
35. ***P.hasitata*** (Kuhl, 1820). **BLACK-CAPPED PETREL or DIABLOTIN.**
   a. *P.h. hasitata* Mountains of Hispaniola; Cuba; Guadeloupe; Haiti; Dominica; Martinique (?).
   b. *P.h. caribbea* Blue Mountains (Jamaica) ? Extinct.
36. ***P.externa*** (Salvin, 1875). **WHITE-NECKED PETREL.**
   a. *P.e. externa* Mas Afuera and Juan Fernandez Is.
   b. *P.e. cervicalis* Kermadec Is.
37. ***P.baraui*** (Jouanin, 1964). **BARAU'S PETREL.**
Reunion and Rodriguez Is.
38. ***P.phaeopygia*** (Salvin, 1876). **DARK-RUMPED PETREL.**
   a. *P.p. phaeopygia* Galapagos Is.
   b. *P.p. sandwichensis* Maui, Hawaii and Kauai Is. (Hawaii).
39. ***P.neglecta*** (Schlegel, 1863). **KERMADEC PETREL.**
   a. *P.n. neglecta* Lord Howe I.; The Kermadecs; islands of the Austral and Pitcairns Groups; Easter I.
   b. *P.n. juana* San Ambrosio and San Felix Is.; Mas a Tierra and Santa Clara Is. (Juan Fernandez).
40. ***P.arminjoniana*** (Giglioli & Salvadori, 1869). **TRINDADE or HERALD PETREL.**
   a. *P.a. arminjoniana* Trindade I. and Martin Vas Rocks (Atlantic); Round I. (Mauritius).
   b. *P.a. heraldica* Raine I. (Queensland); Chesterfield Reefs (New Caledonia); Tonga; The Marquesas; The Tuamotus; Gambier Is.; Pitcairn Group; Easter I.; Cook Is. (?); Rarotonga (?).
41. ***P.alba*** (Gmelin, 1789). **PHOENIX PETREL.**
Phoenix, Line and Marquesas Is.; Tonga; The Pitcairn Group.
42. ***P.inexpectata*** (Forster, 1844). **MOTTLED or PEALE'S PETREL.**
Islands off Stewart I. and the south-west coast of the South I. New Zealand; on an islet in Lake Hauroka, Southland; The Snares.
43. ***P.hypoleuca*** (Salvin, 1888). **BONIN PETREL.**
Bonin and Volcano Is.; islands of the Hawaiian Leeward Chain.
44. ***P.nigripennis*** (Rothschild, 1893). **BLACK-WINGED PETREL.**
Norfolk and Lord Howe Is.; The Kermadecs; New Caledonia; Three Kings and Portland I. (North I., New Zealand); South-East I. (Chathams Group); Rapa I. and Bass Rocks (Austral Group).
45. ***P.axillaris*** (Salvin, 1893). **CHATHAM ISLAND PETREL.**
South-East I. (Chatham Is.).
46. ***P.cookii*** (Gray, 1843). **COOK'S PETREL.**
Little and Great Barrier Is. (North I., New Zealand); Codfish I. (Stewart I.)
47. ***P.defilippiana*** (Giglioni & Salvadori, 1869). **MAS A TIERRA PETREL.**
Mas a Tierra I. and (?) Santa Clara I. (Juan Fernandez); San Felix and San Ambrosio Is.
48. ***P.longirostris*** (Stejneger, 1893). **STEJNEGER'S and PYCROFT'S PETRELS.**
   a. *P.l. longirostris* Mas Afuera I. (Juan Fernandez Group).
   b. *P.l. pycrofti* Islands off the north-east coast of the North I. (New Zealand).

49. ***P.leucoptera*** (Gould, 1844). **GOULD'S PETREL; COLLARED PETREL.**
 a. *P.l. leucoptera* Cabbage Tree I. (New South Wales).
 b. *P.l. caledonica* New Caledonia.
 c. *P.l. brevipes* Fiji and (?) Samoa.

Genus *Halobaena* Bonaparte, 1856.

50. ***H.caerulea*** (Gmelin, 1789). **BLUE PETREL.**
Diego Ramirez, Cape Horn I. and other islands in the region; S. Georgia; Prince Edward and Marion Is.; The Crozets; Kerguelen; Macquarie I.

Genus *Pachyptila* Illiger, 1811. **PRIONS.**

51. ***P.vittata*** (Forster, 1777). **BROAD-BILLIED and SALVIN'S PRIONS.**
 a. *P.v. vittata* Islands in Foveaux Strait and off Stewart I.; The Snares and Chatham Is.; Tristan da Cunha Group and Gough I.
 b. *P.v. macqillivrayi*[3] St Paul and (?) Amsterdam Is.
 c. *P.v. salvini*[3] The Crozets; Prince Edward and Marion Islands.
52. ***P.desolata*** (Gmelin, 1789). **ANTARCTIC or DOVE PRION.**
 a. *P.d. desolata* The Crozets, Kerguelen and Macquarie Is.
 b. *P.d. alter* Auckland and Heard Is.
 c. *P.d. banksi* S. Georgia; S. Sandwich and S. Orkney Is.; Elephant I.; Scott I. (off Antarctica).
53. ***P.belcheri*** (Mathews, 1912). **NARROW-BILLED PRION.**
Isla Noir (Chile); The Falklands; The Crozets and Kerguelen.
54. ***P.turtur*** (Kuhl, 1820). **FAIRY PRION.**
S. Georgia; Beauchene I. (Falklands); Marion and Prince Edward Is.; The Crozets and Ile Croy (Kerguelen); St. Paul I.; Macquarie I.; islands off Victoria (Australia) and off Tasmania; islands off both North and South Is. of New Zealand; at The Snares, Antipodes and Chatham Is.
55. ***P.crassirostris*** (Mathews, 1912). **FULMAR PRION.**
 a. *P.c. crassirostris* Bounty, Snares and Chatham Is.
 b. *P.c. eatoni* Heard and Auckland Is.

Genus Bulweria Bonaparte, 1843.

56. ***B.bulwerii*** (Jardine & Selby, 1828). **BULWER'S PETREL.**
In the Atlantic at the Azores, Salvages, Porto Santo (Madeira), Desertas, Canary and Cape Verde Is.. In the Pacific on Taiwan (?); Fujian and other islets off South-east China; Bonin, Volcano & Ryuku Is.; islands of the Hawaiian Leeward Chain; Kuala and islets off Oahu, Maui & Hawaii (Hawaii); Johnston Atoll; Phoenix I. and the Marquesas.
57. ***B.fallax*** Jouanin, 1955. **JOUANIN'S PETREL.**
Breeding place unknown — in desert of S. Oman?

Genus *Procellaria* Linnaeus, 1758.

58. ***P.aequinoctialis*** Linnaeus, 1758. **SHOEMAKER, WHITE-CHINNED PETREL, SPECTACLED PETREL.**
 a. *P.a. aequinoctialis* The Falklands; S. Georgia; Prince Edward and Marion Is.; The Crozets; Kerguelen; Auckland, Campbell and Antipodes Is.
 b. *P.a. conspicillata* Inaccessible I. (Tristan da Cunha).

59. ***P.westlandica*** Falla, 1946. **WESTLAND BLACK PETREL.**
Coastal range of the north west of the South I. New Zealand.
60. ***P.parkinsoni*** Gray, 1862. **PARKINSON'S or BLACK PETREL.**
Little and Great Barrier Is. off North I. New Zealand.
61. ***P.cinerea*** Gmelin, 1789. **PEDIUNKER or GREY PETREL.**
Tristan da Cunha Group; Gough I.; Marion and Prince Edward Is.; The Crozets, Kerguelen, Amsterdam I. (?); Campbell and Antipodes Is.

Genus *Calonectris* Mathews & Iredale, 1915.

62. ***C.diomedea*** (Scopoli, 1769). **CORY'S and MEDITERRANEAN SHEARWATERS.**
    a. *C.d. diomedea* Balearic Is.; islets off Mediterranean France and Spain; off Italy, Sardinia, Corsica, Crete, Malta, Algeria, Morocco, Tunisia; islets in the Adriatic and off the Balkans and Turkey.
    b. *C.d. borealis* Berlenga Is. (Portugal); Madeira; Porto Santo; The Desertas, Salvages, Canaries and Azores.
    c. *C.d. edwardsi* Cape Verde Is.
63. ***C.leucomelas*** (Temminck, 1835). **STREAKED SHEARWATER.**
Islands off Korea and East China and off mainland Japan from Hokkaido south to the Ryuku, Daito, Senkuku and Izu Groups; The Pescadores.

Genus *Puffinus* Brisson, 1760. **SHEARWATERS.**

64. ***P.pacificus*** (Gmelin, 1789). **WEDGE-TAILED SHEARWATER.**
In Indian Ocean on islets off Madagascar, at the Seychelles, Amirantes, Mascarenes, Cargados Carajos, Chagos and Cocos Keeling Is.; Houtman Abrolhos and other islands off Western Australia from Forestier I. to Fremantle and in Shark Bay. In the Pacific at the Pescadores, Bonin and Volcano Is.; the Marianas, Carolines and Solomon Is.; Marcus I.; Wake I.: Chesterfield Reefs; New Caledonia; Norfolk I.; Lord Howe I.; islets off Papua New Guinea; islands off eastern Australia from Torres Strait to Montagu I. (N.S.W.); on the Marshall & New Hebrides Is.; Fiji; Johnston Atoll; islands of the Hawaiian Leeward Chain and off the main Hawaiian Group; the Phoenix Is; Samoa; Tonga; Kiribati; the Society, Austral, Marquesas and Pitcairn Groups and at the Revillagigedo Is. (Mexico).
65. ***P.bulleri*** Salvin, 1888. **BULLER'S or NEW ZEALAND SHEARWATER.**
Poor Knights Is. off North I. New Zealand.
66. ***P.carneipes*** Gould, 1844. **FLESH-FOOTED SHEARWATER.**
    a. *P.c. carneipes* St. Paul I. (Indian Ocean); islands off the south coast of Western Australia from Cape Leeuwin to Recherche Archipelago.
    b. *P.c. hullianus* Lord Howe I.; islands off the east coast of the North I., New Zealand, off Taranaki and in Cook Strait.
67. ***P.creatopus*** Coues, 1864. **PALE-FOOTED SHEARWATER.**
Mas a Tierra and Santa Clara Is. (Juan Fernandez); Mocha I. (Chile).
68. ***P.gravis*** (O'Reilly, 1818). **GREATER SHEARWATER.**
Tristan da Cunha Group; Gough I.; Kidney I. (Falklands).
69. ***P.griseus*** (Gmelin, 1789). **SOOTY SHEARWATER.**
Islands off Cape Horn and Tierra del Fuego; Chiloe, Guafo and Guamblim Is.

(Chile); Kidney I. (Falklands); islands off N.S.W. from Broughton to Montagu I. and off south east Tasmania; Macquarie I.; islands round New Zealand and Stewart I.; and at The Chathams, Snares, Auckland, Antipodes and Campbell Is.

70. ***P.tenuirostris*** (Temminck, 1835). **SHORT-TAILED SHEARWATER.**
Islands off southern Australia from the Recherche Archipelago to Bass Strait and N.S.W.; off eastern Australia as far north as Broughton I.; many islands around Tasmania and a few mainland sites in that state.

71. ***P.nativitatis*** Streets, 1877. **CHRISTMAS SHEARWATER.**
Islands of the Hawaiian Leeward Chain; the Marshall, Line, Phoenix, and Marquesas Is.; Johnston Atoll; Rapa I. (Austral Group); the Gambier Is.; Ile de Paques (Society Is.); the Pitcairn Group and Easter I.

72. ***P.puffinus*** (Brunnich, 1764). **MANX, LEVANTINE and BALEARIC SHEARWATERS**
   a. *P.p. puffinus* Westmann Is. (Iceland); the Faroes; islands off the west coast of the British Islands, off Ireland and Brittany; Madeira; The Desertas, Azores and Canary Is.; islets off Massachusetts and Newfoundland.
   b. *P.p. yelkouan*[4] Islets off Mediterranean France; off Italy, Sardinia and Corsica; in the Adriatic and in the Aegean off Greece and Turkey.
   c. *P.p. mauretanicus* Pityusae and islets off the Balearics.

73. ***P.auricularis*** Townsend, 1980. **TOWNSEND'S & NEWELL'S SHEARWATERS.**
   a. *P.a. auricularis* Revillagigedo Is. (Mexico).
   b. *P.a. newelli* Kauai I. and (?) Maui (Hawaii).

74. ***P.opisthomelas*** Coues, 1864. **BLACK-VENTED SHEARWATER.**
Islands off the Pacific coast of Baja California.

75. ***P.gavia*** (Forster, 1844). **FLUTTERING SHEARWATER.**
Islands off the North Island and in Cook Strait and the Marlborough Sounds, New Zealand.

76. ***P.huttoni*** Matthews, 1912. **HUTTON'S SHEARWATER.**
Kaikoura Mountains, South I., New Zealand.

77. ***P.lherminieri*** Lesson 1839. **AUDUBON'S SHEARWATER.**
   a. *P.l. lherminieri* Bermuda; Bahamas; Virgin Is.; Guadelope I.; Desirade I.; Leeward Is.; Rocher de Diamant; Martinique; Bird Rock (Barbados).
   b. *P.l. loyemilleri* Islets off Caribbean coast of Panama; Crab Cay (Providentia I. east of Nicaragua); Les Roques (off northern Venezuela).
   c. *P.l. subalaris* Galapagos Is.
   d. *P.l. dichrous* Central Pacific at Palau; Marianas; Carolines; Phoenix Is.; Samoa; Fiji; Society Is.; Tonga; The Marquesas, Line and Gambier Is.
   e. *P.l. gunax* Banks Group (New Hebrides).
   f. *P.l. heinrothi* — ? probably around New Britain.
   g. *P.l. bannermani* Bonin and Volcano Is.
   h. *P.l. bailloni* Mascarene Is.
   i. *P.l. nicolae* Aldabra; Seychelles; Amirante; Maldive and Chagos Is.
   j. *P.l. persicus* Al Hallaniyah I. (Kuri Muria Is. in Arabian Sea).
   k. *P.l. boydi* Cape Verde Is.
   l. *P.l. temptator* Moheli I. (Comoro Group).

78. ***P.assimilis*** Gould, 1838. **LITTLE SHEARWATER.**
    a. *P.a.assimilis* Lord Howe and Norfolk Is.
    b. *P.a.baroli* The Azores; islands off Porto Santo (Madeira); The Desertas, Salvages and Canary Is.
    c. *P.a. tunneyi* Abrolhos Group and other islands off Western Australia from Dongara to Fremantle and from Cape Leeuwin to Chatham I. (near Albany) and the Recherche Archipelago.
    d. *P.a. kermadecensis* Kermadec Is.
    e. *P.a. haurakiensis* Islands off the east coast of the North I. of New Zealand.
    f. *P.a. elegans* Tristan da Cunha Group; Gough I.; Antipodes and Chatham Is. [Birds of Roche Quille, St. Paul., may belong to this form].
    g. *P.a. myrtae* Rapa I. (Austral Group).

Family HYDROBATIDAE Mathews, 1912. **STORM PETRELS.**

Genus *Oceanites* Keyserling & Blasius, 1840.

79. ***O.oceanicus*** (Kuhl, 1840). **WILSON'S STORM PETREL.**
    a. *O.o. oceanicus* Islands off Tierra del Fuego; The Falklands; S. Georgia; The Crozets; Bouvet I. (?); Kerguelen and Heard Is.
    b. *O.o. exasperatus* Coasts and offshore islands of Antarctica; S. Shetland, S. Sandwich and S. Orkney Is; Elephant I.
80. ***O.gracilis*** (Elliot, 1859). **ELLIOT'S STORM PETREL.**
    a. *O.g. gracilis* Only known place Chungungo I. off Chile (once).
    b. *O.g. galapagoensis* unknown: presumably in the Galapagos Is.

Genus *Garrodia* Forbes, 1881.

81. ***G.nereis*** (Gould, 1841). **GREY-BACKED STORM PETREL.**
    Falkland Is.; S. Georgia; Gough I.; The Crozets; Kerguelen; Prince Edward I.; Auckland, Antipodes, Campbell and Chatham Is.

Genus *Pelagodroma* Reichenbach, 1853.

82. ***P.marina*** (Latham, 1790). **WHITE-FACED STORM PETREL.**
    a. *P.m. marina* Tristan da Cunha Group and Gough I.
    b. *P.m. hypoleuca* Salvage Is.
    c. *P.m. eadesi* Cape Verde and Canary Is.
    d. *P.m. dulciae* Houtman Abrolhos Is. and coastal islands southwards off Western Australia, east to Bass Strait, Victoria and N.S.W. as for north as Broughton I. Islands off north-east Tasmania.
    e. *P.m. maoriana* Islands off both islands of New Zealand; in Cook Strait; off Stewart I. and at the Chatham and Auckland Is.
    f. *P.m. albiclunis* Kermadec Is.

Genus *Fregetta* Bonaparte, 1855.

83. ***F.tropica*** (Gould, 1844). **BLACK-BELLIED STORM PETREL.**
    a. *F.t. tropica* S. Shetland, S. Orkney, and S. Sandwich Is.; Elephant I.; S. Georgia; Bouvet I. (?); The Crozets; Kerguelen; Auckland and Antipodes Is.

b. *F.t. melanoleuca* Tristan da Cunha Group; Gough I. (?); Prince Edward Is.

84. ***F.grallaria*** *(Vieillot, 1817).* **WHITE-BELLIED STORM PETREL.**
   a. *F.g. grallaria* Admiralty Is. (Lord Howe I.); The Kermadecs.
   b. *F.g. leucogaster* Tristan da Cunha Group; Gough and St. Paul Is.
   c. *F.g. segethi* Juan Fernandez Is.
   d. *F.g. titan* Rapa I. (Austral Group).

Genus *Nesofregetta* Mathews, 1912).

85. ***N.fuliginosa*** (Gmelin, 1789). **WHITE-THROATED STORM PETREL.**
   Fiji; New Hebrides; Kiribati; Phoenix, Austral, Marquesas and Gambier Is.; Sala y Gomez; Samoa(?).

Genus *Hydrobates* Boie, 1822.

86. ***H.pelagicus*** (Linnaeus, 1758). **BRITISH STORM PETREL.**
   a. *H.p. pelagicus* Westman Is. (Iceland); Lofoten Is.; British Is.; islets off Brittany, Biarritz and Northern Spain; The Canary Is.
   b. *H.p. melitensis* Islets off Mediterranean Spain and France; Balearic Is.; islets off Corsica; Sardinia; Sicily; Malta; Elba; Italy; Yugoslavia; Greece and (?) Turkey; Lipari Is.

Genus *Halocyptena* Coues, 1864.

87. ***H.microsoma*** Coues, 1864. **LEAST STORM PETREL.**
   San Benito I. (Pacific Baja California); Consag Rock, San Luis and Partida Is. (Gulf of California).

Genus *Oceanodroma* Reichenbach, 1853.

88. ***O.tethys*** (Bonaparte, 1852). **WEDGE-RUMPED or GALAPAGOS STORM PETREL.**
   a. *O.t. tethys* Galapagos Is.
   b. *O.t. kelsali* Pescadores and San Gallan Is. (Peru).

89. ***O.castro*** Harcourt, 1851. **MADEIRAN STORM PETREL.**
   In Atlantic at Farilhoes Is. (Portugal); Madeira; islands off Porto Santo; Desertas, Salvages, Azores, Canary Is. (Anaga Rocks) and Cape Verde Is.; Ascension and St. Helena Is.; Sao Tome (?); In Pacific at Hidejama I. ((Honshu, Japan); Kauai (Hawaii); Galapagos Is.

90. ***O.monorhis*** (Swinhoe, 1867). **SWINHOE'S STORM PETREL.**
   Islands off S-W Honshu, Skikotu and Kyushu (Japan); off Korea and Shantung (China); Verkhovski Is. (Vladivostock).

91. ***O.leucorhoa*** (Vieillot, 1818). **LEACH'S STORM PETREL.**
   a. *O.l. leucorhoa* In Atlantic on islands off Newfoundland, Maine, Nova Scotia and Massachusetts; Westmann Is. (Iceland); Lofoten Is. (Norway); Faroe Is.; Scotland. In Pacific on Daikokujima I. (E. Hokkaido); Kuriles, Commanders and Aleutian Is.; islands off Alaska and off western North America south to the Farallon Is.
   b. *O.l. chapmani* San Benitos and Los Coronados Is. (Mexico).
   c. *O.l. socorrensis* Islets off Guadalupe I. (Mexico). [Summer Breeding].
   d. *O.l. cheimomnestes* Guadalupe I. (Mexico). [Winter Breeding].

92. ***O.macrodactyla*** Bryant, 1887. **GUADALUPE STORM PETREL.**

Guadalupe I. (Mexico) — extinct.

93. ***O. markhami*** (Salvin, 1883). **MARKHAM'S STORM PETREL.**
Breeding places unknown — off Chile and/or Peru?

94. ***O.tristrami*** Salvin, 1896. **SOOTY or TRISTRAM'S STORM PETREL.**
Bonin (?) and Izu Is.; Hawaiian Leeward Chain.

95. ***O.melania*** (Bonaparte, 1854). **BLACK STORM PETREL.**
Islands off southern and Baja California and in Gulf of California.

96. ***O.matsudairae*** Kuroda, 1922. **MATSUDAIRA'S STORM PETREL.**
Kita-iwojima and Minami-iwojima (Volcano Is).

97. ***O.homochroa*** (Coues, 1864). **ASHY STORM PETREL.**
Islands off California; San Miguel and Santa Cruz (Channel Is.) & Los Coronados (Baja California).

98. ***O.hornbyi*** (Gray, 1864). **HORNBY'S or RINGED STORM PETREL.**
Coastal mountains of northern Chile.

99. ***O.furcata*** (Gmelin, 1789). **FORK-TAILED STORM PETREL.**
   a. *O.f. furcata* Commander, Kurile and Aleutian Is.
   b. *O.f. plumbea* Islands off western North America from Alaska south to northern California.

Family PELECANOIDIDAE Gray, 1871. **DIVING PETRELS.**

Genus *Pelecanoides* Lacepede 1799.

100. ***P.garnotii*** (Lesson, 1828). **PERUVIAN DIVING PETREL.**
Islands off the west coast of South America from Lobos de Tierra (Peru) to Mocha I. (Chile).

101. ***P.magellani*** (Mathews, 1912). **MAGELLANIC DIVING PETREL.**
Islands off southern Chile and probably off Isla Grande and other islands of Tierra del Fuego.

102. ***P.georgicus*** Murphy & Harper 1916. **GEORGIAN DIVING PETREL.**
S. Georgia; Marion Is.; The Crozets; Kerguelen; Heard I.; Codfish I. (off Stewart I.) and Auckland Is.

103. ***P.urinatrix*** (Gmelin, 1789). **COMMON DIVING PETREL.**
   a. *P.u. urinatrix* Islands off south east Australia and Tasmania; off North I. and in Cook Strait, New Zealand.
   b. *P.u. chathamensis* The Solanders; islands off Stewart I. and in Foveaux Strait; The Snares and Chatham Is.
   c. *P.u. exsul* S. Georgia; Marion and Prince Edward Is.; The Crozets; Kerguelen, Heard I.; Macquarie I.; Auckland, Antipodes and Campbell Is.
   d. *P.u. dacunhae* Tristan da Cunha Group; Gough I.
   e. *P.u. berard* Falkland Is.
   f. *P.u. coppingeri* Breeding places unknown — Straits of Magellan?

---

1. More usually called *Pterodroma brevirostris* (c.f. Imber, 1985).
2. Included in *Pseudobulweria* by Imber (1985).
3. Often called *P.salvini salvini* and *P.s. macqillivrayi*.
4. Full specific status has been proposed for *yelkouan* by Bourne *et al*. 1988.

APPENDIX 2

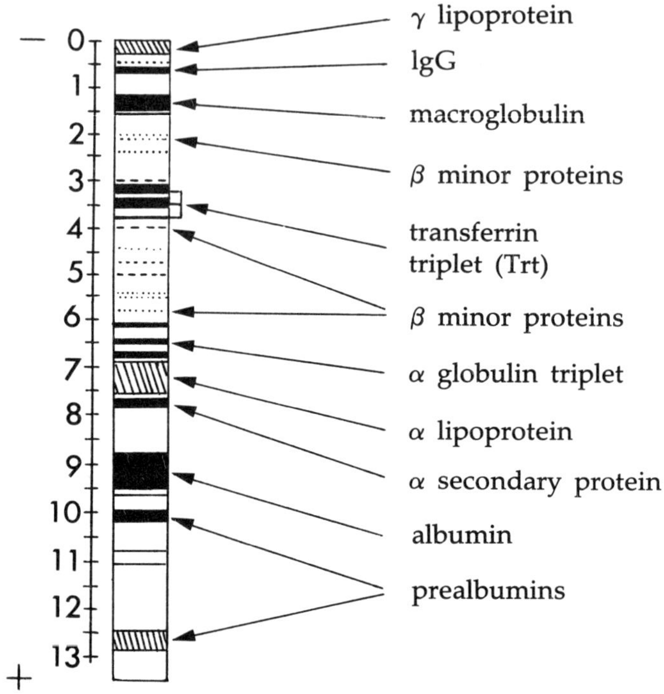

APPENDIX 2: Protein bands in the electrophoresed blood plasma of *Procellaria westlandica* used as a 'reference' plasma for other Procellariiformes. From Harper (1978), courtesy of *New Zealand Journal of Zoology*.

Species Index

Subject Index